Information Centre & Library

IMI *information service*
Sandyford, Dublin 16
Telephone 2078513 Fax
email library

Marketing of High-Technology Products and Innovations

SECOND EDITION

Jakki Mohr
University of Montana

Sanjit Sengupta
San Francisco State University

Stanley Slater
Colorado State University

PEARSON
Prentice
Hall

Pearson Education International

Acquisitions Editor: Katie Stevens
Editorial Director: Jeff Shelstad
Assistant Editor: Melissa Pellerano
Editorial Assistant: Rebecca Cummings
Marketing Manager: Michelle O'Brien
Managing Editor: John Roberts
Production Editor: Kerri M. Tomasso
Manufacturing Buyer: Indira Gutierrez/Michelle Klein

Cover Design: Kiwi Design
Photo Researcher: Teri Stratford
Image Permission Coordinator: Debbie Latronica
Text Permission Researcher: Jane Scelta
Composition/Full-Service
   Project Management: PineTree Composition
Printer/Binder: Phoenix Color Corp

Photo Credits: Chapter 2, page 59, John Gress Photography; Chapter 3, page 99, AP/Wide World Photos; Chapter 4, page 118, Getty Images Inc. – Stone Allstock; Chapter 7, page 223, AP/Wide World Photos; Chapter 9, page 292, Courtesy LG Electronics; Chapter 10, page 325, Courtesy Go2mobile Solutions, Ltd.; Chapter 10, page 327 (all 3 photos), Samsung Electronics America, Inc; Chapter 12, page 404, Getty Images Inc. – Image Bank.

Credits and acknowledgments borrowed from other sources and reproduced, with permission, in this textbook appear on appropriate page within text.

If you purchased this book within the United States or Canada you should be aware that it has been wrongfully imported without the approval of the Publisher or the Author.

Pearson Education LTD.
Pearson Education Singapore, Pte. Ltd
Pearson Education, Canada, Ltd
Pearson Education–Japan
Pearson Education Australia PTY, Limited

Pearson Education North Asia Ltd
Pearson Educación de Mexico, S.A. de C.V
Pearson Education Malaysia, Pte. Ltd
Pearson Education, Upper Saddle River, New Jersey

10 9 8 7 6 5 4 3 2 1
ISBN 0-13-123023-9

To the lights in my life that make me smile: Willy and Claire.

Jakki Mohr

To Baba and Ma
For the gift of life and the values to live it well.

Sanjit Sengupta

To my wife, Paula Galloway, who has been the bedrock in my life for over twenty years, and to
my mother, Anne Fant Slater, who instilled in me a deep love of learning and a sense of
intellectual curiosity.

Stan Slater

To the people who see the possibilities,
and who, sometimes with courage, sometimes with faith and hope—
but always with effort, perseverance, and energy
(despite self-doubt)—
strive to make the possible become reality.

Dreams can come true.

Contents

Preface

With this revised edition of the *Marketing of High-Technology Products and Innovations,* we invite you to consider the following ways in which the last three years have witnessed enormous changes in the technology arena:

- The economy has gone from one that characterized technology as a panacea, the driver of the economic engine, to a major source of the economic downturn in 2000–2002.
- Organizations and enterprise customers have tightened their enthusiasm for tech spending, monitoring more carefully the returns from investments in technology.
- In the area of consumer electronics and technology, the pace of innovation that offers a multiplicity of new products for customers continues unabated.
- In the Internet arena, businesses have more widely adopted electronic business technologies, designed to streamline business processes for enhanced effectiveness and efficiency.
- Companies and retailers are moving to harmonized distribution channels, offering customers bricks-and-clicks-models for a seamless shopping experience via a multitude of channel choices.
- Consumers continue to use the Internet for a wide array of activities, including searching for information, shopping, and joining virtual communities. The wider adoption of broadband technologies has allowed the sharing of more data, file-swapping of music, downloading of videos. Other bandwidth-intensive activities have also brought about a concomitant need to examine intellectual property rules, digital piracy, and legitimate use.

These are but a few of the many changes the technology arena has witnessed during the three years since the first edition of *Marketing of High-Technology Products and Innovations* was released. Our intent with this revised edition is to continue the development and synthesis of decision frameworks and strategies that reflect best practices in the area of high-technology marketing. This edition offers a cutting-edge treatment of research and practice related to the marketing of technology and innovations, supported with a plethora of examples and applications.

Thriving in the high-tech marketplace requires mastery of a diverse set of skills and capabilities. From adroitly reading market trends; investing wisely in future technologies; leveraging the skills and capabilities of technical and marketing personnel in a dynamic, interactive fashion; understanding customers intimately; offering a compelling value proposition; developing astute marketing campaigns; pricing with an eye to customer value; and harmonizing distribution channels and supply chains, high-tech marketing managers must be versatile, yet focused, flexible yet determined, tenacious yet open-minded.

In light of this complicated and challenging environment, we offer a systematic, thorough overview of the issues high-tech marketers must address. More specifically, based on feedback and reviews of the first edition, this second edition:

- Offers more in-depth treatment to the topical coverage in the chapters.
- Brings more examples to the book.
- Streamlines the presentation of the material.
- Updates the material to reflect new developments both in the area of high-technology marketing research and practice (for example, in the area of micro-payments and subscription pricing), as well as new technological developments.
- Adds coverage of under-developed areas in the first edition, such as the marketing of high-technology services.

Importantly, the addition of Sanjit Sengupta of San Francisco State University and Stan Slater of Colorado State University as new co-authors on this edition brings new expertise and perspectives. Their biographies appear at the conclusion of this preface; in brief, Professor Sengupta brings experience in e-business and executive education in high-tech marketing, and Professor Slater brings an extensive background in business strategy and market orientation.

As in the first edition, our primary aim is to show how marketing strategies must be modified and adapted for the high-tech environment. Marketing high-technology products and innovations is not the same as marketing more traditional products and services. For example, the marketing of a familiar product, say Coca-Cola, is very different from marketing products with which customers may be unfamiliar, say, new computer hardware or software such as a Pentium chip, customer relationship management software, or even a new computer video game. Customers' fear, uncertainty, and doubt about how to use and attain the full benefits of using the product contribute to the need for different marketing considerations. In addition, the competitive environment found in high-tech industries is different from that found in more traditional contexts: Innovations are often introduced by industry outsiders that industry incumbents are not aware of. Another factor contributing to high-tech marketing challenges is the velocity of change. Due to technological breakthroughs, products change so rapidly that standard marketing concepts may not be sufficient.

While a standard approach to marketing, such as the "4 Ps" of marketing (product, price, place, and promotion) is still relevant, the standard approach must be modified to account for the inherent uncertainty in high-tech environments. Using a framework to manage the marketing decision-making process fosters greater understanding of the common characteristics in high-tech environments, and helps manage the risk of marketing in a high-tech context.

ORGANIZATION

This revised edition of the *Marketing of High-Technology Products and Innovations* is organized into twelve chapters. Although the material and concepts in each chapter are treated rather distinctly, much of the material, by its very nature, is interrelated and requires consideration from a multiplicity of angles. As a result, concepts are cross-referenced with treatment in other chapters. For example, the material on strategy development in high-technology organizations (Chapter 2) is part and parcel of the product development and management chapter (Chapter 7). Similarly, the discussion of customer relationship management, introduced in Chapter 3, is linked explicitly to pricing strategies (Chapter 9) and advertising and promotion strategies (Chapter 10) as well. This approach at thematic coverage of the underlying drivers of high-tech marketing strategy offers a more comprehensive view of how these issues play themselves out in the different areas of the marketing toolkit.

More specifically:

- **Chapter 1** provides an *overview of the high-technology environment and its key characteristics.* Special attention in this edition is given to the issue of *network externalities* and the need to *develop industry standards.* Chapter 1 also introduces the notion that high-tech marketing strategies should be tailored to the *type of innovation (incremental or radical).* This notion of the contingent effects of the type of innovation on marketing strategies is carried through subsequent chapters.

- **Chapter 2** addresses strategy and corporate culture in the high-tech firm. First, it presents the *strategic planning process* in high-tech companies; in particular, it focuses on the need to develop *competitive advantage* and how to do so. The second part of this chapter focuses on *culture and climate in innovative companies,* noting the forces that can lead established companies to become complacent (i.e., the innovator's dilemma), and the characteristics of companies that overcome such forces—characteristics such as creative destruction, corporate imagination, "unlearning," and reliance on product champions. The final section in this chapter addresses *challenges for small high-tech startups,* including the vexing problem of resources, as well as ways to leverage their unique strengths.

- **Chapter 3** recognizes the vital role that *partnerships and alliances* play in effective high-tech strategies. Special attention is given in the second half of this chapter to *customer relationships.* This material has been strengthened with the addition of recent research linking customer loyalty to company profitability.

- **Chapter 4** covers the domain of *market orientation,* a particularly important consideration for high-tech firms, yet a skill or mindset that is particularly difficult to implement in technology-driven organizations. The second half of this chapter addresses the difficult dynamic between *R&D and marketing personnel.* The incorporation of best practices and cutting-edge research offer insights to manage this important dynamic.

- **Chapter 5** presents an overview of the *research tools* high-tech marketers can use to gather information about their customers. Concept testing, conjoint analysis, customer visit programs, empathic design, lead users, quality function deployment, prototype testing, and beta testing are covered, with examples provided for the types of insights each can offer. This chapter concludes with sections on *gathering competitive intelligence* and *forecasting customer demand* for high-tech products.

- **Chapter 6** addresses a particularly challenging aspect of high-tech marketing: *understanding customer behavior,* including customer decision making for high-tech products, and how marketing to early adopters must be different than marketing to late adopters. This chapter draws heavily on the work of Everett Rogers (*Diffusion of Innovations*) and Geoffrey Moore (*Crossing the Chasm* and *Inside* *the Tornado*). The *market segmentation process* is reviewed. Finally, ways in which high-tech marketers can *manage customers' migration decisions* (in moving to subsequent generations of new technology) wrap up the final section of this chapter.

Chapters 7 through 10 provide coverage of the 4 Ps of marketing: product, place, price, and promotion:

- **Chapter 7** presents the framework of a *technology map* to guide product development. Pertinent considerations include decisions about *technology transfer and licensing, product modularity, platforms and derivatives, and protection of intellectual property.* This chapter has been strengthened with the addition of material on new product development teams, as well as the *intersection of technology and services* and the marketing issues surrounding that intersection.

- **Chapter 8** provides a framework for making *distribution decisions.* Focus is given to the use of the *Internet as a distribution channel* and the need to manage the transition, resulting conflict with existing channels, and how to harmonize/integrate across channels. The final section of this chapter navigates the complex world of *supply chain management* for high-tech products.

- **Chapter 9** provides a framework for pricing decisions, with heavy emphasis on *customer-oriented pricing.* Moreover, in light of the rapid price declines in many high-tech industries, focused attention is given to strategies used to generate profits in light of the *"technology paradox"* (how to make money when the price of product is declining rapidly). New sections related to the *pricing of after-sales services, moving from "free to fee," and price bundling* offer new insights in this revised edition.

- **Chapter 10** emphasizes the importance of using advertising and promotion tools to *develop a strong brand name* (as one mechanism to allay customer anxiety), the need to manage *product preannouncements,* and communication tools used in *managing customer relationships.* This chapter now includes material on using the Internet for advertising and promotion (covered in the prior edition in Chapter 11).

- **Chapter 11** provides a focused lens on *electronic business and e-commerce.* It provides the context for e-business and e-commerce by providing brief highlights of the past three years. Key take-aways in terms of success factors and barriers to online success are highlighted. The chapter is organized into sections related to the Internet and Consumers and the Internet and Business Customers. New material related to Web services and other new Internet technologies are covered in this revision.

- **Chapter 12** concludes with consideration of snafus and other concerns that can inhibit the ability of a new innovation to realize its potential. Issues include *unintended consequences of technology,* the *paradoxes customers face* in adopting and using technology, *ethical controversies* arising from technological advances, *social responsibility* of business in technology arenas, and *government regulation.* Updated coverage of digital piracy, the digital divide, and other timely issues reflect recent trends in this area.

Marketing does not occur in isolation in any firm, but rather, it is cross-functional in nature. This book brings marketing together with other business disciplines (for example, research and development, legal, and management and strategy) to offer insights on how marketing is interrelated and dependent upon interactions with other disciplines. Issues for both small and large businesses are addressed. The book provides a balance between conceptual discussions and examples, startup and established business, products and services, and consumer and business-to-business marketing contexts. Using examples from a wide variety of industries and technologies to illustrate the marketing tools and concepts covered in the book not only captures the richness of the high-tech environment, but also proves the utility of the frameworks presented; the variety also gives the reader experience in applying the frameworks to diverse situations. Some of the industries and contexts covered include telecommunications, information technology (hardware and software), biotechnology, and consumer electronics such as high-density TV and digital video disks.

SPECIAL FEATURES

The following special features bring the material in each chapter to life:

Opening Vignette: Each chapter begins with an opening vignette, highlighting a particular company and how it has grappled with the issues in the coming chapter. The intent is to demonstrate the relevance of the chapter material and to provide a real-world example to facilitate understanding.

Technology Expert's View from the Trenches: Each chapter has one or two technology experts sharing their views about specific issues pertinent to the chapter. These are insights from people working in the field in a variety of high-tech positions. For example, Judy Mohr, Patent Attorney, offers her insights in the section on intellectual property in the product management chapter; Tami Syverson, former Competitive Intelligence Analyst at Sun Microsystems, offers her insights in the market research chapter. Other experts include Wade Sikkink, Supply Chain Manager at Intel; Darlene Solomon, Vice President and Director of Agilent Labs (Agilent Technologies), Daria Schuster, Director of Pricing Strategy at IBM, and Eric Zarakov, Vice President of Marketing, Foveon Corporation.

Technology Tidbits: Each chapter has a one- to two-paragraph summary about cool, cutting-edge technology of which the reader may be unaware, just to stimulate thoughts and knowledge about radical innovations coming down the pike. This material could provide an extemporaneous application of class concepts for the adventurous, motivated reader.

End-of Chapter Discussion Questions: The Discussion Questions at the end of each chapter are designed to assess the reader's knowledge of the material covered, offer additional opportunities to apply the chapter's concepts, and allow students to generate additional insights about the concepts.

SUPPLEMENTS

Companion Website: The text has a Companion Website (www.prenhall.com/mohr) that features current articles on a chapter-by-chapter basis, a password-protected area for professors for sharing resources and pedagogical information (including the instructor's manual with PowerPoint and answers to the discussion questions), and updated/current information on available cases and resources (such as recommended readings from the trade press for books and related information). Links to useful related Web sites are also provided for each chapter.

Instructor's Manual: The instructor's manual includes:

- Answers to the end-of-chapter discussion questions
- Powerpoints for each chapter
- A test bank

TARGET AUDIENCE FOR THIS BOOK

The book will prove useful in a variety of venues, including:

- Upper-level undergraduate and graduate courses on the marketing of high-technology and innovation
- Technology institutes, engineering management programs, biotechnology centers, and/or telecommunications programs
- Executive education courses
- Managers in high-tech firms
- Training programs in high-tech firms
- Technology incubators

Note that this book is not meant to be the only marketing reference for the high-tech marketer. Rather than addressing marketing fundamentals, the book's primary focus is on the unique characteristics of the high-tech environment and the challenges those characteristics pose for marketing. Because of the more advanced nature of the material (for a marketing novice), a book of marketing fundamentals should be used as a reference as well.

In addition, because this book is focused primarily on the *marketing* of technology and innovation, related books on the *management* of technology and innovation might also be useful complements. The website can be consulted for a list of suggested supplementary readings and books.

CONCLUSION

High-tech products and services are introduced in turbulent, chaotic environments where the odds of success are often difficult to ascertain at best and stacked against success at worst. This book is designed to provide frameworks for systematic decision making about marketing in high-tech environments. In doing so, it offers insights about how marketing tools and techniques must be adapted and modified for high-technology products and services. The text highlights possible pitfalls, mitigating factors, and the *how-to*s of successful high-tech marketing.

ACKNOWLEDGMENTS

Although the authors' names are listed on the front of a book, it is only through the efforts of many people that a book actually is completed. First, we sincerely appreciate the time that the reviewers put in to provide us with direction for this revision:

David Corkindale, *The University of South Australia*

John Durham, *University of San Francisco*

Sean M. Hackett, *Vanderbilt University*

Gary Lynn, *Stevens Institute of Technology*

Salvatore J. Monaco, *University of Maryland, University College*

Kenneth A. Saban, *Duquesne University*

Richard Spiller, *California State University, Long Beach.*

Also, one of the special features of this book—the "Technology Experts' Views from the Trenches"—exists solely because of the generosity of the experts and their companies. We appreciate the many willing contributions of dear friends, and friends of friends, who have written excellent contributions throughout the text. The first author, Jakki Mohr, gives special thanks to her sister, Judy Mohr, for her willingness to work with us to develop a prototype for these contributions.

We've also had much helpful assistance along the way in compiling information. Jenny Mish, Lou Fontana, and Jennifer Moe (graduate students at the University of Montana) assisted in researching examples and assisting with organization.

We'd also like to thank the following Prentice Hall team: Katie Stevens, Rebecca Cummings, Larry Armstrong, Bruce Kaplan, Michelle O'Brien, Jeff Shelstad, John Roberts, Kerri Tomasso, the International division, and Patty Donovan from Pine Tree Composition.

Jakki Mohr acknowledges the supportiveness of Dean Larry Gianchetta, Department Chair Nader Shooshtari, and her many thoughtful colleagues at the University of Montana.

Sanjit Sengupta acknowledges that many people have made valuable contributions to his thinking on technology and marketing over the years. His first exposure to technology issues was at the Indian Institute of Technology, Kanpur, through interactions with professors and classmates. Tarun Gupta introduced him to marketing at the Bajaj Institute of Management in Mumbai and got him excited about the subject. He learned a lot about the practical aspects of marketing computers and software services from his colleagues at HCL, especially Amit Dutta Gupta, and at CMC Limited during his early business career. Berkeley provided a vibrant, nurturing environment to incubate his early research ideas. He is especially grateful to Professor Louis P. Bucklin, his mentor at Berkeley, for helping refine research ideas, many of which they published together. His colleagues and students at the University of Maryland and San Francisco State University have played an important role in his intellectual development during his academic career. Finally, he wishes to acknowledge the support of his wife, Amrita, and children, Ishaan and Ila, who make all of his labor worthwhile.

Stan Slater acknowledges his long-time friends and research colleagues, John C. Narver, professor emeritus of marketing at the University of Washington, and Eric M. Olson, professor of strategic management and marketing at the University of Colorado–Colorado Springs.

About the Authors

Jakki Mohr (Ph.D. 1989, University of Wisconsin, Madison) is the Ron and Judy Paige Faculty Fellow and Professor of Marketing at the University of Montana–Missoula. Prior to joining the University of Montana in the fall of 1997, Dr. Mohr was an assistant professor at the University of Colorado, Boulder (1989–1997), where she earned both the Frascona Teaching Excellence Award (1992) and the Susan Wright Research Award (1995). Before beginning her academic career, she worked in Silicon Valley in the advertising area for both Hewlett-Packard's Personal Computer Group and TeleVideo Systems. Dr. Mohr's research has won awards and been published in the *Journal of Marketing,* the *Strategic Management Journal,* the *Journal of Marketing and Public Policy,* the *Journal of Retailing,* the *Journal of High Technology Management Research, Marketing Management,* and *Computer Reseller News.* Her interests lie primarily in the area of marketing of high-technology products and services, including a broad range of technologies, including but not limited to the Internet and e-commerce. She teaches courses in the Marketing of High-Technology Products and Services, Business-to-Business Marketing, Electronic Commerce and Internet Marketing, and Marketing Management.

Sanjit Sengupta (Ph.D. 1990, University of California, Berkeley) is Professor and Chair of the Marketing Department at San Francisco State University. He teaches courses in Strategic Marketing, Business-to-Business Marketing, and e-Business Marketing Strategy. Prior to joining San Francisco State in the fall of 1996, Dr. Sengupta was an assistant professor at the University of Maryland, College Park, where he received two teaching awards. He has taught in many executive development programs in the United States, Finland, and South Korea. His research interests include new product development and technological innovation, strategic alliances, sales management, and international marketing. His research has won awards and been published in many journals including *Academy of Management Journal, Journal of Marketing,* and *Journal of Product Innovation Management.* Prior to his academic career, Dr. Sengupta worked in sales and marketing for Hindustan Computers Limited and CMC Limited in Bombay, India.

**Stanley Slater** (Ph.D. 1988, University of Washington) is a Professor of Marketing at Colorado State University. From 1996 to 2002, he was a Professor and the Director of the Business Administration Program at the University of Washington's Bothell Campus where he was instrumental in launching an MBA Program designed specifically for professionals in technology-oriented businesses. Dr. Slater's major research interests are in the areas of the role of a market orientation in organizational success and marketing's role in business strategy implementation. He has published more than forty articles on these and other topics in the *Journal of Marketing,* the *Journal of the Academy of Marketing Science,* the *Journal of Product Innovation Management,* the *Strategic Management Journal,* and the *Academy of Management Journal,* among others. He has won "Best Paper" awards from the *International Marketing Review* and the Marketing Science Institute. Dr. Slater serves on the editorial review boards of the *Journal of Marketing, Industrial Marketing Management,* and *Business Horizons.* Prior to his academic career, he held professional and managerial positions with IBM and with the Adolph Coors Company. Dr. Slater has consulted with units of Hewlett-Packard, Johns-Manville, Monsanto, United Technologies, Cigna Insurance, Qwest, Philips Electronics, and Weyerhaeuser.

CHAPTER 1

Introduction to High Technology

*In 1981, it was predicted that "'personal V/STOL aircraft' (vertical/short takeoff and landing)
will come into widespread use to supplement automobiles."*
—JOHN P. THOMAS,

*Director of tourism studies at the Hudson Institute,
in* The Book of Predictions

The Excitement is Back in Tech

Hour-long hotel check-in lines at 1 a.m. aren't usually cause for celebration. Nor is the sight of one-time pop princess Debbie Gibson performing a live infomercial for a wireless karaoke microphone. But in this case, these were two scenes from the Consumer Electronics Show (CES) that prove the tech business isn't dead yet... it just went mainstream.

There was definitely an air of excitement at the CES. That's because it's no longer about the computer itself but about putting computer technology into making consumers' home life more comfortable and their car trips more pleasurable. A few products that stood out at the show:

- Real multi-purpose handhelds. The market has seen several "hyphenate" handheld computers in the last few years (combining phone/PDA/MP3/camera capabilities into a single unit), and "clunky" is probably the most charitable term for most of them. So, Samsung impressed the audience with its new phone, the i500, which includes a Palm device but is (happily) *not* the size of a Pop Tart that you have to hold up to your face when you want to make a call.

  In addition, Garmin's new Palm device offers the first handheld with an integrated global positioning system. Punch in an address or a calendar entry and the iQue 3600 smartly pulls up a map, gives turn-by-turn directions, and even voice commands. It will ship with maps of the U.S. and Canada, can serve as an MP3 player as well, and should be a boon for business travelers.

- Remember the Clapper? Hook it up to the lamp, clap your hands, and on or off it goes? Controlling home appliances—everything from lights to your home entertainment system—without having to be right in front of them or using 25 different remotes holds lots of appeal. That technology is now getting a makeover. Leviton's approach doesn't require installing an automated system into your house; homeowners can merely replace light switches and outlets with "smart" ones and use existing wiring. Philips' "home dashboard" iPronto looks impressive, allowing the creation of scenarios for your house, so when you want to watch a movie, press one button and the lights dim, the blinds close, and the DVD player fires up—all automatically.

- Twenty years ago, celebrated serial entrepreneur Nolan Bushnell, founder of Atari, created a big stir at CES demonstrating a personal robot. Since a robot didn't bring you breakfast in bed this morning, you've correctly surmised that the idea didn't take off quite as planned. So fast forward to 2003 and there's serial entrepreneur Bill Gross interacting on stage with a robot that has as much grace as, well, Debbie Gibson testing her karaoke mike. Evolution Robotics' primary goal is to license its technology, and its demos were neat if a bit rough around the edges.

Overall, it's good to see some real excitement back in tech!

Excerpted from: Lidsky, David (2004), "The Buzz is Back," www.fortune.com.

Since the first edition of this book was written in 2000, the high-tech world has changed dramatically. The Nasdaq Composite Index, home to major tech stocks, including Cisco Systems Inc., Microsoft Corp., and Intel Corp., dropped almost 80 percent from its peak of 5049 in March 2000 to 1108 in October 2002. Possibly more troublesome was the reduction in numbers of people employed in the technology sector due to bankruptcies and layoffs. According to the American Electronics Association, high-tech employment fell by 236,000 between January 2002 and December 2002.[1] However, by fall 2003, the technology sector seemed poised for a rebound. From October 2002 to October 2003 the Nasdaq was up by over 60 percent, and the computer and electronics products industry experienced the second largest decrease in initial jobless claims due to layoffs.[2]

Companies like Dell, Cisco, Google, and Salesforce.com seem to have figured out this new business environment and are succeeding in it. The key for tech companies is to think beyond their products—to get in their customers' doors to help remake their work, their ways of doing business, and their industries. As Andy Grove, the Chairman of Intel and one of technology's big thinkers, said, "You know that old saw that 'railroads are not in the railroad business, they are in the transportation business'? We are facing something similar to that." Hewlett-Packard CEO Carly Fiorina explained, "Tech is truly becoming part of the fabric of life. Think about the big problems we have to solve now—health care, homeland security, synchronizing the world's information systems to facilitate the flow of goods and services and to prevent the flow of undesirables—all of those are technology opportunities."[3]

Today's global economy is, to a large extent, driven by technological innovation. Advances in microchip technology are finding applications across a wide range of industries, well beyond traditional computer applications. For example, computer chips are being used in everyday household appliances, such as toasters, and even being implanted in pets and farm animals for identification and health monitoring. The field of biotechnology has taken off and is spawning innovations not only in medical applications but also in waste cleanup and crop biology. The advances in crop biology are one reason that chemical giants such as Monsanto Company, Dow Chemical, and DuPont are acquiring or investing heavily in food-technology firms, including seed and soybean companies such as Pioneer Hi-Bred International Inc., and forming partnerships with grain- and meat-processing companies such as Cargill and ConAgra Inc.[4]

The list of even basic industries that technology is changing is vast and includes automobiles, oil and gas, and consumer foods. Although some might believe that these industries are more low tech than high tech, innovations are revolutionizing them. The traditional mechanical engineering used in automobile design is migrating to electrical engineering. The auto industry has coined the term *mechatronics* to describe this combination of mechanical and electrical principles.[5] For example, brake-by-wire and accelerate-by-wire—where pressing the pedal sends an electronic signal rather than activating a physical connection to the engine or brakes—will become common. All Mercedes models have used brake-by-wire since 1994, and the Chevy Corvette and all recent Audis already have electronic gas pedals.[6] The oil industry is closer to a high-tech industry than to the commodity business it once was, now driven by companies that lead in the key technologies that drive down the costs of exploration and production.[7] Procter & Gamble receives about 5,000 patents each year globally. This makes P&G among the world's largest holders of U.S. and global patents, putting it on a par with Intel, Lucent, and Microsoft.[8] Clearly, technological innovations are revolutionizing many industries, creating a high-tech environment very similar in flavor and feel to more standard high-tech industries of computers, telecommunications, and so on.

As these examples indicate, the scope of high-technology applications is no longer limited to computers, telecommunications, or consumer electronics—the "traditional" high-tech industries; it encompasses a broad cross section of industries in today's business economy. To understand the benefits of technology and innovation, we must also look at the best companies using technology—those that are adopting new technology-based solutions that help create new business models. Take three-year-old JetBlue, which in a time of deep airline trouble is solidly profitable and grew 63 percent in the first quarter of 2003. The company uses technology to automate every aspect of its operation that it can, which is partly how it has kept its cost per seat-mile to 6.25 cents, far lower than American's 11.39 cents or even Southwest's 7.5 cents. Another example is Amazon. Amazon's major competitors are primarily brick-and-mortar companies such as Wal-Mart, Sears, and Barnes & Noble. Amazon uses the Internet and state-of-the-art tech tools to turn over its inventory at a rate of 19 times a year, vs. Wal-Mart's 7.6.[9]

Given the prominence of technological developments in our economy, categorizing particular industries as low- or high-tech may not be as easy as one would expect. Simply drawing a continuum ranging from low-tech industries on the one end to high-tech industries on the other and placing industries on the continuum based on common perceptions might, in fact, be misleading. Agriculture, heavy industry (steel mills, etc.), and services might not be as low tech as some might believe. The next section details various approaches to defining *high technology*.

DEFINING *HIGH TECH*

If high tech is permeating even basic industries, just what is high tech? Is it an industry that produces technology? Or is it one that intensively uses technology? Just what is technology? *Technology* is the stock of relevant knowledge that allows new techniques to be derived and includes both product and process know-how.[10] *Product technology* covers the ideas embodied in the product and its constituent components. *Process technology* encompasses the ideas involved in the manufacture of a product.

If technology is useful know-how, what, then, is *high* technology? There are nearly as many definitions of high tech as there are people studying it. For example, one definition characterizes high-technology industries as

> [those] engaged in the design, development, and introduction of new products and/or innovative manufacturing processes through the systematic application of scientific and technical knowledge.[11]

This section overviews various governmental definitions of high technology, as well as definitions found in research on high-technology marketing.

Government Definitions of High Technology

Most government definitions of high technology classify industries as high tech based on certain criteria such as the number of technical employees, the amount of research and development outlays, or the number of patents filed in a given industry. For example, the U.S. Bureau of Labor Statistics classified industries based on their proportion of R&D employment.[12] Thirty R&D-intensive industries—those with 50 percent higher R&D employment than the average proportion for all industries surveyed (Level I)—were identified; another 10 R&D-moderate (Level II) industries were identified, or those whose proportion of R&D employment was at least equal to the average proportion for all industries (but less than 50%). These 40 industries, and current and projected employment data for them, are shown in Table 1-1. They include both manufacturing and service industries, as well as defense and civilian industries.

The Organisation for Economic Cooperation and Development (OECD) uses a similar definition, defining high tech in terms of the ratio of R&D expenditures to value added of a particular industry.[13] The National Science Foundation examines the R&D intensity, or R&D spending-to-net-sales ratios.[14]

However, definitions based on these specific criteria do have shortcomings. The range of technical innovation in industries classified by the Bureau of Labor Statistics as R&D intensive is extremely wide,[15] including some industries whose products are modified only

TABLE 1-1 High-Technology Industry Employment 2000		
	% of High-Tech Employment 2000	*% Change in Employment 2000–2010*
Total	100.0	
Level I industries[a]	86.3	
Crude petroleum and natural gas operations	1.1	−22.7%
Cigarettes	0.3	−14.6%
Industrial inorganic chemicals	0.9	−16.3%
Plastics materials and synthetics	1.4	−15.8%
Drugs	2.8	23.8%
Soap, cleaners, and toilet goods	1.4	6.0%
Paints and allied products	0.5	7.5%
Industrial organic chemicals	1.1	−9.8%
Agricultural chemicals	0.5	8.9%
Miscellaneous chemical products	0.5	2.2%
Petroleum refining	0.8	−23.2%
Miscellaneous petroleum and coal products	0.4	10.7%
Nonferrous rolling and drawing	1.6	−1.8%
Special industry machinery	1.5	−8.2%
Computer and office equipment	3.2	−3.2%
Electrical industrial apparatus	1.3	−15.6%
Communications equipment	2.4	5.0%
Electronic components and accessories	6.0	17.3%
Motor vehicles and equipment	8.9	8.6%
Aircraft and parts	4.1	23.2%

(Continued)

TABLE I-I (Continued)		
	% of High-Tech Employment 2000	% Change in Employment 2000–2010
Guided missiles, space vehicles, parts	0.8	−3.9%
Search and navigation equipment	1.4	−9.3%
Measuring and controlling devices	2.7	−0.6%
Medical instruments and supplies	2.5	17.4%
Photographic equipment and supplies	0.6	−21.6%
Computer and data-processing services	18.5	86.2%
Engineering and architectural services	9.0	30.8%
Research and testing services	0.4	37.9%
Management and public relations	9.6	42.2%
Services, n.e.c.[b]	0.5	35.9%
Level II industries[a]	13.7	0.0%
Miscellaneous textile goods	0.5	6.2%
Pulp mills	1.8	−11.5%
Miscellaneous converted paper products	2.1	0.0%
Ordnance and accessories, n.e.c.[b]	0.3	−8.4%
Engines and turbines	0.8	−2.2%
General industrial machinery	2.2	3.5%
Industrial machines, n.e.c.[b]	3.3	10.0%
Household audio and video equipment	0.7	−3.3%
Miscellaneous electrical equipment and supplies	1.3	8.6%
Miscellaneous transportation equipment	0.7	19.0%

[a] *See text for definition of Level I and Level II industries.*
[b] *n.e.c. = not elsewhere classified.*
SOURCE: http://www.bls.gov/emp/empocc1.htm

incrementally (e.g., cigarettes) and in which new technological breakthroughs have not been seen in years. The classification may include industries in which most output is standardized and produced in large volume by relatively unskilled workers. These industries have a proportion of scientific or engineering workers high enough to make them R&D intensive or moderate, but the bulk of this talent may be used to alter incrementally the characteristics of established products in slowly growing, advertising-intensive markets.[16] One example would be the cigarette industry. Part of the reason for this ostensible misclassification stems from the underlying basis: the number of employees engaged in R&D.

Moreover, this classification may exclude the development of new products or processes by skilled workers in an industry whose score on R&D employment does not qualify it for high-tech status. For example, in one R&D center funded by the textile industry (generally not considered a high-tech industry), engineers and computer scientists are working to automate the design, cutting, and fitting of garments for retail customers. This project uses the latest in laser and computer technology.[17]

Finally, many low-cost manufacturers of electronic computers (SIC 3571) now use mass-produced components assembled in highly routine settings with minimal engineering and scientific input. Even within the semiconductor industry, high-volume chip manufacturing can involve high capital-to-labor ratios and relatively low scientific labor requirements.[18] Although such industries are generally classified as high tech, the innovations at this stage of the industry development may be fairly incremental.

As these examples, as well as the others noted at the outset of this chapter, show, the effects of technological change can be seen in almost every industry. Hence, rather than taking an industry-based approach to defining high tech, some advocate a definition based on common underlying characteristics. This approach is found in research on the marketing of high-technology products and innovations.

Defining High Tech in Terms of Common Characteristics

As shown in Figure 1-1, another view of high technology is based on common characteristics that all high-technology industries share,[19] most notably, market uncertainty, technological uncertainty, and competitive volatility.[20]

Market Uncertainty

Market uncertainty refers to ambiguity about the type and extent of customer needs that can be satisfied by a particular technology.[21] There are five sources of market

FIGURE 1-1   Characterizing High-Tech Marketing Environments

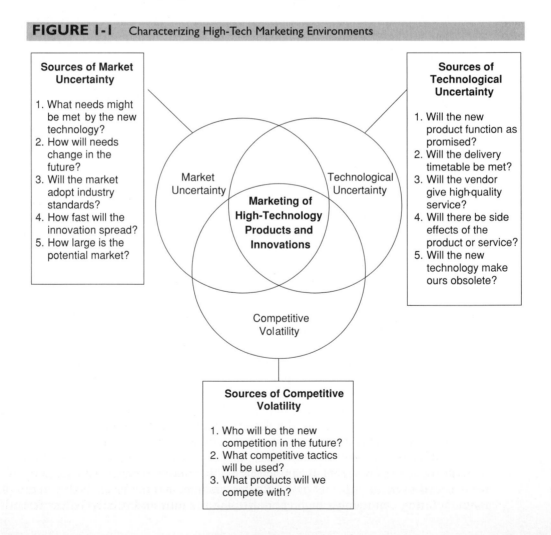

Sources of Market Uncertainty

1. What needs might be met by the new technology?
2. How will needs change in the future?
3. Will the market adopt industry standards?
4. How fast will the innovation spread?
5. How large is the potential market?

Sources of Technological Uncertainty

1. Will the new product function as promised?
2. Will the delivery timetable be met?
3. Will the vendor give high-quality service?
4. Will there be side effects of the product or service?
5. Will the new technology make ours obsolete?

Market Uncertainty

Technological Uncertainty

Marketing of High-Technology Products and Innovations

Competitive Volatility

Sources of Competitive Volatility

1. Who will be the new competition in the future?
2. What competitive tactics will be used?
3. What products will we compete with?

uncertainty. Market uncertainty arises, first and foremost, from consumer fear, uncertainty, and doubt (or the FUD factor[22]) about what needs or problems the new technology will address, as well as how well it will meet those needs. Anxiety about these factors means that customers may delay adopting new innovations, require a high degree of education and information about the new innovation, and need postpurchase reassurance and reinforcement to assuage any lingering doubt. For example, when a business decides to automate its sales force with computers, employees are bound to have some apprehension about learning new skills, wondering if the new mode of working will be better than the old one, and so forth. Hence, marketers must take steps to allay such apprehension both before and after the sale.

Second, customer needs may change rapidly, and in an unpredictable fashion, in high-tech environments. For example, customers today may want to treat their illnesses with a particular medical regimen but next year may desire a completely different approach to the same health problems. Such uncertainties make satisfying consumer needs a moving target.

Third, customer anxiety is perpetuated by the lack of a clear standard for new innovations in a market. The lack of accepted standards for the use of *DVD* technology in long-term data storage has hampered its acceptance. The DVD Forum (www.dvdforum.org), the official DVD standards body, has struggled to set a format standard for the industry. However, there is no single DVD storage standard today. Instead, there are the rival DVD+R and DVD−R write-once, read many times formats. The outcome thus far has been chaos, with rival formats and confusing terminology.[23] Questions about the dominant design of the future will hamper customer adoption, as buyers delay purchase to minimize the odds of making a "wrong" choice.

As another example, third-generation (3G) wireless technology relies either on a standard called CDMA 2000 (code division multiple access) or on W-CDMA (wideband code division multiple access). However, the two standards are not totally compatible. Hence, customers who adopt one technology may not be able to use it in other regions or countries. The lack of a universal standard has further perpetuated consumer fear, uncertainty, and doubt and slowed adoption of 3G technology.[24]

These examples highlight that one important role of high-tech marketing is to recognize the market uncertainty customers face in adoption decisions of new technology. Coalescing disparate product development efforts around common standards can help reduce the perceived risk for customers in terms of making a bad choice. Reducing fear and uncertainty can help serve as a catalyst for adoptions. Moreover, setting standards in an industry is intimately related to the product development process and involvement of business partners. If a firm chooses to use a unique or proprietary system in its product development, that is a very different process (with very different consequences) than choosing to develop a system based on open standards available to multiple players in an industry. For example, some have argued that Apple's decision to retain a proprietary operating system rather than collaborating on development of an open standard was a key weakness in its early strategy. Having a common industry standard allows the various players in an industry to collaborate on developing the complementary infrastructure required for products to perform. This topic of setting industry standards is so important that a later section of this chapter is devoted to it.

Fourth, due in large part to the prior three factors, uncertainty exists among both consumers and manufacturers over how fast the innovation will spread. For example, Figure 1-2 shows that ten years after color TVs were introduced, only 3 percent of U.S.

FIGURE I-2 Technology Adoption (percentage of households 10 years after introduction)

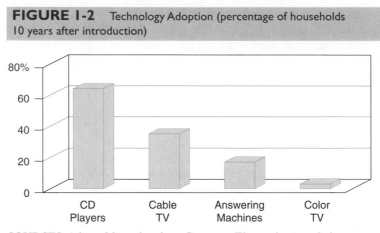

SOURCES: Adapted from data from Consumer Electronics Association, Electronic Industries Alliance, National Cable TV Association, and Encyclopedia Britannica.

households had purchased one. In many cases, the market for high-tech innovations is slower to materialize than most would predict.[25]

Finally, uncertainty over how fast the innovation will spread contributes to an inability for manufacturers to estimate the size of the market. Obviously, market forecasts are crucial for cash flow planning, production planning, and staffing. However, the other sources of market uncertainty contribute to the very real possibility of error in forecasting. For example, in 1988, Zenith forecast that in the year 1992, 66,000 high-definition televisions would be sold (10% of all TV sales), reaching 790,000 (100% of all TVs sold) by the year 1997.[26] However, not until January 1998 were the first HDTV sets sold and broadcasts aired. By year end 2002, HDTV sets were in fewer than 5 million U.S. homes—with a forecast penetration of 29 million homes by 2008.[27]

Indeed, Geoffrey Moore[28] refers to the "chasm" that high-tech products must cross in appealing to a mainstream market. When radically new innovations appear in the marketplace, they appeal to "visionaries" in the market (or innovators and early adopters), who are willing to adopt the new technology despite the often high price tag such items carry. For example, the earliest adopters of electric cars in California paid nearly 25 percent more to lease their cars than did people who leased a year later. These visionaries were willing to make do with the higher price and hassles that can accompany being an early adopter. For early adopters of electric cars, the "hassle factor" came in the limited number of stations that were capable of recharging batteries and the limited mileage range (90 miles between charges). For software adopters, the hassle factor might come in glitches and incompatibilities with other system components. The visionaries are willing to accept such inconveniences for the psychological and substantive benefits they do receive.

However, such benefits oftentimes are not sufficient for the majority of the market to adopt new technology. Typically, "pragmatists" comprise the majority of the market, and they require a different set of benefits and inducements than do the visionaries in their decision to adopt. The chasm represents the gulf between two distinct marketplaces for technology products. Visionaries are quick to appreciate the new development, but the pragmatists need more hand-holding. The transition between these two markets can be rocky at best, with many high-tech firms never crossing the chasm. Many high-tech firms find it hard to abandon their "techie" roots and talk to this group in customer-friendly terms. The inability to predict whether, and the degree to which, the mainstream market will adopt the product—and the rate of such adoption—given the presence of the chasm, makes it extremely difficult for manufacturers to estimate the size of the market.

In summary, market uncertainty arises from not knowing what customers want from the new technology, how their needs and desires will change, and how that uncertainty affects market development and size.

Technological Uncertainty

Technological uncertainty is "not knowing whether the technology—or the company providing it—can deliver on its promise to meet specific needs."[29] Five factors give rise to technological uncertainty. The first comes from questions about whether the new innovation will function as promised. For example, when new pharmaceuticals are introduced, patients may experience anxiety about whether the new treatment will be as effective for them as currently available treatments. As another example, the first online brokerage firms, the "guinea pigs of the Internet age" on the bleeding edge of technology,[30] faced major disabling computer glitches. After embarrassing failures early on (1997), the major online brokerages beefed up their Internet connections and added more computers to handle more transactions. But the crucial software (known as "middleware") knitting the Internet servers that display Web pages and the back-office computers that process the trades was vulnerable. Coupled with the inability to forecast the volume of trades that would occur, even better testing might not have revealed the problems.

The second source of technological uncertainty relates to the timetable for availability of the new product. In high-tech industries, product development commonly takes longer than expected, causing headaches for both customers and firms. For example, Storage Technology Corporation, a manufacturer of tape backup systems for computer data, announced a new technology to back up massive amounts of data on tape library systems, with the code name Iceberg. However, the product was delayed so long that a competitor beat Storage Tech to market, the company nearly went bankrupt, and customers who had delayed purchase of backup systems in anticipation of this new technology were extremely frustrated.

Third, technological uncertainty arises from concerns about the supplier of the new technology: If a customer has problems, will the supplier provide prompt, effective service? When (if?) a technician arrives, will the problem even be "fixable"?

Fourth, the very real concern over unanticipated consequences or side effects also creates technological uncertainty. For example, many companies have invested in information technology with the expectation that such investments would make their businesses more productive. Although the productivity boost from investments in information technology recently became apparent,[31] statistics indicate that 85% of employees use the Internet an average of 3.7 hours per week for personal purposes while at work.

Finally, in high-tech markets, technological uncertainty exists because one is never certain just how long the new technology will be viable—before an even newer development makes it obsolete. As a new technology is introduced, its performance capacity improves slowly and then, because of heavy R&D efforts, improves tremendously, before reaching its performance limits. For example, some people predict that future improvements in microchip performance are limited by the use of semiconductor technology. Japan's Ministry of International Trade and Industry sponsored a $30 million research program focused on technologies that could replace conventional semiconductors. These technologies are based on quantum physics and neural networks, rather than on electrical engineering. In the tape backup market, some are predicting that lasers and optic devices may make developments in tape media obsolete.

Competitive Volatility

A third characteristic that underlies high-tech markets is competitive volatility. **Competitive volatility** refers to changes in the competitive landscape: which firms are one's competitors, their product offerings, the tools they use to compete. There are three sources of competitive volatility.

First, uncertainty over which firms will be new competitors in the future makes it difficult for firms to understand high-tech markets. Indeed, the majority of the time, new technologies are commercialized by companies *outside* the threatened industry.[32] These new players are viewed as disruptive and frequently dismissed by incumbents.

Second, new competitors that come from outside existing industry boundaries often bring their own set of competitive tactics, tactics with which existing industry incumbents may be unfamiliar. However, these new players end up rewriting the rules of the game, so to speak, and changing the face of the industry for all players.[33] The most visible example of this type of volatility is the emergence of the "dot.com" players, which revolutionized the consumer retail business. Amazon.com and Expedia.com, among others, were totally unknown to retail booksellers, and airlines and travel agents, respectively, just a short time ago.

Third, new competition often arises as product form competition, or new ways to satisfy customer needs and problems. Manufacturers and marketers of video cassette recorders are threatened by marketers of digital video recorders (DVRs) such as TiVo and ReplayTV. Today, manufacturers and marketers of DVRs are being threatened by producers of set-top cable TV boxes that have a built-in DVR.[34]

As another example of competitive volatility, witness the personal computer and server operating systems market that has been dominated by Microsoft's proprietary *Windows* software. *Windows* is now being challenged by Linux, which is open-source software that allows users to inspect, modify, and freely redistribute its underlying programming instructions. This has forced Microsoft to begin imitating the ways of the open-source community. In 2002, the firm launched a "shared source" initiative that allowed certain approved governments and large corporate clients to gain access to most of the *Windows* software code, though not to modify it.[35] This was a radical departure from Microsoft's definition of an operating system based on proprietary or closely guarded software code.

Innovations by both new entrants and incumbents can render obsolete older technologies, and hence, the mortality rate of businesses in high-tech industries can be high, further contributing to competitive volatility. Ongoing industry leadership hinges on the notion of *creative destruction.* Paradoxically, a firm must proactively work to develop the next best technology, which is likely to destroy the basis of its current success and make its sunk investments in the prior technology obsolete. However, if the firm does not commercialize a new technology, rivals will surely do so. Even with successful products, rather than being overly focused on experience curve effects and economies of scale in production, firms should instead strive to develop even better technologies.

Figure 1-1 shows how the marketing of high-technology products and innovations occurs at the intersection of these three variables: market uncertainty, technological uncertainty, and competitive volatility. Although one of three characteristics may exist, or even a combination of two of the three, if all three factors do not exist simultaneously, then the uniqueness that such an environment poses for marketers will be less pronounced.

So, for example, in the case of a decision involving customer anxiety (say, a high-involvement–low-tech decision such as which home to purchase), if buyers aren't also simultaneously considering a radically new way of meeting their needs, then it would not be characterized as high tech. Similarly, customer needs may change rapidly in some areas (such as clothing styles, music, etc.), in which new styles and preferences may make obsolete older ones (technological uncertainty). However, although fads may indicate rapid change in an industry, such purchase decisions generally do not include both (1) a high degree of anxiety and (2) totally new ways of meeting customer needs. Finally, although competitive turbulence may be present in many industries (e.g., restaurants in college towns), the issue in a high-tech environment is whether the new competitors offer a radically new way of meeting customer needs. These situations highlight the fact that the *intersection* of these three characteristics typifies a high-tech marketing environment.

Network Externalities and the Importance of Industry Standards

In addition to the prior three characteristics, one of the most important aspects of many high-tech industries is the presence of **network externalities**. Also referred to as demand-side increasing returns or a "bandwagon effect," network externalities exist when the value of the product increases as more users adopt it. Examples include the telephone, portals on the Internet, and so forth: The first telephone was worthless, the second made the first more valuable, and so on.

Also known as Metcalf's Law, this concept illustrates the power that comes from the number of users who have adopted a particular technology. Metcalf's Law states that the value of a network scales as n^2, where n is the number of persons connected. According to Metcalf's law, as the number of users doubles, the value of the network quadruples. In other words, the utility received from an innovation is a function of the square of the number of users (installed base); the rapid take off is where the utility or value of the innovation increases exponentially because a critical mass of users has adopted it. This explains, in part, why some firms are willing to give their products away for free in order to rapidly grow an installed base.

This characteristic is an important contributing factor to the development of "de facto" monopolies. As one enterprise attains dominance in a market, the value of its user base is difficult for another enterprise to match. The self-reinforcing effects of the installed base can lead to a very few firms, sometimes only one, controlling nearly all of the market share in a product category. A challenger may attempt to displace the incumbent by introducing a radically improved technology and bypassing the current generation. However, a technological advantage alone is often not enough. To lure customers away from the existing standard, the new technology must somehow yield more value than the combination of value yielded by the incumbent technology's functionality, installed base, and products or services that make the incumbent more valuable.[36]

The concept of network externalities also explains why it is so vitally important for high-tech industries to set industry standards.

Why are Standards Important?[37]   Different technologies are often incompatible, and yet, due to the idea of network externalities, the first company to have its technology widely adopted may well set the technology standard for all. Indeed, the originator of a new technology has a clear advantage: the first product on the market is the standard. However, when competitors come on the scene, they may have developed alternative technologies and, with the right advantages (say, in product, production, or marketing), may flood the market with their own products.[38]

So, the more successful a firm is at getting its technology accepted as a standard, the more successful it will become in the future. Interestingly, this self-reinforcing cycle exists even when the emerging technological standard is inferior to other designs. One of the earliest examples of a market clinging to an inferior technology is the QWERTY format for typewriter keyboards, which gets its name from the first six characters on the upper left side of the keyboard. Because the type bars in the typewriters of the 1860s had a tendency to clash and jam when keys whose letters frequently appeared next to each other in words (e.g., "t" and "h") were struck in rapid succession, it was necessary to separate those letters. The result was that the QWERTY design slowed typing speed. By the 1890s, better engineering had alleviated the problem of clashing type bars, and a new keyboard format that enabled faster typing was developed. However, these superior keyboards did not do well in the market; typists were so comfortable with the QWERTY design that attempts to use a new design seemed cumbersome to them. As a result, QWERTY is still the standard, even today.

Because new technologies in an industry are often incompatible with each other, adopters of the new technology face increased fear, uncertainty, and doubt. To the extent that companies offering a new technological product can agree on industry-wide standards that result in a common, underlying architecture for products offered by different firms in the market, it can lessen customer's fear, uncertainty, and doubt. The industry standards allow *customers to gain compatibility* across the various components of a product—say, across hardware and software—and across product choices in an industry—say, across different types of computers. Moreover, because the complementary products share a common interface, the customers' hardware and software interface seamlessly. For example, in the cellular telecommunications industry, compatibility allows base stations, switches, and handsets to work with each other across service areas.

Compatibility, achieved when many different companies produce their offerings based on a common set of design principles, in turn, increases the value a customer receives from owning a product. This increase in value should facilitate the adoption decision for customers. As noted previously, this can be particularly helpful when the value of the product to the customer increases as more customers adopt products based on the same technology. In other words, the more customers that adopt products sharing a common, underlying technological standard, the greater the value each of them receives.

Second, the availability of complementary products is largely determined by the size of the *installed base* (or the number of users) of the given product. For example, software developers are more willing to write applications programs around technology platforms that have wider penetration in the market. Standards are one tool that can be used to ensure a greater availability of complementary products. The greater the availability of complementary products, the greater the value a customer derives from the base product.

Each of these factors works in a self-reinforcing manner. A larger installed base leads to greater availability of complementary products, which increases the value of the product to the customer, which in turn increases demand for that product by others, which translates into a larger installed base.

To reiterate: The logical conclusion from the self-reinforcing nature of standards is that a critical success factor in industries where demand-side increasing returns exist is *how quickly a firm can grow its installed base of customers using its design.* Some lawyers

BOX 1-1

STRATEGIES TO USE FOR SETTING
AN INDUSTRY STANDARD[41]

A firm has four main strategies to set an industry standard and grow its installed base of customers.[42]

1. Licensing and OEM agreements. By licensing its technological design to others, a firm can help to grow the market quickly using that design. A spinoff of this strategy is known as an OEM ("original equipment manufacturer") strategy, in which a firm sells subcomponents to other companies (OEMs) that compete in the same market. For example, Matsushita licensed its VHS format to Hitachi, Sharp, Mitsubishi, and Philips NV, who produced their own VHS-format videocassette recorders and tapes. In addition, Matsushita also provided to GE, RCA, and Zenith, on an OEM basis, the components needed to assemble their companies' equipment.

The licensing strategy ensures a wide initial distribution for the technology, which helps to build the installed base. In addition, this strategy co-opts competitors that might have had the capabilities to produce their own competing technology. It also limits the number of technologically incompatible product choices that customers face, reducing their confusion and doubt and hastening market acceptance. For example, Philips NV's decision to license the VHS format from Matsushita eliminated its push into a different VCR format (the V2000). Licensing also signals to suppliers of complementary products the possibility of a larger installed base, providing an incentive for them to pursue development.

The main drawbacks of a licensing strategy are that

- Licensees may attempt to alter the technology and avoid paying licensing fees or royalties.
- By increasing the number of suppliers in the market, the original developer loses a possible monopoly position, having to share revenues derived from the market with its licensees. Moreover, competition may result

in lower prices in the market (lowering profits for the original developer as well).

2. Strategic alliances. By entering into a cooperative agreement with one or more actual or potential competitors, firms can jointly sponsor development of a particular technological standard. For example, during the development of the digital audio technology used in compact disc players, at least four companies were pursuing incompatible designs. Although Philips NV was closest to commercialization, it worried about the issue of compatibility. So it partnered with Sony to cooperate on the commercialization of the first compact disc system. Philips NV contributed a superior basic design and Sony provided the error correction system. This alliance threw momentum behind the Philips–Sony standard, and 18 months prior to product introduction, over 30 firms had signed agreements to license the Philips–Sony technology.

Again, strategic alliances help to ensure a wide initial distribution for the technology, they can co-opt competitors, and they help to build positive expectations for the market demand, inducing other companies to develop complementary products. Alliances can help reduce confusion in the marketplace and build momentum behind the jointly sponsored standard, persuading potential competitors to commit to the standard. A particularly compelling advantage of this strategy for standards development is the fact that alliances may be able to produce a superior technology by combining the best aspects of two companies' know-how, which can also increase the probability of the joint development becoming the industry standard.

Of course, the risk exists that a partner might appropriate the firm's know-how in an opportunistic fashion. This risk might be mitigated by crafting terms to structure and manage the alliance in such a way that provides a disincentive for doing so. For example, by structuring the alliance as a joint venture, in which each company has a stake, each party then has a credible commitment to avoid undermining the alliance. Alternatively, the alliance might be structured in such a way that, after the standard has been developed, each party might be free to go its own way with regard to future extensions of the technology.

3. Product diversification. Customers are hesitant to adopt a new technology unless complementary products from which its value is derived are available. For example, customers won't adopt DVD players if there are no movies available to rent or buy in DVD format. Yet, suppliers of complementary products have no incentive to develop them when no installed base of customers for the new technology exists. Given this chicken-and-egg situation, a firm just may have to diversify into producing the complementary products whose wide availability is crucial to the success of a new technology. For example, Matsushita had its in-house record label, MCA, develop an extended offering of digital compact cassette (DCC) recordable audiotapes in order to jump-start the adoption of its DCC technology. Similarly, Philips issued a wide selection of prerecorded DCC tapes under its in-house PolyGram Records label when it introduced its DCC digital audio recording equipment in 1992. In addition to providing the impetus in the market to start an increasing-returns process, this strategy allowed the company to realize revenue from not only the sales of the base product but also the sales of the complementary products.

The downside risks to this strategy can be significant, however. If the company is starting the development of complementary products from scratch, it can entail a significant capital commitment, as well as possibly straying from its core competencies. If the technology fails to become the standard, the costs of failure are that much greater. However, if no potential suppliers of complementary products will respond rapidly to the firm's introduction of a new technology, then the firm may have no choice but to diversify into complementary products.

4. Aggressive product positioning. Positioning to maximize the installed base of customers relies on penetration pricing, product proliferation, and wide distribution. *Penetration pricing,* including pricing below current costs, makes sense in high-tech markets with demand-side increasing returns. This is why, for example, cellular phone service companies are willing to give handsets away for free and why companies such as Nintendo charge a relatively low price for the console to their games. In each case, the company plans to grow the installed base in order to establish the company's product as the standard in the market. Moreover, in these cases, the companies try to recoup the base product's costs in sales of the complementary products (phone services or the games themselves). *Product proliferation* attempts to serve as many customers as is feasible by developing a product offering to appeal to different segments in the market. Finally, *wide distribution* can help to build the initial market base for a new technology and ensure its place as the market standard. Companies can gain distribution by raising expectations in the market regarding the likely success of the new standard, by participating in alliances, and by licensing the technology. Each of these will signal to the distribution channel the clout that is behind the new technology. Aggressive positioning requires considerable investments in production capacity, product development, and market share building, all of which are sunk costs if the strategy fails.

Some believe that poor implementation of this latter strategy was the reason behind the failure of the DCC technology. Although DCC technology would replace cassette tapes (which are based on analog technology) in much the same way that CDs and CD players replaced analog record players and albums, customers were confused over the benefits of the digital recording technology. Philips did not mention the fact that the DCC tape decks would play both existing analog and new digital tapes and did not highlight the benefits of the new recording technology. Moreover, the initial market price of $900 to $1,200 per DCC tape player was very high. Consumers were also worried about the presence of another, incompatible digital recording standard, Sony's minidisc system, and so adopted a wait-and-see attitude. Retailers were unable to move their initial inventory of DCC players and tapes and were wary about trying again. Philips's initial offer was limited to home entertainment centers and did not include portable players or car players. This lack of proliferation further limited the potential installed base.

WHICH STRATEGY MAKES SENSE?

The question of which of these four strategies to use in growing an installed base in hopes of coalescing customers and competitors around a particular standard depends upon three factors:

- The barriers to imitation, such as those found in patents or copyright protections, for example.

- A firm's skills and resources, such as those found in manufacturing and marketing capabilities, financial resources, company reputation, etc. (Note that technological skills alone are insufficient for establishing a technology as a standard.)

- The existence of capable competitors, which puts a premium on the firm's ability to build an installed base of customers rapidly, despite the risks.

Given consideration of these factors, a firm can decide which strategy to pursue in setting an industry standard, as summarized in Table 1-2. Importantly, a firm should move quickly to establish itself as the *sole supplier* of a standard-defining technology (by avoiding licensing agreements and alliances, developing key complementary products if needed, and adopting an aggressive positioning strategy) when barriers to imitation are high; when the firm possesses required skills and resources to establish

the technology as the standard; and when there is an absence of capable competitors who might develop their own, possibly superior technology.

On the other hand, *passively licensing* the technology to all comers and letting the licensees build the market for the technology makes sense when barriers to imitation are low, the firm lacks requisite skills and resources for market development, and there are many capable competitors. For example, Dolby has adopted this strategy, licensing its high-fidelity sound technology to all players in the audio player market. Its low licensing fee has prevented competitors from developing a superior technology. Despite the low licensing fee, the large market volume provides Dolby a good revenue base.

A firm might undertake a more *aggressive, multiple licensing* strategy (licensing to as many firms as possible to build momentum for a standard), while at the same time adopting an aggressive positioning strategy, in order to become the dominant supplier of the technology. So, while a firm is trying to persuade rivals to adopt its technology (building market expectations that it will become the standard and providing the incentive for development of complementary products) and facilitate their entry into the market, it also tries to preempt its licensees in the marketplace through aggressive positioning. This strategy makes sense when barriers to imitation are low, the firm has the needed skills and resources to establish its technology as the standard, and there are many capable competitors.

TABLE 1-2   Strategies to Set Industry Standards

	Barriers to Imitation	Firm has Requisite Skills	Are there Capable Competitors
Aggressive Sole Provider	High	Yes	No
Passive Multiple Licensing	Low	No	Yes
Aggressive Multiple Licensing	Low	Yes	Yes
Selective Partnering	High	No	Yes

Finally, a firm is best served with a *selective partnering* strategy (the firm partners with one or a few other companies in order to jointly develop the firm's technology as a new industry standard) when there are high barriers to imitation, the firm lacks critical resources and skills, and there exist capable competitors who could develop a competing technology. For example, when IBM was developing the first personal computer, it partnered with Microsoft and Intel. IBM lacked an operating system for the PC, and the operating system was dependent upon the chip. (Note the risk in this approach, in that Microsoft seized industry leadership from IBM as a result of this alliance.)

argue that dominant technology suppliers, such as Intel and Microsoft, have become *de facto* monopolies because their products are so widely used as the basis for the industry standard.[39] The combined standard is referred to as the "Wintel" (Windows–Intel) duopoly. These two products are critical to the success of many of their customers—companies such as Dell, HP, and others. As a result, legal experts call such products "essential facilities"[40] as uneven access to such products may be a basis for unfair competition and, hence, antitrust suits. Box 1-1 presents four strategies companies use to set industry standards.

Additional Characteristics Common to High-Tech Markets

Additional features that are common to technology markets include the following:[44]

- *Unit-one costs.*   **Unit-one costs** refer to the situation when the cost of producing the first unit is very high relative to the costs of reproduction. This type of cost structure is likely to exist when know-how, or knowledge embedded in the design of the product, represents a substantial portion of the value of the products and services. For example, the costs of pressing and distributing software on a CD-ROM are trivial compared to the cost of hiring programmers and specialists to develop the content recorded on it.

- *Tradeability problems.*   When underlying know-how represents a substantial portion of the value of the products and services in question, buyer–seller exchanges are effectively transformed into intellectual property transactions. So, **tradeability problems** occur when it is difficult to value the knowledge, especially when it is tacit and resides in people and organizational routines. This becomes particularly important when a firm is considering a licensing strategy and must decide how much to charge for the license.

- *Knowledge spillovers.*   **Knowledge spillovers** exist when synergies in the creation and distribution of know-how further enrich a related stock of knowledge. Simply, every innovation creates the opportunity for a greater number of innovations. For example, it was once estimated that the Human Genome Project (used to map all human genes) would take at least forty years to complete, but it took only a fraction of that time, due to knowledge building on knowledge.[45]

    These spillovers create increasing returns in the development of related technologies, resulting in modularity in which products are broken down into components and subsystems, or "modules." Such modularity gives users more choice over time, based on compatibility and common standards.

Even when an industry is subject to market, technological, and competitive uncertainty, innovations are a question of degree and occur at different levels in the supply chain. The next section explores where the innovations occur across a supply chain.

A SUPPLY CHAIN PERSPECTIVE ON TECHNOLOGY

A supply chain depicts the flow of product from producer to consumer. A sample supply chain—albeit a simplistic one—for the automotive industry is shown in Figure 1-3. Customers who buy cars—both consumers for personal household consumption as well as businesses who need fleets for salespeople, company cars, and so forth—purchase cars from car dealers (either through a physical location or on the Internet). Car dealers replenish their inventories from the car manufacturers. To produce the cars, manufacturers must either make or buy all the requisite components: glass, metal, tires, electrical assemblies, drivetrains, and so forth, as well as production equipment (i.e., CAD-CAM systems for car design, etc.).

In many cases, technological innovations occur at the supplier level in the supply chain, rather than in the product (i.e., cars) itself. For example, a major innovation in the auto industry is the hybrid gas-electric car. This innovation required major changes in the car's design and componentry. Engines, batteries, wiring, and the frame all had to be reworked. Each of these changes occurred at levels in the supply chain rather distant from the end user, who is still using the same mode of transportation—a car—despite the innovation.

To be sure, innovations can occur in the nature of the product itself; for example, maybe for commuting in the year 2050, we'll be using jetpacks to propel ourselves through the air. (Or, as the Technology Tidbit predicts, we might have flying cars!) However, more often than not, innovations occur at higher levels in the supply chain, affecting the design, componentry, and production processes of the product, rather than revolutionizing the nature of the product itself. Other examples of the prevalence and predominance of technological innovations occurring at higher levels in the supply chain can be found across many industries, both high- and low-tech:

- In the oil industry: Major innovations have occurred in exploration and extraction that have revolutionized that industry.
- In the computer industry: Major innovations have occurred in the chips that power the computers.
- In the food industry: The fat substitute olestra is an ingredient that can be used in many foods, such as potato chips and ice cream.
- In hairstyling: Stylists can use 3-D images to view what a particular style would look like on a specific client prior to the cut.

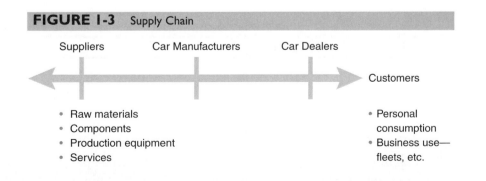

FIGURE I-3   Supply Chain

DREAM MACHINE
THOMAS LUNDELL

Have you ever found yourself stuck in a traffic jam thinking, "If only my car could lift from the ground and fly away . . ."? Well, with the M400 Skycar, you can. The M400 can take off from your backyard and fly at 350 mph, racing over the heads of land-bound commuters stuck in traffic below.

The M400 takes off using the thrust of four sets of shrouded, ducted fans. It is quieter and less dangerous than a helicopter, and the user-friendly control system makes flying the M400 about as easy as driving a car. Once in the air, you log on to a satellite tracking system and the Skycar flies by itself to your destination of choice.

Paul Moller, inventor of the Skycar, predicted in 2000 that we were about five years

Photograph reprinted with the permission of Moller International, Davis, California

from large-scale controlled airways. But if you have one million dollars to spend today, you don't have to wait five years. You can throw condescending looks on your stuck co-workers by rush hour tomorrow morning.

SOURCE: McCosh, Dan (2000), "Dream Machine: M400 Skycar," *Popular Science,* February 28, www.popsci.com/scitech/features/skycar/index.html; Waldman, Peter (1999), "Great Idea . . . If It Flies," *Wall Street Journal,* June 24, pp. B1, B4.

Many innovations occur at levels in the supply chain that are removed from end users. Hence, one cannot assume that an industry is *not* high tech simply because the products don't change much at the end-use level. Has the way in which we gas up our cars changed due to the innovations in oil exploration and extraction? Not really. Do we use our computers differently because they have a new chip design? Not substantially. A supply chain can be a useful tool to help understand and define high tech in broader terms than just the nature of the products we use.

Clearly, some innovations are more significant in terms of the nature of the breakthrough represented. The next section explores the various types and patterns innovations can take.

A CONTINUUM OF INNOVATIONS

As shown in Figure 1-4, innovative developments can be placed on a continuum ranging from radical, breakthrough developments on the one hand to more incremental, modest developments on the other.

Radical/Breakthrough Innovations

Radical innovations are "so different that they cannot be compared to any existing practices or perceptions. They employ new technologies and create new markets.

Incremental	Radical

Incremental	Radical
• Extension of existing product or process	• New technology creates new market
• Product characteristics well defined	• R&D invention in the lab
• Competitive advantage on low-cost production	• Superior functional performance over "old" technology
• Often developed in response to specific market need	• Specific market opportunity or need of only secondary concern
• Demand-side market	• Supply-side market
• Customer pull	• Technology push

FIGURE 1-4   Continuum of Innovations

Breakthroughs are conceptual shifts that make history."[46] In standard marketing parlance, they are discontinuous innovations. Others refer to breakthrough innovations as revolutionary,[47] and they are developed in supply-side markets.[48] Supply-side markets are characterized by innovation-driven practice, in which a company's goal is to achieve profitable commercial applications for laboratory output; R&D is the prime mover behind marketing efforts, and specific commercial applications or targets are considered only after the innovation is developed. For these reasons, these markets are sometimes referred to as "technology-push" situations.

Most radical innovations are developed by R&D groups (in companies, in universities, in research laboratories), who often haven't specifically thought about a particular commercial market application during the development process. For example, Tim Berners-Lee, a software engineer, assembled a network of interconnected computers to share and distribute information easily and cheaply in 1980, well before Marc Andreessen developed a Web browser.[49] In the United States, the Internet emerged as an outgrowth of the need for the U.S. military computer system to survive a Soviet nuclear attack.[50] These innovations were created independently of the vision of the uses they would serve. In the words of Amgen's CEO, Gordon Binder,

> Conventional wisdom says listen to the market. Most pharmaceuticals companies, and quite a few biotech ones as well, are basically market-driven. They see that large numbers of people have a particular disease and decide to do something about it.[51]

However, rather than start with the disease and work back to the science, Amgen does the opposite. It takes brilliant science and finds a unique use for it. The company's immune booster, for instance, helps keep the side effects of chemotherapy from killing cancer patients. And a collaborative arrangement with a professor at Rockefeller University discovered a gene that may yield new treatments for obesity.

In other cases, radical innovations are developed as a new way to meet an existing need, or in response to the identification of an emerging need. Regardless of whether the innovation originates from "pure" science or in response to a need, the new technology then creates a new market for itself. Competitive advantage for a breakthrough technology is based on the superior functional performance that the new innovation has to offer over existing methods or products.

The ways in which radical innovations are developed, the challenges in successful development, and how a company manages its portfolio of innovations is addressed in this chapter's Technology Expert's View from the Trenches.

TECHNOLOGY EXPERT'S VIEW FROM THE TRENCHES

DEVELOPING BREAKTHROUGH INNOVATIONS
DARLENE J.S. SOLOMON, PH.D.
Vice President and Director, Agilent Laboratories
Agilent Technologies, Palo Alto, CA.

Darlene Solomon

Although high-tech companies actively pursue the development of breakthrough innovations, these innovations are neither easily nor predictably achieved. Key success factors include:

- People energized in a work environment that nurtures innovation.
- A conviction to see and impact the future.
- An ability to identify and enable synergies across technical disciplines.

PEOPLE

People are Agilent's most important asset, and it is only though hiring and retaining excellent high performers that we will achieve, or even better, *exceed,* our aggressive goals. Anyone can solve the straightforward problems, but to address the really hard problems that accompany the development of breakthrough innovations requires a *high-performance culture.* High performance has many different meanings; it may encompass aspects such as integrity and respect, access to research tools and support, education and development, empowerment, teamwork, or incentives. In a high-performance culture, everyone works with purpose and knows how his or her efforts and actions are making a difference.

A CONVICTION TO SEE AND IMPACT THE FUTURE

The second success factor is to *envision the future* and solve "problems that matter." While there is no shortage of problems waiting to be addressed, breakthrough innovations preferably focus on the problems that, if solved, will provide clear and compelling customer value and generate attractive financial returns. These are the problems that, when addressed, really "move the needle," enabling 10X improvements in customer value (performance and/or cost), or even better, enable capabilities that couldn't possibly have been envisioned, much less achieved, prior to that innovation.

To begin to see the future, the best starting point is a solid understanding of the present and its inherent and often unstable assumptions. Seeing the future requires an understanding of the current market, including needs, trends, demographics, and emerging or changing paradigms. It also requires understanding the limits of technologies as well as potential changes in assumptions that could develop in the course of decade-long research and development programs.

For commercial entities, seeing the future lies in the ability to go beyond customers' articulated needs and identify a product that exploits a breakthrough innovation with a viable business model with the right timing. Looking back on my twenty years of research experience in HP and Agilent Laboratories, there are several admittedly painful examples of breakthrough technologies that had all of the above factors going for them, but were simply two to five years ahead of their time. As technologists, we are often better at estimating technical risk than market acceptance. But, of course, our best leaders and innovators are able to do both well.

When working at the forefront of research, it is often easy to lose sight of the realities of commercial markets. Working at the leading edge of what is technologically possible in the research environment may be appropriate for very early market adopters, but may not represent the technology that the majority of the market is ready to adopt and purchase.

On a related note, originally envisioned markets and customers for breakthrough innovations may need to change over the course of the research and development process. There is an inherent level of flexibility that is essential when doing longer-range research. This need for flexibility may arise from the continuous evaluation of the technological strengths and limitations revealed in research, the emergence of competitive technologies, and the reality of changing market and industry paradigms.

IDENTIFY AND ENABLE SYNERGIES ACROSS TECHNICAL DISCIPLINES

As we look into the future, many of the high-impact problems that exist and are in need of breakthrough innovation are complex and require *multidisciplinary contributions*. People who have a high level of technical proficiency within a traditional discipline are that much more valuable when they are able to span seemingly unrelated disciplines. This skill often requires levels of communication and metaphor not readily taught in school. Whether it is the convergence of biology and computational analysis, the broad range of applications of nanotechnology across areas as diverse as data storage and therapeutic drug delivery, or the challenges of global industries such as energy or homeland security, research organizations that can attract and harness technical diversification are well positioned for success in addressing these opportunities through breakthrough innovation.

EXAMPLE

Agilent's development of polymer-based microfluidic structures provides an illustration of the success factors and complexities in developing breakthrough innovations. The development of this technology was initiated in HP Labs (now Agilent Labs) in 1993 and is being actively commercialized as of the writing of this book in 2004.

Polymer-based microfluidic structures are tiny pieces (films) of plastic that contain a variety of different patterns of ports, channels, and compartments. Compared to technologies in existence in the early 1990s, these structures can be used to analyze fluids in small samples (say, a water or soil sample from an agricultural application) for the presence of chemicals, waste, and other impurities at the site of interest and at the time of information need.

The biochemist and engineer who proposed this research project championed their approach passionately, despite a significant level of opposition by their undeniably intelligent and experienced colleagues who did not believe that such testing structures could be successfully developed. It was a tough decision to go forward and fund this program against such opposition. The drive and commitment of these researchers to demonstrate that they were "right" were key motivators and were accompanied by milestone-based research objectives to ensure that the early-stage work remained focused on the key issues of technical feasibility.

From a market perspective, the originally envisioned application for polymer microfluidics was in field-based contexts such as water and soil samples for environmental measurement. However, through the 1990s, this application area became a less attractive market. Nevertheless, we continued to invest in the technology foundation as a fundamental enabler of chemical and biochemical analysis. The breakthroughs we were able to achieve in sample size reduction, throughput, and processing integration are fundamental to many scientific applications.

Given the compelling value that microfluidic technology can provide in the biotechnology arena, Agilent's first target for commercialization of this innovation is proteomics (the study of the proteins expressed by genes that make up a cell), a very different market than environmental analysis in field applications. Customers (scientists) in biotech companies will be able to use the breakthrough capabilities of microfluidics to separate complex protein mixtures for subsequent identification by mass spectrometry in order to further the study of proteomics. Ultimately, these studies will lead to the identification of new protein targets and more effective therapeutics to better combat disease and improve the quality and cost of healthcare.

Polymer-based microfluidics is a breakthrough innovation produced by skilled scientists and engineers who envisioned the future, worked across disciplines to solve important problems, tirelessly championed their views, and capitalized on novel combinations of knowledge to develop new ways of analyzing samples. Flexibility to target an emerging market application that was different from the originally envisioned application, while risky, seems to have produced a more compelling application than the original one. Soon, Agilent will know whether the market risk of this new technology will be as skillfully addressed as was the technical risk during its development.

Incremental Innovations

Incremental innovations, on the other hand, are continuations of existing methods or practices and may involve extension of products already on the market; they are evolutionary as opposed to revolutionary. Both suppliers and customers have a clear conceptualization of the products and what they can do. Existing products are sufficiently close substitutes.[52] Incremental innovations occur in demand-side markets,[53] in which product characteristics are well defined and customers can articulate their needs. In contrast to the view of the Internet as a radical innovation, some see it as an evolutionary innovation, "part of a continuum of technologies that drop the cost and improve the distribution of information," comparable to the impact of television.[54]

In an industrial context (manufacturing applications), incremental innovations may be developed by producers of a mature product who have achieved high volume in their production process.[55] Hence, economies of scale may be very important, and pricing may be based on experience curve effects (costs decline by a fixed and known amount every time accumulated volume doubles) that arise from economies of scale and learning curves. Often, because of the importance of scale economies to these firms, innovations may take the form of production *process innovations,* which lower the costs of production. Competitive advantage is frequently based on low-cost production. Firms whose bread-and-butter business comes from a specific product find that they may be less flexible to radical change and are vulnerable to obsolescence.

Some believe that marketing strategy for innovations is complicated by the fact that innovating firms might view an innovation as a breakthrough, whereas customers might view it as incremental (or vice versa). The unique problems this discrepancy in perceptions can cause are explored in Box 1-2, "Suppliers' and Customers' Different Perceptions of Innovation."

BOX 1-2

SUPPLIERS' AND CUSTOMERS' DIFFERENT PERCEPTIONS OF INNOVATION

Figure 1-5 highlights the four possibilities that can occur when considering both the suppliers' and customers' perceptions of the innovativeness of a new product. Obviously, when both parties' perceptions match, the path to marketing is fairly clear—as long as marketers understand that each type of innovation needs to be managed differently, as explained in the chapter. However, when a firm views an innovation as incremental but customers see the innovation as a breakthrough (or vice versa), mistakes can happen.

1. *Shadow products* are developed in the shadow of other, more central products and are not the central thrust of a firm's efforts. An example is the Post-It note at 3M Company, which was developed out of a manager's desire to keep a bookmark from falling out of his hymnbook.

Such innovations appear at the outset to offer a marginal contribution, and very few companies pay attention to marketing them proactively. Hence, such products tend to be marketed within the structure of the existing organization (existing brand manager, sales manager, and manufacturing line). Market segmentation and channel section, if anchored to existing solutions, are typically wrong, presenting a marketing mistake.

The real market might be with new customers in new segments. Imagination and

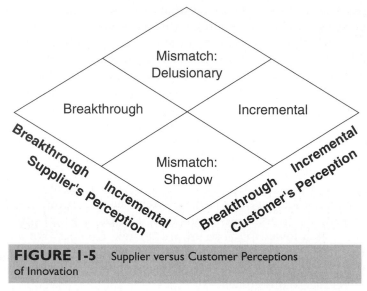

FIGURE 1-5   Supplier versus Customer Perceptions of Innovation

SOURCE: Rangan, V. Kasturi and Kevin Bartus (1995), "New Product Commercialization: Common Mistakes," in *Business Marketing Strategy,* V. K. Rangan et al. (eds.), Chicago: Irwin. Reprinted with permission of McGraw-Hill Companies.

creativity may identify new problems the innovation could solve. Shadowed projects lack urgency and attention, which further undermines their potential odds of success.

2. *Delusionary products* are innovations where the suppliers have grandiose visions for the product but their customers do not share the same euphoria. These might be the typical "lab" projects, wherein the technical team views the innovation as the "next best thing since sliced bread," but consumers simply don't understand it or don't agree that it is so great. For example, some predict that high-definition television may be a "yawner" for the consumer. After all, who wants to watch what many perceive as bad TV programming in higher resolution with better sound quality?

In such a situation, rather than marketing the product as a major innovation, a "new and improved" positioning strategy might be a better option—with the pricing, advertising, and distribution of an incremental extension. To position such products as a revolutionary new category, rather than to position against existing solutions, would be a mistake.

SOURCE: Rangan, V. Kasturi and Kevin Bartus (1995), "New Product Commercialization: Common Mistakes," in *Business Marketing Strategy,* V. K. Rangan et al. (eds.), Chicago: Irwin, pp. 63–75.

It would be overly simplistic to say that high-tech industries are characterized solely by breakthrough innovations. Clearly, many high-tech markets develop incremental innovations. For example, in the area of software, Windows XP was more of an incremental than a breakthrough innovation. It is vitally important that high-tech marketers be aware of the two different types of innovations, because they have very different implications.

Implications of Different Types of Innovations:
A Contingency Model for High-Tech Marketing

Knowledge of the different types of innovations has important implications for how marketing is conducted. "Market planning that explicitly recognizes and accounts for the strategic distinction between market-driven and innovation-driven research goes a long way toward yielding better corporate performance."[56] In other words, the two types of innovation must be managed differently.[57] For example, cross-functional teams must be staffed with people having the right kind of skills and perspectives. Short-term, results-oriented line people would be best matched to incremental innovation and be mismatched on a breakthrough project. Visionaries who question the value of a new product concept might hamper an incremental innovation.[58]

By appropriately matching marketing tools to each distinct type of innovation, the odds of success in the market are enhanced. This means that the appropriate marketing strategy is *contingent upon* the type of innovation. Figure 1-6 provides a picture of how **contingency theory** works.

Many implications arise from the difference between breakthrough and incremental innovation.[59] For example, during the 1980s and early 1990s, U.S. and European firms were competitively challenged in many industries. U.S. firms took a beating in memory chips, office and factory automation, consumer electronics, and automobile manufacturing. These behemoth companies were routinely outmaneuvered by new competitors from other corners of the earth. Kodak watched as videotape camcorders reduced its home movie business to cinders. Xerox's lock on the photocopier business was broken by Canon, Sharp, and others. Consumer electronics products made by Motorola, Zenith, and RCA were largely displaced by better, cheaper, faster versions offered by Sony, Panasonic, and Toshiba. On the automotive front, Toyota, Honda, and Nissan expanded their inroads in the North American market, winning virtually every kudo for quality and reliability. Effective incremental innovation and dramatic improvements in operating efficiency were the two keys to success of these new leaders.

In response, U.S. firms increased their competencies in managing the development of incremental innovation in existing products and processes, with an emphasis on cost competitiveness and quality improvements. Extensive study of incremental innovation by both business managers and academic researchers led to a variety of prescriptions for improvement: six sigma quality in manufacturing, concurrent engineering, just-in-time inventory management, and stage-gate management systems for managing the

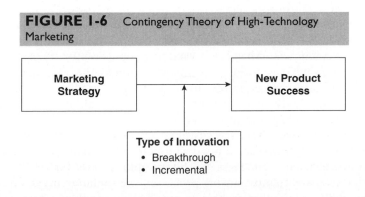

FIGURE 1-6   Contingency Theory of High-Technology Marketing

new product development process. These prescriptions were widely adopted and helped many U.S. companies gain their competitive positions in the world marketplace.

All of these prescriptions are based on the fundamental premise that the firm understands its market's needs and wants and is able to leverage its current technological base to fulfill those needs quickly, cheaply, and reliably. All aspects of the product development project are managed simultaneously by a team comprised of representatives from every function in the business: engineering, production, marketing, cost accounting, and, often, suppliers and customers. Having all constituents present on the team, the argument goes, ensures that decisions are not made with tunnel vision (i.e., that products are not conceived of that cannot be designed) or that products are not designed that cannot be manufactured, etc. This is the world of incremental innovation.

Managers' attention to incremental innovation, however, came at a price. It diminished the focus and capacity of America's largest companies to engage in truly *breakthrough* innovation, which offers the promise of growth through whole new lines of business and the development of new markets. Central R&D labs, traditionally the source of radical innovation ideas, were redirected to serve the immediate needs of business units. Those business units, always under pressure to maximize short-term financial performance, were reluctant to invest in high-risk, long-term projects. The consequences can be disastrous, as Polaroid found. In this case and many others like it, the message is clear: To remain successful over the long haul, firms must be adept at managing both incremental and radical innovation.

However, the processes firms have adopted so well for incremental innovation are not only *not applicable* but may be *detrimental* to the management of radical innovation. The challenges to managing these two types of innovation differ widely and call for different tools, organizational structures, processes, evaluative criteria, and skills. The firm's challenge is to be able to manage both types of innovation simultaneously, because both are needed for the long-term health of the organization.

More specifically, four specific implications of the contingency theory of high-tech marketing are explored here. First, the nature of the interaction between research and development groups and marketing departments depends on the type of innovation. Because technological prowess is key in supply-driven markets, the role of R&D is critical. Research and development is likely to give direction to marketing people in seeking commercial applications for technological advances. A critical issue is the original market that the firm chooses to pursue. The role of marketing is to identify markets.[60]

A second implication relates to the type of market research tools that are appropriately used. Gathering market data to guide the development and marketing of breakthrough products can be difficult. Often the customer doesn't understand the new technology. It can be difficult to articulate performance criteria for the product.[61] For example, if a person has never used jet propulsion for transportation, how will he or she specify what is good performance? Hence, the value of customer feedback through standard marketing research can be questionable. But the voice of the customer remains vitally important; more commonly, qualitative research is used to guide breakthrough product developments.

Or, in situations where the customer might understand the technology, as in the case of lead users who face a need well before the majority of other customers in a market do, the users themselves may be the innovators.[62] For example, Eric von Hippel finds that in many manufacturing processes, the manufacturers who face a particular problem in the production process innovate a solution themselves. An example he cites is Lockheed Martin, which pioneered a new machining technique to speed the removal of titanium metal by up to 20 times with a new face-milling tool that shears rather than chips the metal. The tool was later introduced commercially and expanded to other applications including stainless steels and other hard-to-cut alloys.

A third implication relates to the role of advertising. For breakthrough products, once a viable commercial market is identified, marketers must educate customers, stimulating primary demand for the product class as a whole. Finally, with respect to pricing, because the breakthrough technology may offer a significant advantage over the former mode of doing things, customers may be willing to pay a premium for the new technology.

In contrast to the four implications above for radical innovations, R&D–marketing interaction, market research tools, advertising, and pricing must be managed differently for incremental innovations. For more incremental innovations, the role of marketing is critical. In these situations, customers can play a major role in product development. They can confidently articulate their desires and preferences. In such cases, firms can use standard marketing research tools to identify customer needs, passing the information to R&D, which then develops the appropriate innovations to satisfy those needs. Marketing takes a lead role in such cases. It is more common in incremental innovation to see more standard management controls and formal planning groups.[63] Advertising typically stimulates selective demand, building preference for the firm's specific brand or product. Pricing is more competitive.

It is important to note that the resource allocations of a firm should match the long-term financial attractiveness of the project. Some breakthrough projects might not have a large market potential to start with. And many incremental innovations may absolutely require a major investment (i.e., if a firm's product is hopelessly out of date in a very large market). Hence, marketers should not confuse the nature of the innovation with potential payoff and wrongly assume that breakthrough innovations will have a large payoff![64]

DOES MARKETING NEED TO BE DIFFERENT FOR HIGH-TECHNOLOGY PRODUCTS AND INNOVATIONS?

In light of this discussion, it is clear that the nature of the marketing must be tailored to the type of innovation. But is high-tech marketing all that different from its low-tech counterpart? Or, will standard marketing tools suffice for high-tech markets? Are high-tech marketing disasters caused by the use of a standard marketing approach, when a unique set of tools is necessary to handle the market, technological, and competitive uncertainties? Or, are high-tech marketing disasters merely the result of flawed execution of basic marketing?[65]

Given the high degree of uncertainty, the margin for error for high-tech marketers is likely smaller than for conventional markets. In that sense, high-tech firms must execute basic marketing principles flawlessly.[66] For example, selecting a receptive target market, being able to communicate clearly the benefits the innovation offers relative to other solutions, having an effective/efficient distribution channel, and using solid relationship-building skills cannot be ignored—or overlooked—by high-tech marketers. Because of the importance of following a basic marketing plan in conceptualizing and implementing marketing strategy for any product, be it high- or low-tech, the Appendix to this chapter presents an outline to follow in developing a basic marketing plan.

However, it is all too common that small high-tech startups lack marketing expertise or relegate the role of marketing to second-class status (i.e., beneath the role of engineering/R&D) in the organization. Technical people often have a hard time becoming market focused. Cross-functional collaboration between engineers and marketers is a necessity but extremely difficult to implement well. Further complicating the issue is the fact that many people hired to do "marketing" lack an understanding of how to market in high-tech industries. These organizational realities, as well as the level of market, technological, and competitive uncertainties, mean that although a standard approach to marketing, such as the four Ps (product, price, place, and promotion), is still relevant, the standard approaches must be modified to account for the inherent uncertainty in high-tech environments.[67]

Framework for Making High-Technology Marketing Decisions

This book's primary aim is to provide a framework for making marketing decisions in a high-tech environment. Using a framework to manage the marketing decision-making process will foster greater understanding of the common characteristics in high-tech environments and help manage the riskiness of marketing in a high-tech context.

Figure 1-7 provides the conceptual framework used for making high-technology marketing decisions. On the left side of the figure are the internal considerations that a firm must address and understand as the foundation to effective marketing. The management of high-tech firms has some unique considerations compared to management of traditional companies. Larger high-tech firms that begin to function as a corporate bureaucracy can struggle with how to remain innovative. Smaller high-tech firms wrestle with how to move from a technology-driven, engineering mindset to a market focus. For both sizes of firms, resolving conflicts between R&D and marketing is of paramount importance. Moreover, whereas all marketing is premised upon relationships, the management of relationships and strategic alliances in high-tech industries requires special considerations. For example, strategic alliances often necessitate collaboration with competitors, where protection of intellectual property is even more

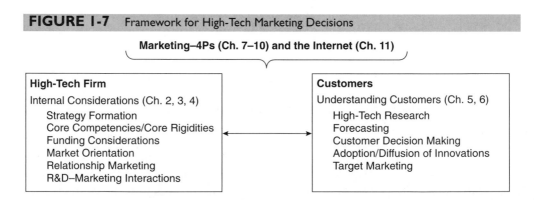

FIGURE 1-7   Framework for High-Tech Marketing Decisions

Marketing–4Ps (Ch. 7–10) and the Internet (Ch. 11)

High-Tech Firm
Internal Considerations (Ch. 2, 3, 4)
    Strategy Formation
    Core Competencies/Core Rigidities
    Funding Considerations
    Market Orientation
    Relationship Marketing
    R&D–Marketing Interactions

Customers
Understanding Customers (Ch. 5, 6)
    High-Tech Research
    Forecasting
    Customer Decision Making
    Adoption/Diffusion of Innovations
    Target Marketing

Societal, Ethical, and Regulatory Concerns (Ch. 12)

important than in more traditional strategic alliances—particularly when the innovative firms are collaborating on cutting-edge research.[68] Hence, Chapters 2 through 4 address issues related to strategy and corporate culture in high-tech firms: strategy formation, competitive advantage, core competencies/core rigidities, relationship marketing (partnering and alliances), market orientation, and R&D–marketing interactions. Coverage of these issues lays a foundation for marketing effectiveness.

On the right side of the figure are the customer considerations. One of the particularly challenging aspects of high-tech marketing is understanding customers and markets. For example, in conducting marketing research in high-tech industries, users often cannot articulate their needs very clearly because they simply cannot envision what the technology can do or how it can benefit them. In addition, conducting marketing research to forecast the size of the market can be extremely difficult. Research tools used in high-tech markets, such as empathic design, customer visits, lead user research, quality function deployment, competitive intelligence gathering, and forecasting, are explored in Chapter 5.

Chapter 6 addresses customer behavior considerations, including customer decision making for high-tech products, issues associated with adoption and diffusion of innovation unique to high-tech markets, and how marketing to early adopters must be different than marketing to late adopters. This chapter draws heavily on the work of Geoffrey Moore.[69]

The link between the firm and its customers is enacted through the marketing mix. Chapters 7 to 10 provide coverage of the four elements (4 Ps) of marketing: product, place, price, and promotion.

- *Product*   High-tech product development and management may not follow standard marketing practice. For example, some argue that the use of the product life cycle to manage products in high-tech markets is flawed; competitive volatility means that products may never reach maturity.[70] Market uncertainty means that the progression from introduction to growth may be greatly disrupted—some call this "the chasm."[71] Hence, Chapter 7, "Product Development and Management Issues in High-Tech Markets," begins with the framework of a technology map to guide product development. Pertinent considerations include decisions about technology transfer and licensing, product platforms and derivatives, protection of intellectual property, and so forth.

- *Place*   Decisions about the role of distribution channels and management of relationships along the supply chain in high-tech markets can be very complicated. For example, consider the case of Intel. When Intel develops chip upgrades, this implies that customers will forgo buying a new computer and buy a replacement chip instead. But such a strategy has the strong possibility of alienating Intel's largest customers—the computer manufacturers who buy the majority of Intel's chips. Hence, it is important that marketers consider the impact of their decisions on all members of the supply chain. In addition, the role of channel members in selling technology is crucial. One study[72] cites two revealing statistics:

  - Seventy-seven percent of computer purchasers who visit a dealer do not have a specific brand in mind; 90 percent of those customers purchase a brand recommended by the dealer.

- Of the 23 percent of computer purchasers who do have a specific brand in mind, 53 percent of them switch to an alternative brand recommended by the dealer.

  Hence, the role of channel members in selling technology can be very powerful. Chapter 8, "Distribution Channels and Supply Chain Management in High-Tech Markets," provides a framework for making distribution decisions. Focus is given to the use of the Internet as a new channel and the need to manage the transition and resulting conflict.

- **_Price_**   Standard pricing practices can be very misleading in high-tech markets. For example, pricing based on experience curve effects—where costs decline by a fixed and predictable amount every time accumulated volume doubles—is simply not feasible. Competitive volatility means that the firm may never see large volumes on any given product. In addition, in order to survive, high-tech firms must constantly introduce new innovations that obsolete their current product line; this means that firms who rely on economies of scale and production costs as a source of competitive advantage are likely doomed.[73] Chapter 9, "Pricing Considerations in High-Tech Markets," provides a framework for pricing decisions, with heavy emphasis given to the need to be customer oriented in managing this element of the marketing mix. Moreover, in light of the rapid price declines in many high-tech industries (the most extreme of which is a price of $0 for many digital products available for free on the Internet), focused attention is given to strategies used to generate profits in light of the technology paradox (how to survive when the price of product is declining rapidly).[74]

- **_Promotion_**   Communicating with customers via advertising and promotion for high-tech products can be difficult. "Preannouncing"[75] is a useful strategy to persuade customers to delay purchase until the new technology is available, but there are issues of timing: When does a firm announce that it is working on a revolutionary new technology? Although preannouncements can be used in both high-tech and more traditional marketing contexts, their effects may be more pronounced in a high-tech environment for several reasons.

  First, given the uncertainty and fears of obsolescence that customers face, preannouncements may encourage customers to forgo a current-generation technology in anticipation of the future one. Some refer to such behavior as "leapfrogging." Second, technological uncertainty implies ambiguity in the delivery timetable. There are many examples of high-tech products being months, if not years, behind promised delivery dates. Indeed, in the software arena, such preannounced products are referred to as "vaporware," products that are never produced. Third, when preannouncements are used, issues of intent arise. In fact, in cases investigated by the Justice Department, if a firm's intent in using preannouncements is specifically to discourage customers from buying a competitor's product, anticompetitive penalties may be applied. Hence, firms in the high-tech arena have strong reasons to use preannouncements but must be especially cautious in doing so. It logically follows, therefore, that future high-tech marketers must have an understanding of such issues.

  In addition to understanding issues surrounding preannouncement communication strategies, high-tech marketers must also effectively use marketing communications to allay customer fear, uncertainty, and doubt. High-tech marketers are turning to one strategy used frequently by consumer goods marketers: the use of communications to build a strong brand name. A strong brand is a heuristic that is used by consumers to reduce perceived risk. Strong brand names in the high-tech arena "stand as a beacon in the confusing and quickly changing world of technology. . . . They offer reassurance about a purchase that is fraught with confusion and anxiety."[76] In technology arenas where products change rapidly, the brand can be more important than in packaged goods industries where a product is more understandable because it has stayed the same for a long time. Microsoft, Intel, Hewlett-Packard, and even smaller firms are using strong brand strategies, in part to assuage consumer anxiety.

To be sure, advertising and promotion strategies used to allay consumer anxiety in a high-tech context may also be used in more traditional marketing contexts. The critical points of distinction are that

1. It is a difficult cultural shift for people in technology-oriented firms to become marketing- or customer-oriented. Even Microsoft sees itself as "doing technology for technology's sake rather than based on customer needs."[77]
2. The combination of customer, technological, and competitive uncertainty surrounding high-tech purchases implies more riskiness for consumers, and hence, the need for a strategy to reduce such riskiness.

Future high-tech marketers must understand how these more traditional marketing tools are especially crucial in their environment, despite the lack of a customer focus in many high-tech firms. Hence, among other topics, Chapter 10, "Advertising and Promotion in High-Tech Markets: Tools to Build and Maintain Customer Relationships," emphasizes the importance of using advertising and promotion tools to develop a strong brand name (as one mechanism to allay customer anxiety), the need to manage product preannouncements, and communication tools used in managing customer relationships.

Although not one of the 4Ps of marketing, the new abilities offered via the Internet to interact with customers and its vital role in serving both high-tech firms and customers warrant special attention. Indeed, a book on the marketing of high-technology products and innovations would not be complete without explicit coverage of how the Internet is revolutionizing the face of business in general (and marketing in particular). Hence, Chapter 11 focuses exclusively on e-business and e-commerce.

Chapter 12, "Realizing the Promise of Technology," concludes with regulatory and ethical considerations high-tech marketers face. Despite the Justice Department's investigations of Microsoft and Intel, there is a heated debate about whether antitrust is even relevant in today's high-tech era.[78]

In light of the high degree of market, technological, and competitive uncertainties surrounding high-tech products and markets, the need for effective marketing in high-tech industries is paramount. High-tech products and services are introduced in turbulent, chaotic environments, in which the odds of success are often difficult to ascertain at best and stacked against success at worst. This book is designed to provide frameworks for systematic decision making about marketing in high-tech environments. In doing so, it offers insights about how marketing tools and techniques must be adapted and modified for high-technology products and services. Effective high-tech marketing includes a blend of marketing fundamentals and the unique tools explored in this

---

BOX 1-3

ODE ON A GRECIAN SCUZZI[a] INTERFACE
M. Pathic and D. Sign (With Apologies to Keats)

THOU chip of silicon, oh please some answers can you give?
Was your birth deemed radical or just derivative?
Will your purchase price afford us not an economic care?
Or will we sell you at a loss to increase market share?

The techies want to dress you up with all the functions they can cite.
But if you do what you do well, I'm sure you will delight.
The mix of FUD and FIT for you should have this implication
That better products are derived from cross-functional collaboration.

Will you cross successfully the high-tech market chasm?
Or will we drop you down the drain when we get a panic spasm?
Oh chip of silicon, just one more answer can you give?
Is as a high-tech marketer a decent way to live?

[a]Scuzzi is a technical term for Small Computer Systems Interface (SCSI).

SOURCE: Written by Scott MacDonald, MBA, University of Montana.

---

book. The text highlights possible pitfalls, mitigating factors, and the how-tos of successful high-tech marketing. The poem in Box 1-3 provides a flavor of these unique tools. Before leaving this opening chapter, readers may be wondering about specific job opportunities in high-tech industries.

JOB OPPORTUNITIES IN HIGH TECH

Due to the differing definitions of high tech, employment statistics vary. A report from the *American Electronics Association*[79] identified a total of over 5.1 million jobs in the key high-tech industry segments of software and computer-related services, high-tech manufacturing, and communications services at year-end 2003. Taking a broader view of what constitutes high-tech industries, the Bureau of Labor Statistics (BLS) identified more than 11 million jobs in 2000 in the core high-tech industries of computers, software, and communications; ancillary management consulting work; and programmers, technicians, and the like in the rest of the economy.[80] The BLS projected high employment growth in the key computer and data processing services, electronic components and accessories, and medical instruments industries. Table 1-1 (page 4) contains employment growth projections for all high-tech industries. Salaries in high tech continue to be attractive for professionals in sales, marketing, finance, engineering, and information technology.[81]

High-tech marketing can include a variety of positions and job titles, including product management, sales, advertising, and research, to name a few. For those with a technical background, product marketing or account management can be a career path out of the lab, if so desired. Many MBA programs report high demand for students with a technical undergraduate degree combined with a Master of Business Administration degree. The many online job sites have no paucity of marketing jobs in high-tech companies.

For individuals with nontechnical backgrounds (say, liberal arts or business majors), it is still possible to find a high-tech job, despite the lack of technical training. As cited by a *Wall Street Journal* article, "examples abound of nontechnical folk thriving at high-tech companies."[82] For example, it is possible to get jobs in marketing communications (advertising and promotion, public relations, trade show management) or marketing research without a technical background. And sales and management experience is a great asset because "it implies that, unlike most techies, the candidate understands how to communicate with fellow humans, including customers."[83] To find employment in a high-tech company without technical experience, the *Wall Street Journal* article offers the following suggestions:[84]

- Get in the door any way you can. Work as a temporary worker or an intern, for example. Once in the door, expand your duties. Take on additional projects; volunteer to work on teams.
- Read industry trade publications; experiment with products; attend meetings.
- Work for a high-tech firm's customer or supplier to learn the industry and make contacts.
- Find a company that will provide training on technology. For example, many consulting firms hire liberal arts majors because they know how to think critically.

It is also important to subscribe to computer and trade magazines so that one understands the jargon and industry issues. Last, but not least, it would be very helpful to understand high-tech marketing and the unique demands high-tech marketers face. Becoming acquainted with the strategies and tools presented in this text would be a step in that direction.

SUMMARY

This chapter has provided an introduction to high-technology marketing. In addition to providing an in-depth examination of the various ways to define and characterize high-tech industries and companies, it has also shown that many innovations frequently occur at levels in the supply chain far removed from the end user. The chapter also showed that innovations are a question of degree, ranging from incremental to radical. In order to be effective, marketing strategies must be tailored to the type of innovation. This notion of matching marketing to the type of innovation is known as the contingency theory of successful high-tech marketing, which will be a common theme running throughout this book. Finally, the chapter concluded with an examination of job opportunities in high-tech markets.

DISCUSSION QUESTIONS

1. What are the pros and cons of the various definitions of high tech? Of these definitions, which do you think is the most useful? Why? Based on that definition, draw a continuum of low- versus high-tech industries.

2. What three characteristics are common to high-tech industries? Provide examples of each of the specific dimensions of the characteristics.

3. Define and give an example of each of the following: unit-one costs, network externalities, tradability problems, and knowledge spillovers.

4. Explain why standards are important to both marketers and customers. Under what conditions are the different strategies for establishing standards appropriate?

5. Think of some examples in which low-tech industries have been transformed by high-tech innovations. Where in the supply chain have these innovations originated?

6. How are radical innovations different from incremental innovations? Provide examples of the two types of innovations.

7. What is a contingency theory of new product success? What marketing tools are appropriately used for incremental innovations? What marketing tools are appropriately used for radical innovations?

8. Does high-tech marketing need to be different from marketing of traditional products? Why? How?

GLOSSARY

Competitive volatility.   Refers to rapid changes in the competitive landscape: which firms are one's competitors, their product offerings, the tools they use to compete.

Contingency theory.   The effects of one set of variables on another (say, marketing variables on new product success) depend on yet a third variable (say, type of innovation).

Incremental innovation.   Continuations of existing methods or practices; may involve extension of products already on the market.

Knowledge spillovers.   When synergies in the creation and distribution of know-how further enrich a related stock of knowledge, which creates increasing returns in the development of related technologies.

Market uncertainty.   Ambiguity about the type and extent of customer needs that can be satisfied by a particular technology, arising from consumer fear, uncertainty, and doubt about the needs or problems the new technology will address and meet.

Network Externalities.   Also called demand-side increasing returns or a bandwagon effect; when the value of the product increases as more users adopt it. In other words, the utility received from an innovation is a function of the number of users.

Radical innovation.   Breakthrough innovation that cannot be compared with any existing practices or perceptions; technology is so new that it creates a new product class.

Technological uncertainty.   Skepticism about whether the technology will function as promised or be available when expected by the company providing it.

Tradability problems.   When underlying know-how represents a substantial portion of the value of the products and services, it is difficult to value the knowledge, especially when it is tacit and resides in people and organizational routines.

Unit-one costs.   The cost of producing the first unit is very high relative to the costs of reproduction; this type of cost structure is likely to exist when know-how, or knowledge embedded in the design of the product, represents a substantial portion of the value of the products and services.

ENDNOTES

1. Available at www.aeanet.org.
2. "Mass Layoffs in January-February 2003 and Annual Averages for 2002," www.bls.gov/mls/.
3. Kirkpatrick, David (2003), "Some in Silicon Valley Have Learned to Stop Worrying and Love the Bust. Here's Why," *Fortune*, April 28. Available at www.fortune.com.

4. Kilman, Scott (1998), "If Fat-Free Pork Is Your Idea of Savory, It's a Bright Future," *Wall Street Journal,* January 29, p. A1.

5. Frost & Sullivan (1997), "Mechatronics for Automobiles Is the Buzzword When It Comes to Electronic Control Modules," www.frost.com/verify/press/transportation/pr556218.htm.

6. Peterson, Thane (2003), "21st Century Cars Hit the Road," *Business Week,* Sept. 4. Available at www.businessweek.com.

7. McWilliams, Gary (1997), "Technology Is What's Driving This Business," *Business Week,* November 3, pp. 146–148.

8. Available at www.pg.com.

9. Kirkpatrick, David, op cit.

10. Capon, Noel and Rashi Glazer (1987), "Marketing and Technology: A Strategic Coalignment," *Journal of Marketing,* 51 (July), pp. 1–14.

11. *Technology, Innovation, and Regional Economic Development* (1982), Washington, DC: U.S. Congress, Office of Technology Assessment, September 9.

12. Hadlock, Paul, Daniel Hecker, and Joseph Gannon (1991), "High Technology Employment: Another View," *Monthly Labor Review,* July, pp. 26–30.

13. Hatzichronoglou, Thomas (1997), "Revision of the High-Technology Sector and Product Classification," OECD STI working paper.

14. *Science and Engineering Indicators* (1996), National Science Foundation, chapters 4 and 6.

15. Luker, William and Donald Lyons (1997), "Employment Shifts in High-Technology Industries, 1988–1996," *Monthly Labor Review,* June, pp. 12–25.

16. Ibid.

17. Lipkin, Richard (1996), "Fit for a King," *Science News,* May 18, pp. 316–317.

18. Luker, William and Donald Lyons, op. cit.

19. Moriarty, Rowland and Thomas Kosnik (1989), "High-Tech Marketing: Concepts, Continuity, and Change," *Sloan Management Review,* 30 (Summer), pp. 7–17.

20. See also Gardner, David (1990), "Are High Technology Products Really Different?" Faculty working paper case #90-1706, University of Illinois at Urbana-Champaign.

21. Moriarty, Rowland and Thomas Kosnik (1989), op. cit. Moriarty, Rowland and Thomas Kosnik (1987), "High-Tech vs. Low-Tech Marketing: Where's the Beef?" Harvard Business School, case #9-588-012.

22. Moore, Geoffrey (2002), *Crossing the Chasm, Marketing and Selling Technology Products to Mainstream Customers,* New York: Harper Business.

23. Schuchart Jr., Steven (2003), "Long-term Care for Your Data," *Network Computing,* 14 (12), pp. 85–87.

24. Aley, James (2003), "Qualcomm: Heads We Win, Tails We Win," *Fortune,* February 18. Available at www.fortune.com.

25. Moore, Geoffrey (2002), op. cit.

26. Sultan, Fareena (1990), "Zenith: Marketing Research for High Definition Television," Harvard Business School, case #9-591-025.

27. Greenspan, Robyn (2002), "HDTV Future Unclear," cyberatlas.internet.com/big_picture/hardware/article/0,,5921_1491461,00.html.

28. Moore, Geoffrey (2002), op. cit.

29. Moriarty, Rowland and Thomas Kosnik (1989), op. cit.

30. Thurm, Scott (1999), "For Frazzled Online Brokers, Technology Is the Problem," *Wall Street Journal,* March 4, p. B6.

31. Karlgaard, Rich (2003), "CEOs Talk Tech," *Forbes,* 172 (1), p. 033.

32. Cooper, Arnold and Dan Schendel (1976), "Strategic Responses to Technological Threats," *Business Horizons,* February, pp. 61–69.

33. Hamel, Gary (1997), "Killer Strategies That Make Shareholders Rich," *Fortune,* June 23, pp. 70–84.

34. Schiesel, Seth (2003), "Can Cable Fast-Forward Past TiVo?," *New York Times,* October 20. Available at www.NYTimes.com.

35. "Microsoft at the Power Point," *Economist,* 9/13/2003, 368 (8341), p. 59.

36. Schilling, Melissa A. (2003), "Technological Leapfrogging: Lessons from the U.S. Video Game Console Industry," *California Management Review;* 45 (3), pp. 6–33.

37. Except where noted, this section is drawn from Hill, Charles (1997), "Establishing a Standard: Competitive Strategy and Technological Standards in Winner-Take-All Industries," *Academy of Management Executive,* 11 (May), pp. 7–25.

38. Ford, David and Chris Ryan (1981), "Taking Technology to Market," *Harvard Business Review,* 59 (March–April), pp. 117–126.

39. Takahashi, Dean and Jon Auerbach (1997), "Digital Files Antitrust Suit Against Intel," *Wall Street Journal,* July 24, p. B5.

40. Gundlach, Gregory and Paul Bloom (1993), "The 'Essential Facility' Doctrine: Legal Limits and Antitrust Considerations," *Journal of Public Policy and Marketing,* 12 (Fall), pp. 156–177.

41. Except where noted, this section is drawn from Hill, Charles (1997), op. cit.

42. Note that this discussion does not address the use of government intervention, standards-setting bodies such as the International Standards Organization, or trade associations' attempts to establish an industry standard.

43. John, George, Allen Weiss, and Shantanu Dutta (1999), "Marketing in Technology Intensive Markets: Towards a Conceptual Framework," *Journal of Marketing,* 63 (Special Issue), pp. 78–91.

44. Mandel, Michael (2000), "The New Economy," *Business Week,* January 31, pp. 73–91.

45. Rangan, V. Kasturi and Kevin Bartus (1995), "New Product Commercialization: Common Mistakes," in *Business Marketing Strategy,* V. K. Rangan et al. (eds.), Chicago: Irwin, p. 66.

46. Abernathy, W. and J. Utterback (1978), "Patterns of Industrial Innovation," *Technology Review,* June–July, pp. 41–47.

47. Shanklin, William and John Ryans (1984), "Organizing for High-Tech Marketing," *Harvard Business Review,* 62 (November–December), pp. 164–171.

48. Maney, Kevin (1999), "The Net Effect: Evolution or Revolution?" *USA Today,* August 8, p. B2.

49. Gross, Neil and Peter Coy with Otis Port (1995), "The Technology Paradox," *Business Week,* March 6, pp. 76–84.

50. Hamel, Gary (1997), op. cit.

51. Rangan, V. Kasturi and Kevin Bartus (1995), op. cit. pp. 63–75.

52. Shanklin, William and John Ryans (1984), op. cit.

53. Maney, Kevin (1999), op. cit.

54. Abernathy, W. and J. Utterback (1978), op. cit.

55. Shanklin, William and John Ryans (1984), op. cit.

56. Rangan, V. Kasturi and Kevin Bartus (1995), op. cit.

57. Ibid.

58. Leifer, Richard, Christopher M. McDermott, Gina Colarelli O'Connor, Lois Peters, Mark Rice, and Robert W. Veryzer, Jr. (2000), *Radical Innovation: How Mature Companies Can Outsmart Upstarts,* Cambridge, MA: Harvard Business School Press.

59. Shanklin, William and John Ryans (1984), op. cit.

60. Abernathy, W. and J. Utterback (1978), op. cit.

61. von Hippel, Eric (1986), "Lead Users: A Source of Novel Product Concepts," *Management Science,* July, pp. 791–805.

62. Abernathy, W. and J. Utterback (1978), op. cit.

63. Rangan, V. Kasturi and Kevin Bartus (1995), op. cit.

64. Moriarty, Rowland and Thomas Kosnik (1987), op. cit, p. 18.

65. Moriarty, Rowland and Thomas Kosnik (1987), op. cit.

66. See also Gardner, David (1990), op. cit.

67. Hamel, G., Y. Doz, and C. K. Prahalad (1989), "Collaborate with Your Competitors—And Win," *Harvard Business Review,* January–February, pp. 133–139.

68. Moore, Geoffrey (2002), op. cit.; Moore, Geoffrey (1995), *Inside the Tornado,* New York: Harper Business.

69. Shanklin, William and John Ryans (1987), *Essentials of Marketing High Technology,* Lexington, MA: DC Heath.

70. Moore, Geoffrey (2002), op. cit.

71. "How Technology Sells," (1997), Dataquest, Gartner Group, and CMP Channel Group, CMP Publications, Jericho, NY.

72. Shanklin, William and John Ryans (1987), op. cit.

73. Gross, Neil and Peter Coy with Otis Port (1995), "The Technology Paradox," *Business Week,* March 6, pp. 76–84.

74. Eliashberg, J. and T. Robertson (1988), "New Product Preannouncing Behavior: A Market Signaling Study," *Journal of Marketing Research,* 25 (August), pp. 282–292.

75. Morris, Betsy (1996), "The Brand's the Thing," *Fortune,* March 4, p. 82.
76. Moeller, Michael (1999), "Remaking Microsoft," *Business Week,* May 17, pp. 106–114 (p. 112).
77. Murray, Alan (1997), "Antitrust Isn't Obsolete in an Era of High-Tech," *Wall Street Journal,* November 10, p. A1.
78. Platzer, Michaela (2003), *Tech Employment Update,* American Electronics Association, Washington, DC.
79. Available at www.bls.gov/emp/empocc1.htm.
80. Ibid.
81. Lancaster, Hal (1997), "You Can Get a Job in a High-Tech Firm with Low-Tech Skills," *Wall Street Journal,* February 4, p. B1.
82. Fisher, Anne (1999), "Job Hunting in Tech Land," *Business Week,* April 12, p. 172.
83. Lancaster, Hal (1997), op. cit.

APPENDIX A

Outline for a Marketing Plan

The purpose of this section is to present an outline of the steps to be systematically considered and developed in the course of a marketing plan. The remainder of the book will elaborate on the content of each section. Supporting detail can also be found in any basic marketing textbook. Examples of marketing plans can be found at http://www.mplans.com/spm/.

1.0 EXECUTIVE SUMMARY – A one- to two-page summary of the market environment and business resources, financial and nonfinancial objectives, and marketing strategy including target market(s), value proposition, and marketing mix elements.

2.0 MARKET ANALYSIS

Market demographics
Market needs
Market trends
Market growth
Buyer behavior
Customer segments
Competition
Collaborators
Macroeconomic forces

3.0 COMPANY ANALYSIS

Tangible assets
Intangible assets
Capabilities
Areas of advantage
Key success factors

4.0 OBJECTIVES

Financial
  Revenues
  Margins
  Growth rate
Non-financial
  Customer satisfaction
  Perceived quality
  Loyalty
  % of sales from new products

5.0 VALUE PROPOSITION

Target market
Functional, emotional, and/or self-expressive benefits
Price

6.0 MARKETING STRATEGY

Positioning - The process of designing the company's image and value offering so that customers in the target market understand and appreciate what the company stands for in relation to competitors.

Product and/or service attributes - New product development processes, decisions about what to sell and what features to include, branding strategies, packaging decisions, warranty, and ancillary services.

Distribution - Locations at which product or service is made available to customer, and the channel members that offer it.

Promotion:

   a. Advertising strategies, regarding both the message content in ads and media used to communicate the message. May include direct media.

   b. Sales promotion strategies, regarding any short-term incentives for both trade members and consumers (coupons, rebates, premiums, etc.).

   c. Public relations and publicity strategies, regarding the generation of news articles, community relations, event sponsorships, and goodwill.

   d. Personal selling/trade shows.

Price - What to charge for specific products, features, or services, as well as discount structures and payment plans.

People – The marketing specialists that are required to execute the marketing strategy. Systems for recruiting, motivating, and retaining them.

7.0 BUDGETING AND CONTROL

Financial resources required to execute the marketing strategy.
System for comparing results to objectives
Processes for taking corrective action

CHAPTER 2

Strategy and Corporate Culture in High-Tech Firms

If cars worked the way computers work, they would have the following characteristics:

- For no reason whatsoever, your car would crash twice a day
- Occasionally, your car would die on the freeway for no apparent reason; you would accept this, restart, and drive on.
- Occasionally, executing a maneuver such as a left turn would cause your car to shut down and refuse to restart—in which case, you would have to reinstall the engine.
- The airbag system would say "Are you sure?" before going off.
- Every time a car company introduced a new model, car buyers would have to learn how to drive all over again, because none of the controls would operate in the same manner as the old car.
- Apple would make a car that was powered by the sun, more reliable, five times as fast, and twice as easy to drive, but would run on only 5 percent of the roads.

Turnaround at Hyperion[1]

The view from the Fountain Room at San Francisco's Fairmont Hotel was spectacular—a sweeping vista of the city's gleaming skyline. It was July 16, 2001, and Jeffrey Rodek, CEO of software maker Hyperion Solutions Corp., had invited 24 of his top managers to dinner at a place that seemed to sit on top of the world. All the more reason why what he told them came as a shock. After playing Elton John's dirgelike "Funeral for a Friend," Rodek announced that the evening's repast would consist of bread and water. "We don't deserve any better," he said.

Rodek's shock treatment kicked off a remarkable turnaround at Hyperion. The company had lost money in two of the previous three quarters, and its stock was trading at 14. By June 2003, it had been profitable for seven straight quarters and its shares traded at 35. Hyperion's software is now used by 88 of the Fortune 100 businesses, 66 of the Nikkei 100, and 53 of the Financial Times Europe Top 100. Hyperion's market share in this category is three times that of its nearest competitor. Rodek focused the company's strategy in the arena of business performance management solutions.

Hyperion software enables companies to translate strategies into plans, monitor execution, and provide insight to manage and improve financial and operational performance. Hyperion gathers data from multiple transactional systems and transforms it into one insightful picture of a company's financial and operational performance. As a result, all managers have the same facts and a common understanding of what affects

---

overall business performance. Hyperion supports event-driven planning, linking goal setting and strategy to models, plans, and execution and aligning individual and corporate goals. As a result, planning with Hyperion software is an agile process, enabling managers to collaborate in modeling scenarios, set new directions or make mid-course corrections, and communicate changes across the enterprise. Hyperion ensures that all managers have on-demand access to relevant information that enables them to understand their company's past and current performance and anticipate future performance. As a result, managers can make fast and informed decisions that lead to immediate and continuous performance corrections and improvements.[2]

As a symbol of Hyperion's turnaround, in February 2003 Rodek treated 100 managers to a dinner at the Mark Hopkins Hotel in San Francisco, where they dined on Australian lamb chops and grilled swordfish.

M any large firms struggle mightily with the task of creating really new products that change the competitive landscape of a given industry. The characteristics of large firms (bureaucratic, focused on economies of scale, and so forth)—characteristics that are useful for developing incremental innovations—can seriously inhibit these firms' ability to develop breakthrough products. It is no small wonder that one of the hottest consulting areas in the past decade was how to develop a culture of innovativeness in Corporate America. Indeed, the Sloan School of Management at MIT offers a two-day program called "Building, Leading, and Sustaining the Innovative Organization" for which participants pay a $2,600 fee. Some of the topics include how companies get the right mix of people and skills they need to generate innovative ideas in a timely manner, develop the processes required to support these people, build cultures that encourage innovative behaviors, and decide which new product opportunities they ought to pursue.

At the same time that many large firms struggle with becoming more innovative and nimble, many small firms struggle with their own unique marketing problems. Small firms often are created on the basis of the superior technology they can bring to a marketplace. The roots of such firms are often found in their founders' sophisticated technical leadership. The technical orientation is a necessary ingredient for success, but not the only ingredient. Many such firms die for a variety of reasons. Some lack access to vital resources including funding and management expertise. Others find it difficult to blend their technical insights with marketing sophistication. Technical leaders often do not recognize that market savvy is a key ingredient for success. Such a difficulty is further exacerbated by the sense that technical people have the innate ability to be marketers as well as technical gurus. These beliefs often prove to be fatal flaws.

Moreover, in many large and small technology-driven firms, a perceived status differential exists between the technical folks and marketers (with marketers coming out on the short end of the stick). This perceived difference in many firms translates into a systemic bias that minimizes the contributions marketers make to the firm, diminishes their voices, and downplays the information they bring to the table. Such perceptions detract from the close R&D–marketing interaction that must occur for successful marketing programs.

Moreover, the marketing of high-technology products requires a discussion of strategic market planning. Because many managers in technology-intensive industries can be overwhelmed with the complexities they face, an overview of strategic market planning tailored to this demanding environment is imperative.

The purpose of Chapters 2, 3, and 4 is to discuss some of these basic issues all high-tech firms face, be they large or small. As shown in the organizing framework in Figure 2-1, these issues represent the building blocks that enhance the odds of marketing

FIGURE 2-1   Internal (Firm) Considerations for High-Technology Marketing Effectiveness

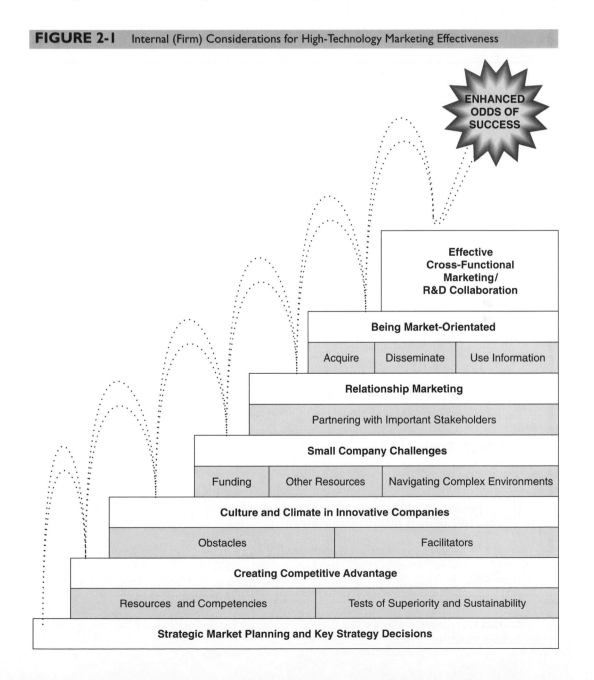

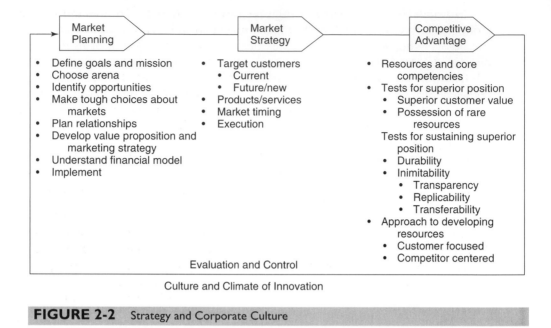

FIGURE 2-2   Strategy and Corporate Culture

success. Moreover, these issues, which provide the foundation for later topics, require that the high-tech firm assess its capabilities in a genuine, wholehearted manner.

While strategy and culture are distinct considerations in the high-tech firm, we consider them both in this chapter since they both strongly influence the capability of the firm to develop innovative products and/or services. Figure 2-2 illustrates the flow of topics in this chapter.

STRATEGIC MARKET PLANNING IN HIGH-TECH COMPANIES

Effective high-technology marketing is built on a foundation of resources and ideas that are the product of strategic market planning.[3] Two fundamental questions are:

- Do traditional marketing planning processes work in a high-tech environment?
- Are the characteristics of the context described in Chapter 1 (market uncertainty, technological uncertainty, and competitive volatility) amenable to traditional strategic market planning processes?

The tendency is to answer the above questions with a *No*. Many managers in technology-intensive situations are simply overwhelmed by the complexities they face and, sadly enough, confront these complexities with strategic market planning processes more complex and unmanageable than the situations themselves. Current planning processes can tie up enormous amounts of management time with extensive analysis, documentation, and a lack of cross-functional involvement. The processes produce plans that are quickly obsolete in the face of competitive actions and reactions.

Confronted with complexity, strategic market planning processes must become simpler, faster, iterative, opportunity-based, team-based, and functionally integrated. Market planning and strategic action must be closely coupled. In *Winning Market Leadership: Strategic Market Planning for Technology-Driven Businesses,* Ryans et al.[4] describe a systematic and highly integrated process for evaluating market opportunities and for developing strategies to win market leadership, focusing on the key issues and tough choices faced by executives in these very demanding, technology-intensive markets. Most of the examples in their book are drawn from the experiences of large, multinational, high-technology companies, such as Intel, Compaq, Hewlett-Packard, Glaxo-Wellcome, and General Electric. Following are the steps in the process:

1. *Define the company's goals and mission.*
2. *Choose the arena.* The business must tentatively identify the markets in which it will compete. Each arena of opportunity should be defined by potential customer segments that could be served, potential benefits that could be provided to these customers, possible technologies or competencies that could deliver those benefits, and possible value-adding roles for the business in the market chain.
3. *Identify potentially attractive opportunities.* This requires understanding: a) customer needs, b) suitability of the business's resources to satisfy those needs, c) competitive threats, and d) profit potential.
4. *Make tough strategic choices.* In this step, the business's management team must identify key strategic issues and determine how it should confront each issue. These issues might include markets to target, technologies to develop, or choice of a value-creating strategy.
5. *Plan key relationships.* By this time, many strategic decisions will have been made. However, in complex and fast-moving markets, it is becoming more and more rare that a business has all of the resources to execute a strategy by itself. At this stage, key collaborators, such as suppliers, distributors, or complementors, are identified and a working relationship is defined.
6. *Complete the winning strategy.* After the broad outlines of the strategy have been determined, several additional issues must be resolved. These include articulation of a clear value proposition, development of a marketing strategy, and recruitment and placement of key personnel.
7. *Understand the profit dynamic.* Once the detailed strategy has been established, the team should assess the profit implications of each market strategy. This requires development of a financial model of the strategy and testing of the financial implications of changes to key variables.
8. *Implement the chosen strategy.* Implementation is concerned with having appropriate structure, systems, personnel, and skills to complement the strategy. Importantly, implementation is not something that follows strategy development. Rather, strategy development and strategy implementation should be tightly integrated to maximize the likelihood of strategy success.

This eight-step process is a guide, around which companies can structure their process. Below is a description of the planning process at *Medtronic, Inc.*, a leader in the medical device industry.

Planning at Medtronic[5]

Medtronic, Inc., based in Minneapolis, Minnesota, is the world's leading medical technology company. A *Fortune* magazine cover story named Medtronic the "Microsoft of the medical device industry" and featured the company's innovative strategy and culture. Founded in 1949 as a producer of pacemakers for heart patients, Medtronic has expanded internally and through acquisitions into producing devices that do everything from controlling pain to reducing the tremors of Parkinson's disease. In 1989, Medtronic was a $1 billion company, with most of its sales coming from pacemakers. As of July 1, 2003, Medtronic had a market value in excess of $60 billion, and its $7.7 billion in revenues for the fiscal year ending on April 25, 2003 came from a variety of cardiovascular devices, neurological stimulators, drug-delivery systems, and spinal implants.

Medtronic's record of innovation is the result of several factors, including a rigorous strategic planning process, shown in Figure 2-3, tied to a clear mission and well-defined goals. Medtronic's financial goal is to generate a minimum of 15 percent revenue and profit growth over any five-year period.

Medtronic develops a strategic plan every year, but the nature of the plan changes from year to year. In alternating years, the company develops a *bottom-up strategic plan,* then a *top-down strategic plan.* Both types of strategic plans are guided by and tested against the company's mission statement to ensure that the strategy and the mission are consistent. Medtronic's management believes that when a company consistently offers employees a sense of purpose—without deviating and without vacillating—then they will buy into the company's mission and make the commitment to fulfill it. They will go the extra mile to serve customers. They may work late into the night or accelerate the timetable for a crucial new-product introduction. The mission statement takes priority, and strategic plans may be reformulated to be consistent with the mission. Essentially, the mission statement and the strategy are inseparable.

Every other year the company conducts a *bottom-up planning process* where all business units follow a common outline but develop their own strategy and the programs to implement that strategy. Bottom-up planning takes place in Medtronic's five key

---

FIGURE 2-3   Medtronic's Planning Process

- Articulate mission and goals
- Develop either bottom-up (driven by business units) or top-down (driven by CEO) strategic plan in alternating years; five-year planning horizon; focus on being visionary and creative
- Examine fit of the plan to the mission
- Develop annual operating plan

businesses: cardiac rhythm management; neurological and diabetes; spinal, and ear, nose, throat surgery; cardiac surgery; and vascular products. The bottom-up planning outline varies across planning cycles, but typically includes:

- An environmental assessment.
- An assessment of disruptive technologies.
- The identification of key business trends such as the shift to the Internet.
- A financial outlook.

The bottom-up strategic planning process begins in the June/July time frame and by late fall is presented to the Medtronic Executive Committee, which challenges the plans of all of the businesses. The Executive Committee is comprised of Chairman/CEO/President, the five business presidents, and key staff members including the CFO, the Chief Legal Officer, the Chief Human Resource Management Officer, the Chief Information Officer, and the Chief Medical Officer. Each business's strategic plan is laid out in thirty to forty slides, of which about one-third are common in format for all businesses with the remainder being specific to a particular business. The actual written plan for a business is likely to be eight to fifteen pages long. The individual business plans are then rolled up into a summary document by the CEO.

The rolled-up plan and the individual business plans are presented to the board of directors in a dedicated corporate strategy session in the last quarter of its fiscal year. A major goal is to keep the plan concise. The primary purpose of the written plan is to provide board members with background for their questions and discussions. Presentations and discussions are the primary way that Medtronic communicates strategy. There is substantial oversight of and challenge to the plan.

The strategic focus shifts in alternating years to a *top-down process*. Top-down planning is driven by the CEO. The intent of top-down strategic planning is to be visionary and creative. The objective is to identify new areas for growth, such as new disease states and new technology platforms that will allow Medtronic to expand into new markets or businesses. At this level the CEO and the senior management team identify a small number of issues and potential new market and business opportunities that they think could be fundamental to the company's growth. The planning template is adjusted to recognize the nature of the issues that have been raised during a planning cycle. For example, in 2000 then-CEO Bill George embarked on an initiative called "Vision 2010." He had a number of teams composed of people from throughout the company look at factors that he thought would impact how the company would evolve. These factors included the role of the patient in consumer marketing and patient advocacy, the impact of the Internet on access to information, the future of biotechnology, and the role of information technology in Medtronic's medical devices. The result was a new vision and new strategies for Medtronic.

While both types of strategic plans look out over a broad five-year horizon, an *annual operating plan* is also developed for each fiscal year. The operating plan is much more detailed and contains specific objectives, milestones, a budget, and clear delineation of responsibility. Medtronic separates the strategic and operating plans in that the company's managers do not carve specific details from one year of the strategic plan and force that on top of the annual operating plan. The intent is to keep a robust strategy intact and not subtly encourage managers to manipulate numbers so that they can make their targets. With this in mind, the financials in the five-year strategic plan reflect only the current, third, and fifth years and do not include the annual operating plan year specifically. In this regard, the strategic plan and the operating plan are only loosely coupled. The belief is that tightly coupling the operating plan to the strategic plan could stifle creativity and innovation. The result is that the strategic plan remains broad and visionary while the operating plan is much more detailed and focused.

STRATEGY IN THE HIGH-TECH FIRM

Key Strategy Decisions

A firm's strategy should answer four key questions:

- Who are our target customers?
- What products or services should we offer them?
- When should we enter a market?
- How can we execute our strategy efficiently and effectively?

Customers

"In making strategy . . . first comes painstaking attention to the needs of customers. . . . Strategy takes shape in the determination to create value for customers."[6] Customer-focused approaches to analysis, value creation, and strategy implementation begin by assessing what customers' articulated and unarticulated needs are and how they are likely to change in the future; they focus on creating the resources that will most effectively satisfy their customers' needs. However, this approach to strategy addresses only the needs of the "served market," or current/existing customers. A narrow focus on the served market may inhibit innovation and blind the firm to the emergence of new segments in a rapidly changing market. Thus, managers must ask both "Who are our current customers?" and "Who should our customers be in five years?" Some refer to this as "bifocal vision," with a focus on both near-term and longer-term customers.

Customer-focused businesses escape the **tyranny of the "served market"** *(current customers)* and avoid head-to-head competition by searching for new **market space**.[7] New market space represents potential—those who might be customers. Businesses in search of new market space look "across substitute industries, across strategic groups, across buyer groups, across complementary product and service offerings, across the functional-emotional orientations of an industry, and even across time."[8] New products and new market space are the foundation for organizational renewal in the customer-focused business.

Hyperion has more than 6,000 customers worldwide representing a broad spectrum of industries. Customers include Cable & Wireless, Compaq, Daimler-Chrysler, General Electric, Hewlett Packard, Mitsubishi, Nissan, Nortel, Southwest Airlines, Sun Microsystems, Unisys, and Wal-Mart. Hyperion has addressed a need that is common to many business customers, thus we would say that it is broad rather than narrow in its scope of market coverage.

Products and Services

Products, services, and technologies should be seen as vehicles for value creation, not as something that has intrinsic value. Explains John Chambers, CEO of Cisco Systems, "I have no love of technology for technology's sake—only solutions for customers."[9] The first question from the mouth of a very successful divisional R&D manager in a Fortune 50 company whenever an engineer brought up a new product concept was consistently, "How will this create value for our customers?" Over time, this simple question reoriented engineers to think about customers first, instead of technology, thus bringing much-needed discipline to the division's R&D efforts.

To continue with the introductory example, Hyperion provides software for performance scorecarding, business modeling, planning, budgeting and forecasting, performance monitoring, analytics, management reporting, as well as financial consolidation.

Timing

Timing of market entry is one of the most vexing issues facing managers today. Should a firm speed the time-to-market cycle and attempt to be the market pioneer? Many studies report that pioneers have a strong first-mover advantages in the marketplace, including higher market share and stronger differential advantage. Yet, one can also find many examples of later entrants having a stronger position than the market pioneer. By allowing competitors to enter the market first, later entrants do not have the market development costs that a pioneer faces in terms of educating customers about the new technology. In addition, later entrants can introduce a product with a higher performance level or a better price–quality ratio.

Pioneering Advantages. What are the arguments in support of being a market pioneer?[10] First movers are thought to have competitive advantages due to entry barriers established by their market entry. Such entry barriers include economies of scale, experience effects, reputational effects, technological leadership, and buyer switching costs. These barriers can lengthen the lead time between a firm's head start and the response by followers. During the time when there is no competition, the first mover is, by definition, a monopolist who can gain higher profits than in a competitive marketplace. In addition, even after competitors enter, the first mover has the established market position, which may allow it to retain a dominant market share and higher margins than later entrants. First movers are also able to "skim off" early adopters, whereas later entrants are left with potential customers who are less predisposed to purchasing new products.

Moreover, if customers know little about the importance of product attributes or their ideal combinations, a first mover can influence how attributes are valued and define the ideal attribute combination to its advantage. The first mover becomes a prototype against which all other entrants are judged, making it harder for later entrants to make competitive inroads. First movers have a higher degree of consumer awareness, which lowers perceived risk and information costs.

For example, the Internet is not a world that tolerates caution or deliberation. In a medium where brand-name recognition is everything, losing the first-mover advantage can be a handicap. Barnes & Noble was blindsided by Amazon.com. And despite its huge expenditures and massive advertising since then, Barnes & Noble remains barely more than one-tenth of Amazon's size online.[11]

Are these arguments so compelling, and the research findings so supportive, that the pioneering advantage is automatically conferred? Not necessarily. A close examination of the evidence shows that for a firm to definitively establish a pioneering advantage, it must have certain competencies and capabilities, including technological foresight, perceptive market research, skillful product and process development capabilities, marketing acumen, and possibly even luck. Being a market pioneer *per se* does not directly produce enduring competitive advantage. Indeed, it also comes with distinct risks.

Pioneering Disadvantages.   In their study of the personal digital assistant (PDA) industry, Bayus, Jain, and Rao[12] found that speeding a product to market is not always a good idea. Apple's Newton is merely a footnote in the history of the handheld computer (PDA) industry. Other pioneers that not only lost out to later entrants but eventually diappeared include VisiCalc, the first personal computer spreadsheet program, and Osborne, the first portable computer. And, in the online grocery business WebVan invested more than $1 billion before closing shop in July 2001. Today, FreshDirect, a later entrant, is thriving with a more focused, lower cost business model.[13] Based on a historical analysis of 500 brands in fifty product categories during the period 1856 to 1979, Golder and Tellis found that the failure rate of market pioneers is 47 percent, while their average market share is 10%. The eventual market leaders entered the market an average of thirteen years after the pioneer. However, their failure rate was only 8 percent and their average market share was 28 percent.[14]

Firms must evaluate tradeoffs among time to market, product performance, and development costs. Pioneers face huge development costs against a high degree of market uncertainty. In fact, Boulding and Christen[15] found that, over the long haul, pioneers are considerably less profitable than later entrants. Although pioneers do enjoy sustained revenue advantages, they also suffer from persistently high costs, which eventually overwhelm the sales gains.

How should marketing managers in a high-tech firm decide whether to pioneer the market? The success of a pioneering strategy depends on how well a firm understands the market (both the market size and market needs) and how well it understands its competitors' strengths and weaknesses. For example, in the PDA market, Apple overestimated the potential market size and underestimated the product performance desired when it launched the Newton. Later entrants, such as Palm Pilots™, were more profitable. They entered the market with higher performance products targeted to narrower target markets.

Late movers can overcome pioneers in at least two ways:[16]

1. A late mover can identify a superior but overlooked market position, undercut the pioneer on price, or outadvertise/outdistribute the pioneer, thereby beating it at its own game.

2. A late mover can innovate either superior products or strategies that change the rules of the game. In particular, *innovative* late entrants, relative to pioneers, grow faster and have higher market potential. Moreover, innovative late entrants slow the pioneer's growth and reduce its marketing spending effectiveness.

The research suggests that by reshaping the category, late entrants can redefine the game in such a way that benefits the late mover and disadvantages the pioneer. So, for any given firm, the question of whether early or late entry is more advantageous depends on the market's and the firm's particular characteristics. Thus, it may be prudent for technology companies to take the time to evaluate the opportunity, assess the competitive threat, then enter the market with a better offering supported by adequate allocation of resources.

Execution

The fourth of the key strategy decisions is execution. It has been said that a mediocre strategy that is *brilliantly executed* will always beat a brilliant strategy that is executed in a pedestrian way. One implication of this statement is that the separation of strategy formation and strategy execution is an artificial distinction. Strategy and execution should be tightly linked in the creative process.

To a large degree, execution is concerned with getting value from the company's mind into customers' hands. Execution requires having the right competencies, an appropriate structure and systems (including compensation systems that reinforce the strategy), plus making good decisions in the distribution, pricing, and promotion arenas. At the same time, executives must be careful to not develop an implementation program that is not flexible. Rapidly changing market conditions and strategy that emerges through learning mean that execution requirements will change as well.

A concluding thought on this topic is that managers should not become enamored with a specific strategy, but be prepared to adapt and change it based on developments in the marketplace—and to do so rapidly.

Strategy Innovation

Merely answering the preceding questions is not sufficient to develop sound strategy. The questions must be answered in such a way that the company brings a unique and innovative perspective to customer value creation. Thus, it would be an oversight not to address how a firm can be innovative in its business strategies as a way to gain competitive advantage. Recent writings on this topic highlight the fact that firms that are able to sustain a high rate of growth do so by radically changing the basis of competition in their industries to create new wealth.

The idea of being innovative in their approach to strategy is vital for today's firms. Digitalization, deregulation, globalization, and new economic business models are profoundly changing the industrial landscape. Indeed, unless today's established corporations learn to reinvent themselves and their industries, new wealth will be created by newcomers. For example, between 1986 and 1996, just seventeen companies of the Fortune 1,000 grew total shareholder return by 35 percent or more per year.[17] Although quality concerns, cost awareness, time to market, and process improvements are vital, they are no longer a source of large gains. These tools and the resulting incremental improvements in strategy will keep profits from eroding and prolong the usefulness of current business strategies, but they won't create new wealth.

To create new wealth, companies must learn how to unleash the spirit of "strategy innovation," the idea of creating a revolution within their own companies. What does it take to identify killer strategies? It is really no different than what it takes to develop breakthrough innovations: Take risks, break the rules, be a maverick. Hamel[18] offers the following insights:

- Rather than focusing on industry analysis as the key to strategy formulation, firms that create new wealth and play by new rules recognize that industry boundaries in today's environment are fluid rather than static. Hence, innovative firms do not focus their analysis on existing industry boundaries.

- Nor do firms with innovative strategies focus on direct competitors. With new business models, it is difficult to distinguish competitors from collaborators, suppliers from buyers. Rivalry is no longer easy to identify, and it is hard to know who is friend and who is foe. It is vital to look at product form competition and partners as potential competitors.

- Strategy formulation must recognize that today's business boundaries (in addition to industry boundaries) are fluid. With the rise of contract workers, outsourcing, and supply relationships, the firm no longer has control of all its critical assets.

A good strategy creation process is a deeply embedded skill; it is a way to understand what is going on in an industry, turn it on its head, and envision new opportunities. And it is based on the paradoxical notion that one can make serendipity happen. How?[19]

- Bring new voices into the strategy formulation dialogue. "Companies miss the future not because they are fat or lazy but because they are blind."[20] Many companies are unequipped to see where the future is coming from and lack the lens to know that. So bringing new voices into the process, outside of the company's normal comfort zone, can provide new vision.

- Bring new connections between new voices, across boundaries of function, technology, hierarchy, business, and geography. Such new conversations can offer a rich web of insights.

- Offer new perspectives, based on a new vantage point in viewing the business. Rather than based on analysis and number crunching, often innovative strategies emerge from novel experiences that yield novel insights.

- Exude passion for discovery and novelty, which engenders an emotional attachment of employees who are committed enough to reduce the time between an idea and its implementation.

- Be willing to experiment. With innovative strategies, the end target may be known, but the route to it may be unknown. The best way to approach such a situation is to be willing to move in the right direction and to refine strategy and process as the firm learns from its experiments. Although such a notion is an anathema to efficiency, it is a must in redefining radical business strategies.

What companies have followed this model in pursuit of "killer strategies"? Intel chose not to follow the conventional wisdom that it pays to extend a product's shelf life. It keeps making its own chips obsolete with better designs and showed that with an effective advertising campaign, one can "brand" a component within another product. Chevron Corp. "mined" seismic data to discover a 1.45-billion-barrel oil field. Other companies, such as Amgen, Oracle, and Iomega, to name a few, have all shown their willingness to look from the outside in and, in doing so, create new rules in established industries.

COMPETITIVE ADVANTAGE—THE OBJECTIVE OF STRATEGY

The purpose of the strategic planning process is to create competitive advantage. **Competitive advantage** exists when the firm possesses resources and competencies that enable it to provide superior benefits to customers or give it a cost advantage, are rare, and difficult to imitate.

Resources and Competencies

The firm's resources are the foundation for the creation of superior customer value. *Resources* may be physical assets, intangible assets, or competencies. Physical assets include such things as manufacturing plants, information systems, distribution facilities, and products. Intangible assets include brand equity, customer loyalty, distribution channels, market knowledge, and the firm's beliefs about customer needs or their responsiveness to pricing, promotional, or distribution changes. *Competencies* are the bundles of skills that enable the firm to achieve new resource configurations as the firm and the markets it competes in evolve. Marketing competencies in high-tech firms include processes for gathering, interpreting, and using market information; the ability to manage customer relationships and establish collaborative relationships with distributors to serve customers more effectively; service delivery; product/service development; new product commercialization; and supply chain management, among others.[21]

It appears that competencies, while important to all businesses, may be more important to high-tech businesses than to low-tech ones. Let's compare Wal-Mart and Dell to get some insight into this phenomenon. In January 2003, Wal-Mart was America's Most Admired Company and the world's largest corporation with revenues of $245 billion on total assets of $95 billion. In contrast, Dell Computer was America's fourth Most Admired Company and had revenues of $35.5 billion on total assets $15.5 billion. Thus, Wal-Mart generated $2.60 in revenues for every $1 of assets and Dell generated $2.30 in revenues for every $1 of assets. This might lead one to believe that each dollar of Wal-Mart's assets would be valued more highly than Dell's. However, each dollar of Dell's equity financed assets has a stock market value of $12.60, while the comparable figure for Wal-Mart is $5.30. This difference cannot be explained by differences in profit margin or growth rate. It can be explained by the fact that Dell possesses even more valuable competencies than does Wal-Mart. Thus, we need to consider core competencies in greater detail.

Core competencies are the set of skills at which a company excels.[22] Such competencies can be identified based on their three underlying characteristics.

- True core competencies make a disproportionate contribution to customer-perceived value. They enable the firm to deliver a fundamental customer benefit such as reliability, user-friendliness, exceptional service, or enhanced productivity.
- To qualify as a core competence, a capability must be competitively unique. Furthermore, core skills and capabilities are very difficult for competitors to imitate, because they are *embedded* deeply in the organization's routines, procedures, and people.
- Core competencies also allow a firm to access a wide variety of very disparate market opportunities. In other words, the firm should be able to leverage its core competency into new arenas, by applying its skills and competencies in product markets where it has not previously competed.

In the high-tech arena, Hewlett-Packard serves as a good example of leveraging core competencies. One of Hewlett-Packard's core competencies is in the area of transferring digital images to paper with superior quality (clarity, detail, and color). This core competency was exhibited in its resounding success in the laser printer business. While other companies also made laser printers, HP's superior technology and production skills made the high quality very difficult to imitate. Moreover, the skill in transferring digital images to paper in a high-quality fashion was significantly related to the benefits customers were seeking in printing their computer images. Hewlett-Packard leveraged this core competency into a very different market: It entered the digital photography business with a digital photography package consisting of a camera, scanner, and printer. The digital photography business taps into essentially the same skills and capabilities that made HP successful in the laser printer business: transferring high-quality images to paper.[23]

Figure 2-4 shows a diagram of core competencies, using the analogy of a tree,[24] applying the analogy to Honda. The branches or canopy of the tree represents the widely different product markets to which the core competencies have provided access. In Honda's case, this would be the end markets in which it competes: small cars,

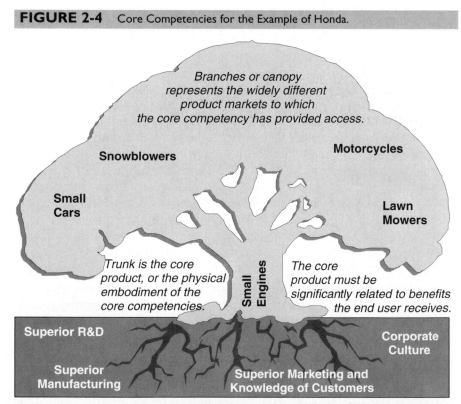

FIGURE 2-4   Core Competencies for the Example of Honda.

Roots are underlying skills and capabilities that represent core competencies.

SOURCE: Adapted from Prahalad, C. K. and Gary Hamel (1990), "The Core Competence of the Corporation," *Harvard Business Review,* May-June, pp. 79–91.

snowblowers, motorcycles, and lawn mowers, to name a few. The trunk represents the core product, or the physical embodiment of the core competencies. The core product must be significantly related to the benefits the end user receives. The roots of the tree represent the underlying skills and capabilities that form the basis of the core competencies. In this case, Honda's superior research and development, manufacturing techniques, marketing, knowledge of customers, and its corporate culture give rise to its success in small engine technology.

Using a core competencies approach to resource allocations can result in decisions that may seem to defy conventional logic. For example, a basic, underlying business tenet is that firms evaluate possible investments in new projects on criteria such as return on investment or payback period. However, using a core competency model of strategy, firms' investments may often defy the use of traditional criteria.

An excellent example can be found at Amazon.com.[25] Jeff Bezos established Amazon based on the concept that he could give customers access to a giant selection of books without going through the time and expense of opening stores and warehouses and dealing with inventory. However, he soon discovered that the only way to make sure customers get a good experience and that Amazon gets inventory at good prices was to operate his own warehouses.

Building warehouses was met with skepticism from Wall Street. At about $50 million each, warehouses are expensive to set up and expensive to operate. However, a visit to one of Amazon's six warehouses today makes it quite clear why Bezos believed he would defy financial logic. Amazon's warehouses are models of efficiency. They are so high tech that they require as much computing power to run as Amazon's Web site does. Computers send signals to workers' wireless receivers, telling the workers what items to pick off the shelves; then the computers determine everything, from which item gets picked first to whether the weight is right for sending.

Along the way the computers generate data on everything from misboxed items to backup times—and managers are expected to study the information religiously. The result is that the warehouses are extremely efficient. For example, by redesigning a bottleneck where workers transfer orders arriving in green plastic bins to a conveyor belt that automatically drops them into the appropriate chutes, Amazon has increased the capacity of one warehouse by 40 percent. Today, Amazon's warehouses can handle three times the volume they could in 1999, and in the past three years the cost of operating them has fallen from nearly 20 percent of Amazon's revenues to less than 10 percent. The warehouses are so efficient that Amazon turns over its inventory roughly twenty times a year. The rate for virtually every other retailer is under fifteen. Based on Wall Street's initial skeptical reaction, it is unlikely that Amazon's investment in this logistics competency would have met traditional investment criteria.

Finally, Amazon leveraged its investment in this competency into new market arenas by becoming a service provider to other e-tailers. For example, Amazon uses its logistics and Web-based skills to run the websites for other companies, such as Target and Borders Books.

Tests of Competitive Advantage for Superiority and Sustainability

Strategy, competencies and assets are the foundation for the creation of superior customer value, but they are only the foundation. Any strategist or planner would be remiss in not formally assessing whether the strategy and its supporting competencies and assets lead to a position of superiority. It is only through this assessment that the strategist can determine whether the strategy is likely to be successful or whether it needs to be adjusted. Recall from Figure 2-2 at the outset of this chapter that there are two tests a resource must pass to lead to a position of superiority—customer *value* and *resource rareness*—and two more tests—*durability* and *inimitability*—that the resource must pass to be a source of sustained superiority.[26] We now consider each of these tests.

Customer Value

Customer value is the difference between the benefits that a customer realizes from using a product and the total life-cycle costs that the customer incurs in finding, acquiring, using, maintaining, and disposing of the product. A resource is valuable if it enables the firm to develop and implement strategies that enhance its customers' effectiveness or efficiency. An effective strategy provides additional customer benefits while increasing life-cycle costs at a slower rate than the increase in benefits.

An efficient strategy focuses on the cost side of the value equation. Its objective is either to reduce life-cycle costs while maintaining benefits so as to increase demand, or to reduce costs internally but not pass the cost reductions on to buyers. As long as benefits and costs are competitive, the firm will achieve "normal" market share but will see its margins and ROI increase.

Different types of buyers will have different perspectives on the worth of different benefits or life-cycle costs. About the only element in the value equation on which there will be agreement about worth is price. Competition based on a low-price strategy, though, is the last strategic marketing tactic that should be considered. Consequently, the firm's managers must conduct careful cost-benefit analysis among members of the target market before making substantial adjustments to the value equation.

How does Hyperion's business performance management software create value? Hyperion's strategy is to create performance-accountable enterprises by offering a Performance Scorecard that links business strategy with actionable goals. Such a scorecard should enhance implementation, speed organizational learning, and facilitate adaptation.

Resource Rareness

The second test for a resource to provide a position of superiority requires that the firm's resources are sufficiently rare that competitors or producers of substitutes are not able to offer the same, or similar, set of benefits and life-cycle costs. If many firms possess the same valuable firm resource, then each firm has the ability to deploy that resource in a similar way. In this case, each firm can implement the same strategy with the result that no firm achieves superiority. This does not mean that rather common resources, such as managerial talent, are unimportant. Indeed, such common resources may be necessary to exploit other, rare resources. However, possessing these resources will not lead to a superior competitive position. Rareness also does not mean that only one firm can possess the valuable resource for it to be a source of superior value. As long as the number of firms in the industry that possess the resource is less than the number required for the resource to approach commodity status, the resource may be a source of advantage.

For example, within the ceramics industry there are many firms that can produce fine china capable of withstanding from 5,000 to 15,000 pounds of pressure per square inch (352–1055 kilograms per square centimeter). However, producing ceramic armor that can withstand 140,000 pounds of pressure per square inch (9843 kilograms per square centimeter) is another matter. It is this type of advanced and highly specialized knowledge that has propelled Japan's Kyocera Corporation to the top spot in that industry. Of course, this leads to the question, how can the firm develop a valuable and rare resource base to achieve a position of superiority? The simple answer, although it is quite difficult to realize, is to create a bundle of complementary resources—including physical assets, intangible assets, and competencies—that produce customer value. Dell has such a combination of unique competencies in production and delivery that allow it to provide a high quality product at a low price, and it possesses a highly regarded brand name.

Achievement of a *sustainable* position of superiority is the Holy Grail of strategic marketing and is based on the last two tests: durability and difficulty of imitation. That a position of superiority is sustainable does not mean that it will last forever. Recent research shows that fewer than 5 percent of firms are able to generate superior profits for ten years.[27] Changes in customer needs or in other elements of market structure may make a resource that once was a source of value no longer valuable. Thus, in one sense, sustained superiority requires continuous improvement in the resource punctuated by the regular, if infrequent, development of new resources. However, durability and inimitability enhance the likelihood of sustainability.[28]

Durability

Durability is concerned with how rapidly a valuable resource becomes obsolete due to innovation by current or potential competitors. The longer it takes for a resource to be rendered obsolete, the more likely it is to be a sustained source of value. Resource durability depends, in large part, on the nature of the industry. Slow-cycle industries, such as many low-tech industries, have a very slow rate of change due to low market and technological turbulence. Many consumer brands such as Coca-Cola, Ivory soap, Campbell's soup, and Kellogg's cereals have maintained strong customer loyalty, even at relatively high prices, for long periods of time. Customers have had good experiences with these brands and are reassured by the brand name. The durability of these brand names is a major reason why they are valuable as a basis for brand extensions.

Fast-cycle industries are often based on a technology or on an idea. In these industries, technology is a rapidly depreciating resource as the list of companies that have led the video game industry illustrates: Atari→Nintendo→Sega→Sony→Microsoft. Each of these firms had a strong position in the industry only to see it eroded as technology advanced. Their positions were not durable.

Intel has been the leader in the microprocessor industry for three decades. However, in 2003, Intel's dominance in the microprocessor industry was challenged by the switch to 64-bit microprocessors in the highly profitable server market. In 2002 Intel's Itanium chip held less than a 1 percent share of the 64-bit market with Sun and IBM holding the dominant positions.[29] Can Intel regain its leadership position in this rapidly changing industry or have its resources become obsolete? Betting on the durability of competitively valuable resources is risky because competitors or new entrants will attempt to develop a new generation of technology that renders the existing technology obsolete.

Inimitability

Inimitability is concerned with how easily a competitor can obtain a valuable resource either through internal development or purchase in the market. Possessing a resource that is easily imitated creates only temporary advantage. Three factors affect inimitability: transparency, replicability, and transferability.

Transparency.   Transparency of the firm's resources, or the ease with which others can observe the source or basis of the superior position, enables competitive imitation to take place more rapidly. Transparency is reduced and inimitability is enhanced when the resources that underlie the position of superiority are not apparent. Furthermore, a superior position that is built on the coordination of several resources is more difficult for a competitor to understand than a position that is based on a single resource. Thus, the greater is the uncertainty within a market regarding why some firms are successful, the greater is the potential that competitors will be discouraged and the longer that successful firms can sustain their superiority.[30]

Replicability.   Replicability is concerned with how easily a firm can develop a valuable resource through internal investment. For example, elements of a product's design may be easily reverse-engineered and replicated by a competitor, thus eroding pioneering advantage.[31] Resources based on complex organizational routines such as production processes, interpersonal relationships among a firm's employees, a firm's culture, or its reputation among suppliers and customers are difficult to replicate. The creation of values, attitudes, norms for behavior, and relationships is quite difficult and takes a long time. Furthermore, complex organizational routines and processes are not a particularly transparent source of superiority. Thus, organizationally complex resources resist imitation.[32]

Other barriers to imitation by competitors include patents, brand names, corporate reputations, specialized assets, and financial resources. All of these inhibit a rival's ability to imitate a valuable resource or, if the resource can be imitated, they inhibit the ability to duplicate the customer's perception of the value that is created. To the extent that effective imitation of a valuable resource can be discouraged or stopped, the greater is the sustainability of the first mover's position.

Transferability.   Transferability is concerned with whether a competitor can acquire, in the market, the resources that are necessary for a superior position. To the extent that resources are easily acquired, then the first mover's position of superiority will be short lived. Impediments to transferability include the difficulty of geographically relocating specialized equipment and employees; failure of a resource, such as brand or personnel with special skills, to maintain its productivity when transferred to a new organization or setting; and the difficulty of transferring a capability without transferring the entire team that is responsible for it.[33]

Approach to Developing Resources and Competencies

The questions raised now are how can the firm's managers determine which resources are necessary to achieve a position of sustainable superiority and how can the firm develop them? The extreme positions are to take a *customer-focused approach* or a *competitor-centered approach*. A customer-focused approach starts with an analysis of customer needs and then develops the resources to satisfy those needs. This approach is consistent with the notion of market orientation, which we will explore in detail in Chapter 4. The competitor-centered approach is based on a comparison of the firm's resources with the resources of a few key competitors.[34]

We believe that it is most appropriate to start with the customer and work backwards. As John Young, former CEO of Hewlett-Packard, said, "If we provide real satisfaction to real customers, we will be profitable."[35] In contrast, a competitor-centered approach may obscure customer value-creating opportunities, blind the firm to changes in customer needs, and lead the firm to emulate competitors' strategies. Imitative strategies reduce the likelihood of experimenting with innovative strategies that could give the firm an advantage in providing superior customer value.[36] Businesses that adopt a competitor-centered perspective typically end up competing on incremental improvements in price, quality, or service. The traditional market segmentation, targeting, and positioning approach to developing customer-focused strategy is useful for identifying the resources upon which to build competitive advantage.

CULTURE AND CLIMATE IN INNOVATIVE COMPANIES

We now turn from the concepts and processes of strategy innovation to the characteristics of culture and climate that are necessary for strategy innovation to occur. Recall from Figure 2-2 at the outset of this chapter that a firm's culture and climate provide the backdrop against which strategy is developed and implemented.

Culture is the deeply rooted set of values and beliefs that provides norms for behavior in the company. Culture helps us understand why things happen the way they do. Climate is the set of behaviors that are expected, supported, and rewarded. Climate is what actually happens in the company.[37] To a large extent, climate is the observable manifestation of culture. Because cultural values are so deeply ingrained, they are difficult to change—which means that dysfunctional behaviors may also be difficult to change. We now explore some cultural obstacles in companies that wish to be innovative and *facilitators* that can overcome these obstacles, as shown in Figure 2-5.

FIGURE 2-5   Culture and Climate Considerations for High-Technology Firms

Cultural Obstacles to Innovativeness	Cultural Facilitators of Innovativeness
• Core Rigidities	• Creative Destruction
• Innovator's Dilemma	• Leveraging Firm Dominance Effectively
	• Unlearning
	• Corporate Imagination
	• Expeditionary Marketing
	• Nurturing Innovation

Cultural Obstacles to Innovativeness

Core Rigidities

Although core competencies are an essential ingredient for success, they might also become **core rigidities** and possibly hinder new product development. For example, new product ideas built on familiar skills and capabilities are more likely to be embraced by a firm than those built on unfamiliar technologies. However, when market conditions are changing—for example, when new technologies are being developed by other firms in the industry—it may be important for a firm to examine closely the viability of a new technology. But ingrained routines, procedures, preferences for information sources, and existing views of the market—all of which can be related to underlying core competencies—can become barriers to a realistic assessment of new market opportunities. In such a situation, core competencies can become core rigidities, which strangle a firm's ability to act on novel information.[38]

Core rigidities are straitjackets that inhibit a firm from being innovative and can include:

- Cultural norms in the firm.
- Preferences for existing technology and routines.
- Status hierarchies that give preference to, say, technical engineers over marketers.

It is understandable that the cultural norms, technologies, routines, and company leaders' beliefs are valued, because they are the basis for the success many companies enjoy. At some point, however, such skills, values, and routines may not be as well suited to the changing business environment and necessitate reexamination and change themselves. Firms that are able to reevaluate such skills and capabilities on a regular basis, and update and modify them as needed, are not as burdened with rigidities as other firms may be.

For example, in the 1920s, New York–based oil-service company Schlumberger virtually established the oil-service industry by using electrical resistance to detect oil deep underground.[39] Recently, however, competition developed something called smart wells, which challenge Schlumberger's most profitable business, Wireline and Testing. Schlumberger has run the risk of losing its technological lead and being poorly prepared to meet customers' demands. It seems as if past financial performance has created a sense of invincibility, which can result in underperformance and lack of innovation. In this case, its core competencies in one set of technologies may have prevented it from recognizing opportunities existing outside of its skill set.

The Innovator's Dilemma

A consistent pattern in business is the failure of leading companies to stay at the top of their industries when technologies or markets change. Apple Computer was an early leader in personal computing and established the standard for user-friendliness but lagged years behind its competitors in introducing a portable computer. The **innovator's dilemma**[40] says that market leaders have great difficulty introducing radical innovations because it is very difficult for them to build a persuasive case for diverting resources from making incremental innovations that address known customer needs in established markets to new markets and customers that seem insignificant or do not yet exist. A radical innovation provides customer value in a very different way than does the traditional solution. An excellent example of a radical innovation that disrupted an entire industry is the digital camera. There is no film and, thus, no film processing. Kodak, which has been an innovator in photographic film, has had a very difficult time adjusting because its business model is focused on camera film. Kodak lost 45 percent of its market value from the beginning of 2000 to the beginning of 2003. It is common for leaders who stay on top of wave after wave of incremental innovations to fail when a radical innovation disrupts their way of doing business.

So, how can a firm avoid falling prey to core rigidities and the innovator's dilemma? In addition to other ideas and insights presented throughout this text, firms in high-tech industries can use the following tools and techniques.

Cultural Facilitators of Innovation

Creative Destruction

To avoid being trapped by core rigidities and the innovator's dilemma, firms must recognize that products in high-tech environments typically sustain only a finite spell at the technological frontier before being made obsolete by better products. Given this reality, firms must proactively attempt to develop that next-generation technology—despite the fact that such developments may alienate some current customers, make obsolete its sunk investments in the prior technology, and render any economies of

---

TECHNOLOGY TIDBIT

THE THIN-AIR DISPLAY

Thin, flat-panel displays are the status symbols du jour. But the displays of tomorrow may be thinner still—so slim, in fact, that they'll literally be made out of thin air. One such display already being tested is the Heliodisplay, invented by MIT researcher Chad Dyner and being developed by IO2 Technology. It projects a video image—or any standard computer image—that appears to float in midair. No special goggles are required.

The developers of the floating monitor say that the projector "modifies the properties of the air within a localized environment" to create a full-color video picture. The image is two-dimensional and can be used like a touch screen by pointing at the objects floating in space. You can watch several video examples of how the Heliodisplay works at www.io2technology.com.

Although the images seem to hover in the air, don't expect to see huge interactive projections like those used in the movie *Minority Report* any time soon. At the moment, the largest Heliodisplay is 27 inches with a 1,024-by 768-pixel display. But the company is working on several 42-inch prototypes it hopes will eventually be priced about the same as comparable plasma displays. "Most commercial interest has come from trade show display firms, amusement parks, and the military," says Bob Ely of IO2 Technology.

SOURCE: John R. Quain (November 4, 2003), "The Thin Air Display," PC MAGAZINE.

scale and experience curve advantages useless. This notion of constantly innovating to develop the next-generation technology, despite the potential drawbacks of doing so, is known as **creative destruction**.[41]

The paradox of such a model of competition is that the firm itself must work to find the next best technology, which is likely to destroy the basis of its current success. The reason for such a model is that if the firm does not commercialize a new technology, rivals will surely do so. Hence, firms must not hold back a new technology that makes its existing products obsolete. Even with successful products, firms should not be too enamored with the technology that forms the basis of that success, but instead strive to develop even better technologies.[42] Indeed, ongoing industry leadership hinges upon creative destruction.

An example is Microsoft's continued desire to ride the wave of the success of its operating systems software (Windows 95, 98, XP, etc.). However, when rivals Netscape (first) and Sun Microsystems (second) introduced new technologies (Internet browsers and Java scripts, respectively), Microsoft had to make a major turnaround in strategy to participate in these technologies[43]—even though they may radically change the world of desktop computing and, at the extreme, render proprietary operating systems obsolete. Rather than information being accessed by desktop computers, these technologies allow consumers to access information via a range of consumer electronic devices and "information appliances" that don't require the Windows operating system.

An additional example is found in the Internet realm. Companies whose past successes have been based heavily in a brick-and-mortar world with tangible distribution channels are finding that to compete with dotcom startups, they must offer sales through an Internet channel. Yet, doing so undermines the very basis for their success. However, this is the nature of creative destruction: If the company itself does not offer customers the access avenues they desire, a competitor surely will. More ties to the Internet will be developed later in this chapter.

Leveraging Firm Dominance Effectively

There has been a significant debate regarding the relationship between firm size and the firm's ability to be innovative. On the one hand, larger firms are said to suffer from inertia; they tend to be more bureaucratic and have more to lose than other firms when they develop radical new innovations that may obsolete their existing product lines. However, contrary to suffering this "incumbent's curse," recent research shows that since World War II, more radical innovations have been introduced by larger, established firms than smaller firms and new entrants.[44] Two variables are related to the innovativeness of large firms: the source of their dominance and managerial expectations about obsolescence.

Sources of Dominance.   There are three variables that give rise to a firm's dominance in the marketplace: (1) investments in the existing product generation, (2) stronger market share, and (3) greater wealth. Although some aspects of a firm's dominance in the marketplace—for example, its *greater investments in the existing product generation* and its *stronger market position* based on sales of existing products—reduce the firm's motivation to invest in radical innovation, dominant firms have a *greater ability to invest* in expensive radical innovations by virtue of their greater wealth.[45] Simply, radical innovations and the technology necessary to generate them have become increasingly complex and require substantial resources. Dominant firms have such resources and enjoy economies of scale and scope in both R&D and in marketing.

Managerial Expectations about Obsolescence. Another variable that allows dominant firms to remain innovative, and overcome the potential negative effects of size (arising from inertia and escalation of commitment), is the managerial expectation that the new technology will make their existing products obsolete. When managers of dominant firms believe that a new technology may make their existing products obsolete, the fear of obsolescence causes firms to invest aggressively in radically new technologies. On the other hand, dominant firm managers who believe that the new technology is likely to increase sales of their existing products actually invest less aggressively in the new technology than do managers who subscribe to the former position. Thus, the fear of loss as a result of obsolescence appears to be a much stronger motivator of investments in radical innovation among such firms than is the lure of gains from enhancement. The lesson is that managers should use fear of obsolescence rather than desire for enhancement as the motivation to engage in creative destruction.[46]

The importance of managerial expectations about obsolescence, or their willingness to cannibalize, begs the question of what factors in the firm make managers predisposed to undertake such creative destruction.[47] Organizations that have strong, autonomous SBUs that compete internally for resources, strong product champion roles, and a focus on future markets more than current markets have a stronger willingness to cannibalize and are more likely to introduce radically new innovations.

This sub-section on leveraging firm size as a facilitator of innovativeness indicates that large firms are not necessarily less innovative than smaller, more nimble firms. For both sizes of firms, one benefit of being innovative is the financial gains to be realized. The more innovative products are, the greater their financial value. In addition to getting greater gains from more radical innovations, the financial value from innovation is also a function of the firm's resources that can support the innovation. Dominant firms have greater marketing resources to sustain the innovation and increase the adoption rate of the new product. They also have "market-based assets," such as brand equity and customer trust, that can reduce the perceived risk that consumers associate with radical innovations. Research shows that the stock market values the introduction of a radical innovation more when it comes from a dominant firm than a nondominant firm.[48]

Thus, this suggests that the introduction of a radically new technology may be more beneficial for a dominant firm than a nondominant firm because it reinforces the market position of the dominant firms by generating greater cash flows than would be generated by a nondominant firm. One mitigating factor is the level of product support of the firm that introduces the product. A high level of product support at introduction, including marketing support (sales) and technology support (R&D expenditures and patent protection), can allow even small firms to benefit financially from radical innovations at a level comparable to that of large firms.

Unlearning

Another cultural facilitator of innovation is the ability to **unlearn**. Many times, companies must "unlearn" traditional but detrimental practices. To do this, managers must surface and challenge their own assumptions and mental models about the market and the business and encourage employees to do the same. Faced with turbulence in traditional markets and an entrenched culture, Ed Artzt, former CEO of Procter and Gamble, found himself in the peculiar situation of having to make "rules that give (employees) intellectual permission to make changes."[49] As John Seely Brown,[50] the former Chief Scientist of the Xerox Palo Alto Research Center, explained, "Unlearning is critical in these chaotic times because so many of our hard-earned nuggets of knowledge, intuitions, and just plain opinions depend on assumptions about the world that are simply no longer true."

General Electric uses "Workout" sessions to "challenge every single piece of conventional wisdom, every book, every rule." In these sessions, executives—including the CEO—take the floor of GE's management development center to respond to tough questions from managers who are taking classes at the center. This has created an environment where difficult issues can be raised without fear of retribution and where executives must respond with plans and solutions.[51] Encouraging unlearning could be the single most important task of the CEO for sustaining innovation momentum.

Corporate Imagination

Another facilitator to overcome the liability of core rigidities and the inertia inherent in a currently successful system is to attempt to develop what is known as **corporate imagination**,[52] creativity, and even playfulness. It is important for established organizations to replenish their stock of ideas continuously. For many innovative firms, a key measure of success at this activity is the percent of revenue derived from recently released products.

Corporate imagination requires the ability to create a vision of the future that consists of markets that do not yet exist and is based on a horizon not confined by the boundaries of the current business. Creative corporate imagination includes four important elements, shown in Table 2-1.

1. *Overturn price–performance assumptions.* Many established firms spend their time making incremental improvements to existing technologies. This is understandable, because existing customers want performance improvements in the products they are using. However, incremental improvements to existing technologies are based on improving performance on existing standards. Alternatively, really new products are more likely to be based on entirely different performance assumptions.

One tool that helps firms understand how to overturn such price–performance issues is the **technology life cycle**. The technology life cycle refers to improvements in product performance relative to the investments in effort in a particular technology.[53] As Figure 2-6 shows, as a new technology is introduced, its performance capacity improves slowly and then, because of heavy R&D efforts, improves rapidly, before reaching its performance limits. When a newer technology is introduced, the two technologies will compete with each other for a time period, until the new technology eventually supercedes the former.

TABLE 2-1   Four Elements of Corporate Imagination

1 Overturn price-performance assumptions, using the tool of technology life cycle curves.
2 Escape the tyranny of the served market.
3 Use new sources of ideas for innovative product concepts.
4 Get out in front of customers: Lead them where they want to go before they themselves know it.

SOURCE: Reprinted by permission of Harvard Business Review. "Four Elements of Corporate Imagination" from "Corporate Imagination and Expeditionary Marketing" by G. Hamel and C.K. Prahalad, July–August 1991. Copyright © 1991 by the Harvard Business School Publishing Corporation; all rights reserved.

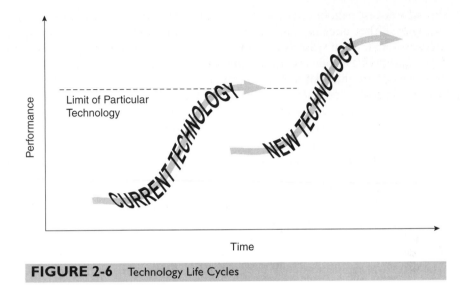

FIGURE 2-6   Technology Life Cycles

For example, advances in semiconductor speed and processing power, relative to price, have been formalized in **Moore's Law**, which states that semiconductor performance doubles every eighteen months, with no increase in price. Stated differently, every eighteen months or so, improvements in technology cut price in half for the same level of performance. However, some predict that future improvements in microchip performance are limited by the use of semiconductor technology and that Moore's Law is reaching its natural limit. In that sense, further investments in enhancing semiconductor performance may result in only incremental improvements in technology, improvements that may have reached a plateau. This possibility is one reason that Japan's Ministry of International Trade and Industry has sponsored a $30 million research program focused on technologies that could replace conventional semiconductors. These technologies are based on quantum physics and neural networks rather than on electrical engineering. Although such research is at a preliminary stage, and the nascent performance of these new technologies is unclear, the technologies hold the possibility of obsoleting semiconductors as we know them today.

Unfortunately, the majority of the time, new technologies are commercialized by companies *outside* the threatened industry,[54] suggesting that it is difficult for industry incumbents to be imaginative in envisioning new technologies. Established firms must realize that all products have performance limits; as an existing technology approaches its limit, it becomes more expensive to make improvements. Therefore, firms must look to develop new technologies.

However, many established firms try to hedge their bets with new technology. At worst, they do not strive to develop new technology. At best, they invest both in improving the current-generation technology and developing new technology. For example, a company may try to make investments to make its manufacturing facility more efficient in producing the current-generation technology. Yet, being the low-cost producer of an obsolete product is not worth much.[55]

Moreover, established firms tend to underestimate the firms introducing new technologies, either because such firms are small or the new technology appears crude. Because of all these reasons, technology substitutes can creep up slowly on established firms and then explode in terms of market performance. This was definitely true of the Internet, where much of the development was initiated by companies unknown to the major computer industry market players. Hence, established firms must aggressively pursue new technologies early on.[56]

How can a firm recognize when a current-generation technology is in danger of obsolescence? The technology life cycle curve demonstrates that one can't rely solely on economic signals: Based on incremental improvements, the revenue of the current technology can reach a peak even after the new technology is introduced. Hence, relying on economic signals may result in the firm moving too late into the new technology, and the competition will have established a stronghold. Underlying technology life cycles, based on diminishing performance returns to increasing investments in current-generation technology, are the crucial indicator.[57]

2. *Escape the tyranny of the served market.*[58] Firms have both expertise in matching their product offering to a particular market segment and a competitive advantage with customers in that segment. As mentioned previously, established firms have a vested interest in maintaining their "bread-and-butter" line in major segments. Hence, they tend to develop more incremental innovations for these existing customers. The tyranny of the served market refers to the tendency for firms to focus very specifically on solving customers' needs with a current technology. Such a myopic focus obscures the possibility that customer needs may change over time and may be solved in radically different ways.

For example, in the Internet world, a firm's best customers may be the last to embrace a disruptive technology because it doesn't provide the service and performance they prefer. Those who embrace it first are typically the customers a company pays least attention to, and the innovation creeps up like a stealth attack.[59]

The tyranny of the served market is one of the main reasons that new innovations are typically introduced by firms that are new to the market or industry outsiders/newcomers. And, these newcomers frequently alter the rules of the game, jumping to a new technology life cycle with new price–performance ratios. Corporate imagination requires that firms look for market opportunities across or between the areas of a firm's competence.

3. *Use new sources of ideas for innovative product concepts.* Many large firms are familiar with standard approaches to market analysis that are often premised on existing market boundaries. For example, most large firms have used the standard marketing research techniques of running focus groups, administering surveys, using conjoint analysis to help decide the optimal combination of attributes for new products, and so forth. However, these tools are generally inadequate for identifying really new opportunities or assessing customer attitudes toward radical innovations. Corporate imagination requires firms to rely on different types of marketing research to open new doors of opportunity. These marketing research techniques, especially useful in a high-tech marketing context, include ethnographic observation ("empathic design"), customer visits, and lead users. These tools are discussed in Chapter 5, "Marketing Research in High-Tech Markets."

4. *Get out in front of customers.* The fourth aspect of corporate imagination is to actually lead customers where they want to go before customers themselves know it. The ability to be a market leader, based on envisioning the future, requires profound insights that are not saddled with existing rules and procedures. Technical wizardry must be based on understanding customer needs for it to be successful. To be in front of customers requires multidisciplinary product teams and procedures to inform those closest to the customers about emerging technological possibilities.

Each of these aspects of corporate imagination leads to another important element of not letting core competencies become core rigidities: expeditionary marketing.

Expeditionary Marketing

Because creating markets ahead of competitors is so risky—sometimes the hoped-for market does not develop at all, or if it does, it emerges more slowly than expected—companies must use strategies to minimize the risks. Successful new products include the right combination of functionality, price, and performance targeted to the correct market. There are two ways to improve the success ratio.

One way is to try to improve the odds on each individual product introduction, or to improve the "hit" rate. To have a successful new product launch, a company tries to gather as much information as possible to tailor the product as needed in terms of functionality, price, and performance, so that when it is delivered to the market, the odds of a successful hit are as high as possible.

The other route to success is to try many "mini-introductions" in quick succession and, by learning from each foray into the marketplace, to incorporate that learning into each successive "time at bat," such that over time, the firm has accumulated loyal customers and higher market share. This strategy is known as **expeditionary marketing**.[60] The objective here is not to improve the hit rate but to increase the number of times at bat in the market. The underlying strategy is based on learning: The firm wants to learn about the marketplace and customer needs by placing many small bets in the marketplace. These low-cost, fast-paced incursions allow the firm to learn and recalibrate its offerings each time, such that the combination of speed and learning enhances the odds of success.

Which of the two ways do most companies focus their attention on? Most companies focus their attention on the first way, trying to improve the hit rate. Through careful market research, competitive analysis, and the use of stage-gate procedures that specify the hurdles a new idea must overcome at each stage of development, they try to maximize the odds of success. However, such a strategy is very time-consuming, and in high-tech markets, the accuracy of the information can be sketchy at best. Moreover, by the time the product is introduced to the market, the marketplace (customer needs and competitors) may have changed. Indeed, this approach might be characterized as "Ready. Aim. Aim. Aim."[61]

Hence, in high-tech markets, it may make more sense to undertake a series of fast-paced market incursions—expeditionary marketing. Such a strategy has several advantages. First, it allows the company to learn more accurately, through successive approximations, about what customer needs are. Second, through fast-paced market incursions, it maximizes the odds that the product actually delivered to the market meets customers' needs. Fast-paced incursions imply that the time-to-market cycle is faster, and therefore, the odds of the customers' needs changing in that time period are lower. Under such a model, what counts the most is not being right the first time, but how quickly a company can learn and modify its product offerings, based on its accumulated experience in the marketplace.[62]

As an example, Storage Technology Corporation, maker of tape backup systems for large data installations (such as banks, insurance companies, etc.), uses the model depicted in Figure 2-7. The idea is that, rather than introducing the most advanced model possible based on a new technology (Model 3), the company attempts to introduce in rapid succession a series of models based on the new technology.

Nurture a Culture and Climate of Innovation

The final strategy discussed here for established companies to retain an innovative culture is to recognize the nature of innovation itself. Innovation is the process, within organizations, of developing and commercializing radical or breakthrough new products, services, or business models. Creativity is the cognitive foundation of innovation. Unfettered creativity, by itself, may not be good for business. It takes *disciplined creativity* to produce useful innovation for a company. When Bob Herbold joined Microsoft as Chief Operating Officer (COO) in 1994, he experienced a freewheeling culture bordering on chaos.[63] Implementing the discipline of new financial, purchasing, and human resource information systems helped Microsoft reduce operating expenses from 50 percent of net revenue to 40 percent, a savings of about $2.7 billion, over a six-year period. It also provided more time and resources for managers and programmers to focus on what they did best—large, complex software projects.

One of the major threats to the innovation process is the boom-bust nature of business cycles. When companies do well, they invest more in innovation and product development. On the other hand, when the economy deteriorates, investments in research and development tend to go down. One way to cushion innovation against cyclical funding

FIGURE 2-7   Expeditionary Marketing: Many Fast-Paced Incursions Into the Market.

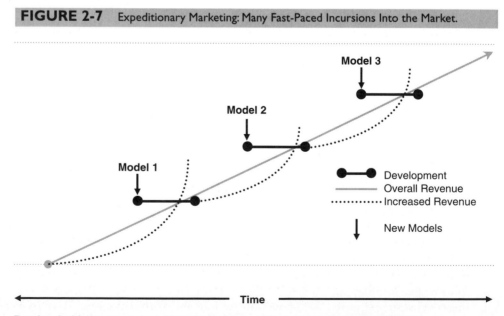

Reprinted with the permission of Storage Technology Corporation, Louisville, Colorado.

cuts and spurts is to *make R&D an indispensable part* of everyday business.[64] If the output of innovation is to be reliable, it must be worked on systematically every day, even during downturns, even if blockbuster results are years away.[65] Consider the Korean conglomerate, Samsung. In 1997 Samsung was in dire straits as a result of the Asian financial crisis. Since then, it has made continuous, significant investments in R&D and product development, even through the global economic downturn of the early 2000s. Today it is the global leader in plasma displays, big-screen TVs, CDMA mobile handsets, DRAM memory chips, and microwave ovens.[66]

Enlightened experimentation involves the use of new information technologies such as computer simulation, rapid prototyping, and combinatorial chemistry to reduce the cost of testing new ideas as well as increase the opportunities for innovation.[67] It has been applied in the automotive, software, and pharmaceutical industries. For example, Millennium Pharmaceuticals uses genomics, bioinformatics, and combinatorial chemistry to test drug candidates for their toxicological profile. Unpromising candidates are eliminated before millions of dollars are spent on further clinical testing and development. This also enables Millennium to test many more potential drugs in search of the real blockbuster innovations.

Bureaucratic processes are also at odds with an innovative culture. Bureaucracy is often manifested in a new product or competency development process that is based on formal plans and procedures. The process is often rational and analytical and bears little resemblance to the nature of innovation, which is often tumultuous, nonlinear, and serendipitous.[68]

Rather than a stage-gate, step-by-step process through which new product ideas must pass, "managing" innovative activity in corporations as an entrepreneurial process can maintain a culture of innovativeness. Referred to variously as "autonomous strategic behavior",[69] "emergent processes",[70] or "intrapreneuring",[71] the idea is to create an internal environment that fosters innovation and an entrepreneurial spirit. What characteristics does a firm exhibit if it wants to foster an innovative climate?

- The market need identified is one that diverges from (rather than converges on) the organization's concepts of strategy.
- The roles and responsibilities of the key players are poorly defined at the outset but become more formalized as the strategies evolve.
- Rather than a formal screen based on established administrative procedures, the screening process for the idea is done via an informal network that assesses technical and market merit.
- The communication between personnel tends to flow less along prescribed organizational decision-making channels and more through an informal network.
- Commitment to the idea emerges largely through the sponsorship efforts of a product champion.

Product champions are integral to the success of any breakthrough innovation. Product champions are the people who create, define, or adopt an idea for an innovation and are willing to assume significant risk to make it happen.[72] Often referred to as iconoclasts, mavericks, or crusaders, these people break the rules, take risks, transform companies, and turn organizations upside down. They work tirelessly and lobby behind the scenes for organizational resources to help their ideas take off. The product champion is a person with drive, aggressiveness, political astuteness, technical competence, and market knowledge. Influential product champions can overcome firms' natural reluctance to cannibalize and motivate radical product innovations.[73]

Although product champions are found in both innovative and noninnovative companies, those in noninnovative companies wield less influence and are frustrated and demoralized. In successful companies, they have the power to make their ideas happen. Product champions in innovating firms wield substantially more influence than in less innovative firms. The reward systems and cultures promote the influence of product champions, and top management actively supports them.[74]

Skunk works are an organizational means for creating an innovative climate by isolating new venture groups in a location removed from the normal corporate operations.[75] The thought is that when large, established companies develop new innovations, they do so *despite* the corporate system, not *because* of it.[76] Hence, in order to protect imaginative individuals from corporate orthodoxies, senior managers isolate them in new venture divisions, or corporate incubators, often at a remote site from the parent organization.

These new venture groups removed from the normal corporate operations are sometimes referred to as skunk works. The etymology of this term comes from the Skonk Works, an illicit distillery in the comic strip "Li'l Abner" by Al Capp, around 1974. Because illicit distilleries were bootleg operations, typically located in an isolated area with minimal formal oversight, the term has been adopted in organizational settings to refer to a usually small and often isolated department or facility (say, for research and development) that functions with minimal supervision or impediments from the normal corporate operating procedures.

Many firms have relied on skunk works operations to develop new lines of business. For example, IBM isolated its PC group in Boca Raton, Florida, away from corporate headquarters in New York and away from any other established IBM locations. Dow Chemical also relies on skunk works for its new venture groups.

Despite the potential advantages of such isolation, some critics argue that trying to leverage corporate competencies into new businesses while at the same time protecting new ventures from the corporate culture is a contradiction in terms.[77] For an established company to become and remain innovative, it must allow individual creativity within the normal corporate operating procedures. A corporate culture that allows innovation to flourish shouldn't have to put up special protection mechanisms for it to happen. Indeed, Gary Hamel refers to such incubators as "orphanages" that isolate the creative conversations and make it hard for new ideas to emerge in the corporate hierarchy.[78]

So, the idea of isolating new product development groups from standard corporate procedures poses a dilemma. Ideally, it would probably be best if the corporation nourished innovation and did not need to have separate units to develop innovative ideas. On the other hand, if normal corporate operating procedures stifle creativity, a skunk works may be something of a temporary or "band-aid" solution; addressing the systemic reasons for the problem may be also necessary.

A recent review of the literature on stimulating innovation coupled with a survey of business executives[79] identified several additional guidelines for encouraging innovation.

1. The company and its top managers must clearly support innovation through words and actions.
2. The company must maintain close relationships with its most innovative customers so that they can jointly determine what those customers will want next.

3. There should be metrics and a procedure for frequent evaluation of project progress.

4. The company should provide considerable freedom of action, substantial resources to educate employees about emerging technologies, and use teams of employees that possess many skills among them.

5. Employees who innovate successfully should be celebrated and rewarded. Also, the company should not rigidly penalize people for ideas that don't work out. Tolerating risks and a certain number of mistakes is part of the entrepreneurial spirit. Indeed, learning from such ostensible "mistakes" may be the basis of the company's next new success.

6. The company should have a process for rapidly communicating new ideas across the company.

Applying Lessons of Innovativeness on the Internet

What can the world of Internet businesses tell us about the preceding "lessons"? As an application of the concepts in this chapter, it is instructive to examine companies' responses and forays into the online world of e-commerce.

Online competitors posed huge threats to incumbents in established industries, because they changed the rules of the game and operated under a radical new business model: fewer requirements for working capital, inherent speed and efficiency, and direct access to customers and suppliers. The obvious threat of this new business model has been that of *extinction*.

Companies that don't move fast to capitalize on the Internet risk being out-competed by tiny rivals who may have barely existed just a few years ago. In industry after industry, fledgling Net companies transformed the way business is done and snatched market share from their much bigger, established rivals.[80]

This is the nature of *competitive turbulence* in a high-tech environment. Companies that don't jump in despite the risks may be cut out altogether. "Big [established] companies are likely to be laggards at spotting fresh ways to do business—and equally ponderous at mobilizing to take advantage of those situations. For startups, the greatest strength may be an ability to react quickly to new opportunities."[81]

Despite the threat of extinction by newcomers, established competitors have a tendency to *underestimate their competition*. Jack Welch of General Electric initially derided Net stocks as "wampum," and IBM's Lou Gerstner said Net companies are "fireflies before the storm:" They shine now but will eventually dim out.[82] And, each time the business model of retailing changed, a new group of leaders emerged, from the Woolworth's of Main Street in the 1950s to the Sears of the malls in the 1970s to the superstores of Wal-Mart and Costco in the 1990s.[83] As Christensen and Tedlow[84] argue,

> Of the four dimensions of the retailer's mission—product, place, price, and time—Internet retailers can deliver on the first three remarkably well. The right products? In categories ranging from books to chemicals, Web stores can offer a selection that no bricks-and-mortar outlet can match. The right price? Internet retailers enjoy unparalleled margin flexibility. To earn a 125% return on inventory investment, an Internet retailer such as Amazon.com, which can turn its inventory up to 25 times each year, needs to earn only 5% gross margins. And the right place? It is here—location—that the Internet is most revolutionary.

The Internet negates the importance of location. Anyone, at any time, can become a global retailer by setting up a Web page. With such advantages, it's no wonder electronic commerce is attracting so much attention.

Another risk companies face in going online is the potential *cannibalization* of existing revenue streams. Going online requires that companies be willing to cannibalize their own core franchise, or engage in *creative destruction*. Often, the initial move to an online environment can result in a decrease in revenues, earnings, and stock prices. Indeed, the decision of how much an existing business should invest in the online world, with its huge up-front costs and only theoretical profits, is an anguishing decision. "In getting from A to C, B is hell," said Intel's VP of business development in discussing Internet investments. B is where revenues decline and profits go down, but there is no way for large companies to get to where they need to be in the future without going through this "valley of death."

Another aspect of being willing to engage in creative destruction is the willingness to risk upsetting partnerships with distributors and retailers. Conflict with existing sales channels has been identified as the biggest impediment to online selling.[85]

How else can the Internet world highlight examples of innovation? In the offline world, where it is sometimes difficult to get rapid feedback when some part of the marketing mix isn't working, companies can be slow to learn from customers and slower to act on what they learn. However, in the online world, that changes. The Web lends itself to immediate customer feedback and rapid adjustment. Learning cycles are much shorter online than offline. Companies that are quick to try, quick to learn, and quick to adapt will win. Those that learn fastest and keep learning will stay ahead. Zipping through the learning cycle creates positive feedback effects. The faster a company learns and adapts, the more customers it wins; the more customers it wins, the faster it can learn and adapt.[86] This learning process highlights the fact that effective Internet companies intuitively understand the notion of *expeditionary marketing*.

Indeed, early on, being a market pioneer on the Web translated into huge advantages in site traffic and stock valuations. The Internet is all about speed, but to be fast necessarily implies that a company can't do it all. Rather, based on a *core competencies* model of strategic planning, companies have to outsource some (or many) aspects of getting online. Companies who can help others get online are seeing huge increases in volume.[87] Designing, hosting, maintaining the site, handling email, and logistics and distribution can stretch many firms well beyond their capabilities.

As a final example of the innovation concepts at work in the Internet arena, when Toys "R" Us decided to begin its online operations, it set up a separate unit in Northern California. The firm decided that being located in corporate headquarters would slow it down and hinder its innovativeness.[88] This is the classic setup of a *skunk works*.

Although large companies do have a number of liabilities in an online environment and need a number of strategies to cope, they also have some important advantages. Household brand names, larger budgets, and tested leadership each provide resources that many smaller companies do not have.[89] Issues faced by small high-tech startups are addressed next.

CHALLENGES FOR SMALL COMPANIES

Although they would rarely, if ever, be accused of not being innovative or nimble enough, small high-tech startups face their own set of unique difficulties, as shown in

TABLE 2-2   Strategic Planning Issues for Small High-Tech Start-Ups

- Sources of Funding
- Other Resources
- Navigating Complex Environments

Table 2-2. One of the most salient concerns is how to fund the budding company prior to the time that real sales volume and positive cash flow materializes. Other concerns have to do with securing other resources and the manner in which small companies deal with a complex environment, demanding customers, and large competitors.

Funding Concerns

Small high-tech startups have three key ways to finance their ventures:

- Friends and family.
- Bootstrapping, or funding the business through early customer revenues.
- Angels and venture capitalists (discussed below).

Gary Schneider is an entrepreneur who successfully dealt with the funding challenge. He has devoted his life to the development and marketing of software that will help small farmers in optimizing their crop selection, an idea he got while studying agriculture and engineering at the University of Arizona.[90] After a decade of refining his idea, Gary finally set up his company, called AgDecision, in 1996 to commercialize his concept. To conserve cash, Gary acted as his own patent counsel, relied on his wife for help with marketing and administration, and drew no salary. He traveled extensively on consulting jobs to make money. While looking for investors, Gary and his wife were selling the product one at a time through networking and conventions. Gary made a cold call on a major crop insurer in Council Bluffs, Iowa: American Agrisurance. A senior researcher at the company found Gary's product very interesting and considered making investments to finance further development of the software.

Absent independent wealth or wealthy family and friends, many high-tech entrepreneurs seek funding from venture capitalists (VCs). One area where the United States excels is the ability to fund innovative companies at an early stage. According to a study by the Harvard Business School, a dollar of venture capital produces three to five times more patents than a dollar of research and development spending. Companies usually need venture capital in early stages in the life cycle, when they begin to commercialize their innovations. At this stage, internally generated cash flows aren't enough to sustain growth, and banks fear the risk of issuing new loans because high-tech companies usually don't own much traditional collateral. Therefore, many emerging high-tech companies consider approaching a venture capitalist.

Simply put, venture capital is money invested in rapidly growing emerging and startup companies.[91] The capital is provided by venture capitalists, who usually spend their days meeting with industry leaders, meeting with other venture capitalists, listening to proposals from excited entrepreneurs, and pondering the future of technology.

There are two types of venture capitalists: formal and informal. *Formal venture capitalists* are professional investors, such as venture capital firms and some banks. These investors often look for companies that have moved beyond the infant stage, where risks are the highest.

Informal venture capitalists are usually referred to as "angels." They are probably called angels not only because of their salvaging qualities but also because they are so hard to find. Angels are usually part of an informal network of investors, who hear about promising startups through acquaintances or friends of friends. However, angels can also act more like professional institutions. They have pooled resources, built explicit networks, and when investing they demand nothing less than a formal venture capitalist; that is, they want a business plan, clear and precise goals, a large equity stake, and very often a seat on the board.

When venture capitalists screen high-tech companies for investment opportunities, they look at four key factors: management, marketing, technology/product, and anticipated return on investment. Venture capitalists don't merely invest in a superior software code or specific technological innovation. Rather, they invest in the talented people who can transform that technology into a profitable product.

The Technology Expert's View from the Trenches is written by a venture capitalist in Silicon Valley.

TECHNOLOGY EXPERT'S VIEW FROM THE TRENCHES

CONSIDERATIONS IN THE VENTURE CAPITAL PROCESS
CHARLIE WALKER

Managing Partner, BlackWolf Partners
Silicon Valley, California

The challenge for an entrepreneur is to always understand his or her audience. Be it customers, employees, partners, or financing sources, each has its own unique set of needs. Nowhere does this apply more than in the area of raising money from venture capitalists.

During the late 1990s, the venture capital industry grew and matured into a permanent fixture in the U.S. capital markets. Long the province of professional money managers, venture capital has gone retail, and with that, an entirely new set of financ-ing dynamics has arisen. Woe to those entrepreneurs who come to this game without understanding the changing rules!

The new dynamics contain elements of the traditional venture capital business, including such requirements as presenting a quality management team and scalability of the proposed business. Newly arising considerations relate to the time constraints being placed on venture capitalists, the readiness of the public markets to provide funding earlier in a business's life cycle, and the ever-shifting requirements of operating a public company—assuming the entrepreneur gets there.

But first, the traditional components: The single most important consideration for obtaining venture capital is the *quality of the management team*. Rare is the deal that obtains professional financing (as opposed to the "aunts and uncles network" or "angel financing") without a management team that has a proven track record of starting and scaling a business or division. Not that prior success is necessary; often a failure helps season a team. Given the time constraints of the industry, however, venture capitalists much prefer situations that are well managed, for the simple reason that they have all learned how draining (emotionally and financially) poorly managed companies can be. Therefore, having "walked the talk" is a powerfully important ingredient in attracting risk capital. Throw in a healthy degree of self-promotion, to ensure ongoing access to capital and customers, and one begins to have the makings of an attractive venture deal.

Scalability of the business plan is also important in generating financial returns. Because most venture capitalists are measured on *how fast* they make money, in addition to *how much* they make (summarized mathematically by a deal's "IRR," or Internal Rate of Return, projections), it helps to propose a business that can grow very rapidly. "Making money while you sleep" is an oft-quoted phrase in the venture capital world. Notably, this is not a reference to 24-hour trading in the public markets, but rather to business models that are capable of generating revenues 'round the clock. Businesses that do not require large numbers of people are ideal examples of this proposition.

Newly emerging considerations include the importance of a *compelling-yet-simple financing pitch,* or "elevator pitch," to rapidly explain your proposal. (The reference to an elevator is intended to limit one's proposition to that amount of time it takes to ride an elevator.) With the ascendancy of venture capital and entrepreneurial activity in America, the industry is awash with business proposals. Time management is critically important, forcing venture capitalists to have the attention span of a gnat! Be quick and to the point, provide compelling reasons (financially *and* intellectually) to focus your audience, and cover all the bases. What are those bases?

Whatever is relevant to your underlying business. On this point, *you* get to make the rules.

Do not characterize the raising of money as a waste of your time. Venture capitalists have the view that the business of raising money is critically important (after all, it *is* their line of work), and these people will add more value to the underlying enterprise than most anything else you can do during the entire life cycle of the business. And, in a sense, they are correct. Done properly, a financing can elevate the valuation of a business more rapidly in the span of three months than the underlying business can in three years. Therefore, treat the process with the respect it deserves.

Be an expert in the venture capitalist's business. Concepts don't sell a financing—although they help; expertise of management does. Know your industry, thoroughly, and be prepared to explain *exactly* how you will deal with all the expected challenges, particularly in the capital markets. Provide confidence to the financier.

And, because new financing options are always emerging, become an expert in financing alternatives that will be available to you as you grow the business, beyond simply assuming the business can "go public." Matters of lease financing, secondary offerings, subordinated debt, and strategic investors can all provide the financial leverage that will further fuel returns for the early investors.

Finally, quite apart from the venture capital industry, understand that ultimately, the venture capitalist will cash out, and you will end up wanting and/or needing to still operate a business. What the entrepreneur needs to do is to propose a business that will endure far beyond the venture capital life cycle. Oddly enough, the more irrelevant venture capitalist money seems in the context of the business' overall life cycle, the more attractive the proposition becomes to a venture capitalist. In the final analysis, everyone wants to be associated, however briefly, with enduring companies!

Other Resources

Small high-tech startups are also in need of general assistance in getting their budding businesses off the ground. **Technology incubators** can be a nice resource for many small startups. A business incubator is an economic development tool designed to accelerate the growth and success of entrepreneurial companies by offering an array of business support resources and services. A business incubator's main goal is to produce successful firms that will leave the incubation program financially viable and free-standing. Critical to the concept of an incubator is on-site management, which develops and orchestrates business, marketing, and management resources that are tailored to a company's needs. Incubators may also provide clients access to appropriate rental space and flexible leases, shared basic office services and equipment, technology support services, and assistance in obtaining the financing necessary for company growth.[92] The incubator concept has several benefits. For example, sharing office space allows entrepreneurs to swap ideas, and inexperienced entrepreneurs will have access to qualified consultants.[93]

Small businesses looking to develop new ventures and markets have many other options for assistance, both public and private. At a basic level, startups may want to get in touch with the Senior Corp of Retired Executives (SCORE), a volunteer group available to counsel budding entrepreneurs (www.score.org). The volunteers are identified by the Small Business Administration (SBA), and there are local SCORE chapters throughout the country. The nature of the expertise available may be geographic-specific, so high-tech startups may find this service more valuable if they are based in areas with a plethora of high-tech businesses. The SBA also offers loan guarantee programs that may be useful for a new business (www.sba.gov).

For those firms that are looking to international markets, the U.S. Department of Commerce has multiple tools for identifying markets and finding partners. First, the U.S. government has commercial sections in most U.S. embassies around the world. These offices, along with a network of export assistance centers and offices throughout the United States, are available to provide counseling, market intelligence, and contacts. Indeed, the staff in these offices will help a firm develop a customized strategy to export to a particular country by identifying possible partners and trade events. Moreover, the Commerce Department conducts an annual analysis for U.S. products in export markets. These reports are made available via the Internet. Being listed on the online trade show maintained by the Department of Commerce (www.eexpousa.doc.gov) is also possible. In addition to market intelligence, the commercial sections employ locals who are experts in targeted industries. These trade experts have the ability to identify potential partners and introduce U.S. firms. Using the introduction by the embassy can expedite the search process and add a prestige factor that can be critical in gaining access to some countries. Each of these services can be found at the U.S. Department of Commerce's Commercial Service's Web site: www.usatrade.gov.

When a firm sells products overseas, a critical issue for high-tech products is knowing what standards and certification are necessary for different countries. The National Institutes of Standards and Technology (www.NIST.gov) compiles and maintains such information for marketers and others.

Another option for small high-tech startups facing resource constraints is to partner with other firms who offer complementary skills and resources. Strategic alliances and partnerships fall under the domain of relationship marketing, the topic of the next chapter.

Navigating a Complex Environment

How can the small- to medium-sized firm stay ahead of rapid technological, economic, and market change? There seem to be three requirements for meeting this challenge: speed, flexibility, and time orientation.[94]

Speed

As the pace of innovation accelerates, so too does the speed at which new knowledge is created and existing knowledge becomes obsolete. The speed of learning matters: Organizations that learn slowly may find their innovation performance rapidly deteriorating. Speed of learning is only one piece in the speed puzzle, though. Speed in other processes can improve effectiveness and/or efficiency. A speedy product development process may reduce the number of person-hours required for product development and may reduce the likelihood that the company will miss a market window. A rapid manufacturing process reduces investment in raw materials, work in progress, and finished goods inventory while a fast fulfillment process speeds cash flow.

Flexibility

The small company's strategy, competencies, and products must fit with its current environment while remaining flexible enough to respond to environmental change. Flexibility implies three things: The firm must be able to sense or anticipate changes in its environment; it must have the cultural readiness to accept change; and it must have the requisite skills and competencies to compete in the new environment.[95] Thus, flexibility has a cultural aspect (i.e., being willing to change) and a competence element to it (i.c., being able to change). This is a very difficult challenge for the smaller company, but there are some processes by which flexibility can be maintained, such as:

- Develop a market sensing system. A market sensing system embodies the ability of the firm to learn about customers, competitors, and channel members and act on events or trends in present and potential markets. The processes for gathering, interpreting, and using market information are systematic, thoughtful, and proactive.[96]
- Monitor fringe customers. Fringe customers are not members of the firm's core market. Their needs may be more advanced than those of members of the core market or they may require product or service customization to satisfy their needs. They often provide insight into untapped markets.
- Institutionalize a learning orientation. Firms with strong learning orientations encourage employees to question the organizational norms that guide intelligence generation and dissemination, to understand the cause and effects of their actions, and to detect and correct errors in their organizational routines.[97]
- Undertake low cost market experiments.
- Encourage decentralized decision making within clear parameters.

Time Orientation

One of the most pressing issues confronted by top management of small companies in developing markets is the need to recognize new opportunities and act on them. Nowhere is this issue of greater importance than in companies competing in technology-based markets. The challenge is particularly great for younger and smaller companies as they typically suffer from a lack of experience and accumulated resources.

Managers' time orientation is one means to avoiding these problems. Time orientation refers to managers' ability to anticipate and understand future events and to share their insights across the organization. Anticipation, dissemination, and shared interpretation should produce proactive and coordinated change.[98]

SUMMARY

This chapter, as the first of three chapters on considerations internal to the firm that affect high-tech marketing success, has addressed three broad topics. First, it has addressed the strategic planning process in high-tech firms, including market planning, development of market strategy, and the ultimate purpose of creating competitive advantage through the development and possession of resources and competencies.

Second, it has addressed the need for large firms to remain nimble and innovative and to not allow their core competencies to become core rigidities. Also known as the innovator's dilemma, firms must contend with the tension that disruptive innovation creates in their organizations. Strategies and techniques for maintaining a culture and climate of innovativeness include:

- Being willing to engage in creative destruction.
- Having corporate imagination (using technology life cycles to jump to next-generation technology, escaping the tyranny of served markets, using new sources of ideas, and leading customers before they themselves know where they want to go).
- Engaging in expeditionary marketing by making many quick incursions into the market in order to apply learning to successive versions of the product.
- Maintaining a culture of innovation (allowing product champions to flourish, giving time and incentives to innovate, using skunk works if necessary).

Third, the chapter addressed the issue of resources needed by small firms in order to successfully compete. Venture capital, incubators, and cultural and strategic dimensions were discussed. In the continued exploration of the internal factors that affect marketing success, the next chapter looks at the role of establishing partnerships and alliances with key stakeholders.

DISCUSSION QUESTIONS

1. Provide an overview of the strategic marketing planning process in high-tech firms. How does it differ from the traditional strategic planning process?
2. What are the key questions that a company's strategy should answer?
3. What are the requirements for competitive advantage?
4. What are core competencies? Give an example of a firm's core competencies. Explain how your example stacks up on each of the criteria for a core competency.
5. Why is it hard for large firms to be innovative?
6. How can core competencies become core rigidities? Give an example.
7. What are the causes of the innovator's dilemma?
8. What is creative destruction?
9. How can a firm effectively leverage its dominance to be innovative?
10. Why is unlearning important?
11. What are the four elements of corporate imagination?

12. What are technology life cycles? How can companies use them to be innovative? Give an example.
13. What is expeditionary marketing? What are the implications in terms of bringing products to market?
14. How does a firm nurture a culture of innovation?
15. What are skunk works? Do they make sense to you? Why or why not?
16. Who are product champions? What are their characteristics? What are the pros and cons of taking on such a role?
17. What are the issues faced by small high-tech startups?

GLOSSARY

Competitive advantage. A position where a firm is able to create more value for customers than its competitors, while earning a superior return on investment. Competitive advantage requires possession of superior tangible assets, intangible assets, or competencies.

Core competency. Underlying skills and capabilities that give rise to a firm's source of competitive advantage; typically based in embedded knowledge, which is hard to imitate.

Core rigidity. Well-rehearsed skills and competencies that are so entrenched that they prevent a firm from seeing new ways of doing things; might include cultural norms giving status to engineers over marketers, preferences for existing technology, and so forth.

Corporate imagination. A characteristic of a firm that allows its culture to exhibit creativity and playfulness, such that it (1) is willing to overturn existing price–performance assumptions, (2) does not slavishly serve current markets, (3) uses new sources of ideas for innovative products, and (4) is willing to lead customers where they want to go.

Creative destruction. The notion that in order to remain viable, a firm must be willing to destroy the basis of its current success. If a firm doesn't constantly innovate and reinvent itself, it will find its market share eroded by new competitors who are willing to do so.

Expeditionary marketing. A strategy for new-product success based on trying many mini-introductions in quick succession and, by learning from each foray into the marketplace, incorporating that learning into each successive time "at bat," such that over time, the firm has accumulated loyal customers and higher market share than firms that have made fewer incursions into the market.

Innovator's dilemma. The conflict between continuing to allocate resources to serve current customers with incrementally improved products or allocating resources to develop new products that might disrupt current customers operating processes.

Market space. The arena in which the firm competes. New market space is represented by customers who are underserved by current offerings in the market or by previously unidentified segments. Either represents an opportunity for the firm to avoid head-to-head competition.

Moore's Law. Every eighteen months or so, improvements in technology double product performance at no increase in price. Stated a different way, every eighteen months or so, improvements in technology cut price in half for the same level of performance.

Product champion. A person who is so committed to a particular idea that he or she is willing to work tirelessly advocating the idea, to work outside normal channels to pursue it, and to bet future successes on the idea. Often, champions are iconoclasts, mavericks, risk takers.

Skunk works. New venture teams that are isolated or removed from normal corporate operations in order to foster an innovative culture that allows the team to think out of the box.

Technology incubator. An economic development tool designed to accelerate the growth and success of technology startups by offering an array of business support resources and services in a facility that houses new businesses for a brief period of time.

Technology life cycle. A graph depicting investments made in a particular technology relative to improvements in performance, typically an S-shaped curve. Radical innovations "jump" technology life cycles and begin a new S-shaped curve.

Tyranny of the served market.   A narrow focus on serving the needs of current customers at the expense of identifying possible new customers with new needs.

Unlearning.   The process of surfacing knowledge and assumptions that are the basis for strategic action, testing them for their validity, and discarding those that have become barriers to proactive change.

ENDNOTES

1. Hamm, Steve (2003), "Reversals of Fortune,"*Business Week,* June 23. Available at www.businessweek.com.

2. www.hyperion.com.

3. Cooper, Lee (2000), "Strategic Marketing Planning for Radically New Products," *Journal of Marketing,* 64 (January), pp. 1–16; Eisenhardt, Kathleen and Shona Brown (1999), "Patching: Restitching Business Portfolios," *Harvard Business Review,* 77 (May–June), pp. 72–82.

4. Ryans, Adrian, et al. (2000), *Winning Market Leadership,* Toronto: John Wiley & Sons Canada, Ltd.

5. Based on an interview with Robert Guezuraga, President of the Cardiac Surgery business, and Jan Shimanski, General Manager of Biologics and Therapeutics, on June 3, 2003.

6. Ohmae, Kenichi (1988), "Getting Back to Strategy," *Harvard Business Review,* 66 (November–December), pp. 149–156.

7. Hamel, Gary and C. K. Prahalad (1994), *Competing for the Future,* Boston: Harvard Business School Press; Kim, W. Chan and Renee Mauborgne (1999), "Creating New Market Space," *Harvard Business Review,* 77 (January–February), pp. 83–93.

8. Kim, W. Chan and Renee Mauborgne (1999), op. cit.

9. Serwer, Andy, Irene Gashurov, and Angela Key (2000), "There's Something About Cisco," *Fortune,* May 15, pp. 114–127.

10. Kerin, Roger A., P. Rajan Varadarajan, and Robert A. Peterson (1992), "First-Mover Advantage: A Synthesis, Conceptual Framework, and Research Propositions,"

11. Byrnes, Nanette and Paul Judge (1999), "Internet Anxiety," *Business Week,* June 28, pp. 79–88.

12. Bayus, Barry L., Sanjay Jain, and Ambar G. Rao (1997), "Too Little, Too Early: Introduction Timing and New Product Performance in the Personal Digital Assistant Industry," *Journal of Marketing Research,* 34 (February), pp. 50–63.

13. Kirkpatrick, David (2002), "The Online Grocer Version 2.0 ," *Fortune,* November 25, pp. 217–222.

14. Golder, Peter N. and Gerald J. Tellis (1993), "Pioneer Advantage: Marketing Logic or Marketing Legend?" *Journal of Marketing Research,* 30 (May), pp. 158–171.

15. Boulding, William and Markus Christen (2001), "First-Mover Disadvantage," *Harvard Business Review,* 79 (October), pp. 20–21.

16. Shankar, Venkatesh, Gregory S. Carpenter, and Lakshman Krishnamurthi (1998), "Late Mover Advantage: How Innovative Late Entrants Outsell Pioneers," *Journal of Marketing Research,* 35 (February), pp. 57–70; see also Zhang, Shi and Arthur B. Markman (1998), "Overcoming the Early Entrant Advantage: The Role of Alignable and Nonalignable Differences," *Journal of Marketing Research,* 35 (November), pp. 413–426.

17. Hamel, Gary (1997), "Killer Strategies That Make Shareholders Rich," *Fortune,* June 23, pp. 70–84.

18. Ibid.

19. Ibid.

Journal of Marketing, 56 (October), pp. 33–52.

20. Ibid.
21. Day, George S. (1994), "The Capabilities of Market-Driven Organizations," *Journal of Marketing,* 58 (October), pp. 37–52; Srivastava, Rajendra K., Tasadduq A. Shervani, and Liam Fahey (1998), "Market-Based Assets and Shareholder Value: A Framework for Analysis," *Journal of Marketing,* 62 (January), pp. 2–18; Srivastava, Rajendra K., Tasadduq A. Shervani, and Liam Fahey (1999), "Marketing, Business Processes, and Shareholder Value: An Organizationally Embedded View of Marketing Activities and the Discipline of Marketing," *Journal of Marketing,* 63 (Special Issue), pp. 168–179.
22. Prahalad, C. K. and Gary Hamel (1990), "The Core Competence of the Corporation," *Harvard Business Review,* 68 (May–June), pp. 79–91.
23. Gomes, Lee (1997), "H-P to Unveil Digital Camera and Peripherals," *Wall Street Journal,* February 25, p. B7.
24. Prahalad, C. K. and Gary Hamel (1990), op. cit.
25. Vogelstein, Fred (2003), "Mighty Amazon," *Fortune,* May 26, pp. 60–67.
26. Barney, Jay (1991), "Firm Resources and Sustained Competitive Advantage," *Journal of Management,* 17 (1), pp. 99–120.
27. Wiggins, Robert and Timothy Ruefli (2002), "Sustained Competitive Advantage: Temporal Dynamics and Persistence of Superior Economic Performance," *Organization Science,* 13 (1), pp. 82–105.
28. Collis, David and Cynthia A. Montgomery (1995), "Competing on Resources: Strategy in the 1990s," *Harvard Business Review,* 73 (July–August), pp. 118–128; Grant, Robert M. (1991), "The Resource-Based Theory of Competitive Advantage: Implications for Strategy Formulation," *California Management Review,* 33 (3), pp. 114–135; Williams, Jeffrey R. (1992), "How Sustainable is Your Competitive Advantage?" *California Management Review,* 34 (3), pp. 29–51.
29. Kirkpatrick, David (2003), "See This Chip?" *Fortune,* February 3. Available at www.fortune.com.
30. Collis, David and Cynthia A. Montgomery (1995), op. cit. Grant, Robert M. (1991), op. cit.
31. Golder, Peter N. and Gerald J. Tellis (1993), op. cit.
32. Barney, Jay (1991), op. cit.
33. Grant, Robert M. (1991), op. cit.
34. Day, George S. and Robin Wensley (1988), "Assessing Advantage: A Framework for Diagnosing Competitive Superiority," *Journal of Marketing,* 52 (April), pp. 1–20.
35. Collins, James C. and Jerry I. Porras (1994), *Built to Last,* New York: HarperBusiness.
36. Day, George S. and Robin Wensley (1988), op. cit.
37. Deshpande, Rohit and Frederick E. Webster, Jr. (1989), "Organizational Culture and Marketing: Defining the Research Agenda," *Journal of Marketing,* 53 (January), pp. 3–15.
38. Leonard-Barton, Dorothy (1992), "Core Capabilities and Core Rigidities: A Paradox in Managing New Product Development," *Strategic Management Journal,* 13, pp. 111–125.
39. McWilliams, Gary (1999), "Schlumberger Digs Deeper," *Business Week,* July 27, pp. 48–49.
40. Christensen, Clayton M. (1997), *The Innovator's Dilemma,* Boston, MA: Harvard Business School Press.
41. Schumpeter, Joseph (1942), *Capitalism, Socialism and Democracy,* New York: Harper & Row.
42. Shanklin, William and John Ryans (1987), *Essentials of Marketing High Technology,* Lexington, MA: DC Heath.
43. Moeller, Michael (1999), "E-Commerce May Be One Race Microsoft Can't Win," *Business Week,* March 22. Available at www.businessweek.com.
44. Chandy, Rajesh K. and Gerard J. Tellis (2000), "The Incumbent's Curse? Incumbency, Size, and Radical Product Innovation," *Journal of Marketing,* 64 (3), pp. 1–17.
45. Chandy, Rajesh K., Jaideep C. Prabhu, and Kersi D. Antia (2003), "What Will the Future Bring? Dominance, Technology Expectations, and Radical Innovation," *Journal of Marketing,* 67 (3), pp. 1–18; Sorescu, Alina B., Rajesh K. Chandy, and Jaideep C. Prabhu (2003), "Sources and

Financial Consequences of Radical Innovation: Insights from Pharmaceuticals," *Journal of Marketing,* 67 (4), pp. 82–102.

46. Chandy, Rajesh K., Jaideep C. Prabhu, and Kersi D. Antia (2003), op. cit.

47. Chandy, Rajesh K. and Gerard J. Tellis (1998), "Organizing for Radical Product Innovation: The Overlooked Role of Willingness to Cannibalize," *Journal of Marketing Research,* 35 (November), pp. 474–487.

48. Sorescu, Alina B., Rajesh K. Chandy, and Jaideep C. Prabhu (2003), op. cit.

49. Saporito, Bill and Ani Hadjian (1994), "Behind the Tumult at P&G," *Fortune,* March 7, pp. 74–80.

50. Brown, John Seely (1991), "Research That Reinvents the Corporation," *Harvard Business Review,* 69 (January–February), pp. 102–111.

51. Potts, Mark (1992), "Toward a Boundary-less Firm at General Electric," in *The Challenge of Organizational Change* by Rosabeth Moss Kanter, Barry A. Stein, and Todd D. Jick, New York: The Free Press, pp. 450–455

52. Hamel, Gary and C. K. Prahalad (1991), "Corporate Imagination and Expeditionary Marketing," *Harvard Business Review,* 69 (July–August), pp. 81–92.

53. Shanklin, William and John Ryans (1987), op. cit.

54. Cooper, Arnold and Dan Schendel (1976), "Strategic Responses to Technological Threats," *Business Horizons,* February, pp. 61–69.

55. Shanklin, William and John Ryans (1987), op. cit.

56. Shanklin, William and John Ryans (1987), op. cit., chapter 7.

57. Ibid.

58. Leonard-Barton, Dorothy, Edith Wilson, and John Doyle (1995), "Commercializing Technology: Understanding User Needs," in *Business Marketing Strategy,* V. K. Rangan et al. (eds.), Chicago: Irwin, pp. 281–305.

59. Byrnes, Nanette and Paul Judge (1999), op. cit.

60. Hamel, Gary and C. K. Prahalad (1991), op. cit.

61. MacDonald, Elizabeth and Joann Lublin (1998), "In the Debris of a Failed Merger:

Trade Secrets," *Wall Street Journal,* March 10, p. B1.

62. See also the notion of waste in Gross, Neil and Peter Coy with Otis Port (1995), "The Technology Paradox," *Business Week,* March 6, pp. 76–84.

63. Herbold, Robert (2002), "Inside Microsoft: Balancing Creativity and Discipline," *Harvard Business Review,* 80 (January), pp. 72–79.

64. Prahalad, C. K. and Gary Hamel (1990), op. cit.

65. Leonard-Barton, Dorothy (1992), op. cit.

66. Mehta, Stephanie (2003), "Samsung's New Play," available at www.fortune.com, September 17.

67. Thomke, Stefan (2001), "Enlightened Experimentation: The New Imperative for Innovation," *Harvard Business Review,* 79 (February), pp. 67–75.

68. Quinn, James (1985), "Managing Innovation: Controlled Chaos," *Harvard Business Review,* 63 (May–June), pp. 73–85.

69. Burgelman, Robert (1983), "Corporate Entrepreneurship and Strategic Management: Insights from a Process Study," *Management Science,* 29 (December), pp. 1349–1364.

70. Hutt, Michael, Peter Reingen, and John Ronchetto, Jr. (1988), "Tracing Emergent Processes in Marketing Strategy Formulation," *Journal of Marketing,* 52 (January), pp. 4–19.

71. Pinchot, Gifford (2000), *Intrapreneuring: Why You Don't Have to Leave the Corporation to Become an Entrepreneur,* San Francisco, CA: Berrett-Koehler Publishing.

72. Maidique, Modesto (1980), "Entrepreneurs, Champions, and Technological Innovations," *Sloan Management Review,* 21 (Spring), pp. 59–70; see also Howell, Jane (1990), "Champions of Technological Innovation," *Administrative Science Quarterly,* 35 (June), pp. 317–341.

73. Chandy, Rajesh K. and Gerard J. Tellis (1998), op. cit.

74. Ibid.

75. Tabrizi, Behnam and Rick Walleigh (1997), "Defining Next-Generation Products: An Inside Look," *Harvard Business Review,* 75 (November–December), pp. 116–124.

76. Hamel, Gary and C. K. Prahalad (1991), op. cit.

77. Ibid.

78. Hamel, Gary (1997), op. cit.

79. McGosh Andrew, Alison Smart, Peter Barrar, and Ashley Lloyd (1998), "Proven Methods for Innovation Management: An Executive Wish List," *Creativity & Innovation Management,* 7 (December), pp. 175–193.

80. Byrnes, Nanette and Paul Judge (1999), op. cit.

81. Anders, George (1999), "Buying Frenzy," *Wall Street Journal,* July 12, pp. R6, R10.

82. Byrnes, Nanette and Paul Judge (1999), op. cit.

83. Hamel, Gary and Jeff Sampler (1998), "The e-Corporation," *Fortune,* December 7, pp. 80–92.

84. Christensen, Clayton M. and Richard S. Tedlow (2000), "Patterns of Disruption in Retailing," *Harvard Business Review,* 78 (January–February), pp. 42–46.

85. Hof, Robert (1998), "The Click Here Economy," *Business Week,* June 22, pp. 122–128.

86. Hamel, Gary and Jeff Sampler (1998), op. cit.

87. Sager, Ira (1999), "Go Ahead, Farm Out Those Jobs," *Business Week e.biz,* March 22, p. EB35.

88. Byrnes, Nanette and Paul Judge (1999), op. cit.

89. Sager, Ira (1999), op. cit.

90. "Selling a 'Killer App' Is a Far Tougher Job Than Dreaming It Up," (1998), *Wall Street Journal,* April 13, p. B1.

91. Mandel, Michael (2000), "The New Economy," *Business Week,* January 31, pp. 73–91.

92. www.nbia.org.

93. Bransten, Lisa (1999), "Seeking More High-Tech Home Runs," *Wall Street Journal Interactive,* October 15. Available at http:interactive.wsj.com.

94. Riolli-Saltzman, Laura and Fred Luthans (2001), "After the Bubble Burst: How Small High-Tech Firms Can Keep in Front of the Wave," *Academy of Management Executive,* 15 (3), pp. 114–125.

95. Markides, Constantinos C. (1999), *All the Right Moves: A Guide to Crafting Breakthrough Strategy,* Cambridge, MA: Harvard Business School Press.

96. Day, George S. (1994), op. cit.

97. Baker, William E. and James M. Sinkula (1999), "The Synergistic Effect of Market Orientation and Learning Orientation of Organizational Performance," *Journal of the Academy of Marketing Science,* 27 (4), pp. 411–428.

98. West III, G. Page and G. Dale Meyer (1997), "Temporal Dimensions of Opportunistic Change in Technology-Based Ventures," *Entrepreneurship: Theory & Practice,* 22 (2), pp. 31–53.

CHAPTER 3

Relationship Marketing: Partnerships and Alliances

In 1967, Olaf Helmer predicted that by the year 2000 a permanent colony would exist on the moon, and that people would have landed on Mars.
—OLAF HELMER, RAND INSTITUTE

IBM's Big Partner Agenda

As IBM rolled into 2002, the company hoped to build on the strengths it established in 2001, particularly partner relationships. To that end, Peter Rowley, general manager of IBM's Global Business Partners unit, had been working on plans for the channel, while taking some time to look back at his first year on the job.

The three reasons IBM had a relatively good year in a tumultuous economy were its strategy around e-business, strong and fair relationships with its business partners, and excellent products. "Lou Gerstner was very focused on using the technology around customers and making sure that if they do something out on the Web, that it's integrating into the customers' legacy systems [older, existing computer technology in use]," Rowley said.

IBM's *alliance strategy* with its partners, particularly software vendors, also paid off. In the summer of 1999, IBM executives decided against pursuing the application-software business inside the company. Rather than developing applications for various vertical markets serving key industry sectors, IBM aligned itself with software vendors. "It's such a fragmented marketplace—it's so specialized and it moves so quickly—that we're better off allying ourselves with the software vendors," Rowley said. Such a move certainly helped the sale of WebSphere and DB2 (two software programs used in e-business applications), and that's why WebSphere is outgrowing its competitors and DB2 is outgrowing Oracle.

"We're trying to become the world's best provider of middleware (the software that connects two otherwise separate applications; for example, linking a database to a Web server), but we are not trying to become the world's best provider of Enterprise Resource Planning (ERP) systems or Customer Relationship Management (CRM) systems or business intelligence systems," Rowley claimed.

While alliances with software companies are one way to generate revenue for IBM, they can also lead to more opportunities. For example, if Siebel (an ERP software provider) was to license a new product, opportunities would abound for IBM in hardware, other software, and services.

The challenge, Rowley said, is that it's up to him and other top IBM executives to determine which areas will grow. He noted that the CRM market, despite some levels of customer dissatisfaction, is still growing, as is ERP integration. "SAP [another ERP software provider] is doing well around the world, and despite its recent faltering from excessive growth rates, Siebel is still growing."

IBM's relationships with its partners remain strong, mainly because of open communication regarding any changes within the program and the identification of conflict within the distribution channel.

IBM's business partners tend to deal with various groups of the large organization, and Rowley hopes his work creates cohesion among those groups. There are, according to Rowley, four groups (or tracks): software, systems and services, personal systems, and developers. "I think you're going to find the relationships with the four tracks are going to get closer and closer, so the business partner who's a member of multiple tracks will see more consistency from IBM." And solution providers are noticing the changes, as proven each year by *VARBusiness* Annual Report Card (ARC) (*VARBusiness* is a trade publication serving value-added resellers in the computer industry). In recent ARC surveys, IBM has steadily improved in "ease of doing business."

"The biggest thing we tried to do [last] year was extend our portfolio of partners to include software vendors, systems and Web integrators and software resellers, who, I think, are the true generators of demand.[1]

Relationship marketing refers to the formation of long-term relationships with customers and other business partners, which yield mutually satisfying, win–win results. In high-tech fields, several forces exist that necessitate the use of partnerships and alliances. Because the time-to-market cycle is short and development costs and risks are high, firms can find it faster and more cost efficient to develop products jointly than alone. For example, GM and Toyota teamed up in the development of alternative fuel vehicles, seeing the race to produce a viable alternative-technology auto as a battle "not of individual auto makers, but of corporate—and often transcontinental—alliances."[2]

Moreover, the need to bring a complete, end-to-end solution to the market may mean that partners are needed to develop different aspects of the product. For example, in the development of enterprise resource planning (ERP) software—which companies use to integrate and improve the various business activities such as manufacturing, sales, finance, human resources, supply chain management, and customer relationship management—the major players have found that they need to partner in order to offer customers a complete solution. A leading provider of this software, SAP AG (Germany) works with a host of other companies to round out its offering in order to provide customers an integrated set of features. And, as noted in Chapter 1, the importance of setting industry standards necessitates collaborating with other companies.

This chapter explores important issues in the formation and management of the partnerships, alliances, and relationships that are vital to success in high-tech markets.

PARTNERSHIPS AND ALLIANCES

Types of Partnerships

The synergies of partnering can lead to both parties becoming more competitive through a win–win situation, possibly even strengthening both companies against outsiders. For example, scientists and research executives at most of the large pharmaceutical companies believe that open access to a map of genetic breakthroughs will be central to the way drugs are developed and tested in the future. They have formed a nonprofit venture in which the usually highly competitive companies are working together to build the biological blueprint for all human life. This new venture will also collaborate with competing biotech companies, which currently lead the race in genetic mapping. The idea is to store valuable data in a public database to which drug companies, academic researchers, and biotech firms will all have free and equal access. Executives at the major drug companies realize that progress would be much slower if every company had to create its own map—especially given the lead the biotech companies hold.[3] Sixteen labs joined in April 2003 to release the full results of the Human Genome Project.

A wide variety of different types of partnerships can be formed at all levels of the supply chain, as shown in Figure 3-1.[4]

Vertical Partnerships

Firms may form **vertical partnerships** with members at other levels in the supply chain (either suppliers, distribution channel members, or customers). Relationships with *suppliers* are often formed to gain efficiencies in accessing parts and materials. Collaborative relationships built around common procedures and intensive information sharing mean that the supplier's operations can be more closely fitted to the customer's needs. Moreover, early supplier involvement (ESI) is useful in developing innovations in supplies that can help differentiate the customer's product in downstream markets. For example, a firm may choose to form partnerships with key suppliers whose skills and experiences complement its strengths to develop next-generation

FIGURE 3-1   Possible Alliance Partners Along the Supply Chain

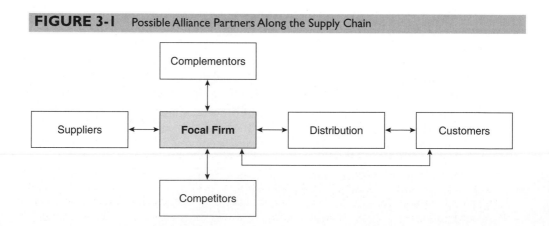

technology. Chip manufacturers and computer manufacturers are working together to develop next-generation computers. Because chips are a supply used in computers, this is an example of a supplier–OEM (original equipment manufacturer, or a company that buys components such as disk drives from suppliers that it integrates in a manufacturing process into a finished product such as a computer), or vertical, relationship.

Relationships with *channel members* are used to gain efficiency and effectiveness in accessing downstream markets. For example, collaborative relationships with channel members provide both a source of competitive advantage that can be used in more effectively implementing marketing programs and a conduit of market information back to the manufacturer. Relationships with channel intermediaries are discussed in Chapter 8.

Another source of important relationships referenced in discussions of relationship marketing are relationships with *customers,* be they end users of the product or business customers who use the product in their businesses. Close, long-term relationships with customers are vital in many types of markets, and particularly in high-tech markets. Because of the need to rely on customers for beta-test sites and ideas for innovations, firms that have close relationships with customers have a strong source of market-based information. Moreover, a focus on establishing long-term relationships with each customer is more likely to produce a long-term revenue stream from that customer, capturing the lifetime value of a customer's purchases in a particular product category. Customer relationships are discussed in detail in a later section of this chapter.

Horizontal Partnerships

In the high-technology arena, a common type of relationship is with a company at the same level of the value chain. **Horizontal alliances** are created with either competing firms or firms that provide jointly used, complementary products. When these relationships provide customers a product that delivers a complete, integrated solution, such alliances are referred to as **complementary alliances**, and the members are referred to as *complementors.*[5] For example, in 1997, Hewlett-Packard and Kodak jointly decided to pursue the digital photography market. The alliance relied on Kodak's thermal dye transfer process to produce prints on HP's printers. And, as the opening vignette for this chapter describes, IBM has a series of relationships with software vendors in order to develop software for particular applications. This form of horizontal partnering (because both firms are producers of some good that, in turn, is resold to customers or channel intermediaries) enables each firm to maintain flexibility and focus on its core competencies, and stimulate demand through synergistic innovation.

Competing firms may choose to join forces to develop next-generation technology, to define standards for new technologies, to provide market access in an area that one firm lacks, or to be a stronger force against a larger competitor. **Competitive collaboration** between firms is also a form of horizontal alliance, because these firms typically compete at the same level of the supply chain. Another name given for this type of collaboration is *co-opetition,*[6] whereby firms compete in some arenas and collaborate in others. For example, in the mid-1980s,[7] semiconductor industry analysts predicted that by 1993, the U.S. market share, once at 85 percent, would shrink to 20 percent. Given the superordinate threat posed by Japanese competitors, U.S. semiconductor firms decided to join forces. In 1987, the Semiconductor Manufacturing Technology Consortium (SEMATECH) of U.S. semiconductor manufacturers and the U.S. government was jointly founded by fourteen firms that then accounted for 80 percent of the U.S. semi-conductor manufacturing industry (including Digital Equipment, IBM, Intel, NCR, Texas Instruments, National Semiconductor, Advanced Micro Devices, and LSI Logic, to name a few). Its mission was to provide the U.S. semiconductor industry the capabil-ity of achieving a world-leadership manufacturing position by the mid-1990s. The U.S. industry had to do two things to remain competitive: increase the number of usable chips that could be manufactured from each wafer of silicon and make each chip capa-ble of doing more. Through SEMATECH, firms could pool resources in the fight to recover market share from Japanese companies.

SEMATECH's main cited achievement was in finding ways to pack features onto chips by reducing the width of the circuit lines etched on chips. Eleven of the original fourteen member companies and the U.S. government agreed to extend their member-ship in SEMATECH for a second five-year period and committed themselves to new and expanded goals. Their continued commitment to SEMATECH indicated that they believed SEMATECH had achieved something worth their investments.

Indeed, to stimulate more collaboration between competing firms in the hopes of generating innovation and enhancing the competitiveness of U.S. businesses, tradi-tional antitrust laws were eased in the 1980s and 1990s to allow more forms of collabo-ration. For example, the **National Cooperative Research Act** (1984) promotes research and development, encourages innovation, and stimulates trade. The act applies to R&D activities up to and including the testing of prototypes. This act was expanded in 1993 with the **National Cooperative Production Amendment**, which allowed for joint production as well. The amendment excludes marketing and distribution agreements, except for the products manufactured by the venture, and also excludes the use of existing facilities by the joint venture.

The nexus of relationships formed may be quite complex, with the same players serving as suppliers in some arenas, complementors in others, and competitors in still others. For example, in the enterprise resource planning software market, key players include Oracle and SAP. Oracle builds databases, which are the core of all ERP sys-tems. SAP is a customer of Oracle, cooperating to bundle Oracle databases into its applications, and it simultaneously competes with Oracle in the ERP market. In a similar vein, Oracle competes aggressively with Microsoft in the database market; Microsoft offers its SQL server with 75 percent of Oracle's functionality at 25 per-cent of the price.[8] But Oracle must cooperate with Microsoft to ensure that its soft-ware works well with Microsoft's operating system. AMR's President Tony Friscia states:

> One side of Oracle considers Microsoft the enemy, and the other side considers SAP the enemy. The side that considers Microsoft the enemy needs a relation-ship with SAP, and the side that considers SAP the enemy needs a relationship with Microsoft.[9]

Reasons for Partnering

In general, partnerships are formed to provide one firm access to resources and skills that, if it had to develop in isolation, would be costly in terms of either money or time. So, by partnering, firms are able to gain *access to such resources and skills* in a *timely, more cost-efficient manner.* Cisco has methodically built its company by strategically investing in or cooperating with other companies whose technology rounds out its product line. For example, Cisco recognized that it needed fiber optic expertise and initially partnered with (and then bought) Cerent, a company whose products made it cheaper to move voice and data over fiber optic lines. Cisco currently has alliances with Motorola, IBM, HP, EDS, KPMG, Peoplesoft, Cap Gemini, and Microsoft that blend complementary strengths to address opportunities created by the growth of the Internet.

Another important reason for collaborating is to *define standards for new technologies*. The GM–Toyota partnership to share research into battery-powered electric, fuel-cell-powered electric, and hybrid vehicles was formed partly in the hopes that through sheer combined size, they would be able to set technical standards regarding what fuel the industry will use for fuel cells. Recall from Chapter 1 that industry-wide standards result in a common, underlying architecture for products offered by different firms in the market. As a result, *customers gain compatibility* across the various components of a product—say, across hardware and software—and across product choices in an industry—say, across different types of computers. Because the complementary products share a common interface, the customers' hardware and software interface seamlessly. For example, in the cellular telecommunications industry, compatibility allows base stations, switches, and handsets to work with each other across service areas.

Compatibility, achieved when many different companies produce their offerings based on a common set of design principles, increases the value a customer receives from owning a product. First, standards reduce customer fear, uncertainty, and doubt about which technology to purchase. This can be particularly helpful when the value of the product to the customer increases as more customers adopt products based on the same technology (demand-side increasing returns). As a result, the more customers that adopt products sharing a common, underlying technological standard, the greater the value each of them receives. Second, the availability of complementary products is largely determined by the installed base of the given product. For example, software developers are more willing to write applications programs around technology platforms that have wider penetration in the market. In turn, standards ensure a greater availability of complementary products, and therefore enhance the value a customer derives from the base product.

Reasons for Partnering over the Stages of the Product Life Cycle

The product life cycle provides a useful way to look at the different reasons for partnering at different points in the life cycle for technology-based products,[10] as illustrated in Figure 3-2. Recall that product innovations are new technologies that will be sold to customers in the marketplace while process innovations are new technologies used in manufacturing and business processes.

During the *emergence* stage of the life cycle, substantial uncertainty surrounds the product. As discussed subsequently in Chapter 6, purchasers at this stage are innovators or technology enthusiasts who are willing to take on substantial risks. As a buyer group, innovators have few requirements but they are very demanding when it comes to those requirements. First, they want an accurate portrayal of both the benefits and the liabilities of the innovation. In their role as gatekeepers in the system, innovators have the ability to undermine the prospects of any product that holds a surprise for them. Second, when they have a problem (and they will have problems due to the early

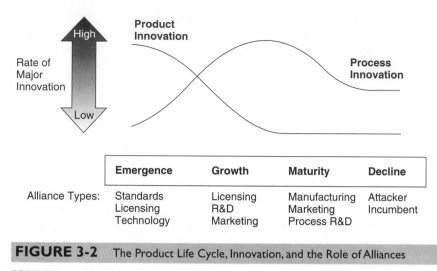

FIGURE 3-2   The Product Life Cycle, Innovation, and the Role of Alliances

SOURCE: Adapted from Utterback (1994) and Roberts and Liu (2001).

stage of development), they require access to the most technically knowledgeable person in the organization to help them work through the problem. Finally, they want the new technology early and they want it at a low cost. They often see themselves as providing a service to the seller, acting as a beta test site by testing early versions of the new technology prior to its full-scale launch. They will refuse to pay a high price to be guinea pigs.[11]

From a competitive perspective, there may be little direct competition at the emergence stage, but there are competing standards. For example, at the emergence of the personal video cassette recorder (VCR) industry there were two competing standards, Betamax, a Sony technology, and VHS, which was developed by Philips NV and Matsushita. Since the technologies were incompatible, Betamax and VHS VCRs were not direct competitors per se, but they did compete to establish the industry standard. Furthermore, barriers to entry may be low during the emergence stage since brand loyalty has not been established, little cumulative learning has occurred, and major investments in production facilities have not been made.

Thus, one of the key reasons for partnering at this stage is that alliances among potential competitors are valuable to establish standards. Managers of companies in this stage pursue licensing agreements in which the company is willing to license its technology to other manufacturers and competitors in the industry in order to establish their technology as the standard. For example, in a major departure from its prior strategy (and in order to more effectively compete with Microsoft's entry into cellular phone technology), Nokia began to license its software to other cell phone makers in 2002; its objective was to try to make the Finnish company's software the industry standard.[12]

Finally, because the product technology is likely to still be immature, it behooves potential competitors to collaborate to improve the technology so that the next buyer group in the diffusion process can be addressed. For example, during the development of the digital audio technology used in compact disc players, at least four companies were pursuing incompatible designs. In a turnabout from their VCR experience, Philips NV and Sony partnered to develop the first compact disc system. Philips NV contributed a superior basic design, and Sony provided the error correction system. This alliance threw momentum behind the Philips–Sony standard, and eighteen months prior to product introduction, over thirty firms had signed agreements to license the Philips–Sony technology.

One of the signals that the life cycle has shifted from emergence to *growth* is that a dominant design becomes the industry standard. Consistent with this, customer needs and wants become increasingly clear. Purchasers in this stage are early adopters to whom technology is important only insofar as it furthers the opportunity to achieve breakthrough performance. Because of the potential that early adopters can envision, they tend to be the least price sensitive of the innovation adopter groups. The lack of price sensitivity is offset by the fact that they are in a hurry to reap the rewards of the innovation before the opportunity window closes. A critical conclusion is that the seller must manage early adopters' expectations by understanding their goals and clearly communicating how the innovation will help attain those goals.[13] The early adopter category is important because it is large enough to generate meaningful revenue and is often quite profitable due to low price sensitivity. While innovation in product technology declines during the growth stage, process technology becomes more important due to the increased emphasis on quality and efficiency.

In the growth stage, the winners of the race to establish the dominant design license their technology to the losers. In addition, R&D alliances are formed to enable companies to improve the dominant design and develop product extensions and new features. If the dominant design is controlled by a small or a technology-oriented company, the company's managers will want to establish a marketing alliance with a larger, well-established firm that has well-developed distribution and other marketing competencies. This is a common strategy in the biotech/pharmaceutical industry.

In the *maturity* stage, sales volume and revenue are high but their growth rate slows dramatically. Purchasers are members of the mass market. Mass-market purchasers seek a demonstrable reason to buy. They utilize a rational decision-making style to evaluate the relative benefits of the innovation. They typically seek a measurable and predictable improvement in performance or productivity compared to the previous solution to this particular need. They put strong pressure on price. To maintain their profits in the face of price pressure, sellers emphasize cost control from purchasing through distribution and post-sale service. Process innovation dominates product innovation and process R&D alliances take center stage.

For example, in 2001 IBM Corp., Infineon Technologies Inc., and United Microelectronics Corp. (UMC) formed an R&D partnership for making integrated circuits (semiconductors). The combined experience of the three companies was anticipated to enable faster process technology development at a lower cost for each company. Jim Kupec, president of worldwide marketing and sales for UMC explained, "With the combined expertise of IBM, Infineon, and UMC, we believe that the processes that we jointly develop will represent a new world standard for quality in logic semiconductor manufacturing."[14]

Also at this stage, many firms form alliances in order to outsource production to contract manufacturers whose competency is in efficient production. The outsourcing capitalizes on the idea of each firm's core competencies, enabling the developer to focus its resources on developing a next generation of the technology and the contract manufacturer to focus on its skills in efficient manufacturing. As in the growth stage, marketing alliances are formed to address new markets.

A major reason that products start to *decline* is that products based on new technologies are developed and introduced. As the transition is made from the old to the new technology, the industry newcomers form technology and marketing alliances to take on incumbents. Because incumbents often have a difficult time developing and introducing a technology that will obsolete or cannibalize their cash cow, they are often in the position of having to license the disruptive technology from a new competitor. And then, the cycle starts again. (Note that this stage of the cycle is consistent with the concepts of technology life cycle, core rigidities, and creative destruction from Chapter 2.)

Because of their ostensible benefits, partnerships and alliances are frequently prescribed as the panacea for success, the *modus operandi* for a successful business model in today's business environment. However, despite the many reasons for partnering, as summarized in Table 3-1, the prescription to partner overlooks the reality that the overwhelming majority of partnerships fail to achieve the objectives set by at least one of the partners.[15] In addition to outright failure, many risks are inherent to partnering efforts. These inescapable realities of strategic alliances highlight the need to understand fully the risks as well as the factors that contribute to the potential success and viability of the partnership.

TABLE 3-1 Reasons, Risks, and Success Factors for Partnering	
Reasons to Partner	Access resources and skills
	Gain cost efficiencies
	Speed time to market
	Access new markets
	Define industry standards
	Develop innovations and new products
	Develop complementary products
	Gain market clout
Risks to Partnering	Loss of autonomy and control
	Loss of trade secrets
	Legal issues and antitrust concerns
	Failure to achieve objectives
Success Factors	Interdependence
	Appropriate governance structure
	Commitment
	Trust
	Communication
	Compatible corporate cultures
	Integrative conflict resolution

Risks Involved in Partnering

Although partnerships do provide the many benefits mentioned previously, firms face serious impediments to realizing those benefits. For starters, working in tandem with another organization increases the project's complexity. More serious is the *potential loss of autonomy and control* that accompanies a joint effort. In teaming with another firm, decisions must be made jointly, and the success of the project becomes, to some extent, dependent on the efforts of another. Sharing decision-making control is a very difficult hurdle for many firms in choosing to partner, and because some are never able to give up their autonomy, their partnerships fail (at worst) or do not function effectively (at best).

Some identify the *loss of trade secrets* as the least noticed, but potentially riskiest, aspect of alliances.[16] Although companies typically sign confidentiality agreements that ostensibly prevent partners from exploiting what they learn about each other, shared secrets are a hazard of the game. Many experts recommend that a firm never forget that one partner may be out to "disarm" the other. A quote from a business manager highlights this fact: "If they were really our partners, they wouldn't try to suck us dry of technology ideas they could use in their own products. Whatever they learn from us they are going to use against us worldwide."[17] Indeed, product managers cited the leakage of information as the greatest risk in joint product development.[18]

The potential leakage of information can lead to one partner learning skills and knowledge from the other. Partners may gain access to knowledge and know-how that can be very valuable. As Oracle notes, "The partners we catch up to in our core applications will cease to be useful partners." Although one wants to learn as much as possible from one's partners in order to maximize the effectiveness and efficiency of the partnership, one also must limit transparency and leakage of information in the partnership so as not to dilute the firm's sources of competitive advantage. The Appendix to this chapter explores more fully the issues involved in inter-firm learning.

Another risk strategic alliances face is that of *legal issues and antitrust problems.*[19] Cooperative ventures may run afoul of U.S. antitrust laws, especially when they involve large firms. Public policy officials wrestle with finding an appropriate balance in the tradeoffs between protecting consumer welfare and maintaining national competitiveness in an increasingly global market. On the one hand, collaborative ventures are necessary to compete globally. On the other hand, collaborative relationships can result in less competition in markets, potentially harming consumer welfare.

To help clarify its position, on October 1, 1999, the Federal Trade Commission, in consultation with the Justice Department, issued its Guidelines for Collaborations Among Competitors.[20] Any partnership that has the potential to directly or indirectly affect pricing is of great concern. Hence, partnerships that affect allocation of customers or supply in a market are likely to be carefully scrutinized because of the indirect impact on pricing. On the other hand, partnerships that arise because firms are unable to pursue projects alone, and that create some common good in the market, are likely to be encouraged. For example, partnerships focused on costly research and development efforts are typically easily justified. Another factor involves the impact of the collaboration on market share. If, together, the partners will not control more than 20 percent of the market being affected, their partnership is unlikely to raise concerns.

Possibly the largest risk of any partnership is the reality in the overwhelming majority of cases, there is failure to achieve objectives set by at least one of the partners.

Reasons for this failure can be attributed to a host of factors: incompatible cultures between the two firms, lack of attention and resources allocated to the ongoing management of the relationship, lack of trust in the other party's motives or ability to deliver its part of the agreement, and so forth. The risks involved highlight the need to understand the factors that contribute to the potential success and viability of the partnership.

Factors Contributing to Partnership Success

Effective partnerships are characterized by the existence of traits that are important in nearly all business relationships, but alliances show a greater amount or intensity of these characteristics.[21] Effective strategic alliances have been shown to exhibit the following characteristics.

Interdependence

To enhance the odds of partnership success, both parties must be dependent on the other for some important resource that is valued and hard to obtain elsewhere. Shared mutual dependencies form the basis for a give-and-take relationship in which both parties are equally motivated to ensure the success of the alliance. Uneven, or asymmetrical, dependence undermines the dual nature of the relationship, can lead to exploitation, and may leave one party more vulnerable than the other. Alliances with low levels of interdependence suffer from a lack of commitment and need.

A special case of interdependence arises with partners of very disparate sizes. Past research has shown that partnerships between relatively equally sized partners are more likely to be successful than those between partners of unequal size.[22] However, in the technology arena, one commonly finds small startups partnering with industry behemoths. In a typical case, the small startup has an exciting new technology, whereas the large company brings needed resources, access to markets, and management and marketing expertise. When small companies partner with large companies, the risks can loom large and special attention should be paid to the governance structure of the relationship.

Appropriate Governance Structure

Governance structures are the terms, conditions, systems, and processes used to manage the ongoing interactions between two companies. At a simple level, governance structures can be *unilateral* in nature, granting one party the authority to make decisions, or *bilateral* in nature, based on mutual expectations regarding behaviors and activities.

Generally, the governance structure should be matched to the level of risk in the partnership. When a partner has assets that are at risk, should the other party behave in an opportunistic fashion, or when uncertainty is high, it is important to have a governance structure that mitigates that risk. Governance structures based on "credible commitments," or mutual investments that both parties place in the relationship such that both are vulnerable should the partnership fail, are one way to achieve interdependence. Alternatively, a vulnerable partner could adopt an interfirm agreement that only loosely couples the organizations,[23] relying on a narrow marketing agreement or licensing agreement rather than a joint venture. However, these looser, arms-length relationships can compromise the ability of the partnership to function effectively. As a last resort, the more vulnerable party might either trust or rely on its partner's reputation as a governance strategy. However, this too can involve a gamble.

A governance structure based on bilateral norms, including expectations for future interactions (commitment), acting in the best interests of the partnership (trust), and intensive information sharing (communication), can also provide a reasonable solution, as discussed in the next three factors.

Commitment

Commitment, or the desire to continue the relationship into the future, is an important element for strategic alliances to succeed. Partners who are committed to the relationship are less likely to take advantage of the other partner or to make decisions that may sabotage the long-run viability of the relationship. Commitment can be demonstrated by making investments in the relationship that are dedicated solely to that relationship. Importantly, commitment should arise from a positive feeling and regard for each other's contributions, rather than from feelings of desperation or economic necessity.[24] When the nature of commitment between partners is of the "have to be committed" variety (rather than because of a positive, voluntary desire), the impact on the alliance is negative.

Trust

Trust refers to the sense that the other partner will make decisions that serve the best interests of the partnership when one party is vulnerable and will act honestly and benevolently. Trust is necessary for the partnership to succeed because it leads to more effective information sharing, a willingness to allocate scarce and sensitive resources to a shared effort, and the sense that both parties will benefit in the long run.

Communication

Effective communication in strategic alliances is absolutely critical to success. Effective communication is characterized by frequent sharing of information, even information that may be considered proprietary. Such communication flows bidirectionally, with both partners participating in the flow of information about their needs and potential problems. The quality of the communication, in terms of its credibility and reliability, is vital. Communication needs to be somewhat structured, with someone accountable for maintaining open lines of information. Informal, unplanned, and ad hoc interactions are also an important component of communication.

Compatible Corporate Cultures

Although two firms may have synergistic skills that could usefully be shared in a partnership, such synergies are difficult to realize if corporate cultures clash. For example, some suspect that one of the contributing factors to the demise of Taligent, the joint venture formed between IBM and Apple, was the very different cultures brought to the venture by the respective companies' employees. The IBMers were unaccustomed to working in an open and nonhierarchical environment, whereas Apple's people were equally uncomfortable in a more formal atmosphere. Conversely, John Chambers, CEO of Cisco, knows that a good fit is important. When Cisco partnered with Cerent, Cerent's sales team had a "high-octane" spirit similar to that of the Cisco sales force, and its offices were as utilitarian as Cisco's.

Integrative Conflict Resolution and Negotiation Techniques

In any relationship in which parties' outcomes are dependent to some extent on the actions and decisions of another, difficulties will arise. For example, SAP's relationships with its partners has been confrontational in some cases. The company has stated, "We will select a handful of partners to work with. If our partners cross us, we will crush them into the dust."[25] Similarly, Microsoft exerted coercive (albeit indirect) pressure on even one of its largest, most important partners, Intel, by threatening to withhold Windows 95 if Intel's largest customers (computer makers) supported some of Intel's new-product development efforts that would undermine Microsoft's clout in the market.[26] In the Intel–Microsoft case, conflict arises because of differing goals: Intel tends to focus on innovating products that will persuade existing users to buy new hardware whereas Microsoft wants to sell products for existing machines.

Some level of conflict is likely functional, because it indicates that problems are being identified and addressed rather than being ignored. The ways in which such conflicts are addressed are more important than the sheer level of conflict in determining the alliance's future success. Parties must be willing to resolve conflict in a way that allows for both partners to have a stake in the outcome, addresses both partners' needs simultaneously, and is mutually beneficial to both. Although this may seem idealistic, many creative problem-solving techniques can identify solutions that do not result in a win–loss outcome, but rather win–win solutions.

In sum, the *spirit of cooperation,* identified in the traits mentioned, signals the likelihood of future success of partnerships and alliances. In the Technology Expert's View from the Trenches, IBM's Melissa Porter offers her insights on managing horizontal alliances.

TECHNOLOGY EXPERT'S VIEW FROM THE TRENCHES

THE ART OF ALLIANCES

MELISSA PORTER

Global Alliance Executive
Telecommunications Industry
IBM Corporation, Oakbrook Terrace, IL

Alliance (*n*) (1) A bond or connection between parties or individuals; (2) an association to further the common interests of the members; (3) union by relationship in qualities.

Companies today are finding that they cannot be all things to all people, so are seeking out other companies to form alliances in order to combine strengths and to get leverage within a market segment. Working together as a team, each bringing to the partnership unique skills, abilities, and experiences, allows for a more powerful resource to meet the needs of mutual customers.

Forming and managing alliances is truly an art, not a science. To bring together different companies and to build a partnership that has lasting power has its challenges. Once a firm puts a stake in the ground to define the relationship and establishes principles to engage in the marketplace, effective partnerships should allow a firm to provide best-of-breed solutions to its customers. I often recommend a Crawl-Walk-Run strategy when first entering into an alliance . . . start slow and get some success in working together. This will generate momentum leading to a more successful partnership.

One place where alliances can be particularly useful is in serving customers facing ever-changing environments. For example, telecommunication companies are being challenged by many issues in a boom-to-bust marketplace across the globe. Faced with competitive market pressure to reduce costs, offer new products and services, and better manage their networks, many telecoms now require information technology solutions to help them compete successfully in a radically different, consumer-driven environment. By partnering with software vendors to develop these solutions, IBM is helping telecom service providers all over the world meet the challenge of change.

The complexity and dynamic nature of the marketplace does continually present barriers to both sides of an alliance. When discussing alliances, I sometimes refer to the term "co-opetition," meaning that in certain instances we cooperate with our partners, and at other times we compete. Enhancing cooperation between parties and offering a support structure will allow for a long-term and productive partnership.

IBM has addressed the need for strong relationships with its alliance partners by offering development, marketing, sales, and financing support. Each year at the PartnerWorld conference, we review the status of key objectives with each business partner and announce new products, programs, and processes to further enhance the alliance. A consistent message to our business partners and IBMers alike is that we are committed to a strong relationship with partners, many of them developing industry software applications, and that cooperation in delivering solutions will make each business stronger. In fact, the percentage of IBM's revenue generated from business with its partners has greatly increased over the years, and the partner firms report similar results.

A current value theme throughout IBM, and one that resonates with partners and customers alike, is that of *on-demand*. Businesses that are on-demand can respond quickly to customer requirements, market opportunities, and external threats. For example, telecom service providers can use on-demand solutions to increase their value and operational flexibility, and many of our alliance partners are enabling their solutions to fit this model.

IBM now enjoys an ecosystem of strategic alliance partners that couple applications from leading software firms with IBM's hardware, software, services, financing, and unmatched industry expertise to meet growing customer demand for end-to-end information technology solutions. As part of these alliances, independent software vendors (ISVs) gain access to new customers and revenue opportunities through IBM's extensive marketing, sales, and solution resources. At the same time, ISVs commit to increased use of IBM's leading hardware (server) platforms, middleware, and services.

Partnerships like these are truly:

1. A connection between parties that
2. furthers the common interests of members and
3. are seen as a relationship in qualities!

The next section provides focused coverage on one type of vital relationship—relationships with customers.

CUSTOMER RELATIONSHIPS

Customer relationship marketing refers to the development of close, long-term relationships with customers that provide mutual benefits, or win–win solutions. Companies that believe in the philosophy of relationship marketing recognize the value of using marketing as a process of building long-term relationships with customers to keep them satisfied and to keep them coming back. Conventional wisdom holds that relationships are valuable in that the cost of retaining customers is only about one-fifth the cost of finding and acquiring new customers. This idea has led many companies to shift their philosophy from managing the enterprise as a portfolio of products and services to managing it as a portfolio of segments and customers. Thus, many businesses are now being managed to maximize customer equity, the net present value of the cash flows associated with a particular customer or segment.

However, managing to maximize customer equity does not mean that all customers or potential customers are equally valuable or are worthy of efforts to retain them. For example, when asked who its unprofitable customers are, executives at one Fortune 500 company responded that they had no unprofitable customers. However, this was a company that was not economically profitable. In a bit of perverse logic, its senior managers seemed to be arguing that the firm earned profit from every customer but destroyed economic value in the aggregate![27] This illustrates the importance of an active rather than a passive approach to customer equity management. It is no longer sufficient for managers to offer the platitude "Customers are our most important asset." Managing for customer equity requires specific marketing activities to acquire customers and to retain them. Underlying customer equity management is a system for assessing profitability at the customer or market segment level. We first describe the principles used to evaluate the financial ramifications of customer acquisition and retention and then the strategies for accomplishing these goals.

Conceptually, the *customer acquisition* decision is like most other investment decisions because customer acquisition is simply an investment in the customer or customer group. As we noted, customer equity is the net present value of the cash flows associated with a particular customer or segment. Its major determinants are acquisition costs (the cash outflow in a net present value calculation) and the profit stream from retained customers (the cash inflows). Acquisition costs include such things as price discounts or rebates, advertising campaigns, direct marketing, and personal selling. To compute customer acquisition investment, the analyst should determine: (1) the total marketing cost of the acquisition campaign, (2) the number of prospects contacted during the campaign, (3) the number of prospects who become customers, and (4) the acquisition cost per new customer. The analyst then should: (5) compute the revenue from the new customer's first set of purchases and (6) subtract the cost calculated in step 4 from the revenue calculated in step 5. The difference is the net acquisition investment per customer.[28]

Two additional points must be made. It is most desirable to compute net acquisition investment at the customer level. While this is realistic for producers of big ticket items such as MRIs (magnetic resonance imaging scanners that are used in the diagnosis of many medical conditions) or ERP software systems and by companies such as Dell Computer and Charles Schwab that have very sophisticated customer databases, it may not be feasible for many sellers of high-volume, low-price goods. In the latter case, the analysis should be done at the segment level and an average acquisition investment per customer computed. Any company that is interested in developing a relatively sophisticated customer equity management system must adapt its accounting and information systems to collect appropriate data. Such a database is interactive, allows reliable measurement of the return on marketing investments, and can lead to opportunities to cross-sell or up-sell customers.[29]

Calculation of customer equity requires estimating the future profit stream generated by the customer and discounting it to its present value at the firm's cost of capital. The profit stream stems from the base profit generated by the customer, increased business from the customer over time, and referrals of new customers from the loyal customer (see Figure 3-3 below for an illustration of this concept).

The analyst should subtract the net acquisition investment for the customer from the present value of the expected future profit stream. If the value is positive, the company should cultivate a relationship. If it is negative, the marketing team should decrease its costs to acquire or serve the customer, or should increase prices to the customer. If these actions cause the seller to lose the customer, that is acceptable since the customer was not profitable. We now consider customer acquisition and retention strategies in more depth. Not only are these strategies distinct; many companies have separate organizations that are responsible for formulating and executing them.

Acquisition Strategy

The firm can choose to pursue either a broad or focused customer acquisition strategy. AOL utilized a broad acquisition strategy through its mass distribution of software for connecting to the Internet. Using this strategy, AOL acquired 500,000 new customers in one month during 1997. Fewer companies have the ability to use a focused customer acquisition strategy, selectively pursuing high potential customers. Dell employs a focused acquisition strategy based on its direct-sales system and limited service that attracts more sophisticated buyers. While more difficult to manage, the focused strategy generally creates greater customer equity because the profiles of the acquired customers are more likely to match the characteristics of the firm's target market. Following are six key steps in the customer acquisition process:[30]

FIGURE 3-3   Computing Customer Equity

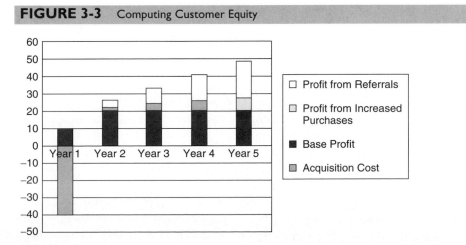

1. Segmentation.   Demographics, the traditional approach to market segmentation, often provides little insight into how the product is used and creates customer value. It is more useful to segment the market based on how the product is used and the value of the product in that application. Instead of identifying homogeneous demographic groupings, the market research task is to identify groupings of customers that use the product in similar ways and realize similar value from the product. For example, Gateway has identified a segment of relatively unsophisticated computer users that it can reach through its retail outlets.

2. Targeting.   The seller must now determine which segments it should target for relationship-building efforts. This is a function of two influences: the *attractiveness* of the segment and the *competitive position* of the seller in the segment. *Attractiveness* is determined by general factors such as rate of market growth, buyer power, competitive intensity, and relationship-potential factors such as recognized or potential need for the firm's offerings, product complexity, role of user input, quality/performance requirements, and presence of switching costs. Generally speaking, the faster a market is growing, the less power buyers have, and the more that competition is based on differentiation rather than price, the more attractive the segment is. And, the higher the product's complexity, the more important are user input, product quality, and performance, and the higher are switching costs, the more appropriate the segment is for relationship building.

Switching costs are a particularly powerful influence in that they make it costly or risky for a customer to switch to a competing vendor's products. Switching costs can arise from investments in equipment, procedures, or people that a customer has made. For example, to implement electronic data interchange between firms, the customer has to purchase dedicated equipment to link to the supplier's computer. As another example, switching online services means finding a new Internet service provider, loading new software, and creating a new email address. Switching costs can also arise from the risk of making a poor choice. If a purchase is important to operations, if the brand is not well known, or if the product is complex, the customer's switching costs will be perceived as higher. A relationship marketing strategy makes the most sense when customers face high switching costs. If a vendor were to follow a relationship marketing philosophy when customers face low switching costs, it may find itself using high-price marketing strategies and then losing customers to low-cost competitors.

Switching costs are also related to whether the customer's purchase is wrapped around a particular technology (say, a Unix operating system) or a particular vendor (say, an IBM e-business solution). In these situations, customers tend to be more long-term focused and experience greater switching costs. On the other hand, firms that focus on a particular product or on a salesperson may not be as committed to a certain company. If a customer can buy a similar product from another vendor or if the customer's sales contact leaves, that customer may move its account to another vendor.

Assessing *competitive position* requires that the seller develop a deep understanding of needs among buyers in the segment and conduct an honest assessment of whether the seller possesses the resources (tangible assets, intangible assets, competencies) to address those needs.

3. Awareness Generation and Positioning.   Once the firm has identified its target market(s), it must create awareness of the offering and clearly communicate its benefits to members of the target market and set customer expectations. Mass communications is the least expensive approach to building awareness when the firm is pursuing a broad acquisition strategy. Many Internet companies use mass media such as television, magazine, and billboards to create awareness and to log onto their websites. Targeted communications such as direct mail are more appropriate for a focused strategy.

Positioning is the process of creating a set of associations in the buyer's mind that create a distinctive and positive image of the product or the company. Through those associations, positioning sets customer expectations about product benefits. The firm must carefully balance the product's promise against its ability to deliver on the promise.

4. Acquisition Pricing.   The firm can use either a skimming or penetration pricing approach. Under the *skimming* approach, a relatively high initial price is set to attract the least price-sensitive customers first. The price can be lowered in stages to attract more and more price sensitive customers. Under the *penetration* approach, the initial price is set relatively low to capture a high share of market. One problem with the penetration approach is that by lowering the revenue from the customers' purchases, it essentially increases the net customer acquisition investment. Moreover, if the firm discovers that the price is too low, it is very difficult to raise prices. One way to manage this is to use rebates or special offers to attract first-time buyers.

5. Trial.   Product trial marks the point at which customers move from assessing alternatives to making purchases. This is a delicate time since customers have generally not committed to repurchasing the product but are actually evaluating product performance against expectations. If the product fails to meet expectations, then the customer will probably not repurchase and customer equity may be negative.

6. Post-purchase Service.   The initial purchase is only a start. Especially in the case of many complex high-tech products, the customer will need help to fully realize the benefits of the product and avoid significant post-purchase dissonance.

RFID: THE NEXT STEP IN INVENTORY CONTROL

Imagine strolling into Wal-Mart to buy the new DVD of *The Matrix*. As you take it off the shelf, a radio signal alerts an employee to restock, telling him where in the backroom to find *The Matrix* and giving a warning ping if he mistakenly slides it onto the *Legally Blonde* shelf. Meanwhile, forget going through the checkout line: An electronic reader scans the items in your cart and automatically charges your debit card.

Sound far-fetched? The future is closer than you think. Called radio frequency identification (RFID), the geeky-sounding technology will revolutionize both the way stores sell and the way consumers buy. Radio-frequency identification (RFID) tags are attached to each item that the retailer wants to track. A transmitter at the door "illuminates" each tag, whose flat antenna picks up the radio-frequency field. The energy from the door's transmitter then powers a tiny radio transmitter embedded in the tag. The tag's transmitter sends out an encoded stream of bits—in essence a radio bar code—for as long as it receives power. So not only can an item with an RFID tag tell the store's information system its color, its size, how long it had been on the shelves, and any other information the reseller or manufacturer chooses to put into the bitstream; it can also alert store managers that it hasn't been paid for.

Imagine a future in which every food item has an RFID, where self-inventorying retail shelves and warehouses will be commonplace. Imagine the convenience of bagging your groceries as you go up and down the aisles, then having the checkout device scan your entire shopping cart. You wouldn't have to handle each item three times before you've even left the supermarket. And if your refrigerator has a scanner, it can remind you to put an item on the grocery list by knowing how many times it goes out and back in before it's exhausted. Imagine that RFIDs are woven into your garments. Your washing machine could choose the best setting for the clothes you put in it. Or it could refuse to wash an item that requires dry cleaning.

Larry Kellam, director of supply-network innovation at P&G, notes that reducing out-of-stock products by 10 to 20 percent could boost its annual sales by anywhere from $400 million to $1.2 billion. But benefits of that magnitude will come only when companies start analyzing their data to reduce inventories and respond better to fluctuations in consumer demand.

SOURCES: Boyle, Matthew (2003), "Wal-Mart Keeps the Change," *Fortune*, 148 (10), pp. 23–26; Machrone, Bill (November 5, 2003), "RFID: Promise and Peril," *PC Magazine*.

Customer Relationship and Retention Strategies

Reinartz and Kumar[31] argue that different relationship strategies are appropriate for different market segments that the firm is serving. Rather than assuming long-term, loyal customers are profitable, their research shows that some loyal customers may be more costly to serve and may demand lower prices than customers who are not as loyal. At a minimum, the decision to invest in customer relationships must include an assessment of both customer profitability and the projected duration of the customer's relationship with the company. Reinartz and Kumar[32] identify four categories of customers and strategies for managing customers in those categories:

- *True friends,* the most valuable customer group, are highly profitable, loyal customers who find a good fit between the company's offerings and their needs. Relationship-oriented customers such as true friends seek to build strong and extensive social, economic, service, and technical ties with the intent of increasing mutual benefit.[33] True friends often have a philosophy of doing business that is similar to the seller's philosophy, have roughly the same power in the relationship as the seller, anticipate conflicts of interest and develop resolution processes, and pay substantial attention to measuring benefits and costs for both parties.[34]

The greatest risk in marketing to true friends is that of overkill, or intensifying the level of contact and potentially drowning them in marketing messages. Because they are already loyal, the overkill strategy can needlessly increase costs or even alienate customers without changing their purchasing habits. Rather, a company should attempt to stimulate these customers to engage in word-of-mouth communications with others and reward them for their loyalty by giving them preferential treatment. These are the customers that the company should strive to retain.

Retention is accomplished by keeping the relationship fresh. All relationships have the potential to go stale or sour. It is essential for both parties to be honest with each other. Although there may be a contractual obligation, a relationship is doomed if the parties feel that they can't trust one another. The larger or stronger partner should wield power cautiously. A relationship should be mutually beneficial and the exercise of power may produce an imbalance in benefits, thus jeopardizing the future of the relationship. The seller should regularly update the offering's value and continuously ask questions that encourage customers to relate problems and sources of dissatisfaction. Open and frequent communications can go a long way toward avoiding conflict.

- *Butterflies,* the second most valuable customer group, are short-term (transient) customers who also are highly profitable. Butterflies are shoppers; they constantly are searching for the best value at a particular point in time. The goal with butterflies is to capture as much of their business as possible in the short time they buy from a particular company. While it is tempting to try and convert butterflies into true friends, these attempts are rarely successful. A common mistake in marketing to this group is to continue to invest in retaining or reacquiring them after they have switched their business to a new company. Absent the ability to increase these customers' switching costs, the company should market intensively during the time the customer buys from them, by following up with marketing efforts after their most recent purchase. If these followup efforts are unsuccessful, stop marketing to these customers altogether.

- *Barnacles* are very loyal in that they desire long-term relationships. However, they are not very profitable either because the size and volume of their transactions are too low or the cost to serve them is too high. In this sense, these customers can be the most problematic: Like barnacles on the hull of a cargo ship, they create drag. Properly managed, however, they can sometimes become profitable. The selling company should attempt to determine whether it can capture a greater share of these customers' business. If not, it should attempt to increase the margins it earns from these customers either by increasing price to cover its costs or by imposing strict cost controls. The reality is that the seller will lose many barnacles, but if these are unprofitable customers, that is acceptable.

- *Strangers* are transaction-oriented customers who tend to focus on price rather than value. They focus on the timely exchange of products for highly competitive prices. Each transaction is treated as a discrete event. These transactions are typically characterized by limited communications between the buyer and seller. Because of this attitude toward exchange, they will have the lowest profit potential. The company should identify these customers early on and not invest in marketing to them. Just as these customers treat every interaction as a transaction that is to be exploited for maximum benefit, the seller must profit on every transaction.

A company that has learned how to segment its customer base for relationship-specific efforts is Tech Data Corp, a computer distributor. During the industry downturn in

2001, it was able to remain profitable through its efforts to apply sophisticated relationship-management techniques. Tech Data's customers include corporations, smaller distributors, and computer retailers. For each customer, Tech Data computes 150 customer service costs, ranging from average order size to freight charges. Based on this data, it calculates gross expenses and margin on every account. CEO Steven Raymund was able to persuade Tech Data's barnacles to order more efficiently, for example, by placing one order of $1 million rather than 10 orders of $100,000 each, or their business was terminated! In addition, it was willing to link its true friends directly to its inventory system as a way to bond relationship-oriented customers through high-value services.[35]

Customer relationship management has important implications for distribution, pricing, and advertising and promotions strategies, which are discussed in Chapters 8, 9, and 10.

SUMMARY

This chapter has explored the role of partnerships and alliances in high-tech markets. Because of the dynamic nature of the high-tech marketplace, most firms find that they cannot go it alone and must partner in some capacity to be effective in their marketing efforts. Understanding the range of partnerships available, the purposes each serves, and the risk factors as well as the success factors is necessary to managing partnerships effectively. Moreover, the focus on customer relationships is consistent with the need to have a customer or market orientation, the focus of the next chapter.

DISCUSSION QUESTIONS

1. What various types of partnerships may a firm form? Provide a current example of each. Draw a supply chain to demonstrate the nature of the relationship.
2. What is the National Cooperative Research Act and its amendment?
3. What are the various reasons for a firm to form partnerships? How do these change over the course of the product life cycle?
4. What are the risks of partnering arrangements? How can each of these risks be mitigated?
5. What factors are associated with the success of strategic alliances?
6. What is customer relationship marketing? How does a firm assess the financial ramifications of customer acquisition and retention (i.e., measure customer equity)?
7. What are the two types of acquisition strategies a firm could pursue? What are the six key steps in the customer acquisition process?
8. Enumerate and describe the various types of switching costs.
9. In terms of customer retention strategies, what are the four categories of customers (based on loyalty and profitability), and what are the marketing implications for each?
10. "Thought" question: How do the unique characteristics of the high-tech environment affect the desirability of and techniques used to build customer relationships?

GLOSSARY

Competitive collaboration, co-opetition. When firms who typically compete in some segment in the marketplace join forces to collaborate on some aspect of their business, say new-product development. These firms typically compete at the same level of the supply chain.

Complementary alliances.   When firms who make products that are jointly used team together in some aspect of their business, say new-product development. The members are referred to as *complementors.*

Horizontal alliances/partnerships. A relationship between firms at the same level of the value chain; may be competitors or firms that provide complementary products.

National Cooperative Production Amendment (1993). Act to allow joint production of products. The amendment excludes marketing and distribution agreements, except for the products manufactured by the joint venture, and the use of existing facilities by the joint venture.

National Cooperative Research Act (1984). Act passed by the U.S. government to promote joint research and development, encourage innovation, and stimulate trade.

The act applies to R&D activities up to and including the testing of prototypes.

Relationship marketing. The forming of close, long-term relationships with business partners to achieve win–win business solutions.

Switching costs. Costs incurred by a customer that make it costly or risky to switch to a competing vendor's products. Switching costs can arise from monetary (tangible) investments in equipment, procedures, or people that a customer has made, or from intangible feelings of risk or exposure.

Vertical partnership. When firms who sell products to another firm at a different level of the value chain team up to collaborate on some aspect of their business strategy.

ENDNOTES

1. Russell, Joy (2002), "IBM's Big Partner Agenda," *VARBusiness,* January 21, p. 19.

2. Ball, Jeffrey (1999), "To Define Future Car, GM, Toyota Say Bigger Is Better," *Wall Street Journal,* April 20, p. B4.

3. Langreth, Robert (1999), "DNA Dreams: Big Drug Firms Discuss Linking Up to Pursue Disease-Causing Genes, *Wall Street Journal,* March 4.

4. Morgan, Robert and Shelby D. Hunt (1994), "The Commitment–Trust Theory of Relationship Marketing," *Journal of Marketing,* 58 (July), pp. 20–38.

5. Brandenburger, Adam and Barry Nalebuff (1996), *Co-opetition,* New York: Currency/ Doubleday.

6. Ibid.

7. Browning, Larry D., Janice M. Beyer, and Judy C. Shetler (1995), "Building Cooperation in a Competitive Industry: SEMATECH and the Semiconductor Industry" (Special Research Forum: Intra- and Interorganizational Cooperation), *Academy of Management Journal,* 38 (February), pp. 113–139.

8. Kirkpatrick, David (1998), "The E-Ware War," *Fortune,* December 7, p. 102.

9. Ibid.

10. Utterback, James M. (1994), *Mastering the Dynamics of Innovation,* Boston: Harvard

Business School Press; Roberts, Edward B. and Wenyun Kathy Liu (2001), "Ally or Acquire? How Technology Leaders Decide," *Sloan Management Review,* 43 (1), pp. 26–34.

11. Moore, Geoffrey A. (1999), *Inside the Tornado,* New York: HarperBusiness.

12. Pringle, David (2002), "Nokia Signs Accord With Samsung," *Wall Street Journal,* September 3, p. B6.

13. Moore, Geoffrey A. (1999), op. cit.

14. Available at www-3.ibm.com/chips/news/ 2000/0127_infineon_umc.html.

15. Mohr, Jakki and Robert Spekman (1994), "Characteristics of Partnership Success: Partnership Attributes, Communication Behavior, and Conflict Resolution Techniques," *Strategic Management Journal,* 15 (February), pp. 135–152.

16. MacDonald, Elizabeth and Joann Lublin (1998), "In the Debris of a Failed Merger: Trade Secrets," *Wall Street Journal,* March 10, p. B1.

17. Hamel, Gary (1991), "Competition for Competence and Inter-Partner Learning with International Strategic Alliances," *Strategic Management Journal,* 12, pp. 83–103 (quote p. 87).

18. Littler, Dale, Fiona Leverick, and Margaret Bruce (1995), "Factors Affecting the Process of Collaborative Product Development,"

Journal of Product Innovation Management, 12, pp. 16–32.

19. Mohr, Jakki, Gregory T. Gundlach, and Robert Spekman (1994), "Legal Ramifications of Strategic Alliances," *Marketing Management,* 3 (2), pp. 38–46; Gundlach, Greg and Jakki Mohr (1992), "Collaborative Relationships: Legal Limits and Antitrust Considerations," *Journal of Public Policy and Marketing,* 11 (November), pp. 101–114.

20. Brady, Diane (1999), "When Is Cozy Too Cozy?" Business Week, October 25, pp. 127–130; www.ftc.gov/bc/guidelin.htm.

21. Mohr, Jakki and Robert Spekman (1994), op. cit.; Mohr, Jakki and Robert Spekman (1996), "Perfecting Partnerships," *Marketing Management,* 4 (Winter–Spring), pp. 34–43.

22. Bucklin, Louis P. and Sanjit Sengupta (1993), "Organizing Successful Co-Marketing Alliances," *Journal of Marketing,* 57 (April), pp. 32–46.

23. Dutta, Shantanu and Allen M. Weiss (1997), "The Relationship Between a Firm's Level of Technological Innovativeness and Its Pattern of Partnership Agreements," *Management Science,* 43 (March), pp. 343–356.

24. Kumar, Nirmalya, Jonathon Hibbard, and Louis Stern (1994), "The Nature and Consequences of Marketing Channel Intermediary Commitment," working paper #94–115, Cambridge, MA: Marketing Science Institute.

25. Kirkpatrick, David (1998), op. cit.

26. Takahashi, Dean (1998), "Microsoft–Intel Relationship Has Become Contentious," *Wall Street Journal,* September 25, pp. B1, B5.

27. Selden, Larry and Geoffrey Colvin (2002), "Will This Customer Sink Your Stock?" *Fortune,* September 30. Available at www.fortune.com.

28. Blattberg, Robert C., Gary Getz, and Jacquelyn S. Thomas (2001), "Managing Customer Acquisition," *Direct Marketing,* 64 (6), pp. 41–55.

29. Solomon, Michael and Elnora Stuart (2002), *Marketing: Real People, Real Choices,* 3rd ed., Upper Saddle River, NJ: Prentice Hall.

30. Anderson, James C. and James A. Narus (1991), "Partnering as a Focused Market Strategy," *California Management Review,* 33 (3), pp. 95–113; Blattberg et al. (2001), op. cit.

31. Reinartz, Werner and V. Kumar (2002), "The Mismanagement of Customer Loyalty," *Harvard Business Review,* 80 (July), pp. 86–95.

32. Ibid.

33. Anderson, James C. and James A. Narus (1991), op. cit.

34. Dwyer, F. Robert, Paul H. Schurr, and Sejo Oh (1987) "Developing Buyer-Seller Relationships," *Journal of Marketing,* 51 (April), pp. 11–27.

35. Available at www.techdata.com.

APPENDIX A

Learning from Partners in Collaborative Relationships[*]

Knowledge can be categorized in terms of its nature or properties. A common typology

---

[*]The information in this appendix is drawn from: Mohr, Jakki and Sanjit Sengupta (2002), "Managing the Paradox of Inter-firm Learning: The Role of Governance Mechanisms," *Journal of Business and Industrial Marketing,* 17 (special issue), pp. 282–301.

categorizes knowledge as either explicit or tacit.[1] *Explicit,* or *migratory, knowledge* can be written down, encoded, and explained. Examples include blueprints, technical specifications, product designs, steps in the manufacturing process, and so forth. Importantly, explicit knowledge is not veiled in organizational routines, practices, and culture. Because such knowledge is transparent—anyone with a comparable knowledge or skills base can understand and decipher it—some firms turn to patent protection and other forms of intellectual property rights (discussed in Chapter 7) for ownership.

On the other hand, *tacit knowledge* is unwritten know-how and "know-why" that is embedded in the organization's skills and routines. Tacit knowledge is less transparent than is explicit knowledge and has a "sticky" quality to it, making it difficult to learn and absorb. It can include a way of approaching and solving problems, imagination, continuous improvement techniques, or artisan-like skills.[2] The *embeddedness* of tacit know-how may arise from *causal ambiguity,*[3] in which the relationship between a firm's knowledge/ actions and outcomes is highly uncertain.

For example, product/market knowledge tends to be explicit and more easily transferred than are skills in technology development, manufacturing, or continuous quality improvement techniques.[4] On the other hand, tacit knowledge forms the basis of core skills and competencies, which (1) are harder to share and imitate than is explicit knowledge due to their deeply embedded nature, but (2) present the greater value.

Learning from partners in strategic alliances has both an upside and a downside. On the one hand, interfirm learning can lead to a win–win situation, in which both parties improve their skills and performance in the marketplace. Further, learning about each other can make the partnership more successful and contribute to positive relationship dynamics. On the other hand, interfirm learning can involve the "deskilling" of a partner, whereby one party absorbs the proprietary information of its partner, ultimately leading to a situation in which the deskilled partner is no longer needed.

These two views of interfirm learning highlight its paradoxical nature. In painting the rosy picture of interfirm learning, cooperative, harmonious ties are viewed as indicators of an effective relationship; the focus is on the value that interfirm learning can offer to the learning organization. The risks involved in the sharing of tacit, embedded information that the partner is attempting to acquire or internalize by having a tightly knit partnership agreement are minimized or overlooked. On the other hand, the risky view of interfirm learning explicitly alerts managers to the potential seepage of information in tightly knit relationships where the partner intends to acquire the other's knowledge. Cooperative, harmonious relational ties are potentially problematic and signal that one partner may be naïve to the other's intent. Rather, less relational characteristics may be desirable. One implication of these risks, which runs directly counter to the implications of the more favorable view of interfirm learning, is that firms must not partner too closely and that they must guard against interpersonal ties that are too collegial.[5]

These two views of interfirm learning are not mutually exclusive, but rather, are complementary sides to the same coin. Hence, the real challenge is to manage the paradox through the use of tools that allow firms to maximize the upside potential of interfirm learning while limiting its downside risks. At the heart of the issue is the reality that two different parties come to the relationship with potentially conflicting goals and objectives. One party may be content with mere access to a partner's knowledge and skills, whereas the other desires acquisition and internalization of the partner's knowledge. Such a situation makes the less aggressive party vulnerable to the risk of the potential opportunism of its partner. In such a situation, how can the more vulnerable party protect itself? One key is the nature of the governance mechanisms, or the tools used to structure and manage the interfirm relationship.

To the extent that the most valuable knowledge is tacit—which is also the most difficult to transfer—firms have an incentive to structure the closest type of partnership agreement possible. Indeed, the only way to learn skills and competencies that are highly embedded in organizational routines is to partner closely.[6] Firms seeking to learn and internalize tacit knowledge must absorb it through an apprenticeship model of learning, which often entails collaborative agreements such as joint ventures and R&D consortia. Close collaborative agreements allow organizational routines to be examined and understood, in terms of what is done, why it is done, and how it is done. Because more tightly linked relationships face greater risks from interfirm learning, the higher the firm's technological innovativeness, the less likely it is to use more transparent types of partnering arrangements such as joint ventures and research

and development agreements, compared to marketing and licensing agreements.[7]

Close interpersonal ties between the parties may be especially risky. The partner with the learning intent would potentially take advantage of the close interpersonal relationships, using them to gain access to crucial information that the more vulnerable partner might share unintentionally or unconsciously. Indeed, studies of leakages of proprietary information find that the most common ones occur through such unintentional sharing.[8]

In any case, an explicit tradeoff must be made between the need to restrict information sharing and the risks of not doing so. Although ideally there would be a combination of both looser, more collaborative mechanisms as well as more rigid procedures, in reality this may be quite difficult for operating personnel to implement.[9] Ultimately, the people involved will determine the success or failure of a plan to limit knowledge transfer, and hence, a governance strategy must be clear to them.

NOTES

1. Badaracco, J. L., Jr. (1991), *The Knowledge Link,* Boston: Harvard Business School Press.
2. Lei, David T. (1997), "Competence-Building, Technology Fusion and Competitive Advantage: The Key Roles of Organizational Learning and Strategic Alliances,"*International Journal of Technology Management,* 14 (2, 3, 4), pp. 208–237.
3. Reed, R. and R. De Fillippi (1990), "Causal Ambiguity, Barrier to Imitation, and Sustainable Competitive Advantage," *Academy of Management Review,* 15, pp. 88–102.
4. Hamel, Gary (1991), "Competition for Competence and Inter-Partner Learning with International Strategic Alliances," *Strategic Management Journal,* 12, pp. 83–103; Inkpen, Andrew C. and Paul W. Beamish (1997), "Knowledge, Bargaining Power, and the Instability of International Joint Ventures," *Academy of Management Review,* 22 (January), pp. 177–202.
5. Hamel, Gary, Yves Doz, and C. K. Prahalad (1989), "Collaborate with Your Competitors —And Win," *Harvard Business Review,* 67 (January–February), pp. 133–139; Dutta, Shantanu and Allen M. Weiss (1997), "The Relationship Between a Firm's Level of

Technological Innovativeness and Its Pattern of Partnership Agreements," *Management Science,* 43 (March), pp. 343–356.

6. Badaracco, J. L., Jr. (1991), op. cit.

7. Dutta, Shantanu and Allen M. Weiss (1997), op. cit.

8. Mohr, Jakki (1996), "The Management and Control of Information in High-Technology Firms," *Journal of High-Technology Management Research,* 7 (Fall), pp. 245–268.

9. Readers who would like more detail on the exact types of governance strategies recommended to navigate this tension can refer to Mohr, Jakki and Sanjit Sengupta (2002), "Managing the Paradox of Inter-firm Learning: The Role of Governance Mechanisms," *Journal of Business and Industrial Marketing,* 17 (special issue), pp. 282–301.

CHAPTER 4

Market Orientation and R&D—Marketing Interaction in High-Technology Firms

A man is flying a hot air balloon and realizes he is lost. He reduces altitude and spots a woman down below. He lowers the balloon further and shouts,

"Excuse me. Can you help me? I promised a friend I would meet him half an hour ago, but I do not know where I am."

The woman below says, "Yes, you are in a hot air balloon, hovering approximately 30 feet above this field. You are between 40 and 42 degrees north latitude and between 58 and 60 degrees west longitude."

"You must be an engineer," calls down the balloonist.

"Yes, I am," replies the woman. "How did you know?"

"Well," says the balloonist, "everything you have told me is technically correct, but I have no idea what to make of your information, and the fact is, I am still lost."

The woman below says, "You must be a manager."

"Yes, I am," replies the balloonist, "but how did you know?"

"Well," comes the answer from the engineer, "you did not know where you are nor where you are going. You have made a promise that you have no idea how to keep, and you expect me to solve your problem. The fact is, you are in exactly the same position you were in before we met, BUT NOW IT IS SOMEHOW MY FAULT."
—FROM THE INTERNET

Market Orientation at General Electric[1]

On a visit to Bentonville, Arkansas, Jack Welch, then CEO of General Electric, was blown away by Wal-Mart's Saturday-morning meetings where two rivers of customer information came together: up-to-the-minute sales data and up-to-their-elbows reports from managers who had spent all week on the road visiting stores and soaking up impressions. GE transformed that into a technique it calls quick market intelligence (QMI)—insisting that all top managers regularly call on at least one customer, then holding regular meetings devoted to nothing but discussing what customers say and do. In typical GE style, the same setup is replicated in many places and at different scales. Cross-functional new-product development teams, for example, visit customers to discuss and test-drive ideas and feed what they learn both to R&D and to account managers. The result: increased two-way talk with customers.

As Welch explained, "We must concede no markets—and no customers—because our competitors do not. I know with turbines and jet engines and CT medical scanners, that there is a value nub, an intersection, where low cost and just-the-right features intersect. That value nub, when hit, causes products to fly off the shelves and out of the showrooms. The consuming passion of each of the divisions in our company must be to become so fast and so lean and so close to that customer that the value nub is always in our sights." To achieve that success, GE strives to be "boundaryless." According to Welch, "boundaryless behavior assumes that there are no customers in the world who don't have something valuable to share with you, so why not hand them a coffee mug and bring them into the room when you sit down to design a new product?" It "laughs at the concept of little kingdoms called finance, engineering, manufacturing, and marketing sending each other specs and memos, and instead gets them all together in a room to wrestle with issues as a team." The creation of superior customer value through cross-functional collaboration is the essence of market orientation.

Probably one of the most daunting challenges in many high-tech firms is to bring a customer-oriented philosophy to the firm's operations. This challenge is particularly daunting because its necessity is significantly less obvious to a firm than is the need for funding and additional resources, for example. High-tech startups usually begin with a great technological idea that offers improvements over existing ways of doing things. Whether this great technological idea is one that customers will actually embrace is a completely different issue. Think of the numerous examples of superior technology that never became the *de facto* standard in a marketplace or lost their position to a better-known rival, for example, Beta-format video cassette recorders or Netscape browsers.

The need for firms in high-tech industries to understand customers, to be customer-oriented, and to be market-driven is vital. No information "is more important to a technology-based firm than information flowing in from the market, as this information shapes science into commercial products or services."[2] The ability of the market-oriented business to facilitate the development of innovative products is supported by Atahuene-Gima's[3] finding that "market orientation is more strongly related to new product performance at the early stage of the product life cycle than at the late stage.... Such an environment seems to warrant greater market intelligence and information sharing within the firm." There is substantial evidence of a strong relationship between market orientation and innovative capability and between market orientation and new product success.[4]

However, high-tech firms often lack an appreciation for the value of developing an in-depth understanding of customers and their needs and, indeed, may even be suspicious of marketers and marketing tactics. Yet, the need to have a strong marketing capability is actually most important for firms that are good technologically.[5] Marketing capability has a disproportionate positive effect on R&D productivity for firms with strong technical skills. The greater are the technological strengths, the greater is the impact of strong marketing capability on R&D productivity. To be sure, being technology- or customer-oriented is not an either/or situation; a firm can benefit from being both technology- and customer-oriented in uncertain markets. Research has compared the impact of a technology orientation versus a customer orientation on performance. In highly uncertain markets, a customer orientation (a set of beliefs that puts satisfaction of customers' needs first) has a positive influence on the commercial performance of an innovation.[6]

Although the smaller firm typically lacks marketing specialists and, indeed, may even be suspicious of marketing tactics, the need to be market driven is vital *for small and large firms alike.* A large body of research and literature on developing a market orientation applies to firms in many industries, not solely those that are high tech. However, given the market and technological uncertainty present in high-tech markets, the need for a market orientation may be even stronger than in more traditional contexts.

This chapter covers of the basic underpinnings of what it means to be market oriented, along with the vital importance of collaborative interactions between marketing and R&D personnel.

WHAT IT MEANS TO BE MARKET ORIENTED[7]

It is important to note that a market orientation is *not* the same thing as a marketing orientation. A marketing orientation might imply that marketers have disproportionate influence or that marketing activities are the source of the firm's competitive advantage. On the other hand, in market-oriented firms, there is no consistently dominant function or coalition of individuals. In fact, any group may take the lead as long as its members are committed to the continuous creation of superior customer value. And, while marketing activities may be the source of competitive advantage, competitive advantage is just as likely to derive from market-focused skills in R&D or in product development. A firm cannot be fully market oriented if the entire organization is not committed to creating customer value.

As shown in Figure 4-1, a firm that is market oriented emphasizes the *gathering, dissemination,* and *utilization* of market intelligence as the basis for decision making.[8] Customer-oriented marketing activities are critical to gathering information to reduce overwhelming uncertainty over demand.

First, market-oriented firms *gather a wide array of information* from the market. **Market intelligence** includes information about current and future customer needs, as well as competitive information and trends in the marketplace. The acquisition of information can be done via customer hot lines, trade shows, customer visits, working with lead users, competitive intelligence, or some of the more high-tech-oriented research tools discussed in Chapter 5.

Second, the market-oriented firm *disseminates the information* throughout the company. Effective dissemination increases the value of information when each piece of information can be seen in its broader context by all organizational players who might be affected by or utilize it. People in the organization must be able to ask questions and augment or modify the information to provide new insights to the sender. For example, to drive new products from concept to launch more rapidly and with fewer mistakes, "all functional interfaces (contacts) are important in the product development process."[9] Effective interfacing is accomplished through greater emphasis on

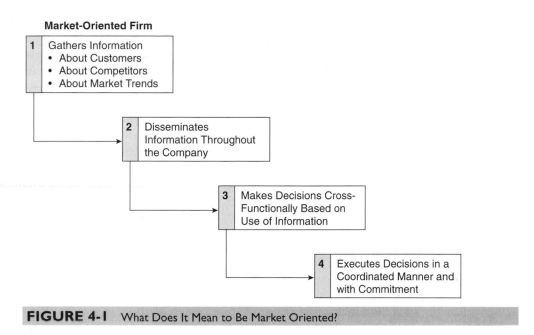

FIGURE 4-1   What Does It Mean to Be Market Oriented?

"multifunctional activities (activities that are the joint responsibility of multiple functions in the business) . . ., multifunctional discussions, and information exchange."[10]

When organizations remove the functional barriers that impede the flow of information from development to manufacturing to sales and marketing, they improve the organization's ability to make rapid decisions and to execute them effectively. For example, Andersen Consulting uses an internal electronic bulletin board so that its consultants can describe problems they have encountered that others may have already seen and dealt with. The use of a market information database can be useful in making the information available in a centrally located manner. People across functions and divisions in the organization should share information with each other and have a dialogue about it. The wise use of information technology can effectively facilitate this process.

Another common approach to encourage information sharing in the development process is to send people from multiple functions on customer visits. Not only does this stimulate real-time information sharing, it generally increases the quality of the information gathered.[11]

Third, the market-oriented firm *uses the information* to make decisions. To ensure that all information is considered before a decision is made, organizations must provide forums for information exchange and discussion. When decisions are made interfunctionally and interdivisionally, greater representation of the information and a closer connection to the market issues will occur. Moreover, interfunctional decision making implies that the people who will be involved in *implementing* the decisions are the ones actually involved in *making* the decisions—the idea being that if one is involved in making the decision, he or she will be more committed to implementing that decision.

Finally, the market-oriented firm *executes the decisions* in a coordinated manner. Commitment to execution is necessary to successful implementation of a market orientation. An organization can generate and disseminate intelligence; however, unless it acts on that intelligence, nothing will be accomplished. Responsiveness to market intelligence involves selection of target markets; development of products/services that address their current and anticipated needs; and production, distribution, and promotion of the products in a way that produces both customer satisfaction and customer loyalty.[12] All functions in a market-oriented company—not just marketing—participate in responding to market needs.

These, then, are the characteristics of firms that value and rely on market information to guide strategic decision making. Distilled to its essence, a **customer** or **market orientation** simply means gathering, sharing, and using information about "the market" (customers, competitors, collaborators, etc.) to make decisions. This philosophy of approaching decision making requires effective management of the *knowledge* that resides in different places within the firm. An even broader concept than market orientation is knowledge management, since it subsumes all forms of knowledge that may be valuable to the firm. **Knowledge management** is the process of creating, transferring, assembling, integrating, and exploiting knowledge assets.[13] It requires the proactive management of the firm's bases of knowledge and the effective use of that knowledge to produce good decisions. Moreover, it requires the tearing down of walls and barriers between departments, functions, and individuals, both inside and outside the company, in order to better share and use information.[14]

Although the sharing of knowledge makes intuitive sense, it really represents a paradigm shift in the sources of competitive advantage within the firm. In the past, labor and capital were the primary determinants of firms' profits, but increasingly know-how is the profit engine. Yet, competitive advantage that resides in know-how is only as strong as the ability to share and use that knowledge across the organization's boundaries. Indeed, some say that effective knowledge management requires a boundaryless learning organization, which takes good ideas from disparate functions and uses them in many areas, as Jack Welch of General Electric noted in the opening vignette. However, freely sharing, and using, information is easier said than done.

Becoming Market Oriented

The logic behind the value of a market orientation is clear, and the evidence of its value is compelling. Why, then, aren't more firms able to become market oriented? It simply boils down to the fact that many firms do not have the facilitating conditions in place. These include, at a minimum: (1) top management advocacy, (2) a flexible, decentralized structure, and (3) a market-based reward system.

If a firm's *top managers* are not unequivocally and visibly committed to its customers, the firm will not bring its resources to bear on developing solutions to its customers' needs. As Michael Dell, CEO of Dell Computers, says,[15] "We have a relentless focus on our customers. There are no superfluous activities here. Once we learn directly from our customers what they need, we work closely with partners and vendors to build and ship relevant technologies at a low cost. Our employees feel a sense of ownership when they work directly with customers."

Market-oriented behaviors thrive in an organization that is *decentralized,* with fluid job responsibilities and extensive lateral communication processes. Members of these organizations recognize their interdependence and are willing to cooperate and share market intelligence to sustain the effectiveness of the organization. The necessity of effective information sharing in the market-oriented organization demands that bureaucratic constraints on behavior and information flow be dismantled. The high uncertainty in high-tech markets requires high frequency and informality in communication patterns among organizational units for effective intelligence dissemination.

A *market-based reward system* is the organizational factor that has the greatest impact on a market orientation.[16] A market-based reward system is one that rewards employees for generating and sharing market intelligence and for achieving high levels of customer satisfaction and customer loyalty. Market-oriented firms seem to place less emphasis on short-term sales and profit goals than do their more financially- or internally-focused competitors.

The Technology Expert's View from the Trenches in this section comes from a senior executive who understands what it takes to make the transition from being engineering-driven to customer-oriented.

TECHNOLOGY EXPERT'S VIEW FROM THE TRENCHES

WHAT DOES IT TAKE TO BECOME CUSTOMER FOCUSED?
JACK TRAUTMAN, SENIOR VICE PRESIDENT AND GENERAL MANAGER,

Automated Test Group
Agilent Technologies
Loveland, Colorado

Jack Trautman

The challenge in any high-tech organization is creating a customer-centered culture. On the surface, it sounds easy—listen to the customer and invent what the customer wants. In practice, it's a complex task to choose a target customer, accurately anticipate the target's exact needs, translate those needs into product features, and generate not only the right product, but also the total customer experience that captures the customer's money today and their loyalty tomorrow—and do it better than your competitors.

For a lot of high-tech startups, the founders are the customers. Their bright new ideas are spawned from their own user experiences. Through their engineering expertise, they solve their own problems with the products they invent, and in so doing, solve the same problem for all of their customers.

Such was the case in the early days of the Hewlett-Packard Company. Engineers developed test and measurement equipment for engineers. When they wanted to know what the customer wanted, they asked their fellow engineers. This "Next Bench Syndrome" served HP well for many years.

But then the business got more complex. HP diversified into computers, printers, and a host of other product types, and even spun off Agilent Technologies to give the test and measurement parts of its business more focus. Suddenly the customer was not sitting at the next bench. In the absence of real customers, engineers will still do their best to anticipate needs. But their instincts are more likely grounded in technical prowess than customer requirements. "I'll give them the most technically elegant solution, and surely they'll love it!" Maybe.

Enter marketing. Through market research, customer segmentation, competitive analysis, focus groups, a customer relationship management process, and a host of other techniques, a surrogate for the customer can be constructed. These are well-understood processes that can yield great results.

But an effective product marketing team is not enough to ensure success. It takes teamwork between marketing and engineering to truly bring the customer into the center of the product generation process. This teamwork comes through mutual respect, working together side by side, and an open sharing of ideas. Enlightened companies are co-locating product marketing and engineering teams beside each other to maximize interaction and a free-flowing exchange of ideas. Insights into ways to best satisfy customers will not occur in a "throw it over the wall" environment between marketing and engineering.

Finally, and most importantly, it takes management leadership to commit the entire organization to a customer-centered culture. Managers must budget the time and money to get product marketing and product development engineers out with customers so they can experience the customer's environment firsthand. They must identify early adopter customers who help complete the final product feature set through hands-on testing of early prototypes. Capturing the customer's insights early in the prototyping phase can often make the difference between the ultimate success or failure of a product or service. It is vital to involve the customer in each of the key phases of product generation: design, develop, test, and launch.

Beyond marketing and engineering, every department and every person must embrace the total customer experience. Managers must be willing to invest in all employees to give them the tools and skills to get close to the customer. And then those managers must insist that every decision in the company be made with the customer in mind.

It's hard work to keep a focus on the customer. But the payoffs are big: increased customer satisfaction and loyalty, high employee motivation and morale, and increased shareholder value.

Barriers to Being Market Oriented

It is all too easy in an organization to give mere lip service to being market oriented; organizational barriers can prevent the implementation of a market orientation. First, people often hoard information to *protect their turf*, having the sense that access to and control of information provide power and status within the organization. As a result, it logically follows that a very difficult part of knowledge management and being market oriented is to convince people to share information rather than to hoard knowledge to protect their standing in the organization.

Second, *core rigidities* may prevent market information from being utilized in decisions. As discussed in Chapter 2, core rigidities are straitjackets that inhibit a firm from being innovative and can include status hierarchies that give preference to, say, technology enthusiasts within the company over customer-focused organizational members. This preference often results in the disregard of information about users—unless it comes from someone with status in the organization.

For example, during the design of the Deskjet printer at Hewlett-Packard, marketers tested early prototypes in shopping malls to determine user response. They returned from their studies with a list of twenty-one changes they believed essential to the success of the product; however, the engineers accepted only five. Unwilling to give up, the marketers persuaded the engineers to join them in the mall tests. After hearing the same feedback from the lips of users—feedback that they had previously rejected—the product designers returned to their benches and incorporated the other sixteen requested changes.[17] Unfortunately, because marketing personnel are often not a part of the distinctive competencies of technology-oriented firms, the information they bring to the design process goes unheeded—which can have very negative consequences.

Other barriers that can prevent a firm from being market oriented are the *tyranny of the current* ("served") *market* (discussed previously in Chapter 2) and users' inability to envision new solutions that technology may have to offer.[18] These firms have **marketing myopia**, a tendency to focus very specifically on solving customers' needs with a current technology. Such a myopic focus obscures the possibility that customer needs may change over time and may be solved in radically different ways. Although a market-oriented culture necessitates that a firm actually solicit information from, and listen to, customers, this can be hard to do when customers may not be able to articulate their needs clearly.

Figure 4-2 below illustrates the relationships among the facilitators of a market orientation, barriers to becoming market oriented, and the results of a market orientation.

The Hidden Downside of a Market Orientation

Even if customers can articulate their needs, some argue that listening to customers may not always be the panacea that advocates of a market orientation indicate:

FIGURE 4-2   Influences on Achieving a Market Orientation

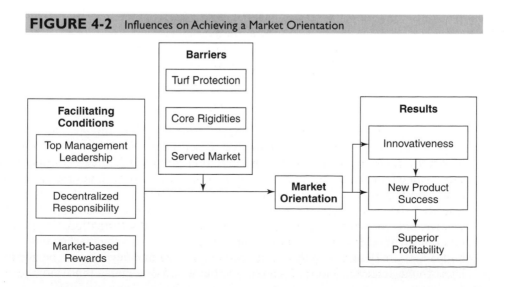

Listening to customers too carefully can inhibit innovation, constraining it to ideas that customers can envision and articulate—which may lead to safe, but bland, offerings. A *Fortune* magazine article suggests that the obsessive, even slavish, devotion to listening to customers has gone too far and that in some cases, it's better to ignore the customer.[19] Why would this be?

First, problems may arise in listening to customers if they actually give market researchers bad information. For example, during a marketing research project, customers may say they love a new-product idea but then not buy the product when it comes out on the market. Spalding tried to market "pump baseball gloves" that tested well but just didn't do well in the marketplace. So, in some cases, marketers may need to ignore customers' assertions of what they want.

Second, marketers may need to ignore feedback about what customers say they *don't* want. For example, some products that met with initial customer resistance included fax machines and overnight express delivery. In addition, people said they'd never give up their mainframe computers. A Motorola marketing executive has stated, "Our biggest competitor isn't IBM or Sony. It's the way in which people currently do things."[20]

Third, customers are not always able to articulate their needs. Customers have needs that they are not aware of. These are referred to as "latent" needs. They are real, but not yet in the customers' awareness. If these needs are not satisfied by a provider, there is no customer demand or response. They are not dissatisfied, because the need is unknown to them. If a provider understands such a need and fulfills it, the customer is rapidly delighted. Managers describe this kind of need fulfillment as having "attractive" quality. It delights and excites customers and inspires loyalty. A simple example is the 3M Post-It Notes, for which a need had long been present, but was not articulated until the product existed. It met a latent need, generated great enthusiasm, and became wildly successful.

So, in light of these concerns, a key question becomes this: How can a firm be market driven and yet not be overly constrained by what customers say (become a "feedback fanatic"[21]) or don't say?

Overcoming Possible Pitfalls in Being Market Oriented

One way to be market oriented without being customer-led is to give more emphasis to how customers behave and less emphasis to what they say.[22] The best information can be gleaned through observation of what customers do under normal, natural conditions. Known as empathic design, this type of research is covered thoroughly in Chapter 5.

Second, it is important to match when a firm chooses to rely heavily on customer feedback to the type of innovation the firm is getting feedback on,[23] or in other words, to use the contingency model of marketing presented in Chapter 1. In the context of *incremental innovation,* relying on customer feedback is important and useful for fine-tuning a product. Customers are adept at providing useful information that can reinforce or refine existing technology. However, in the context of *radical or breakthrough innovations,* customers are less adept at providing useful information. Customers have natural "myopia,"[24] which means that, in gathering data from the market and customers, informants are bounded by their context or environment. They may not be aware of the latest trends in usage, and users cannot see the world through an innovator's eyes; they cannot know what solutions the technology may have to offer. Indeed, in advocating the use of customer observation to gather new insights, some high-tech marketing researchers believe that what customers *can't* tell you might be just what is needed to develop successful new products.[25]

Moreover, with the "tyranny of the served market," firms that focus too narrowly on their established customers may be constrained in the strategies and technologies they choose to pursue. For example, a firm's best customers may be those who are last to embrace a disruptive technology. In such a case, the people who embrace a new technology first are the ones a company pays the least attention to, and the innovation creeps up like a stealth attack.[26] For this reason, successful companies adopt a "future market focus."[27] In addition to being focused on current customers, innovative firms focus on future customers as well, exhibiting bi-focal vision.

Note that customers who adopt a radical new technology are typically on the fringes of an established market or in an entirely new, emerging market. For example, in the early 1980s, the 5.25-inch disk drive technology was out of step with mainframe and minicomputer customer demands but was embraced by the emerging desktop personal computer marketplace.[28] Established disk drive firms did not fail because they were unable to develop innovative technologies. Rather, because established customers were uninterested in new technologies that didn't address their immediate needs, industry leaders did not allocate resources to the new technologies. This decision allowed new entrants to gain leadership in the new market.

Ignoring the direct customer input may be difficult when market researchers and company conformists use reams of consumer research to stifle creative ideas. But a person who believes fervently in the value of an idea—say, a product champion—can overcome a sea of skeptics. To be a visionary requires an informed view of the market and faith in one's ideas and the resulting products, as well as a realistic time frame for perseverance. One shouldn't persevere so long that the signs of imminent failure are totally ignored. On the other hand, one wants to persevere long enough to give the invention a fair chance. A Compaq executive suggests that twelve to eighteen months is a good time frame "to get a good read on whether what you're hearing is surmountable skepticism or a downright lack of market acceptance."[29]

For a firm to be truly market driven, these characteristics must be infused throughout the company, starting from the top down. If anything, these characteristics idealize a culture or philosophy that places very high value on market information. The acquisition, sharing, and dissemination of information require trust in coworkers and collaborative communication among them. In the high-tech arena, close collaboration between marketing personnel and R&D personnel is especially vital.

R&D–MARKETING INTERACTION

High-technology companies must effectively link research and development and marketing efforts in order to be successful.[30] Firms in high-technology markets must excel at three activities: opportunity identification, product and process innovation, and product commercialization. Because one of marketing's tasks is to listen to the customer and define a broad set of opportunities, a strong marketing capability implies that marketing is able to identify a wide range of markets and customer applications for the innovative technology. The voice that marketing brings to the innovation

---

THE BIONIC PERSON!

Researchers all over the world are looking into electroactive polymer (EAP) actuators, artificial muscles possibly 200 percent stronger than those of the average person. EAP actuators could be woven into exoskeletons, which you may be able to wire to your brain someday. For quadriplegics, this could allow them full use of their body again.

We're decades away from a working muscle suit, but before then we could see these actuators used in anti-frostbite socks and gloves, which would increase circulation in extremities. "In theory," says Gordon Giesbrecht, a physiologist at Canada's University of Manitoba, "if you can increase your blood flow, you'll bring oxygen to your muscles, which would bring heat to your extremities and enhance performance."

Until then, the first commercial application of EAP actuators has come from Japanese company Eamex, which developed a toy fish that uses them to swim. So keep training: It's going to be a long time before your Spider-Man suit comes off the rack.

SOURCE: Brian Alexander, October 2003, "Are You Ready? The New Power Suit," *Outside Online*.

---

process must be joined with the knowledge that R&D brings in order to develop an offering that effectively addresses customer needs.

R&D–marketing interaction, where personnel from marketing and R&D interact effectively, is most important during the early stages of a product development project.[31] Marketing should have the knowledge about customer preferences and competitive offerings that is crucial for resolving design and positioning issues. And, since R&D has primary responsibility for translating technology into a design that addresses customers' needs, its knowledge is crucial for resolving design issues also. As a project moves into the production and commercialization stages, product engineering and operations supercede R&D in importance, while the focus for marketing shifts from product definition to development of a marketing program.[32]

Similar to the hurdles in becoming market oriented, high-tech firms face barriers to R&D–marketing integration. For example, the following are jokes from the Internet made by engineers about marketing:

- Marketing Research: When marketing goes down to engineering to see what they're working on.
- Marketing: What you do when your products aren't selling.
- A software manager, a hardware manager, and a marketing manager are driving to a meeting when a tire blows. They get out of the car and look at the problem.

The software manager says, "I can't do anything about this—it's a hardware problem."

The hardware manager says, "Maybe if we turned the car off and on again, it would fix itself."

The marketing manager says, "Hey, 75 percent of it is working—let's ship it!"

- What high-tech salespeople say and what they mean by it:

All new: Parts not interchangeable with previous design.
Field-tested: Manufacturer lacks test equipment.
Revolutionary: It's different from our competitors'.
Breakthrough: We finally figured out a way to sell it.
Futuristic: No other reason why it looks the way it does.
Distinctive: A different shape and color than the others.
Redesigned: Previous faults corrected, we hope.
Customer service across the country: You can return it from most airports.
Unprecedented performance: Nothing we ever had before worked *this* way.

Because of its vital role in high-tech marketing success, this section provides supporting detail on R&D–marketing interactions. First, as shown in Figure 4-3, the nature of the cross-functional interaction must be effectively matched to the nature of the innovation (radical versus incremental). Second, firms must understand the barriers to interaction. Finally, strategies to overcome barriers to interaction and, more specifically, to enhance communication between marketing and engineering personnel are addressed.

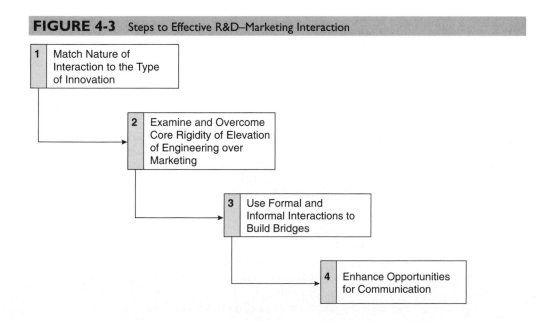

FIGURE 4-3   Steps to Effective R&D–Marketing Interaction

1 Match Nature of Interaction to the Type of Innovation

2 Examine and Overcome Core Rigidity of Elevation of Engineering over Marketing

3 Use Formal and Informal Interactions to Build Bridges

4 Enhance Opportunities for Communication

Nature of R&D–Marketing Interaction: Breakthrough versus Incremental Innovations

Although the initial impetus for success in many high-tech firms comes from engineering breakthroughs, a successful transition to being market driven requires that input from marketing be both heard and responded to. Managing the R&D–marketing interface is vital for the firm to succeed as the market evolves from a supply-side, innovation-driven market to a demand-side, market-driven market.[33] Although R&D tends to play a stronger, more influential role in breakthrough products and marketing tends to play a greater role in more incremental products, for either R&D or marketing to overlook the vital perspective that the other brings to the table is a contributing factor to failure. Although cross-functional interaction is critical for both breakthrough and incremental products, consistent with the contingency model of marketing decision making, the nature of the interaction must be tailored to the type of innovation.

Research findings demonstrate that greater R&D–marketing integration in technical activities is required for *breakthrough products* than for incremental products.[34] Because the possibilities for the application of the new technology tend to be either nonobvious or very numerous, it can be difficult for engineering to proceed in isolation from market-related feedback. Hence, for breakthrough products, much of the early interface efforts between R&D and marketing should address what industry the company should compete in, what the conceivable market opportunities are, and what the market development priorities are.[35] The cross-functional interaction is helpful in determining desired product features and assessing engineering feasibility.

For *incremental innovations,* R&D again should actively participate in the market planning process, especially in setting objectives. R&D can ensure that marketing does not lose sight of R&D's vision for the product. Marketers can offer parameters for the engineers' efforts. Through give and take, the team members can agree on the target market, priorities, expectations, and timing. Moreover, R&D efforts don't end once selling begins; engineers should continue to help with brochures, research, pricing, sales promotion, trade shows, and customer visits. Research findings validate the importance of achieving R&D–marketing interaction for incremental innovations in the new-product development launch to establish the direction for commercialization, designing marketing plans, and implementing the launch.[36] Similarly, marketing should participate during the precommercialization period, bringing the voice of customer and marketplace into the development process.

Barriers to R&D–Marketing Collaboration

Despite the crucial need for effective R&D–marketing collaboration in high-tech product development, managing the reality of the interaction is very different.

One important barrier to R&D–marketing collaboration can be the corporate culture of a high-tech firm that respects and values engineering knowledge more than marketing knowledge. This dominant engineering culture has been identified as a core rigidity in many high-tech firms.[37] High-level executives from firms with a dominant engineering culture typically come out of engineering; they are expected to develop a business orientation and understanding of customers as they advance. This type of technology-driven culture translates into a lack of regard and respect for marketing personnel. Obviously, it is very difficult for marketing personnel to be effective when the prevalent view is that "engineering does its thing and then marketing helps get it out the door."

The disdain for marketers also manifests itself in engineering taking on many tasks traditionally thought of as marketing, such as competitor analysis, product management, and so forth. The engineering culture further reinforces the dominant role of engineering and justifies not listening to marketing.[38] One of the most frequent reasons engineers give for not using market research or input from consumers is that "customers don't know what they want" or that "marketers don't know what they're talking about" (because they lack technical expertise). For example, "marketing wants everything right now at no cost—they have no concept of feasibility—they want a $5,000 Cadillac tomorrow."[39]

Marketing and R&D personnel tend to differ on a number of dimensions such as education, goals, needs, and motivation. More importantly, they tend to differ with regard to their values as illustrated in Table 4-1.[40] Of course, these differences are generalities and do not apply in all situations. However, they do illustrate why marketing and R&D may have a difficult time understanding each others' goals and decision processes.

Even spatial dynamics contribute to the problem: Locating R&D and marketing in different parts of the building further contributes to the lack of collaboration. Yet, studies have found that increased R&D–marketing interaction increases the likelihood of a new-product development project's success.[41] In light of these barriers, how can R&D–marketing interaction be structured to enhance effectiveness?

Achieving R&D–Marketing Integration

Many firms specify a number of formal systems and processes by which marketing groups provide information to engineering groups.[42] For example, during specific review phases of the new-product development project, marketing offers input into the product requirements document and an understanding of the tradeoffs involved in the myriad attributes being considered. In addition, during the annual planning process, marketing groups forecast revenue and profit for their market segments and indicate products and programs needed to achieve their goals. Finally, marketing communicates with engineering via the sales forecasting system.

Although these formal mechanisms exist for R&D–marketing interaction, they tend not to be the primary means by which such interaction occurs; indeed, they offer little in the way of marketing influence in the process. Such formal measures are often a chimera, erecting a façade of R&D–marketing interaction but doing little to make it productive. The question is, how can R&D–marketing interaction be made productive for the firm? Useful techniques fall into the categories of co-optation, cooperation, communication, and constructive conflict, as illustrated in Figure 4-4.

TABLE 4-1   R&D and Marketing Stereotypes

	R&D	*Marketing*
Time Orientation	Long	Short
Projects Preferred	Breakthrough	Incremental
Ambiguity Tolerance	Low	High
Department Structure	Informal	Moderately Formal
Bureaucratic Orientation	Less	More
Orientation to Others	Permissive	Permissive
Professional Loyalty	Profession	Firm
Professional Orientation	Science	Market

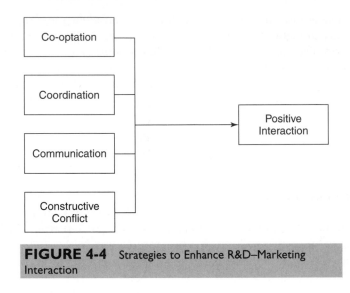

FIGURE 4-4 Strategies to Enhance R&D–Marketing Interaction

Co-optation

The strategy of co-optation has the objective of merging the interests of R&D and marketing. In his study of a high-tech firm, John Workman[43] found that effective marketers use the following methods to gain the support of R&D in the product development process. Effective marketers:

- Use informal networks and build bridges to engineering. They know the right people; they are in close physical and organizational proximity to engineering.
- Understand products and technology, which gives them credibility with engineers. Engineers "don't mind talking to marketers" if they know what they're talking about.
- Don't tell others what to do; rather, they ask questions, tell stories, and build consensus across groups.
- Form strategic coalitions that include high-level managers who push through changes that engineering resisted. However, there is a cost to this strategy, because it can alienate peers in engineering, so this option is the least preferred and should be saved for the most important issues.
- Recognize that it is the minor improvements to the new innovations that can be particularly important. "It's the 5 percent that's uninteresting to the engineering folks that can produce five times the levels of sales you would otherwise have."[44] In such a situation, effective marketers either undertake development themselves or turn to external partners to complete the work.

Cooperation

Based on their extensive review of the research in this area, Griffin and Hauser[45] found that the following cooperative strategies enhance marketing–R&D interaction.

- Co-locate marketing and R&D to overcome the barrier of physical separation and encourage information transfer. To realize the benefits of co-location, it must be complemented with techniques that foster communication and collaboration.

- Move personnel across functions to give them insight into the challenges their counterparts face. Personnel movement blurs distinctions between groups, may decrease market or technological uncertainty, and reduces barriers that stem from differences in values and language.
- Develop informal cross-functional networks to encourage open communication and provide contact across the functions within the team.
- Create a structure and systems that foster cooperation. This would include clear responsibilities and performance standards, decentralized decision making, tolerance of failure that is not the result of poor planning or execution, and joint reward systems.

Even with these strategies, the reality still remains that marketing and R&D have very different world views, which can result in misunderstanding and conflicts in goals and solutions.[46] So, what factors affect the ability of the two groups to interact effectively? Most experts point to the need for enhanced communication.

Communication

One commonly studied factor in enhancing R&D–marketing interaction is communication. Many argue that simply increasing the *frequency of interaction* between marketing and R&D will help improve the understanding and harmony between the two functions; increase their ability to cope with complex, dynamic environments; and lead to greater product success. Indeed, the use of market information provided by a marketing manager to nonmarketing managers (including not only R&D managers but also manufacturing and finance managers) requires a minimum threshold of interactions (approximately 125 interactions in a three-month period).[47] However, too frequent communications can hurt perceptions of information quality (beyond 525 interactions in a three-month period). Hence, increasing frequency of cross-functional communication may be warranted if the minimum threshold has not been reached, but increasing frequency may not always improve perceived quality and use of information.

This study[48] also found that when disseminated through *formal means* (those that are planned and verifiable), market information is used to a greater extent than when disseminated through *informal channels*. Whereas informal channels, which are spontaneous and unplanned communication interactions, may provide greater openness and clarification opportunities, formal interactions are more credible.

Two other factors that affect the nature of communication between marketing and R&D are the presence of information-sharing norms within the organization and the degree to which engineering goals are integrated with marketing goals (in which case the marketing manager's goal attainment depends on actions of the engineering counterpart and vice versa).[49]

Information-sharing norms indicate organizational expectations about the exchange of information between functions and can promote increased communication behaviors. However, the degree to which such norms really influence marketers' communication with engineering depends on on how strongly marketing managers identify with the marketing function. Marketing managers *who identify more strongly with the organization as a whole* (than with the marketing function) communicate more bidirectionally when information-sharing norms are stressed.[50]

Similarly, *integrated goals* suggest that the organization's needs are superordinate to the goals of the individual functional units and again can promote increased collaboration and cooperation. Marketing managers who *identify very strongly with the marketing function* are more likely to communicate more frequently and more bidirectionally when integrated goals are stressed. However, marketing managers who identify very strongly with their functional area of marketing also resort to coercion to ensure that the engineering contacts comply with their functional perspective on organizational issues. In this case, integrated goals increase the coerciveness of influence attempts by high-functional-identification managers because of their emphasis on functional solutions within the organization.[51]

So, how should managers in high-tech firms use this information to facilitate R&D–marketing interaction? First, managers need to establish policies to encourage information-sharing norms when marketing managers identify more strongly with the organization as a whole. Second, managers should set integrated goals for marketing and R&D when marketing managers identify more strongly with the marketing function specifically. Note, however, the increased risk of coercive influence attempts in this latter case.[52]

Constructive Conflict

Additional findings show that when marketers perceive a greater frequency of interaction, they also perceive more conflict with R&D personnel; however, that conflict doesn't necessarily result in a less effective relationship.[53] In fact, although many studies have corroborated the need for close R&D–marketing interaction for new-product development success, when relationships are too close, the desire to retain harmony precludes alternative viewpoints from emerging.[54] When such groupthink takes over, adverse opinions are not expressed, and potential problems are not addressed, resulting in lower product performance. A real question, then, is how to structure the R&D–marketing collaboration for frequent interaction with the simultaneous ability to challenge others' viewpoints. Having formalized roles within the group, with certain members assigned the role of devil's advocate, may be helpful.

A Caveat

Each of these tools can help improve the flow of communication between marketing and R&D. However, the nature of the communication must continue to be solidly grounded in an *understanding of customer needs and wants*. As Figure 4-5 on "The Rock Game" shows, marketing and engineering may think they are doing a good job of interacting, but in reality, they have not accurately captured or conveyed the customer's needs in the process.

As a case in point, a healthcare company adopted a new information system to manage customer records. This new information system included customized software from the vendor, which allowed the healthcare company to customize the screens and menus for its caseworkers, based on certain treatment criteria and standards. Unfortunately, in tailoring the software to its needs, the healthcare company encountered a series of bugs and other issues (i.e., security concerns) that required extensive rework on the part of the software vendor. In order to communicate their needs and

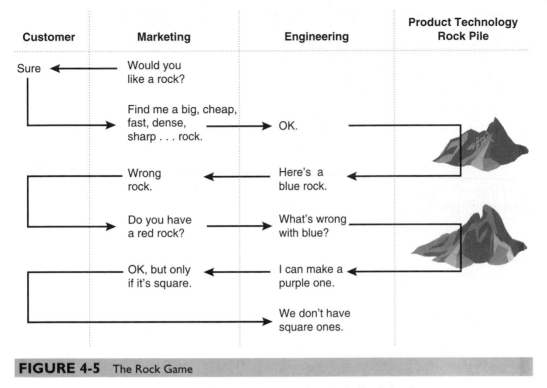

FIGURE 4-5   The Rock Game

Reprinted with the permission of Storage Technology Corporation, Louisville, Colorado.

concerns, the healthcare company personnel had talked to either the salesperson or a customer-support person, who, in turn, expressed the customer's concerns to the software engineers. The filtering and recommunicating of the customer's needs had resulted in several misunderstandings between the engineers and the customer. Despite this, the customer was not allowed to communicate directly with the software engineers. And why not? The reason given by the software vendor was that "its engineers' time was too costly to be spent in talking to customers"!

In summary, cross-functional integration between marketing and R&D is a key driver in diffusing market and customer knowledge among all members of a project team in high-tech firms. This integration ensures that an understanding of market needs, desires, and behavior in the early stages of development constitutes the foundation for technological applications—applications that are valued by customers. This diffusion of customer and market knowledge to the engineering team is enhanced by allowing them to have direct and repeated contact with customers and other outside sources of information—and, as discussed previously, having a strong connection between all internal departments and customers is part of a firm's market orientation. This section's Technology Expert's View from the Trenches is offered by Jennifer Longstaff, whose perspective is influenced by her background in both engineering and marketing.

TECHNOLOGY EXPERT'S VIEW FROM THE TRENCHES

ENGINEERING/MARKETING COLLABORATION

JENNIFER LONGSTAFF

Technical Marketing Engineer
Xilinx Inc., Longmont, Colorado

The relationship between engineering and marketing is often a source of conflict, even in the most cooperative, consensus-driven, proactive companies.

In a market-driven, product-based company, marketing's role is typically to define a product based on input from customers and its knowledge of market needs. Marketing delivers the finished product, and then solicits customer feedback on how well the product has met needed solutions and fared against competition. This feedback gets rolled into the next cycle of product definition as the product is improved. In this way, marketing can assure a consistent, long-term evolution of an improving product.

Engineering's role is to use the marketing definition and create an efficient implementation of the product within a development time-frame. Part of engineering's product design must include extensibility for inclusion of future enhancements within the long-term vision. Engineering must make it known whether marketing's product definitions are feasible and work with marketing to revise the product definitions if not.

In the high-tech environment, new products are often new-to-the-world innovations with no existing models to compare with customer experiences. Marketing receives general requirements from customers, often expressed vaguely as "I need the ability to do something, but I'm not sure how I want that ability to look." Marketing may need to envision a new paradigm to deliver compelling new benefits that satisfy the customer. Engineering can assist marketing with development of the new paradigm by creating a prototype to examine and assess if it will meet customer need. However, engineering may view a prototyping stage as a time-consuming step and a rationale for marketing indecision. Therefore, it is very important for marketing to do sufficient research and study feasibility before requesting prototype work from engineering. Ideally, the prototype will be close to a final product and may need only minor changes. Marketing (as driver of new products) must be responsible and must use engineering's time expeditiously.

In the most common scenario, engineers serve both as *developers* (R&D for new products and feature enhancements) as well as *maintainers* of the existing installed product in the field. Marketing, in contact with customers, is also aware of the issues in the field with existing products; however, engineering takes the brunt of the "resource hit" because maintenance is difficult to schedule, and it is a distraction away from new development. Marketing may not always understand the disruptive nature of the maintenance distraction and the tension it creates with new development.

Engineering and marketing personnel would probably all agree with the concepts and roles listed above. It would seem that both groups, in following their role definitions, should be able to work together productively. Yet, there seems to be an abundance of conflicts and problems between these two groups during the development process.

Inherent problems seem to be:

- Each group assumes it can do the other's job better. Engineering questions marketing's ability to define products, especially when marketing can't immediately finalize a definition and requests prototyping. Engineers don't acknowledge marketing's experience with customers or marketing's ability to make long-term "crystal ball" prediction of future product needs. Marketing doesn't understand why engineering takes so long to develop products, and why engineers react against late definition changes. Each group often sees only a superficial view of the other's tasks.

- Communication between the groups is often poor. Engineers, even those who have moved into marketing, may be introverted and reluctant to take the initiative to contact others, even when definition is needed in order for work to continue. This communication is especially difficult when negative feedback must be given. But the marketing/engineering dynamic is based on a foundation of open and frequent communication. Without this communication, the interaction degenerates quickly, usually due to simple misunderstandings.

The solutions are obvious! Each group must acknowledge the other's skills. This is easier to do if each group consists of competent and hard-working individuals. If either group fails to deliver its commitments in a competent manner, it will cause unnecessary work for the other and feelings of resentment. A key solution to misunderstandings and conflicts is frequent, open communication.

SUMMARY

This chapter has focused specifically on the final two internal considerations for high-technology marketing effectiveness: being market oriented and having effective R&D–marketing interactions. These final two building blocks complement those established in the prior two chapters: relationship marketing (partnerships and alliances), small company challenges, maintaining a culture of innovativeness, and the process of strategic decision making to establish competitive advantage.

Many of the topics addressed in these past three chapters deal with the organizational strategies and culture for being innovative, for valuing market information, and for allowing that information to be shared cross-functionally. In that sense, the topics share the common tie of being housed within the firm itself.

Chapters 5 and 6 take a different perspective and more directly address issues related to customers: how marketers can gather information about customers, not only to develop successful products but also to forecast the market size for the new innovation, and issues customers face in adopting high-tech products.

DISCUSSION QUESTIONS

1. What are the characteristics of a market-driven organization?
2. What are the advantages and pitfalls of being market oriented?
3. What steps might a firm take in order to become more market oriented?

4. How does a firm balance when to listen to customers and when to ignore them?
5. Why is R&D–marketing interaction so important in high-tech firms?
6. What are the barriers to such interaction?
7. How are the barriers to effective R&D–marketing interaction overcome?
8. Under what conditions are market orientation and R&D–marketing interaction most important?

GLOSSARY

Customer/market orientation. A philosophy of doing business that emphasizes the shared gathering, dissemination, and utilization of market intelligence as the basis for decision making.

Knowledge management. The process of creating, transferring, assembling, integrating, and exploiting knowledge assets.

Market intelligence. Information about customer's expressed needs, latent needs, and decision-making processes and about competitors' objectives, strategies, assets, and capabilities.

Marketing myopia (also known as customer compelled). The firm is constrained by traditional and narrow perspectives of its customers and business. This is a common weakness of firms that say they are market oriented.

R&D–marketing interaction. A process in which R&D and marketing work collaboratively to understand customer needs and create customer value. Illustrates the importance of R&D, not just marketing, being market oriented.

ENDNOTES

1. Stewart, Thomas (1999), "Customer Learning Is a Two-Way Street," *Fortune,* May 10, pp. 158–159.
2. Leonard-Barton, Dorothy (1995), *Wellsprings of Knowledge.* Boston: Harvard Business School Press.
3. Atahuene-Gima, K. (1995), "An Exploratory Analysis of the Impact of Market Orientation on New Product Performance: A Contingency Approach," *Journal of Product Innovation Management,* 12, pp. 275–293.
4. Atuahene-Gima, K. (1995), op. cit.; Atuahene-Gima, K. (1996), "Market Orientation and Innovation," *Journal of Business Research,* 35 (2), pp. 93–104; Deshpande, Rohit, John U. Farley, and Frederick Webster, Jr. (1993), "Corporate Culture, Customer Orientation, and Innovativeness in Japanese Firms: A Quadrad Analysis," *Journal of Marketing,* 57 (January), pp. 22–37; Han, Jin K., Namwoon Kim, and Rajendra K. Srivastava (1998), "Market Orientation and Organizational Performance: Is Innovation a Missing Link?" *Journal of Marketing,* 62 (4), pp. 30–45;

Hurley, Robert F. and G. Thomas M. Hult (1998), "Innovation, Market Orientation, and Organizational Learning: An Integration and Empirical Examination." *Journal of Marketing,* 62 (3), pp. 42–55; Slater, Stanley F. and John C. Narver (1994), "Does Competitive Environment Moderate the Market Orientation Performance Relationship?" *Journal of Marketing,* 58 (January), pp. 46–55.
5. Dutta, Shantanu, Om Narasimhan, and Surendra Rajiv (1999), "Success in High-Technology Markets: Is Marketing Capability Critical?" *Marketing Science,* 18 (4), p. 547.
6. Gatignon, Hubert and Jean-Marc Xuereb (1997), "Strategic Orientation of the Firm and New Product Performance," *Journal of Marketing Research,* 34 (February), pp. 77–90.
7. Shapiro, Benson (1988), "What the Hell Is 'Market Oriented'?" *Harvard Business Review,* 66 (November–December), pp. 119–125.
8. Kohli, Ajay K. and Bernard J. Jaworski (1990), "Market Orientation: The Construct, Research Propositions, and Managerial

Implications," *Journal of Marketing,* 54 (April), pp. 1–18.

9. Gupta, Ashok K., S. P. Raj, and David L. Wilemon (1986), "A Model for Studying R&D–Marketing Interface in the Product Innovation Process," *Journal of Marketing,* 50 (April), pp. 7–17.

10. Cooper, Robert G., and Elko J. Kleinschmidt (1991), "New Product Processes at Leading Industrial Firms," *Industrial Marketing Management,* 20 (2), pp. 137–148.

11. McQuarrie, Edward F. and Shelby H. McIntyre (1992), "The Customer Visit: An Emerging Practice in Business-to-Business Marketing," Working Paper #92-114, Cambridge, MA: Marketing Science Institute.

12. Kohli, A. and B. Jaworski (1990), "Market Orientation: The Construct, Research Propositions, and Managerial Implications," *Journal of Marketing,* 54 (April), pp. 1–18.

13. Teece, David J. (1998), "Capturing Value from Knowledge Assets: The New Economy, Markets for Know-How, and Intangible Assets," *California Management Review,* 40 (3), pp. 55–79.

14. McWilliams, Gary and Marcia Stepanek (1998), "Taming the Info Monster," *Business Week,* June 22, pp. 170–172.

15. Mears, Jennifer (2003), "Customer Focus Keeps Dell Productive," *Network World,* April 21, pp. 51.

16. Jaworski, B. and A. Kohli (1993), "Market Orientation: Antecedents and Consequences," *Journal of Marketing,* 57 (July), pp. 53–70.

17. This example is cited in Leonard-Barton, Dorothy (1992), "Core Capabilities and Core Rigidities: A Paradox in Managing New Product Development," *Strategic Management Journal,* 13, pp. 111–125.

18. Leonard-Barton, Dorothy, Edith Wilson, and John Doyle (1995), "Commercializing Technology: Understanding User Needs," in *Business Marketing Strategy,* V. K. Rangan et al. (eds.), Chicago: Irwin, pp. 281–305.

19. Martin, Justin (1995), "Ignore Your Customer," *Fortune,* May 1, pp. 121–126.

20. Ibid.

21. Ibid.

22. DeYoung, Garrett (1997), "Listen, Then Design," *Industry Week,* February 17, pp. 76–80;
Martin, Justin (1995), op. cit.; Leonard-Barton et al., op. cit.

23. Christensen, Clayton and Joseph Bower (1995), "Customer Power, Strategic Investment, and the Failure of Leading Firms," *Strategic Management Journal,* 17, pp. 197–218.

24. Leonard-Barton, Dorothy, Edith Wilson, and John Doyle (1995), op. cit.

25. Leonard-Barton, Dorothy and Jeffrey F. Rayport (1997), "Spark Innovation Through Empathic Design," *Harvard Business Review,* 75 (November–December), pp. 102–113.

26. Byrnes, Nanette and Paul Judge (1999), "Internet Anxiety," *Business Week,* June 28, p. 84.

27. Chandy, Rajesh K. and Gerard J. Tellis (1998), "Organizing for Radical Product Innovation: The Overlooked Role of Willingness to Cannibalize," *Journal of Marketing Research,* 35 (November), pp. 474–487.

28. Christensen, Clayton and Joseph Bower (1995), op. cit.

29. Martin, Justin (1995), op. cit.

30. Song, X. Michael and Mark E. Parry (1997), "A Cross-National Comparative Study of New Product Development Processes: Japan and the United States," *Journal of Marketing,* 61 (2), pp. 1–18.

31. Olson, Eric M., et al. (2001), "Patterns of Cooperation During New Product Development Among Marketing, Operations and R&D: Implications for Project Performance," *Journal of Product Innovation Management,* 18, pp. 258–271.

32. Workman, John (1993), "Marketing's Limited Role in New Product Development in One Computer Systems Firm," *Journal of Marketing Research,* 30 (November), pp. 405–421.

33. Shanklin, William and John Ryans (1984), "Organizing for High-Tech Marketing," *Harvard Business Review,* 62 (November–December), pp. 164–171.

34. Song, X. Michael and JinHong Zie (1996), "The Effect of R&D–Manufacturing–Marketing Integration on New Product Performance in Japanese and U.S. Firms: A Contingency Perspective," report summary #96-117, Cambridge, MA: Marketing Science Institute.

35. Shanklin, William and John Ryans (1984), op. cit.

36. Song, X. Michael and JinHong Zie (1996), op. cit.

37. Leonard-Barton, Dorothy (1992), op. cit.

38. Kunda, Gideon (1992), *Engineering Culture: Culture and Control in a High-Tech Organization,* Philadelphia: Temple University Press.

39. Workman, John (1993), op. cit.

40. Griffin, Abbie and John Hauser (1996), "Integrating R&D and Marketing: A Review and Analysis of the Literature," *Journal of Product Innovation Management,* 13, pp. 191–215.

41. Ayers, Doug, Robert Dahlstrom, and Steven J. Skinner (1997), "An Exploratory Investigation of Organizational Antecedents to New Product Success," *Journal of Marketing Research,* 34 (February), pp. 107–116.

42. Workman, John (1993), op. cit.

43. Ibid.

44. Ibid.

45. Griffin, Abbie and John Hauser (1996), op. cit.

46. Gupta, Ashok K., S. P. Raj, and David L. Wilemon (1986), "A Model for Studying R&D–Marketing Interface in the Product Innovation Process," *Journal of Marketing,* 50 (April), pp. 7–17; Griffin, Abbie and John R. Hauser (1992), "Patterns of Communication Among Marketing, Engineering and Manufacturing—A Comparison Between Two New Product Teams," *Management Science,* 38 (March), pp. 360–373.

47. Maltz, Elliot and Ajay K. Kohli (1996), "Market Intelligence Dissemination Across Functional Boundaries," *Journal of Marketing Research,* 33 (February), pp. 47–61.

48. Ibid.

49. Fisher, Robert J., Elliot Maltz, and Bernard J. Jaworski (1997), "Enhancing Communication Between Marketing and Engineering: The Moderating Role of Relative Functional Identification," *Journal of Marketing,* 61 (3), pp. 54–70.

50. Ibid.

51. Ibid.

52. Ibid.

53. Ruekert, Robert and Orville Walker (1987), "Interactions Between Marketing and R&D Departments in Implementing Different Strategies," *Strategic Management Journal,* 8, pp. 233–248.

54. Ayers, Doug, Robert Dahlstrom, and Steven J. Skinner (1997), op. cit.

CHAPTER 5

Marketing Research in High-Tech Markets

By 2000, atomic-powered zeppelins will speed along at 300 miles per hour.
—RUDYARD KIPLING, 1927, "WITH THE NIGHT MAIL"

Marketing Research at Microsoft

In 1975 Bill Gates and Paul Allen founded Microsoft Corporation with this improbable vision: A personal computer on every desk and in every home. Today, Microsoft is the worldwide leader in software, services, and Internet technologies for personal and business computing. Microsoft reported almost $10 billion in net income on revenues of over $32 billion for its 2003 fiscal year.

Part of Microsoft's general business philosophy has been to use a team-based approach to product development that incorporates direct customer input. Since the late 1980s, the company has formalized its process of learning from current and potential customers. Figure 5-1 below illustrates the pervasive role of marketing research at Microsoft.

FIGURE 5-1 Customer Input during Product Development

Activity-Based Planning	Wish Lines	Calls Data

Analysis and User Needs Definition
Specification Development
Product Prototyping

Usability Lab Testing

Additional Product Development
Internal Alpha Release
Feedback Analysis and Product Refinements

Beta Site Testing

Feedback Analysis and Product Refinements
External Product Release

Surveys	Studies

Activity-Based Planning: With (user-) activity-based planning, the project planning phase focuses on user activities first and features second. Rather than selecting features and crafting a vision statement around them, program managers and marketers determine key user activities and focus the product's vision statement on features that support those activities.

To identify key activities, Microsoft product managers and program managers research exactly how people use a particular piece of application software, such as Word or Excel, to identify the specific activities performed to accomplish a task such as creating a budget. For example, they found about twenty basic activities that were involved in the creation of a budget. Feature selection and feature prioritization in the development schedule were based on this technique. This process produces better decisions about features and synchronization among marketing, user education, and product development.

Wish Line: Another source of information on features that might be added or improved is the customer suggestion database. The main source of input for the suggestion database is the Microsoft Wish Line, a phone system that customers can call with suggestions. Product Support Services (PSS) division staff transcribe these suggestions into a database available throughout the company. It should be noted that Microsoft probably receives more customer calls and electronic messages than any of its competitors simply because it has so many customers.

Calls Data: Microsoft receives hundreds of thousands of telephone calls and electronic messages every day from customers who experience software problems. PSS prepares a weekly report, called *Off-line Plus,* which details problem reports. Aside from helping to set priorities for software fixes, it is a source of important input for next-version product development.

Usability Lab Testing: Microsoft tests the usability of new features by developing a special version or prototype of the feature and asking groups of ten people from "off the street" to try the new feature out. Testers observe and record many things such as problems that users have and the percentage that are able to complete a task without seeking help. The usability lab does not support product planning; rather, it helps developers decide how to make new or existing features simpler to use.

Beta Site Testing: A beta test is a field test of a prerelease version of a new product. Microsoft attempts to get a diverse set of beta testers, who are selected customers, to adequately represent many different hardware, software, and networking combinations. Beta testing detected 27 percent of the total number of bugs in Windows 3.1.

Surveys: Microsoft conducts a wide variety of traditional marketing research surveys on topics such as market segmentation; customer satisfaction with Microsoft products, customer support, and the company overall; and how customers use Microsoft products and competitors' products. In the annual End-User Customer Satisfaction Benchmark survey, Microsoft tries to measure satisfaction by identifying what is necessary to attract and retain a customer. Customers who are very satisfied with the company, its products, and its support and who would definitely buy and recommend Microsoft products are rated as "secure" customers.

Other Studies: Microsoft groups conduct some longitudinal studies of users, visiting them every three or four weeks to see how they are using a product, any problems they are encountering, and any solutions or modifications they may have developed to deal with those problems. They also use a technique they call "contextual inquiry" to study the activities of users working in groups to perform particular tasks. In this process, the project team first chooses a focus for study, such as a group that needs to create a compound (integrated text and graphics) document. The team then visits a site and observes customers performing this task. In the third stage, which occurs during and after the visit, observers take extensive notes and record both the activities and their context. Finally, they hold a session in which they use various techniques to reach a consensus on the nature of the activity and its associations and implications for software development.

As Microsoft states on its Web site,

Great software doesn't happen by chance. By listening to customers and exploring possible solutions to common problems, new technologies can emerge. That's a role of customer research at Microsoft—to connect our products and services with the needs of the industry and, most important, the needs of our customers.

Microsoft has a systematic and comprehensive approach to marketing research. In fact, we have only touched on some of the key pieces of this critical activity. Microsoft continually strives to learn from the marketplace; its focus on understanding the many ways that customers and software relate is indicative of this.[1]

As we've seen in the prior chapters, technology marketers face a paradox. On the one hand, customers find it difficult to articulate what their specific needs are; on the other hand, high-tech firms must keep a finger on the pulse of the market in order to enhance their odds of success. Winning high-tech firms do not first develop new products and then worry later about how to market them, and taking calculated risks does not mean ignoring customers. High-tech firms must incorporate information about customers into the product development process, despite the inherent difficulties and the all-too-common tendency to overlook them.

For example, in the online arena, Amazon.com built its site with the visitor experience in mind. However, because site features are easily copied, Amazon's ability to innovate is based on identifying novel ways to deal with customers and leapfrog its competitors. "We ask customers what they want," says Jeff Bezos, CEO.[2] Amazon.com encourages feedback, sorts through purchase histories to identify customer preferences, conducts focus groups, and collects information in ways that don't impose on customers. Because of this superior customer knowledge, even if other sites offer better prices, customers tend not to switch. As one customer says, in a useful analogy: "I'm happily married, so it doesn't matter how cute the guys are I meet."[3] Other Internet companies, such as REI.com, use customer feedback to make sites easier to use or to add new features.

Ultimately, what separates companies is the kind of information they collect on customers and from whom they collect it.[4] Hence, this chapter focuses on gathering information in high-tech markets, addressing marketing research techniques such as concept testing, conjoint analysis, empathic design, customer visit programs, lead user research, and quality function deployment (a tool to link customer input to product design). Moreover, a look at the gathering of competitive intelligence in high-tech markets is provided. Finally, effective forecasting in high-tech markets is paramount to making good decisions.

GATHERING INFORMATION: HIGH-TECH MARKETING RESEARCH TOOLS

High-tech environments are fraught with change and uncertainty. Customers have difficulty envisioning how technology can meet their needs. They aren't aware of what new technologies are available or how those technologies might be used to solve current problems. They might not even be aware of the needs they have. Moreover, in this environment, firms must accelerate the product development process, closing the time between idea to market introduction. Successful firms in high-tech markets collect useful information to guide decisions.

As Figure 5-2 shows, research methods must be aligned with the type of innovation being developed.[5] This is consistent with the contingency theory of high-technology marketing. For incremental innovations, new-product developments are in alignment with the current market. Customer needs are generally known, and traditional marketing research can help companies understand such needs. Indeed, traditional marketing research techniques are most effective when a product or service is well understood by

FIGURE 5-2   Aligning Market Research with Type of Innovation

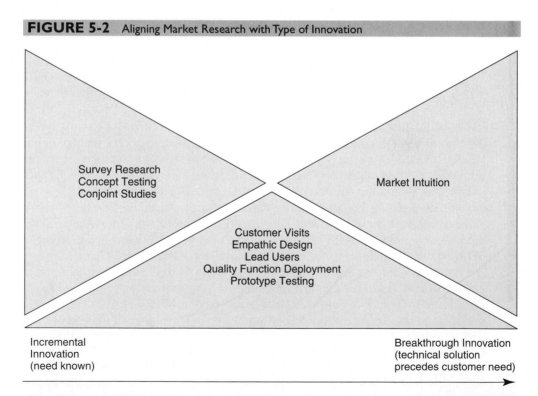

Survey Research
Concept Testing
Conjoint Studies

Market Intuition

Customer Visits
Empathic Design
Lead Users
Quality Function Deployment
Prototype Testing

Incremental
Innovation
(need known)

Breakthrough Innovation
(technical solution
precedes customer need)

SOURCE: Adapted from Leonard-Barton, Dorothy, Edith Wilson, and John Doyle (1995), "Commercializing Technology: Understanding User Needs," in *Business Marketing Strategy,* V. K. Rangan et al. (eds.), Chicago: Irwin, pp. 281–305. Reprinted with permission of the McGraw-Hill Companies.

customers, or when the customer is familiar with possible solutions because of related experience in other contexts. Traditional marketing research techniques such as focus groups, surveys, concept tests, conjoint studies, and test markets can be useful to match new-product characteristics with customer needs. While we touch on some of these, readers interested in pursuing these standard marketing research tools may consult one of the many excellent resources available.[6]

However, standard marketing research tools typically don't address new uses or new benefits and are less effective when customers are unfamiliar with the product being researched. Hence, for breakthrough products or for rapidly changing markets, standard marketing research techniques might not provide useful information. In the extreme, where technical solutions precede customer needs, market research might consist largely of guided intuition. Industry experts may be helpful, and the creation of different future scenarios can be used to guide decision making based on intuition.[7]

In the midrange (between incremental and radical innovation), very useful techniques include customer visits, empathic design, the lead user process, quality function deployment, and prototype testing.

Figure 5-2 also depicts the flow of the chapter in that we first consider traditional tools of marketing research that are most appropriate for identifying opportunities for incremental innovation or for managing existing products. We then consider tools that may be useful for providing insight into opportunities for breakthrough innovations.

Concept Testing

One of the more challenging decisions faced by a new product development (NPD) team is concept selection, the narrowing of multiple product concepts to a single, "best" product concept. The NPD process starts with the generation of ideas for a product that addresses an identified customer need. Many product concepts should be considered since only a small percentage of new product ideas ultimately prove to be profitable.[8] Also, keeping multiple product concept options open and freezing the concept late in the development process affords the flexibility to respond to market- and technology shifts and may actually shorten total product development time.[9] Common approaches to idea generation include: the various observational techniques that we will discuss; brainstorming, where employees from engineering, marketing, sales, and manufacturing are guided through a series of creativity exercises to generate new product ideas; focus groups, where members of the target market are asked to think about how different product or service ideas could satisfy their needs; and depth interviews, where target customers participate in lengthy, nondirective, one-to-one interviews regarding their needs and potential solutions to those needs.

Concept testing then evaluates these early-stage ideas and decides which of them are good enough to be developed further. These concepts are described in one or two paragraphs, sometimes with a name and a price, and potential customers are asked to rate them on dimensions such as interest in trying the product, purchase intent, uniqueness, and perceived value. The results can give the firm a better idea of customer interest, so the new product concept can be refined to improve its chances of success before going to a full-blown, predictive concept test.

In the last stage, the number of concepts is reduced, based on the results from the previous stage, to a manageable set that can be thoroughly assessed. In this stage, a representative sample of potential customers is asked to view a small number of new product concept finalists and complete a battery of questions and diagnostic ratings.

This is commonly done on a secure Internet site due to the speed, power, convenience, and flexibility of the Internet.[10] Conjoint analysis, which we discuss next, is often used at this stage. The goal is for the firm to be able to focus limited R&D and marketing resources on the one or two concepts that have the greatest probability of market success.

Conjoint Analysis

In **conjoint analysis**, respondents are asked to make judgments about their preferences for combinations of product attributes, such as price, brand, speed, warranties, technical services, etc., that involve various levels such as high or low price, premium or value brand, and so forth. The basic objective is to determine the tradeoffs respondents are willing to make within the range of the attributes provided. For example, all other things being equal, a consumer may prefer to have a warranty, but is price sensitive. Conjoint analysis helps to infer whether, on average, consumers in the market would be willing to trade off a less extensive warranty for a lower price or whether the warranty is crucial, despite a slightly higher price. Conjoint analysis accomplishes this by estimating how much each attribute level is valued based on the choices that respondents make about product concepts whose attributes are varied in a systematic way. The value of conjoint analysis lies in the attention it gives to the specifics of each product offering and how the various product features fit together to deliver a complete offering.

One of the first steps in designing a conjoint study is to develop a set of attributes and levels of those attributes that adequately characterize the range of product options. Focus groups, customer interviews, and internal corporate expertise are some of the sources used to structure the sets of attributes and levels that guide the rest of the study. The analyst then develops a set of profiles that cover the full range of attribute levels specified in the study. Respondents then indicate their preference for each profile. The preference can be decomposed into the utility (value) of each level of an attribute and the relative importance of each attribute.[11] The results from this analysis are then used to make product development and positioning decisions. In the Technical Appendix on Conjoint Analysis (see pages 164–167), Bryan Orme of Sawtooth Software provides a managerial overview of how conjoint analysis works and how it can be used to make critical product development, positioning, and pricing decisions.

Customer Visit Programs

A **customer visit program,** or a systematic program of visiting customers with a cross-functional team to understand customer needs, when implemented correctly, can also lead to significant insights and benefits for high-tech marketers.

The idea of using customer visits for market research has developed in response to the challenges faced by managers in many industries. Customer visits are more than a tool to groom customer relationships; they offer a variety of benefits, including the following:

- *Face-to-face communication.* Development of new-to-the-world products benefits from the unique capacity of personal communication to facilitate the transfer of complex, ambiguous, and novel information.
- *Field research.* Doing research at the customer's place of business allows personnel to see the product in use, talk to actual users of the product, and gain a better understanding of the product's role in the customer's total operation.

- *Firsthand knowledge.* Everyone believes his or her own eyes and ears first. When key players hear about problems and needs from the most credible source—the customers—responsiveness is enhanced.
- *Interactive conversation.* The ability to clarify, follow up, switch gears, and address surprising and unexpected insights provides depth to interactions.
- *Inclusion of multiple decision makers.* Many technology products are purchased by groups of people, and customer visits allow all of the players' various needs and desires to be addressed.

To realize these advantages, customer visits are much more than merely talking to people. Good customer visit programs can reveal new pieces of information that may have a direct impact on products or services offered to customers. How should customer visits be structured to maximize the benefits?

1. Get engineers in front of customers.   It is vital that cross-functional teams participate in the customer visit program. Relying solely on marketing personnel to conduct customer visits makes cross-functional collaboration unlikely, and marketing may lack credibility with key technical people. The people who participate in the visits must be the ones who will use the information. Teams should include, at a minimum, an engineer, a product-marketing representative, and the account manager. For cross-functional teams to work smoothly in customer visits, good teamwork must exist between engineering and marketing.

For a customer visit program to be successful, it must be part of the corporate culture and enthusiastically embraced by the technical team. R&D managers who say, "Go see the customers yourself," or "Take the project team out to visit customers" are vital to communicating the appropriate attitude. Having only marketers go out to visit customers does not substitute for a commitment on the part of the entire organization to understand customers. Finally, having only high-level executives on customer visits makes other company personnel question the degree to which a customer focus is real or just window dressing.

2. Visit different kinds of customers.   Ideally, teams should visit multiple customers to get more than just an idiosyncratic reading on customer needs. The common tendency in customer visit programs is to visit only national accounts. Although visiting national accounts may result in increasing satisfaction with these accounts, market share may shrink if the firm falls into the trap of developing products that exactly suit an ever-smaller number of customers. Often the freshest perspectives and greatest surprises come from atypical sources, such as competitors' customers, global customers, lost leads, lead users, distribution channel members, or "internal" customers of the firm's own field staff. Customer councils are another important source of information. They are typically designed to get feedback, share perspectives, and build stronger customer relationships. They offer the potential of synergy through group action.

3. Get out of the conference room.   Because customers often don't realize and cannot vocalize specific needs, it is important to listen and observe what they do. This is especially important for companies that tend to invite customers to their own premises. When a firm hosts its customers' visits on the premises, the visits tend to take place in the company's visit center. Such a policy may cut costs and save time in the customer visit program, but it puts the customers in a passive role; the company is typically showcasing its products and giving VIP treatment to customers.

4. Take every opportunity to ask questions.   Customer visit programs are useful not only for new-product development ideas but also for customer satisfaction studies, identification of new market segments, and a myriad of other issues. Interesting questions to ask include:

- If you could change any one thing about this product, what would it be?
- What aspects of your business are keeping you awake at night?
- What things do we do particularly well or poorly, relative to our competition?
- What things do we do particularly well or poorly, relative to your expectations?

5. Conduct programmatic visits.   A systematic approach including between 15 and 40 visits will yield a depth of understanding and illumination that can go well beyond what a few scattered visits can offer. It is important to coordinate the visits so that customers are not confused and irritated by a series of haphazard visits from different divisions and levels in the firm. Promptly log and review customer visits in a central database. Reviewing all profiles that are kept in a central database allows the firm to spot trends, define segments, identify problems, and glimpse opportunities.[12]

Empathic Design

The process of using empathic design tools is very similar in flavor to the notion of customer visits. Being market oriented in high-tech markets means that observation of customers (what they do) is often more useful in developing novel insights than is asking customers more direct questions (what they say). **Empathic design**, or contextual inquiry in Microsoft's terminology, is a research technique based on the idea that users may not be able to articulate their needs clearly. It focuses on understanding user needs through empathy with the user world, rather than from users' direct articulation of their needs.[13] For example, users may have developed "workarounds"—modifications to usage situations that are inconvenient yet so habitual that users are not even conscious of them. Or customers may not be able to envision the ways new technology could be used. Based in anthropology and ethnography, empathic design allows the marketer to develop a deep understanding of the current user environment, to extrapolate the evolution of that environment into the future, and to imagine the future need that technology can satisfy.[14]

Insights from Empathic Design

Observation of customers can provide illuminating insights when customers find it difficult to articulate their needs. For example, observations of customers using the product allow marketers to identify:[15]

- *Triggers of use.* These are the circumstances that prompt people to use a product or service. For example, when Hewlett-Packard watched customers using its personal digital assistants (PDAs), it found that what it thought was the key reason people used PDAs—to use spreadsheet software—was strongly augmented by using the product for personal organizing functions.
- *How users cope with imperfect work environments and unarticulated user needs.* For example, when engineers from a manufacturer of lab equipment visited a customer, they noticed the equipment emitted an unpleasant smell when being used for certain uses. Customers were so accustomed to the smell that they had never

mentioned it. In response, the company added a venting hood to its product line, which actually became a compelling sales point when comparing the product to those of competitors.

As another example of the use of empathic design to show how users cope with imperfect work environments and unarticulated needs, the design team of the Power Tool Division at Ingersoll-Rand used customer observation to improve the designs of its products. Upon visiting factories where the tools were used, the design team found that half the people using wrenches on an auto assembly line were women, who typically have smaller hands than men. The women found it difficult to grasp the tools properly. As a result, the team developed a two-size variable-grip wrench that was made even easier to hold by using rubberized plastic. An unexpected bonus was the wrench's success in Japan, where hands generally are smaller than in the United States.[16]

- *Different usage situations.* When Intuit's developers observed customers using their Quicken software, they found that many small business owners were using Quicken to keep their books, an important market that Intuit has since targeted more specifically.

- *Customization of products of which marketers are unaware.* For example, in studying consumers' use of beepers and cell phones, researchers observed individuals giving special beeper codes to friends to screen out undesired calls. Based on this observation, companies have creatively implemented the filtering capability on cell phones.

- *The importance of intangible attributes that even customers may not articulate.* These include smell, feel, and so forth, which frequently aren't addressed in traditional surveys.

Empathic design techniques exploit a company's existing technological capabilities in the widest sense of the term. Company observers carry the knowledge of what is possible for the company to do. When that knowledge is combined with what customers need, existing organizational capabilities can be redirected toward new markets. A note of caution: Empathic design techniques do not replace market research; rather, they contribute to the flow of ideas that warrant additional testing before committing to the project.[17]

Process to Conduct Empathic Design

Leonard-Barton and Rayport[18] offer a five-step process to conducting empathic design.

1. *Observation.*   At the first step of undertaking an empathic design study, researchers should clarify the following:

   - *Who should be observed?*   Although "customers" is a logical answer, often noncustomers, customers of customers, or a group of individuals who collectively perform a task may provide useful information.

   - *Who should do the observing?*   Differences in perception and background lead different people to notice very different details when observing the same situation. Hence, it is best to use a small cross-disciplinary team to conduct observational studies. Members should be open-minded and curious, and they should understand the value of observation. For this reason, hiring trained ethnographers to assist in the study is useful. Moreover, as we learned in the prior chapter, those who know the capabilities of a particular technology are often *not* the ones who are in contact with the customer (who knows what needs to be done). Hence, the process of conducting empathic design requires cross-functional collaboration between marketing and R&D.

- *What behavior should be observed?* It is important to observe the "subjects" in as normal an environment as possible. Although some believe that observation changes people's behavior (which is probably unavoidable), some alternatives to observation are experiments in highly artificial lab settings or focus groups, both of which also have limitations. The idea here is to gather new kinds of insights that other research techniques cannot.

2. *Capture the data.* At the second step of the empathic design process, researchers need to establish how to record the information. Most data from empathic design projects are gathered from visual, auditory, and sensory cues. Hence, photographs and videographs can be useful tools that capture information lost in verbal descriptions, such as spatial arrangements.

   Whereas standard research techniques may rely on a sequence of questioning, empathic design asks very few questions other than to explore, in a very open-ended fashion, why people are doing things. Researchers may want to know what problems the user is encountering in the course of the observed activity.

3. *Reflection and analysis.* At the third step, the different team members and other colleagues review the team's observations contained in the captured data. The purpose is to identify all of the customers' possible problems and needs.

4. *Brainstorm for solutions.* At the fourth step, brainstorming is used to transform observations into ideas for solutions.

5. *Develop prototypes of possible solutions.* At the fifth step, researchers need to consider more concretely how possible solutions might be implemented. The more radical an innovation, the harder it is to understand how it should look and function. Researchers can stimulate useful communication by creating some prototype of the idea. Such prototypes, because of their concreteness, can clarify the concept for the development team, allow insights from others who weren't on the team, and stimulate reaction and discussion with potential customers. Simulations and role playing can be useful prototypes when a tangible representation of the product cannot be made.

Increasingly, high-tech firms, such as Hewlett-Packard, IBM, Motorola, and Intel, are using empathic design to augment their traditional marketing research practices. They are hiring social scientists, anthropologists, and psychologists to help them figure out how people use products. By observing customers in their work environments and other natural settings, the research technique helps to close the gap between what people say they do and what they really do. Ethnographers tend to study relatively few subjects, chosen with great care, looking for big insights rather than statistical data.

Example[19]

How does Intel learn about how customers work and use electronic equipment? How does the knowledge it learns help Intel design more effective products in the future?

Intel has hired an eight-person team of "design ethnographers" who go to customer sites to observe customers in their natural settings. Their goal is to learn about how customers navigate their daily environments so that they can then use this information to help Intel design more effective products in the future.

At first, the corporate culture within Intel—particularly the R&D folks—did not take the ethnographers seriously. Indeed, their presence was an acknowledgment that the personal computer had shortcomings, and in the Intel culture, that acknowledgment put them at odds with most other employees. But the success rate when technical engineers design what they think other people want is only 20 percent, says former Intel chairman, Andy Grove. For example, Grove believes that Intel wouldn't have sunk millions of dollars into its Proshare videophone, introduced by Intel in the early 1990s, if it had done more ethnographic research. The quality of the video was slow, jerky, and not synchronized with the sound. However, Intel loved the phone because it required significant computing power, based on the underlying microprocessors. Yet consumers hated it, because the out-of-sync video resulted in miscues when people nodded or shook their heads.

So, Intel has used empathic design in a variety of industries and with different customer groups to gain new insights. For example, its design team has spent time observing people working in the salmon industry off the coast of Alaska. The team was trying to understand how technology, such as satellite-guided locators instead of helicopters to monitor fishing boats, would help. Other insights have come from observing business owners. In observing their often harried schedules, the design team learned that businesspeople needed a tool to capture all the messages and phone numbers they write down on self-stick notes, such as an electronic organizer that recognizes handwriting.

Does the nature of being observed change people's behavior? The Intel team finds that most people love to be observed and eventually lower their guard when they are being studied by the researchers. And ethnographic researchers are masters at getting people to feel at ease under observation. For example, one of the Intel team members has spent hundreds of hours with teenagers in their bedrooms, using videotapes to catalog their behaviors and belongings, from dirty laundry to posters. His goal is to find out more about how they live and what technology they might find useful. Some of his insights: Teens should be able to send pictures to each other instantaneously, over phone lines to computers and into flat-display bedside picture frames. They also need handheld computers that allow them to communicate schedule changes to their parents when they're out and about. The bottom line is that what a user does with a product, rather than what the product can do, ultimately drives its success.

Lead Users

Another research technique helpful in high-tech environments is the lead user process.[20] Used to generate ideas for breakthrough innovations, the lead user process collects information about both needs and solutions from the leading edges of a company's target market and from markets that face similar problems in a more extreme form. The types of customers that tend to innovate are **lead users**—customers that are well ahead of market trends and have needs that go far beyond those of the average user.[21] Lead users may face needs months or years before the bulk of the marketplace and, as such, are positioned to benefit significantly by obtaining solutions to those needs now. In some cases, lead users may have even developed a solution to their needs that marketers can then commercialize for other users. Figure 5-3 shows where lead

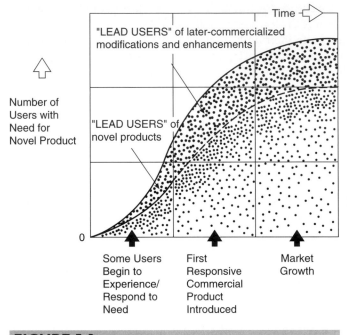

FIGURE 5-3 Lead Users

A schematic of lead users' position in the life cycle of a novel product, process, or service. Lead users (1) encounter the need early and (2) expect high benefits from a responsive solution. (Higher expected benefits indicated by deeper shading.)
SOURCE: von Hippel, Eric (1986), "Lead Users: A Source of Novel Product Concepts," *Management Science,* 32 (July), pp. 791–805.

users exist, relative to the broader target market. Research on lead users[22] shows that many products are initially thought of, and even prototyped by, users rather than manufacturers. For example, Table 5-1 shows that across a variety of industries, the number of innovations conceived of by users is quite high.

Lead user problem solving may apply existing commercial products in ways not anticipated by their manufacturers. Or, lead users may have developed completely new products to solve their needs. For example, Lockheed Martin pioneered a new machining technique in the development of titanium aircraft; the innovation was later commercialized by a machine tool operation that refined the Lockheed tool.[23] Other lead users may not have developed a solution but may simply be aware of the need. Experience with the problem is what makes the lead user's experience so valuable. The lead user process transforms the difficult job of creating breakthrough products into a systematic task of identifying lead users and learning from them. The development team actively attempts to track down promising lead users and adapt their ideas to its customers' needs.

Let's say an automobile company wanted to design an innovative braking system.[24] It might try to identify some users who had a strong need for better brakes—such as in auto racing teams or in an even more technologically advanced field in which users had even greater needs to stop quickly, such as in aerospace. In fact, because military aircraft must stop before reaching the end of the runway, antilock braking systems

TABLE 5-1   Innovations Developed by Lead Users

	Percent of Products Developed By		
	User	*Manufacturer*	*Other*
Computer industry	33%	67%	
Chemical industry	70	30	
Poltrusion-process machinery	85	15	
Scientific instrument with major functional improvements	82	18	
Semiconductor-electronic process equipment with major functional improvements	63	21	16%[a]
Electronic assembly	11	33	56[b]
Surface chemistry instruments with new functional capability	82	18	

[a]Joint user-manufacturer innovation

[b]Supplier innovations

SOURCES: Adapted from von Hippel, Eric (1986), "Lead Users: A Source of Novel Product Concepts," *Management Science,* 32 (July), pp. 791–805; and von Hippel, Eric, Stefan Thomke, and Mary Sonnack (1999), "Creating Breakthroughs at 3M," *Harvard Business Review,* September–October, pp. 47–57.

were first developed in the aerospace industry. The data from lead users consist of examining the solutions they have used to solve their problems.

Eric von Hippel advocates the use of a four-step process to incorporate lead users into marketing research. The process is conducted by a cross-disciplinary team that includes marketing and technical departments. The process can be time-consuming, with each step taking about four to six weeks, and the entire process four to six months.[25]

1. *Identify important market/technical trends.* Lead users are defined as being in advance of the market with respect to an important dimension, which is changing over time. Therefore, before one can identify lead users, one must identify the underlying trend on which these users have a leading position. "One cannot specify what the leading edge of a target market might be without first understanding the major trends in the heart of the market."[26]

The identification of trends is a standard part of most basic marketing courses. In the context of strategic planning, firms undertake an assessment of the external environment in which they operate, examining competitive, economic, regulatory, physical (natural), global, sociocultural, demographic, and technological opportunities and threats. For example, at 3M, the firm determined that a critical trend in the medical industry, particularly in developing countries, was the need to find inexpensive methods of infection control during surgery.

2. *Identify and question lead users.* Customers who are affected early on by significant trends often face product and process needs sooner than do others in a market. As such, they may be positioned to realize a relatively higher benefit from solutions to those needs than are others. In business-to-business markets, manufacturers typically have a better understanding of their key customers than may be possible in consumer markets. Hence, personal knowledge of customers may identify lead users, whereas surveys may be used to identify lead users in consumer goods industries. A very practical method for identifying lead users involves identifying those users who are actively innovating to solve problems present at the leading edge of a trend.

To track lead users down most efficiently, development teams may use telephone interviews to network their way into contact with experts on the leading edge of the target market.[27] People with a serious interest in any topic tend to know of others who may know even more than they do. Such people are research professionals, may have written articles on the topic, or have presented research at conferences. This networking can lead the team to the users at the front of the target market (as shown in Figure 5-3). It is also important to network to identify lead users in markets and fields that face similar problems, but in different and more extreme forms, as in the previous aerospace braking example.

Note that lead users may not be within the firm's usual customer base; they may be customers of a competitor or outside the industry. Moreover, if lead users have already solved a problem, they may no longer articulate the solving of that need as an issue; hence, using a survey to identify them may be unproductive. In such a situation, the use of empathic design in the identification of lead users may be particularly wise.

A final issue in selecting and talking to lead users relates to their willingness to share information. The lead user project team should be up front about its company's commercial interest in the ideas being discussed. In a lead user study devoted to improving credit reporting services, a team found that at least two major users of such services had developed advanced online credit reporting processes. One of the users was unwilling to discuss details because the service was viewed as a significant source of competitive advantage. The other said, "We only developed this in the first place because we desperately needed it—we would be happy if you developed a similar service we could buy."[28] If a customer hesitates to talk, it is better to not pursue that interview due to the intellectual property concerns.

3. *Develop the breakthroughs.* The team may begin this phase by hosting a workshop that includes several lead users who have a range of expertise, as well as a number of representatives from different areas of the company (marketing, engineering, manufacturing, etc.). During the workshop, the group combines insights and experiences to provide ideas for the sponsoring company's needs.[29]

Many users will participate simply for the intellectual challenge. Because they tend to come from other fields and industries, they generally are not concerned about loss of competitive advantage within their field. Moreover, by transferring their knowledge to a willing supplier (either voluntarily or on a licensing basis), they can continue to focus on their own core competencies and have an improved source of supply through transferring the innovation. On the other hand, lead users may not be willing to participate when the advantage they have gained is significant to their competitive position.

It is rare for a firm simply to adopt a lead user innovation "as is." Rather, information gained from a number of lead users and in-house developers leads to adaptations and modifications. The team may assess the business potential of ideas that emerge from the workshop and how they fit with the company's interests.

4. *Project the lead user data onto the larger market.* One cannot assume that today's lead users are similar to the users who make up the major share of tomorrow's market. Firms must assess how lead user data will apply to more typical users rather than simply assume such data transfer in a straightforward fashion. Prototyping the solution and asking a sample of typical users to use it is one way to gather data to make the projection.

Based on a determination of how the new concept fits the needs of a larger target market, the team will present its recommendations to senior managers. This presentation will include evidence about why customers would be willing to pay for the new products.

What are some of the benefits of the lead user process?[30] Ultimately, the lead user process allows a firm to gather and use information in a different way, which leads to new insights. In addition, because the process involves a cross-functional team from the organization, no one person feels like the lone ranger, pushing for change. The process brings cross-functional teams into close working relationship with leading-edge customers and other sources of expertise. But the lead user process is not a panacea for the difficulties in gathering research to develop breakthrough products. Without adequate corporate support, skilled teams, and needed time, the process may not succeed. A detailed example of how the lead user process works at 3M Corporation is addressed in Box 5-1.

BOX 5-1

THE LEAD USER PROCESS AT WORK: 3M CORPORATION

Examples of companies using the lead user process to improve their ability to match product development with customers' needs include Bose (maker of consumer electronics) and Cabletron (designer of fiber optic networks).[a] Another company, 3M Corporation, has relied on the lead user concept extensively.

3M EXAMPLE 1: MEDICAL IMAGING

Step 1: Identify important market/technical trends.   A team focused on medical imaging knew that a major trend was the development of capabilities to detect smaller and smaller features in medical images—very early stage tumors, for example. Its initial goal was to develop new ways to create better high-resolution images.

Step 2: Identify and question lead users. Through networking with research experts in the field, the team identified a few radiologists who were working on the most challenging medical problems. They discovered some lead users in radiology who had developed imaging innovations that were ahead of commercially available products.

Networking to other fields that were even further ahead in *any* important aspect of imaging led the team to specialists in pattern recognition

and people working with images that show the fine detail in semiconductor chips. The lead users in pattern recognition were very valuable to the team. Military specialists relied on computerized pattern recognition in reconnaissance. These users had actually developed ways to enhance the resolution of the best images by adapting pattern recognition software. This discovery of the use of pattern recognition helped the development team refine its initial goal (developing new ways to create better high-resolution images) to finding enhanced methods for recognizing medically significant patterns in images, whether by better imaging or by other means.

Step 3: Develop the breakthroughs.   In the course of a two-day lead user workshop, lead users with a variety of experiences were brought together: people on the leading edge of medical imaging, those who were ahead of the trend with ultra-high-resolution images, and experts on pattern recognition. Together, they created a solution that best suited the needs of the medical imaging marketplace and represented a breakthrough for the company.

3M EXAMPLE 2: INFECTION CONTROL

Another 3M team was charged with developing a breakthrough product for the division's surgical drapes unit, which designs the material that prevents infections from spreading during surgery. Surgical drapes are thin, adhesive-backed plastic films that adhere to a patient's skin at the site of the surgical incision prior to surgery. Surgeons cut directly through these films during an operation. Drapes isolate the area being operated on from the most potential sources of infection: the rest of the patient's body, the operating table, and the members of the surgical team. But drapes don't cover catheters or tubes being inserted into the patient. The drapes' cost prohibited market entry into less developed countries.

Step 1: Identify important market/technical trends.   In looking for a better type of disposable surgical draping, the team first had to learn about the causes and prevention of infections by reading research articles and interviewing experts in the field. Then, the team gathered information about important trends in infection control. During this process, the team realized it didn't know about the needs of surgeons in developing countries where infectious diseases are still major killers, so the team traveled to more hostile surgical environments to learn how people keep infections from spreading in those operating rooms. Some surgeons combated infections by using cheap antibiotics as a substitute for more expensive measures. The team saw a coming crisis with the doctors' reliance on antibiotics: Bacteria would become resistant to the drugs.

Step 2: Identify and communicate with lead users.   The team networked to find the innovators at the leading edge of the trend toward much cheaper, more effective infection control. As is usually the case, some of the most valuable lead users turned up in surprising places. For example, specialists at leading veterinary hospitals were able to keep infection rates very low, despite facing difficult conditions and time constraints. As one vet said, "Our patients are covered with hair, they don't bathe, and they don't have medical insurance, so the infection controls we use can't cost much."

Another surprising source of ideas was from Hollywood: Makeup artists in Hollywood are experts in applying materials that don't irritate the skin and are easy to remove. Because infection control materials can be applied to the skin, those attributes were very important.

Step 3: Develop the breakthroughs.   During the lead user workshop, the participants were invited to brainstorm about revolutionary ideas for low-cost infection control. The outcome of this session were the following ideas:

- An economy line of surgical drapes, made with existing 3M technology, targeted to the increasingly cost-conscious medical world.
- A "skin doctor" line of handheld devices to layer antimicrobial substances onto a patient's skin during an operation and to vacuum up blood and other liquids during surgery; this line again would be developed from existing 3M technology and would offer a new infection prevention tool.
- An "armor" line to coat catheters and tubes with antimicrobial protection, created with existing 3M technology; this line could open up major new markets outside surface infections, including bloodborne, urinary tract, and respiratory infections.
- A revolutionary approach to infection control based on the idea that some people enter the hospital with a greater risk of contracting infections—for example, those suffering from malnutrition or diabetes. Rather than providing every patient the same degree of infection prevention from the same basic drapes, this approach worked on a different philosophy. Through "upstream" containment of infection, treatment before people went to surgery, doctors could reduce the likelihood of these higher-risk patients contracting disease during an operation. This approach, however, would require 3M to radically change its strategy in the market and require new competencies, products, and services. After much debate, 3M decided to fund a "discovery center" service to develop and diffuse the new approach to infection control, and the product lines needed to deploy it are being developed.

[a] DeYoung, Garrett (1997), "Listen, Then Design," *Industry Week,* February 17, pp. 76–80.
SOURCE: von Hippel, Eric, Stefan Thomke, and Mary Sonnack (1999), "Creating Breakthroughs at 3M," *Harvard Business Review,* September–October, pp. 47–57.

In addition to using empathic design and the lead user process to understand customer needs and to innovate breakthrough solutions, high-tech firms must also make tradeoffs among product specifications and product functionality in the product design process. A useful research tool that incorporates a customer orientation into design decisions is quality function deployment.

Quality Function Deployment

Quality function deployment (QFD) is an engineering tool that first identifies what the customer's requirements are (through customer visits, empathic design, working with lead users, etc.) and, second, maps those requirements onto the product design process.[31] The basic idea in quality function deployment is to use the voice of the customer in the new-product development process to ensure a tight correlation between customer needs and product specifications.[32] The process prioritizes and ensures that all design decisions take into account the importance of that design requirement from the customer's perspective. The ultimate outcome is a new product that provides superior value to the marketplace via a customer-informed design team. It requires close collaboration between marketing, engineers, and customers.

The implementation of QFD is a multistage process, including the following:[33]

- *Collect the voice of the customer.* Through customer visit programs or empathic design, identify customer needs, in the customers' own words, regarding the benefits they want the product to deliver. Roughly ten to twelve customers will yield close to 80 percent or more of the customers' needs (assuming a relatively homogeneous market segment). These desired benefits and attributes can be weighted or prioritized to help the product development team in design tradeoffs later (e.g., to trade off processing speed versus price, in the case of a computer chip).[34]

- *Collect customer perceptions of competitive products.* Surveys of customers can be used to assess how well current products fulfill customer needs. These data are an important component in identifying any gaps or opportunities in the market.

- *Transform customer insights into specific design requirements.* Sometimes called customer requirements deployment, the idea in this step is to identify the product attributes that will meet the customers' needs. It is important here to understand the interrelated nature of various attributes. For example, although customers may want more speed in processing, they may also want a lower price. This step is sometimes also referred to as the *house of quality,* or the planning approach that links customer requirements, competitive data, and design parameters.

Kano Diagram

At the heart of the process lies one of the key tools in QFD, the Kano concept (or Kano dimensions/diagram). The **Kano concept** (Figure 5-4) provides a graphical representation of the nature of the relationship between the presence of certain product attributes and customer satisfaction or dissatisfaction.

The graph in Figure 5-4 shows three types of attributes. Attributes that are linearly related to customer satisfaction are deemed *one-dimensional quality* attributes; increasing the performance of these attributes leads to a linear increase in satisfaction. These attributes are typically known and voiced by the customer. For example, in a laptop computer, lengthening the life of the battery would probably lead to a predictable increase in satisfaction.

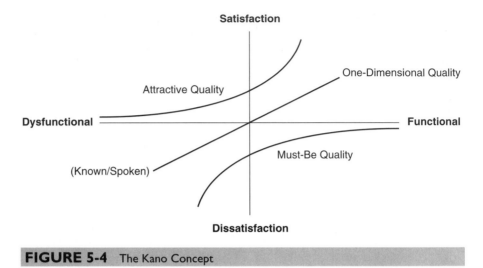

FIGURE 5-4   The Kano Concept

SOURCE: Adapted from Kano, Noriaki, Shinichi Tsuji, Nobuhiko Seraku, and Fumio Takerhashi (1984). "Miryokuteki Hinshitsu to Atarimae Hinshitsu (Attractive Quality to Must-Be Quality)," Japanese Society for Quality Control.

The two other types of attributes have nonlinear relationships with satisfaction. *Must-be quality* attributes must be present in order for the customer to be satisfied. Although the *absence* of the attribute is exponentially related to levels of *dis*satisfaction, increasing the level of that attribute does not increase customer satisfaction with the product. Moreover, these attributes are so essential to product functionality that they may not be explicitly voiced by the customer. For example, in the laptop industry, the computer must be fairly immune to bumps and roughness in handling. If the computer failed upon booting every time the laptop received rough handling, customers would be horribly dissatisfied. On the other hand, allocating significant resources to improve the degree of roughness the laptop can handle likely would not appreciably increase the satisfaction of most laptop users.

The final type of attributes, also exhibiting a nonlinear relationship with satisfaction, is *attractive quality* attributes, or those that exhibit an exponential relationship with satisfaction. When the attribute is lacking, the customer is not dissatisfied, but the presence of the attribute leads to an extremely favorable reaction. These attributes delight the customer and provide the "wow" factor in the product usage experience. Often, customers cannot articulate these attributes, and hence, they must be discovered through some of the techniques mentioned earlier (empathic design and lead users). In the laptop example, the "wow" factor might be found in the presence of a laptop that is decompressable into a pocket size for carrying but expands upon opening. Many experts in product innovation believe that firms that know how to identify those attributes that delight the customer are destined for success.

Rather than culminating in a specific design solution, the QFD process reveals friction points in the design process. It allows the product development team to develop a common understanding of the design issues and tradeoffs, bases resolution of those tradeoffs in customer needs, and enhances the collaborative processes between marketing, manufacturing, and engineering.[35] One study found that using QFD reduces design time by 40 percent and design costs by 60 percent, while enhancing design quality.[36] There are many seminars offered on this process and research firms that can provide expertise in implementing it. Interested readers are referred to additional reading.[37]

QFD and TQM

QFD emerged from the total quality management (TQM) movement in manufacturing and has become closely linked with the notion of market orientation. The total quality management paradigm is originally based on the notion of using the process to create value for customers. Value creation requires excellence in four key areas:[38]

1. *Customer excellence.* Knowing what customers require and delighting them with attractive quality are the hallmarks of customer excellence.
2. *Cycle time excellence.* Shortening a firm's time to market is vital in creating value. When cycle times run longer than expected, cost overruns occur. More importantly, however, when cycle times run longer than expected, the loss of market share contributes even more than do cost overruns to a decline in profitability. Although being fast to market is a necessary condition for excellence, it is not, in and of itself, sufficient: The firm must have the ability to hit customer requirements accurately. Because customer needs and expectations are a rapidly moving target in high-tech markets, faster cycle time also ensures a higher correlation with quality—as defined by the customer.

    Recall the notion of expeditionary marketing: By undertaking a series of low-cost, fast-paced incursions into the market, the firm learns about customer needs and recalibrates its offerings each time; the combination of speed and learning enhances the odds of success. One of the key factors that affects cycle time is *complexity* in the product's features and functionality. As Figure 5-5 shows, in striving to get to the

---

FIGURE 5-5   Relationship Between Entries in the Market and Quality

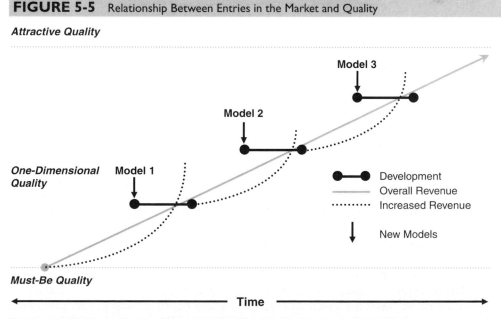

Reprinted with the permission of Storage Technology Corporation, Louisville, Colorado.

market quickly, firms may initially develop a product with a relatively basic combination of attributes, such that the attributes exceed the must-be quality level and are in the desired one-dimensional quality space. As the firm brings additional versions of the product to market, it adds additional features, approaching the attractive quality threshold. Guy Kawasaki refers to this willingness to commercialize an initial version of the product with basic features only as Rule #2 for revolutionaries: "Don't worry, be crappy."[39] Although he uses fairly inflammatory rhetoric, what he means is that it is sometimes acceptable to strive not for perfection but for the minimum level of market acceptability with the first generation of a radical new product. This message does not mean that companies can introduce products that overlook key attributes in customer choice (which would fall below acceptable quality). Rather, it means they must be quick to market with an acceptable level of quality.

Although this prescription might sound counterintuitive, it makes good business sense for at least two reasons. First, many high-tech firms have failed because, in their striving to attain a complicated combination of product attributes—many of which require design time, testing, and debugging well beyond what was initially projected—either customer needs have changed or competitors have beat them to the market with similar products or products that serve the customers' needs in a different way. Storage Technology encountered this situation in its lengthy development of one of its radical innovations in data storage devices. Delays in technological development due to the high level of complexity resulted in a competitor beating it to market.

Second, many high-tech customers are faced with switching costs; as a result, getting to market quickly captures an installed base of customers for later upgrades and versions. Hence, a firm should work incrementally on features and functionality, guided closely by marketing input. Firms must ensure that their first foray in the market is quick and at least must-be quality; over time, additional product extensions can strive for attractive quality.

A final thought on cycle time excellence: Firms that focus on cycle time must realize that merely getting the product to market quickly is not enough. The real cycle time that matters is the time to market acceptance.[40] To hasten the time to market acceptance, involving customers early in the development process with a continuous dialogue is vital. New technologies make this more and more feasible.

3. *Cost excellence.* Cost excellence is found in providing customer value at minimum cost. Supply partnerships are one useful way to work on cost issues. Although many firms in the past focused on downsizing as a way to cut costs, downsizing can also cause decreases in customer value. Downsizing does not inherently lead to customer value unless the assets are somehow added back to enhance value.

4. *Cultural excellence.* Cultural excellence refers to the alignment of individual and organizational goals to respond to business conditions; organizational goals must be supported in order to capitalize on opportunities in the marketplace. A culture of innovation, discussed in Chapter 2, is one way to achieve cultural excellence. Engineering and marketing must also be integrated to focus on customer value.

Prototype Testing

A prototype is a model of the ultimate product or service. As a model, the prototype may provide only the essential elements of the planned final product while ignoring minor or purely supporting elements. The first test in **prototype testing** is against the technical design specifications. If the prototype does not meet specifications, appropriate adjustments are made. When it does meet specifications, the prototype is then evaluated by potential customers.

Beta Version Testing

Beta versions of new products are prerelease versions that the company provides to customers to try. In a **beta test,** a customer agrees to provide feedback on this early version of the new product to the producer so that the product can be improved prior to its commercial release. For example, Symantec provides a broad range of content and network security software and appliance solutions to individuals, enterprises, and service providers. The Symantec External Test Program is designed to expose prerelease software to a wide range of equipment and real-world usage. Participants in the External Test Program receive prerelease software, test scripts, and documentation to review and test. Participants are expected to remain active throughout a project's lifecycle and communicate issues effectively to Symantec team members.[41]

The ways in which Hewlett-Packard uses research to guide the new product development process is highlighted in this section's Technology Expert's View from the Trenches.

TECHNOLOGY EXPERT'S VIEW FROM THE TRENCHES

USING MARKET RESEARCH TO DRIVE SUCCESSFUL INNOVATIVE PRODUCTS

BONNIE GEBHART

Future Product Manager, Hewlett-Packard,
Loveland, Colorado

Ever wonder how *new* products are created? What about *new product categories* that don't exist today? These are very relevant questions to Future Product Managers who work with technologies that *replace* existing products, or radical innovations. Examples of radical innovations include digital cameras, digital entertainment products for the living room, and portable CD/mp3 players. In all these examples, there existed a "current" product that satisfied the customer need for such things as capturing memories, recording a favorite TV show, or listening to music.

Gathering customer insights, understanding consumer behavior, and "use models" are very important steps in all phases of the product development cycle. How can this information be gleaned about products that don't exist in the market? Hewlett-Packard faced this challenge in 1995 when it researched the digital photography market. HP saw a market opportunity in digitally capturing and printing high-resolution images that would compete with the traditional silver-halide photography market. To start, HP had to ask:

- What is consumer behavior regarding photography?

- What would the user like to do with photography, but is unable to do today?

- What problems would digital photography solve?

- What would be barriers to market adoption of digital photography?

These are critical questions to answer during the investigation phase of new product development. HP set out to learn what customers liked and disliked about the traditional silver-halide photography process through a number of worldwide focus groups. These focus groups provided valuable insights about ease-of-use, quality, photo storage and retrieval, getting the "right" picture, red-eye, etc. Could digital photography improve the traditional photography process through offering benefits that customers valued? Would it be a barrier to also need a PC and a printer to consider the new technology? Would the new process be more intimidating than the current point-and-shoot method of photography? Solving customer frustrations and understanding customer concerns were essential for the successful adoption of this new technology.

Another step in the research process involved enlisting the help of "thought leaders"—individuals who are recognized as experts in a particular field. Individuals with backgrounds in computing, photography, anthropology, and psychology were brought together in a moderated session to give their perspectives on digital photography adoption. This provided critical insights into understanding trends, and market readiness, in addition to technology constraints (examples here included camera resolution, costs, etc.). Insights were also gained about the role that photography plays in different cultures in society.

Cross-functional design teams were also involved in additional thought-provoking "ideation" sessions. This phase of work built on the knowledge gained from the previous customer focus groups and thought leader insights. Through creative exercises, "out-of-the-box" ideas and concepts were generated. These ideas were captured through real-time artist renditions, and a series of early product concepts were generated. Using customer needs as the basis for sorting and prioritizing concepts, the concepts were screened, and the remaining ideas were developed in more detail, keeping in mind design feasibility, cost, and schedule. The most promising concepts were then presented to customers at a series of focus groups.

To gain valuable customer input on future "new" products, consumers would need to "envision" the future. Creating a video that easily and quickly allows people to see future use models is extremely helpful—in this case the video would portray how you would take a digital picture, connect the camera to your PC, and easily print it out on your home printer. This seems ordinary today, but looking back to 1995, it was revolutionary. Mocked-up products, artist's drawings, and examples of product output are also very useful tools. The more realistic the "props," the better a focus group would be able to respond to research needs, particularly when they involve radical changes from today's products. Customer feedback on product concepts will also help "flush" out musts versus wants in a new product definition and bring forth new insights and potential pitfalls or concerns.

Often as a result of focus group feedback, real-time product concepts would be created and tested in future focus groups. For this reason, it is very important that the cross-functional team participate in observing the focus groups and listen to customers discuss their likes, dislikes, and needs. The cross-functional team would typically consist of individuals from product marketing, individuals who would be involved in messaging and selling the product, R&D, and manufacturing.

The techniques described above have been used in numerous investigations where a disruptive technology or radical innovation is involved. The challenge for creating these types of products is the ability to carefully understand customers' needs, current use models, customer desires and frustrations, and the key technology enablers.

GATHERING COMPETITIVE INTELLIGENCE

Another vital element in the information arsenol for the high-tech marketer is competitive intelligence. **Competitive intelligence** is information about competitors: who they are, their products, their marketing strategies, and likely responses to the marketing strategies of other firms in the market. Effective competitive intelligence provides solid knowledge of the market, customers, and competitors; quick response time; and superior strategy based on identification of threats and opportunities.[42] Competitive intelligence provides firms with an early warning system to ward off disasters. Indeed, "the essence of smart competitive management is an action that preceded its obvious time."[43]

For competitive intelligence programs to work, they must affect the mindset and decisions of the people whose actions most significantly affect the bottom line—namely, top management. Moreover, effective competitive intelligence programs are much more than mere passive watching of the market (i.e., competitive monitoring); rather, firms that are skilled at reading signals from the market actually develop a core competency in understanding the competition. To do so, they must find it safe to challenge the status quo, to bring an outside perspective, and to be unconventional.

For example, in 1985 Motorola launched a study of Japanese business strategies in Europe. At that time, Motorola-Europe executives saw no signs of a thrust by Japan into Europe. The lack of such entry did not seem to fit the character of their Japanese rivals. So, through a strategic gathering of competitive intelligence, Motorola discovered that the Japanese planned to double their capital investments, going after the European semiconductor market. As a result, Motorola changed strategy, sought joint ventures with European partners, and held off their Japanese rivals.[44]

It can be difficult to gather competitive intelligence in high-tech markets. Sometimes, one doesn't think to look outside the industry for competition. However, Chapter 2 highlighted the fact that industry newcomers often pioneer breakthrough technology. As a result, firms must monitor related industries for competitive moves.

The Web has had a radical effect on the way marketers get the business intelligence they need.[45] Useful websites for gathering competitive intelligence are listed in Table 5-2. For example, The Google Groups news service (www.groups.google.com) shows job notices posted on specific Usenet groups, including details about hardware and software with which applicants must be familiar. The geographical nature of such postings

TABLE 5-2   Useful Web Sites for Competitive Intelligence

Society of Competitive Intelligence Professionals	www.scip.org
Fuld & Co.	www.fuld.com
Hoover's Online	www.hoovers.com
Farcast	www.farcast.com
Individual Inc.	www.individual.com
SEC's EDGAR file	www.sec.gov/edgar/searchedgar/webusers.htm
The CI Resource Index	www.bidigital.com/ci
Competia	www.competia.com/home
Moodys	www.moodys.com
Dun & Bradstreet	www.dnb.com

can also yield insights into possible expansions. To leverage their power to communicate with prospective customers, employees, and other stakeholders, companies post plenty of information on their sites. Some of this information can be available to anyone who seeks it, making it possible to gather more information on competitors than ever, including:

- Customer and client lists
- Detailed product and pricing information, as well as product specifications and technical data
- Specifics about business goals and strategies
- New-product plans and R&D efforts
- Extensive company job postings that shine a spotlight on new business emphases
- Details about manufacturing processes and quality-control efforts
- Company organizational structures and biographical information about managers
- Comprehensive information about business locations, distributors, and service centers
- Information about partnerships and alliances

Interestingly, cultural differences exist between countries in the views of intelligence. In the United States, intelligence is often associated with the military, with related connotations of covert activity and secrecy. In other countries, say Japan, for example, social networks of intelligence transfer are an integral part of society. And in Israel, the vast majority of top executives once served in the military. Given other countries' greater receptivity to and acceptance of intelligence-gathering activities, U.S. companies can probably learn some useful lessons about treating competitive intelligence activities more strategically.

Tami Syverson, former competitive intelligence analyst at Sun Microsystems, provides insight in this section's Technology Expert's View from the Trenches.

TECHNOLOGY EXPERT'S VIEW FROM THE TRENCHES

A DAY IN THE LIFE OF A COMPETITIVE ANALYST

Tami Syverson,

Product Marketing Manager
(and Former Competitive Analyst)
Sun Microsystems, Hillsboro, Oregon

The Competitive Intelligence (CI) Team functions as a service organization within the corporation. The goal of a CI team is to provide market research and competitive intelligence on the industry, both proactively and upon request by the internal customer base consisting of product managers, sales force, public relations, and executive management. Information is provided to assist the development of internal and external materials such as product and market requirements documents, marketing and sales collateral, strategic business cases, RFPs (requests for proposals), press releases, and customer-ready presentations.

Taking a look at daily events will shed light on how the CI Team helps the organization gain a competitive advantage via awareness of the market and the competitive landscape. A typical day starts with a couple hours of reading trade journals and online news subscriptions to remain current on competitive moves and evolving industry trends. An integral service provided to people within the organization is keeping them alerted of current news events. Press releases are a good source of product announcements, service offerings, and pricing. A second source can be a competitor's Web site. On any given day, a major competitor can announce a new product that may compete directly or indirectly. First, notification must be made internally followed by an analysis of the effect on market positioning. A competitive positioning document is necessary to prepare executives and the sales force when responding to customer questions and concerns. This involves working with personnel in public relations, and many other areas of the organization.

Later in the day, a noon lunch meeting with a product manager uncovers a research inquiry. A new product is in development and the marketing manager needs an overview of the competition for the marketing plan. In order to correctly position the service/product within the marketplace, it is important to identify the correct price, complete offering, features, channel of distribution, and competitive positioning.

Typically, most research requests can be categorized into the following three areas:

- Market trends and statistics.
- Competitive offerings.
- End user wants and needs.

Some answers can be found in secondary research. In the event primary research is needed, the project may be outsourced to a market research company. At this time, the CI Team will manage the custom research study. Industry analysts are integral to gathering competitive information as they have insights that may not otherwise be found via standard research efforts. They track daily the industry and market events in their areas of expertise (i.e., computer hardware, software, etc.).

Conducting competitive research is not always easy; there will be challenges along the way. A few challenges to watch out for in this new era of information technology include business rules that are constantly changing, limited information, and new, unknown competitors.

When rolling out new products or services, often there are no set parameters to follow. It is difficult to find market trends and statistics on new innovations. Part of the challenge involves predicting the future. Just when you understand the competition, the market changes. New competitors land on the scene so quickly that your radar screen may not be fixed on them. It is important to widen the range to include competitors not previously thought of that can meet your customers' needs.

News breaks all day—don't miss it! A discontinuous innovation can radically change the competition and the way you do business. Staying abreast of the competitive landscape allows you to answer questions that arise from your customer base. It is impossible to know all the answers, but the key is to know where to find solutions.

Last words of warning, beware of "Web-myopia," a nearsighted focus on Internet research. There are so many sources available beyond the Internet, but it may become difficult to refocus your energy.

To be effective as a Competitive Analyst, it is important to manage your relationships and information carefully. Focus on building and maintaining good relations with internal and external contacts, as they will be a significant source of information. When gathering and analyzing information, be sure to integrate market research tools and techniques.

Note: The information presented here represents the opinions of the author and not those of Sun Microsystems, Inc.

The flipside of *gathering* competitive intelligence is *sending* competitive signals. Indeed, some firms proactively attempt to send signals to competitors in the marketplace via a variety of mechanisms.[46] For example, preannouncing of products, or the announcement of a firm's intention to release a product in the future, is commonly used (and is covered in Chapter 10 on advertising strategies) and can preempt competitors by postponing customers' buying decisions. Firms can send competitive signals by sharing information with industry contacts, customers, or distributors; the information will eventually be disseminated to others.

Moreover, firms have been known to deliberately release several different stories to the marketplace, so as to confuse competitors. As stated by one manager:

> [We try to] keep competitors off guard as to what our "specs" really look like, when we are going to announce, and what the price is going to be. We always have three or four stories out in the marketplace, so people really don't know which is the right one, until we formally announce something.[47]

Hence, firms must carefully scrutinize the competitive information they receive from the marketplace and attempt to gauge its accuracy.

In addition to researching customers and competitors, high-tech firms are faced with the daunting prospect of forecasting demand for new products.

FORECASTING CUSTOMER DEMAND[48]

Forecasting future sales of high-tech products is difficult for many reasons. Quantitative methods typically rely on historical data, but for radically new products, there are no historical data. Moreover, data obtained through traditional techniques are of dubious value, because it is difficult for customers to articulate their preferences and expectations when they have no basis for understanding the new technology.

Although gathering information regarding customers in high-tech markets is difficult, the issue of developing a specific sales forecast is often akin to looking into a crystal ball. The "crystal ball" technique to develop forecasts for high-tech products is imprecise at best and flat-out wrong at worst. "Managers know little about predicting new-product sales, and nothing about the takeoff."[49] For example,[50]

- In reacting to the addition of audio technology to silent movies (circa 1927), Harry M. Warner said, "Who the hell wants to hear actors talk?"
- His later colleague, Darryl Zanuck, head of 20th Century Fox Films in 1946, predicted that "Television won't be able to hold onto any market it captures after the first six months. People will soon get tired of staring at a plywood box every night."
- Ken Olsen, president and founder of the DEC Corporation, said in 1977, "There is little reason for any individual to have a computer in their home."

High-tech marketers must not be daunted by the challenge. Tools are available to help them address the important issue of forecasting. And because the task is fraught with uncertainty and many sources of error, using a systematic process to develop the forecast is more important than ever.

Forecasting Methods

Forecasting tools can be categorized into quantitative and qualitative tools.[51] Basic quantitative tools include moving averages, exponential smoothing, and regression analysis. As noted previously, because of their reliance on historical data—which are often nonexistent in a new high-tech marketplace—quantitative tools may not be available in high-tech markets. Qualitative forecasting methods, such as the Delphi and morphological methods, may be more applicable. Readers interested in these traditional forecasting tools can use one of the many excellent resources available.[52]

The *Delphi method* is probably the most common qualitative method. In this technique, a panel of experts is convened and asked to address specific questions, such as when a new product will gain widespread acceptance. These experts are purposefully kept separate, so that their judgments will not be influenced by social pressures or group influences. The answers to initial questions are sent back to the participants, who are asked to refine their own judgments and to comment on the predictions of the others, in an attempt to find a consensus. Anonymity among the panel members allows for open debate.[53]

Although this method does have limitations, including lack of reliability assessment and potential sensitivity to the experts selected, such limitations also apply—possibly even more so—to other subjective estimates. Selection of the experts also warrants careful attention. Experts from the industry in general, including lead users, can offer their knowledge as a useful benchmark against the estimates generated internally by a firm.

Another useful forecasting tool in high-tech markets relies on *analogous data* to make inferences about the new technology.[54] The basic idea is to use data about another product currently on the market, or one that existed at an earlier time, to forecast a new product's expected growth pattern. For example, in forecasting sales of high-definition TV (HDTV) equipment, forecasts can be based on the history of similar consumer products, such as color TVs or videocassette recorders. In selecting the analogous product, it is critical to establish a logical connection between the two. For example, do the two products serve a similar need or share other important characteristics? One also must take into consideration environmental factors and market conditions that may uniquely affect the new product's growth pattern. Then, based on the sales pattern for the analogous product, the use of intuitive judgment traces the expected pattern of sales for the new product.

This technique is valid only to the extent that the analogy holds true. The degree to which the analogy is appropriate depends on the logical connection between the products involved. For example, in forecasting the demand for personal digital assistants (also known as handheld computers), possible analogous products might include personal computers and cell phones.[55] The degree to which these analogous products are logically connected to handheld computers depends on similarities in the attributes of importance to the consumer in making purchase decisions and in the business factors that contribute to product success. Important attributes include technical support, ease of use, and product form/design considerations. Critical business factors include distribution considerations, brand-name considerations, and model options. Based on consideration of these factors, Handspring, Inc. (Mountain View, CA) concluded that both products served as useful benchmarks but neither alone was entirely appropriately analogous.

Additional techniques might also be useful in making forecasts for high-technology products. The *information acceleration (IA)* technique relies on a virtual representation of a new product to assist in product development and forecasting.[56] Such representations are more vivid and realistic than are traditional concept descriptions and less expensive than relying on actual prototypes. Hence, they provide a useful middle ground between traditional concept descriptions and actual physical prototypes. Feedback from customers is obtained through the use of the virtual representation of the new-product idea.

In addition, merely creating a virtual representation yields several other benefits. First, to simulate a future environment, the design team must agree on the implications of that future environment. This forces the team to carefully define the target group of customers and the core product benefits early in the process. Other issues that are brought to the fore are

- The requisite infrastructure required for product usage (e.g., recharging stations for electric vehicles).
- Technology requirements for future generations of the innovation (e.g., new battery technology for electric vehicles).
- Competitive forecasting of new market entrants.
- Available alternatives to the new technology (e.g., hybrid electric vehicles that combine gas power with electric).

To simulate one product, the team must plan for the entire product line (including vans, two seaters, sedans, etc.) and cannibalization of existing products.

Other Considerations in Forecasting

Whichever forecast method or combination of methods is used, the forecaster must ensure that bias does not enter into the forecast due to personal or organizational desires of success for the technology. Stakeholders in a new technology often inflate predictions of its future success, and "since their bullish statements of technical potential are often misleadingly packaged as precise market forecasts, unwary businesses and investors often suffer."[57] Marketing researchers can avoid bias by studying a new technology's potential buyers, who have less of a stake in its success. However, this is typically not done due to the fact that the group of potential customers can be difficult to reach, making accurate market research expensive and time-consuming.[58]

Another problem with forecasting new technologies is the "cross-competition of current technologies with new technologies serving the same market."[59] Although Chapter 2 addressed the "incumbent's curse," in which existing firms downplay the competitive threat posed by new technologies, new startups also suffer from their own curse: that of overenthusiasm. In developing a forecast for a new technology, managers must consider the entrenched market position of the incumbents. The main advantage of incumbent technologies is that they already have developed markets with established distribution channels and loyal consumers. Also, they have proven production processes and higher production volumes. All of these factors allow established technologies to be marketed with a cost leadership strategy, pushing prices downward and helping them maintain and even increase market share.[60] To avoid forecasting inaccuracies due to this problem, forecasters must fully consider the advantages of the established technologies and, at a minimum, temper their enthusiasm for the success of the innovation while adjusting their predictions of how quickly the new technology will overtake the existing technology.

Many times, decision makers are less than confident in the prepared forecast for a certain technology, and this lack of confidence can sometimes lead to indecisiveness or bad decisions. Although forecasting demand for new technologies is difficult, it is often critical to provide information to decision makers. Forecasters should keep in mind that the success of the forecast is not based on whether it comes true, but on the quality of information provided to the decision makers who are the end users of the forecast.

The Technology Tidbit focuses on the use of information technology for data mining.

TECHNOLOGY TIDBIT

DATA MINING

Effective data mining is all about connecting the dots. Nearly every organization has a database, and most have more than one. If organizations can piece together information from these various databases, they can sell more products or operate more efficiently.

"It's ironic. The more organizations know about themselves and their customers, the less they know what to do about it," says Usama Fayyad, CEO of datamining company digiMine. "Everyone understands the value of data, so they store a lot of it. But they can't always get to their data. And when they get to it, they have trouble making sense of it." Companies like digiMine, Autonomy, ClearForest, and iPhrase Technologies sell software that uses complex algorithms to look for relationships between data points that are either spread across multiple data warehouses or collected into one.

Microsoft's Web site for its popular game console, the Microsoft Xbox, uses digiMine to study Web activity and map it against marketing data. Web logs of Xbox's site visitors are stored in a data warehouse that digiMine hosts, while customers' personal information is stored at Microsoft. Site managers in twenty-seven countries, as well as Xbox marketing staff, can initiate a data query from a Web browser. digiMine then extracts the relevant data, such as players of the game Halo who have read the site's article on "Star Wars: The Clone Wars" and indicated an interest in buying any game within the last ten days by clicking through to one of Xbox online's retail partners. Once digiMine returns the information, Xbox marketers can map the data against its internal database to create targeted email offers. The goal is to sell more products and tailor the site to individuals' tastes. "The information we pull together is central to Xbox marketing," says Scott Pickle, Xbox online's site manager. "It's important for us to be able to segment users and understand their behavior in order to give them what they want."

SOURCE: Grimes, Brad (2003), "Data Mining: The Xbox Files," *PC Magazine,* 22 (11), p. 68.

SUMMARY

A key point to take away from this chapter is the overwhelming need to diligently and assiduously gather information from the marketplace. In addition to collecting information on competitors, high-tech marketers must work with customers—to understand them, to have an ongoing dialogue with them, to study them, and to incorporate their needs into the product development and marketing process. High-tech firms sometimes end up with baffled, frustrated, and unhappy customers, which is itself a threat to the health of the high-tech economy. Technical people find that, much of the time, users don't know what they want; and when they do know, they all want something different. Despite that, users are the customers; developers should delight the customers and be responsive to their needs and anticipate them in designs.[61]

The next chapter takes another step toward understanding customers and explores issues related to customer adoption decisions for high-technology products.

DISCUSSION QUESTIONS

1. Using contingency theory from Chapter 1, identify how marketing research techniques must be matched to the type of innovation to ensure greater success and insight.
2. What is concept testing and how is it used by high-tech marketers?
3. What is conjoint analysis, and how can high-tech marketers use it to refine the product-development process?
4. What is a customer visit program and what benefits does it offer? What are the elements that make a customer visit program successful?
5. What is empathic design? What insights can it generate? What are the steps in the process?
6. How are customer visits similar to and different from empathic design?
7. Who are lead users? What are the four steps in the process of using lead users in marketing research?
8. What is QFD? How does a Kano concept add insights into customer needs? What are the four characteristics necessary for QFD?
9. What are prototype testing and beta testing? How are they different from each other?
10. What are some of the complicating factors in gathering competitive intelligence in high-tech markets?
11. What are some useful tools for forecasting in high-tech markets?

GLOSSARY

Beta test. When a customer is given an early version of a new technology in order to provide feedback to the producer to make needed improvements prior to its commercial release.

Concept testing. The process of evaluating early-stage ideas to determine which of them are good enough to be developed further.

Competitive intelligence. The gathering of information about competitors (who they are, their products, their marketing strategies, and likely responses to the marketing strategies of other firms in the market).

Conjoint analysis. A market research tool in which respondents are asked to make judgments about their preferences for combinations of product attributes; statistical tools are then used to estimate how much each attribute is valued and to use this information in the product design process.

Customer visit program. A systematic program of visiting customers with a cross-functional team to understand the customers' needs, how they use products, and their environment.

Empathic design.   A research technique based on understanding user needs through observation of the customer, rather than through traditional questioning methods (focus groups, surveys).

Kano concept.   A graphical representation of the nature of the relationship between the presence of certain product attributes and customer satisfaction or dissatisfaction. "Must-be quality" attributes are those that, if absent, cause an exponential decrease in satisfaction. "Attractive quality" attributes are those that delight the customer; when present, there is an exponential increase in satisfaction.

Lead users.   Customers who face needs months or years before the bulk of the

marketplace and are positioned to benefit significantly by obtaining solutions to those needs now. In some cases, lead users may have even developed a solution to their needs that marketers can then commercialize for other users.

Prototype testing.   The evaluation of a model of the desired product that replicates the product's critical features while ignoring non-critical features. The first test of a prototype is whether it meets technical specifications. The second test compares the prototype to customer expectations.

Quality function deployment (QFD).   An engineering tool to identify customer requirements and map those requirements onto the product design process.

ENDNOTES

1. Cusumano, Michael and Richard Selby (1995), *Microsoft Secrets,* New York: Free Press.
2. Brown, Eryn (1999), "9 Ways to Win on the Web," *Fortune,* May 24, pp. 112–124.
3. Ibid.
4. von Hippel, Eric, Stefan Thomke, and Mary Sonnack (1999), "Creating Breakthroughs at 3M," *Harvard Business Review,* September–October, pp. 47–57.
5. Leonard-Barton, Dorothy, Edith Wilson, and John Doyle (1995), "Commercializing Technology: Understanding User Needs," in *Business Marketing Strategy,* V. K. Rangan et al. (eds.), Chicago: Irwin, pp. 281–305.
6. Churchill, Gilbert and Dawn Iacobucci (2001), *Marketing Research: Methodological Foundations,* Fort Worth, TX: South-Western College Publisher.
7. Leonard-Barton, Dorothy, Edith Wilson, and John Doyle (1995), op. cit.
8. Stevens, Greg A. and James Burley (1997), "3,000 Raw Ideas = 1 Commercial Success!" *Research-Technology Management,* May–June, pp. 16–27.
9. Iansiti, Marco (1995). "Shooting the Rapids: Managing Product Development in Turbulent Environments," *California Management Review* 38(1): pp. 37–58.
10. For an example, go to http://survey.confirmit.com/wi/p56967573/ctl.asp.
11. The following sources contrast conjoint analysis with the simpler notion of *concept testing* and explain conjoint analysis in great clarity and detail: Moore, William (1992), "Conjoint Analysis," in E. Pessemier, *Product Planning and Management: Designing and Delivering Value,* New York: McGraw-Hill; Green, Paul and Abba Krieger (1997), "Evaluating New Products," *Marketing Research,* 9 (4), pp. 12–21; and Dolan, Robert (1999), "Analyzing Consumer Preferences," *Harvard Business Review,* reprint #9-599-112.
12. McQuarrie, Edward (1995), "Taking a Road Trip," *Marketing Management,* 3 (Spring), pp. 9–21; McQuarrie, Edward (1993), *Customer Visits: Building a Better Market Focus,* Beverly Hills, CA: Sage Publications.
13. Leonard-Barton, Dorothy, Edith Wilson, and John Doyle (1995), op. cit.
14. Leonard-Barton, Dorothy and Jeffrey F. Rayport (1997), "Spark Innovation Through Empathic Design," *Harvard Business Review,* November–December, pp. 102–113.
15. Ibid.
16. Nussbaum, Bruce (1993), "Hot Products," *Business Week,* June 7, pp. 54–57.
17. Leonard-Barton, Dorothy and Jeffrey F. Rayport (1997), op. cit.
18. Ibid.
19. Takahashi, Dean (1998), "Doing Fieldwork in the High-Tech Jungle," *Wall Street Journal,* October 27, pp. B1, B22.

20. von Hippel, Eric (1986), "Lead Users: A Source of Novel Product Concepts," *Management Science,* 32 (July), pp. 791–805; von Hippel, Eric, Stefan Thomke, and Mary Sonnack (1999), op. cit.; Urban, Glen L. and Eric von Hippel (1988), "Lead User Analyses for the Development of New Industrial Products," *Management Science,* 34 (May), pp. 569–582.

21. von Hippel, Eric, Stefan Thomke, and Mary Sonnack (1999), op. cit.

22. von Hippel, Eric (1978), "Users as Innovators," *Technology Review,* 80 (January), pp. 3–11.

23. von Hippel, Eric (1986), op. cit.

24. von Hippel, Eric, Stefan Thomke, and Mary Sonnack (1999), op. cit.

25. Ibid.

26. Ibid.

27. Ibid.

28. Ibid.

29. 3M asks respondents first to sign an agreement granting 3M intellectual property rights for any ideas resulting from the workshop.

30. von Hippel, Eric, Stefan Thomke, and Mary Sonnack (1999), op. cit.

31. Center for Quality Management (1995), *Concept Engineering,* Cambridge, MA.

32. Griffin, Abbie and John R. Hauser (1993), "The Voice of the Customer," *Marketing Science,* 12 (Winter), pp. 1–27; Hauser, John R. and Don Clausing (1988), "The House of Quality," *Harvard Business Review,* 66 (May–June), pp. 63–73.

33. Center for Quality Management (1995), op. cit.

34. Another way to determine the value customers place on various features, attributes, or benefits, to better understand possible tradeoffs, is conjoint analysis.

35. Griffin, Abbie and John Hauser (1992), "Patterns of Communication Among Marketing, Engineering, and Manufacturing: A Comparison Between Two New Product Teams," *Management Science,* 38 (March), pp. 360–373.

36. Hauser, John R. and Don Clausing (1988), op. cit.

37. Clark, Kim and Steven Wheelwright (1992), *Managing New Product and Process Development,* New York: Free Press.

38. Much of this material draws on the insights of Don Kleinschnitz, chief quality officer and former vice president of corporate quality, Storage Technology Corporation.

39. Kawasaki, Guy and Michele Moreno (1999), *Rules for Revolutionaries,* New York: Harper Business.

40. McKenna, Regis (1995), "Real-Time Marketing," *Harvard Business Review,* July–August, pp. 87–95.

41. Available at http://www.symantec.com/corporate.

42. Gilad, Ben (1995), "Competitive Intelligence: What Has Gone Wrong," *Across the Board,* October, pp. 32–36.

43. Ibid.

44. Ibid.

45. Yovivich, B. G. (1997), "Browsers Get Peek at Rivals' Secrets," *Marketing News,* November 10, pp. 1, 6; Graef, Jean (1996), "Using the Internet for Competitive Intelligence," *CIO Magazine,* www.cio.com/CIO/arch_0695_cicol umn.html; Bort, Julie (1996), "Watching Rivals on the Net," *Denver Post,* February 16, p. C1.

46. Mohr, Jakki (1996), "The Management and Control of Information in High-Technology Firms," *Journal of High-Technology Management Research,* 7 (Fall), pp. 245–268.

47. Ibid.

48. This section was written in conjunction with information compiled by Tom Disburg.

49. "Will It Fly," (1998), *Wall Street Journal,* March 19, p. A1.

50. These next examples came from a presentation by Rajesh Chandy, professor of marketing, University of Minnesota.

51. Levary, Reuven R. and Dongchui Han (1995), "Choosing a Technological Forecasting Method," *Industrial Management,* 37 (January–February), p. 14.

52. Makridakis, S., S. C. Wheelwright, and V. E. McGee (1997), *Forecasting: Methods and Applications,* New York: John Wiley (comprehensive and detailed); Kress, G. and John Snyder (1994), *Forecasting and Market Analysis Techniques,* Westport, CT: Greenwood Publishing Group.

53. Krajewski, Lee and Larry P. Ritzman (2002), *Operations Management, Strategy and Analysis,* 6th ed., Upper Saddle River, NJ: Prentice-Hall.

54. Weiss, Allen (1999), *Hitchhiker's Guide to Forecasting,* available at www.marketingprofs. com.

55. This example is based on a presentation by Donna Dubinsky, CEO of Handspring Technologies, to the American Marketing Association Summer Educators' Conference, San Francisco, August 1999.

56. Urban, Glen, John Hauser, William Qualls, Bruce Weinberg, Jonathan Bohlmann, and Roberta Chicos (1997), "Information Acceleration: Validation and Lessons from the Field," *Journal of Marketing Research,* 34 (February), pp. 143–153.

57. Brody, Herb (1991), "Great Expectations: Why Technology Predictions Go Awry," *Technology Review,* 94 (July), p. 38.

58. Ibid.

59. Stevenson, Mirek J. (1998), "Advantages of Incumbent Technologies," *Electronic News,* 44 (July 20), p. 8.

60. Ibid.

61. Wildstrom, Stephen (1998), "They're Mad as Hell Out There," *Business Week,* October 19, p. 32.

RECOMMENDED READINGS

Lead User Process

Donath, Bob (2000), "Irritations Lead Users to Innovations," *Marketing News,* 34 (21), p. 16.

von Hippel, Eric (1988), *The Sources of Innovation,* New York: Oxford University Press.

Lilien, Gary L., Pamela D. Morrison, Kathleen Searls, Mary Sonnack, and Eric von Hippel (2002), "Performance Assessment of the Lead User Idea-Generation Process for New Product Development," *Management Science,* 48 (8), pp. 1042–1059.

Gathering Information on Customers

Barabba, Vincent and Gerald Zaltman (1991), *Hearing the Voice of the Market: Competitive Advantage Through Creative Use of Market Information,* Boston: Harvard Business School Press.

Dickinson, John R. and Carolyn P. Wilby (1997), "Concept Testing With and Without Product Trial," *Journal of Product Innovation Management,* 14 (2), pp. 117–125.

Fink, Arlene and Jacqueline Kosecoff (1998), *How to Conduct Surveys—A Step-By-Step Guide* (2nd ed.), Thousand Oaks, CA: Sage Publications.

Green, Paul E., Abba M. Krieger, Terry G. Vavra (1997), "Evaluating New Products," *Marketing Research,* 9 (4), pp. 12–21.

Guillart, Francis and Frederick Sturdivant (1994), "Spend a Day in the Life of Your Customer," *Harvard Business Review,* January–February, pp. 116–125.

Mello, Sheila (2001), "Right Process, Right Product," *Research Technology Management,* 44 (1), pp. 52–58.

McQuarrie, Edward (1996), *The Market Research Toolbox,* Thousand Oaks, CA: Sage Publications.

Stone, Brad (2002), "Microsoft's Software Snoops," available at www.newsweek.com.

APPENDIX A

Technical Appendix

What Is Conjoint (Tradeoff) Analysis?

BRYAN ORME,

Vice President, Sawtooth Software, Sequim, WA

A great deal of market research commissioned today is descriptive in nature rather than predictive. Descriptive information is useful to characterize demographics, usage patterns and attitudes of individuals. Beyond descriptive information, managers need survey research tools that can predict what consumers will buy when faced with the variety of brands available and myriad product characteristics. It is precisely due to this focus that conjoint (tradeoff) analysis has become so popular over the last three decades.

Humans employ a variety of heuristics when evaluating product alternatives and choosing in the marketplace. Many high-tech products are made up of a dizzying array of features (computers, cell phone calling programs, manufacturing equipment). How does the manager decide what product characteristics, branding strategy or price to charge to maximize profits? And how does the consumer evaluate the offering vis-à-vis other alternatives in the marketplace?

To decide what product to sell, some managers use their own intuition, the recommendations of design engineers, or they look to competitors for indications of what already "works." These strategies are myopic and reactive. In consumer-oriented organizations, potential products are often evaluated through concept (market) tests. Buyers are shown a product concept and asked their purchase interest, or new products are actually placed in test markets. These tests can

be quite expensive and time consuming, and generally investigate just one or a few variations of a product concept. Sometimes survey research has been used, where respondents are asked to check or rate which brands and product features they prefer. None of these approaches by itself has been consistently successful and cost efficient. Conjoint analysis uses the best elements of these techniques in a cost-effective survey research approach.

In the early 1970s, marketing academics applied the notion of conjoint measurement (which had been proposed by mathematical psychologists) to solve these complex problems. The general idea was that humans evaluate the overall desirability of a complex product alternative based on a function of the value of its separate (yet conjoined) parts. In the simplest form, one might assume an additive model. Consider a PC purchase. A consumer browsing the Internet might see the following alternative:

> Dell
>
> 3 GHz processor
>
> 512 MB RAM
>
> 21-inch monitor
>
> $1,399

Again, assuming that this consumer uses some internal, subconscious additive point system to evaluate the overall attractiveness of the offer, the unobserved scores (called part-worths) for each of the attributes of this product for a given buyer might be:

	Part-Worths
Dell	20
3 GHz processor	50
512 MB RAM	5
21-inch monitor	15
$1,399	30
Total Utility:	120

The estimated overall utility (desirability) of this product alternative is equal to the sum of its parts, or 120 utiles. The trick is to somehow reliably obtain these scores from individuals, for the variety of attributes we might include in the product, or that our competitors might include. To do this, one first develops a list of attributes and multiple levels (degrees) within each (see Table 5A-1).

It is easy to see that there are many possible combinations of these attribute levels. In the 1970s, it became popular to print each of many product profiles on separate cards and ask respondents to evaluate them (by either ranking or rating). For example, see Table 5A-2.

By systematically varying the features of the product and observing how respondents react to the resulting product profiles, one can statistically deduce (typically using linear regression) the scores (part worths) for the separate features respondents must have been subconsciously using. In contrast to simpler direct questioning approaches, conjoint survey respondents cannot simply say that all features are important—they must trade off different aspects of the product (as in real life), weighing products featuring both highly desirable and less desirable qualities.

TABLE 5A-1   Combinations of Attribute Levels

Brand	Processor	RAM	Monitor	Price
Dell	2 GHz	256 MB	15-inch	$1,099
IBM	3 GHz	512 MB	17-inch	$1,199
HP	4 GHz	1 GB	21-inch	$1,399
Micron				$1,699

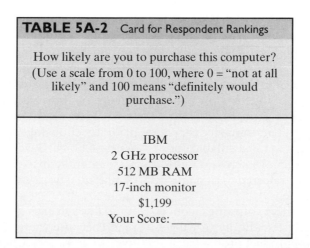

TABLE 5A-2   Card for Respondent Rankings

How likely are you to purchase this computer?
(Use a scale from 0 to 100, where 0 = "not at all likely" and 100 means "definitely would purchase.")

IBM
2 GHz processor
512 MB RAM
17-inch monitor
$1,199
Your Score: _____

Using the attribute list in Table 5A-1, there are 432 possible product profiles (4 brands × 3 processor speeds × 3 memory sizes × 3 monitor sizes × 4 prices) that could be considered! But what makes conjoint analysis work so nicely is that each respondent doesn't have to evaluate all possible product profiles. If we are willing to assume a simple additive model (which tends to work well in practice), each respondent needs to evaluate only a fraction of the total combinations. With our example, only about 20 to 26 carefully chosen product concepts (using experimental design principles of independence and balance) would need to be evaluated to lead to a full set of part-worth scores for each respondent for all 17 attribute levels. The part-worth scores are useful for determining which levels are preferred and the relative importance of each attribute. Once we know these scores, we can simply sum them to predict how each respondent would react to any of the 432 possible product profiles.

Although the scores on the attribute levels provide significant value in and of themselves, the real value from conjoint analysis comes from the "what-if" market simulators that can easily be developed, often within spreadsheets. It follows that if, for each respondent, we can predict the overall desirability for all possible product profile combinations (given the set of attribute levels we measured), we can also predict how each respondent might choose if faced with a choice among two or more competing profiles. For example, we can simulate what percent of the market would prefer each of four PCs (described using the different brands and performance characteristics we measured) if available for purchase. These predictions across a sample of respondents are referred to as shares of choice or preference.

Holding competitive offerings constant, managers can systematically vary the features of their own product profile (such as pricing changes, or performance attributes) and observe what percent of the market would prefer their product under each condition. With conjoint simulators, one can estimate demand curves, substitution effects (e.g., from which competitors do we take the most share if we increase the processor speed?), or cannibalization effects (e.g., what happens to our overall share if we come out with another product with lesser performance at a lower price?). In essence, the manager has the ability to estimate the results of millions of possible concept/market tests based on data collected in a single survey research project among typically 300 to 600 respondents. If additional information is included (such as feature costs), computer search algorithms can find optimal product configurations (holding a set of competitors constant) to maximize share, revenue, or profit.

Since the 1970s, as one might expect, additional improvements and refinements have been made to conjoint analysis. In the 1980s, a computerized version of conjoint analysis called Adaptive Conjoint Analysis (ACA) was developed that could customize the conjoint interview for each respondent, focusing on the attributes, levels, and tradeoffs that were most relevant to each respondent. As a result, even more attributes and levels could be studied effectively than before. In the 1990s, researchers began to ask respondents to simply choose among product profiles rather than rate each profile individually on a numeric scale. The feeling was that buyers in the real world don't actually score each alternative on a rating scale prior to choosing—they simply choose. With Choice-Based Conjoint (CBC), respondents answer perhaps 12 to 24 choice questions such as those shown in Table 5A-3.

Although each question takes longer to read (because there are multiple alternatives to consider), Choice-Based Conjoint

TABLE 5A-3 Choice-Based Conjoint			
If you were in the market to purchase a PC today, and these were your only alternatives, which would you choose?			
Dell 3 GHz Processor 512 MB RAM 21-inch Monitor $1,399	HP 2 GHz Processor 1 GB RAM 17-inch Monitor $1,199	Micron 2 GHz Processor 512 MB RAM 15-inch Monitor $1,099	None: If these were my only choices, I'd defer my purchase.
○	○	○	○

questions seem more realistic and can include a "None" choice that can be selected if none of the products would appeal to the survey respondent. New developments in computationally-intensive statistical methods (hierarchical Bayes estimation) still make it possible to estimate a full set of part-worth scores on each attribute level for each respondent. The results are typically even better than with ratings-based conjoint, and the "what-if"

market simulators are even more accurate in predicting actual market choices.

Today, thousands of conjoint studies are conducted each year, over the Internet, and by fax, person-to-person interviews, or mailed paper surveys. Leading organizations are saving a great deal of money on research and development costs, successfully using the results to design new products or line extensions, reposition existing products, and make more profitable pricing decisions.

CHAPTER 6

Understanding High-Tech Customers

> *Prediction about the year 2000, made in 1975: Cars without wheels will float on air,*
> *bringing about the passing of the wheel.*
> —Arthur C. Clarke,
>
> from *The People's Almanac*

TiVo 1999–2002

TiVo was launched in March 1999 to high expectations. TiVo was the first digital video recorder (DVR). A DVR is like a videocassette recorder except that it records TV shows onto a hard drive instead of a tape. The viewer looks over an electronic programming guide and pushes a Record button on a remote for the shows that she or he wants to record. Another press of the Record button sets the DVR to record every episode of that series automatically. The number of hours that a DVR can record depends on the capacity of the hard drive and the specified recording quality. The benefits of a DVR to the consumer include the ability to pause and rewind live TV, since the DVR maintains a cache of the live program being watched, and the ability to record and retain a program on the hard drive. The net effect is to free the viewer from the networks' program of commercials, fillers, and rigid scheduling.

In 1999, the TiVo set-top box was priced at $499 for a model with up to 14 hours of recording capacity and $999 for a model with up to 30 hours of capacity. The required TiVo service came at an additional charge of $9.95 per month, $99 per year, or $199 for the life of the unit. By May 2000, TiVo had signed up 42,000 subscribers, with a rate of 14,000 new subscribers per quarter. Based on 102 million TV watching households in the United States, that gave TiVo less than a .05 percent penetration rate. This was in spite of TiVo being available in most major consumer electronics stores, generating very high satisfaction ratings, and having 90 percent of customers saying they would recommend it to family and friends.

By March 2002, TiVo was available bundled with satellite television receiver DirecTV and with a special model of AT&T's cable receiver. TiVo's subscriber base was growing by more than 20 percent each quarter and the company had 380,000 customers. However, this number was still far below initial estimates. And, the TiVo customer base did not mirror the general population. The typical TiVo customer was a relatively affluent single male, between 25 and 44 years old, and with at least a college education.

By mid-year 2003, TiVo sales were picking up. In the second quarter of the year, TiVo added 90,000 subscribers, twice the number added in the same quarter of the previous year. Much of this growth was due to a more than 150 percent increase in subscriptions from its relationship with DirecTV. And in September 2003, TiVo dropped the price for a 40-hour recording capacity unit to $199, after a $50 rebate, and the price for an 80-hour recording capacity unit to $299 after rebate. The required TiVo service was priced at $12.95 per month or $299 for the life of the unit. Mike Ramsay, TiVo's CEO, explained, "TiVo's momentum is accelerating. . . . (W)e expect to roll past 1 million subs (subscriptions) during the holiday season" (www.tivo.com.). Nevertheless, 1 million subscribers represents less than a 1 percent penetration rate among TV-watching households almost five years after the product's introduction. This suggests that TiVo has not yet reached beyond the "innovator" or "early adopter" categories of technology adopters. For TiVo to be a success, it had to penetrate the broader consumer market. The questions on the minds of TiVo's executive team were: (1) Why has adoption not been faster? and (2) What needs to be done to penetrate the mass market?

SOURCES: Pogle, David (2003), "For TiVo and Replay, New Reach," nytimes.com, May 29; Wathieu, Luc (2000), "TiVo," Harvard Business School Case 9-501-038; Wathieu, Luc (2002), "TiVo in 2002: Consumer Behavior," Harvard Business School Case 9-502-062.

Business Week states:

> As the consumer market for technology soars, companies that sell stuff ranging from cellular phones and computers to software and Internet services have some surprising blind spots about who their customers are and what motivates them.[1]

In order to develop effective marketing strategies, firms must have a solid understanding of how and why customers make purchase decisions for high-technology products and innovations. For example, take the case of a large company deciding to purchase enterprise resource planning (ERP) software to help manage and integrate a variety of different business functions and applications.[2] Some of the key vendors in the ERP software market include J. D. Edwards, Baan, Oracle, PeopleSoft, SAP, i2 Technologies, Siebel Systems, and Trilogy Software. *Front-office products* are designed both to help companies find and sell to customers and to automate and track data related to sales force management, marketing, and customers. For example, Siebel Systems' programs close the loop between sales, marketing, and customer service. *Back-office solutions* are designed to handle functions that do not interface with customers: functions such as supply chain management, accounting and finance, and so forth.

Effectively implementing enterprise resource planning requires a company to eliminate its stovepipe mentality, which keeps each functional area isolated from the others. Where different functional areas operate autonomously, goals may not be well integrated. For example, salespeople might have been rewarded on volume, operations personnel on cost of products and conformance to specifications, and so forth. The success of ERP planning hinges on integrating and collaborating across functions, such that each is working toward common goals.

Moreover, these programs are not cheap. To install an ERP system in a *Fortune* 500 company may cost tens of millions of dollars in license fees for global rollouts, with an additional expenditure for consulting, typically in the ratio of one to three times the license fees, plus investments in computers and networks. Issues such as application interfaces, compatibility ("plug and play"), interoperability between disparate systems, scalability across the enterprise, and linking new and legacy applications in an enterprise-wide system must be addressed. The time line can take one to three years—or more. The suppliers of these software programs must have an intimate understanding of their customers' operations, concerns, and decision processes in order to market their products effectively.

Firms must look at at least three critical issues in assessing the motivations of customers to buy their products, as shown in the organizing framework in Figure 6-1. Marketing must be tailored to address these issues:

- What affects customers' purchase decisions? What motivates them to buy (or not)?
- Who is likely to buy? Are there categories of customers who are predisposed to adopt an innovation earlier than others? How can technology markets be segmented?
- What affects the timing of customers' purchase decisions? Are they likely to postpone purchases or bypass new generations of technology in anticipation of better options coming in the near future?

As an initial step in understanding the customer behavior of technology buyers, basic models of consumer behavior (business to consumer, or B-to-C) and organizational buyer behavior (business to business, or B-to-B), such as that depicted in Figure 6-2, can be useful. While many companies have concluded that when it comes to high-tech products and services, conventional consumer behavior models "don't go far enough,"[3,4] others believe that the purchase process as depicted in Figure 6-2 is adequate for understanding the broad outlines of buyer behavior in high-tech markets.[5] We utilize this model to illustrate the basic process.

The next sections develop an understanding of customers based on their technology-adoption behaviors. The chapter begins with a brief overview of the purchase process. A discussion follows of the different categories of customers, derived from the

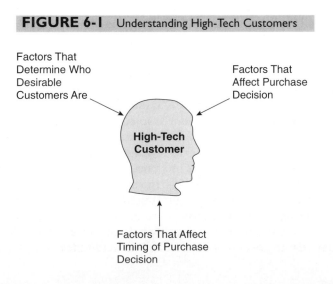

FIGURE 6-1   Understanding High-Tech Customers

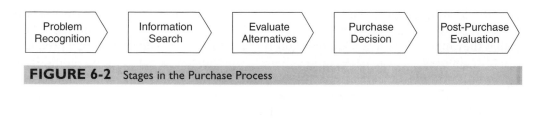

FIGURE 6-2   Stages in the Purchase Process

BOX 6-1

ORGANIZATIONAL BUYER BEHAVIOR IN HIGH-TECH MARKETS

ALLEN WEISS

Marshall School of Business,
University of Southern California,
Los Angeles, California

One defining feature of a high-technology market is the fast pace of technological change. Such change creates high levels of uncertainties for buyers. The uncertainties that buyers face in high-technology markets is really about the *information* in these markets.

Unlike in slower-paced markets, the information in high-technology markets is time sensitive.[a] Time-sensitive information quickly loses its value. For example, computers and the microprocessors on which they are based have been rapidly improving in terms of speed, capabilities, and so forth. Consequently, knowledge associated with a given generation of computer quickly diminishes, and both customers and engineers are finding it difficult to maintain up-to-date knowledge. There are two broad implications for buyers in these markets.

First, as it relates to their purchase behavior, customers who perceive a rapid pace of technological improvements in computer workstations do tend to recognize the short shelf life of the received information. As a result, they tend to search for shorter periods of time because

the information they receive is time bound.[b] In low-technology markets, uncertainties are typically resolved by longer periods of information search.

The short shelf life of information also affects buyers who have an existing relationship with a current (incumbent) vendor, but who decide to consider other vendors for a subsequent purchase. Paradoxically, this causes buyers both to expand their information collection effort and at the same time to restrict the tendency to switch vendors.[c]

For incumbent vendors, these information search characteristics of buyers pose an apparent dilemma. On the one hand, it would appear important for incumbent vendors to convince buyers that they are remaining technologically active. On the other hand, to the extent that such efforts contribute to increasing buyers' perception of technological change, it may create an incentive for buyers to consider the products of competitors. Interestingly, it appears that buyers end up staying with existing vendors. As such, rapid technological change actually buffers incumbent vendors from competition.

Rapid technological change also generates expectations in prospective customers that they may purchase a soon-to-be-obsolete technology. These expectations have been shown to induce prospective customers to *leapfrog* current generations of high-technology products.[d] Presumably, these expectations reduce the perceived benefits of owning a current product generation.

When information is time sensitive and customers anticipate rapid improvements, marketers of high-technology products face several other challenging product decisions. In particular, they must decide when to introduce new generations of a product and whether older generations should be sold concurrently. Although the pressures of producing leading-edge products are high, managers who quickly introduce new generations may both cannibalize their existing products and increase buyers' perceptions that their technology is changing rapidly. This may reduce the benefits of owning a current generation and ultimately encourage customers to leapfrog.

ENDNOTES

a. Glazer, R. and A. M. Weiss (1993), "Marketing in Turbulent Environments: Decision Processes and the Time-Sensitivity of Information," *Journal of Marketing Research,* 30 (November), pp. 509–521.

b. Weiss, A. M. and J. Heide (1993), "The Nature of Organizational Search in High-Technology Markets," *Journal of Marketing Research,* 30 (May), pp. 220–233.

c. Heide, J. and A. M. Weiss (1995), "Vendor Consideration and Switching Behavior for Buyers in High-Technology Markets," *Journal of Marketing,* 59 (July), pp. 30–43.

d. Grenadier, S. and A. M. Weiss (1997), "Investments in Technological Innovations: An Options Pricing Approach," *Journal of Financial Economics,* 44, pp. 397–416; and Weiss, A. M. (1994), "The Effects of Expectations on Technology Adoption: Some Empirical Evidence," *Journal of Industrial Economics,* 42 (December), pp. 1–19.

traditional adoption and diffusion of innovation model. The chapter continues with adaptations to marketing, based on the notion of a "chasm" between early-market adopters and later-market customers. Then a look at segmenting markets presents a process to identify attractive, viable target market opportunities in high-tech markets. The final topic covers the complications in high-tech customer decision making that arise from customers' desire to avoid obsolescence.

CUSTOMER PURCHASE DECISIONS

Process of Customer Purchase Decisions

Problem Recognition

As shown in Figure 6-2, the purchase process begins when the buyer recognizes a need, be it a problem or an opportunity. Need recognition can be stimulated by internal or external stimuli. An example of an internal stimulus is recognition that a bottleneck exists in the order fulfillment process. Advertising might provide an external stimulus, as might insight provided by using the lead user process or from customer complaints.

Information Search

At this stage the buyer actively searches for information about how to solve the problem. This often takes the form of identification of alternatives for solving the problem. The buyer may utilize personal sources such as friends or colleagues, commercial sources such as advertising or a vendor, public sources such as the Internet or reviews in trade publications, or experiential sources such as examining the product. Particularly for distributors or retailers of high-tech products in making decisions about which products to carry, trade shows (such as Comdex, which focuses on computers and computer-related products, and the International Consumer Electronics Show) are important sources of information about new products and cutting-edge technologies. The amount of information required varies by product category and customer type.

Evaluate Alternatives

Evaluation of alternatives for high-tech products and innovations follows Everett Rogers' framework for the evaluation and adoption of innovations.[6] From a customer's perspective, making the decision to adopt a new technology is a high-risk, anxiety-provoking one. The sources of market and technological uncertainty mean that customers are worried about making a bad decision, switching costs involved, training needs, and so forth. Understanding the factors that affect customers' purchase decisions is vital. The critical characteristics that influence a customer's potential adoption of a new innovation are shown in Table 6-1 and discussed here. High-tech marketers must be able to articulate their vision of how their product fares on each of these factors.

1. Relative advantage. Relative advantage refers to the benefits of adopting the new technology compared to the costs. In addition to the price of buying the new technology, the ambiguity of high-tech products can lead to emotional worry, a type of psychic cost. The customer will have fear, uncertainty, and doubt about whether (a) the technology will deliver the promised benefits and (b) the customer will have the skills and capabilities to realize those benefits.

Many high-tech entrepreneurs believe that their invention is the Holy Grail, a better mousetrap, and the next best thing to sliced bread, all rolled into one. However, the factor of relative advantage suggests that it is not sufficient for the inventor to believe that he or she truly has a better product; the improvements must be readily perceived by the customer *and* be worth the monetary and other costs of adoption.

As an example, some question whether high-definition TV (HDTV) really provides a perceived relative advantage to the large majority of consumers. Initially, the relative advantage was discussed in terms of the higher resolution that the digital format provided. The cost of initial sets was in the $2,000 to $3,000 range. Consumers asked themselves whether they needed to see their favorite shows in higher resolution for such a high price tag, relative to standard sets. When this concern was coupled with the reality that TV broadcasters were sending only a portion of their programming in the new digital format, the relative advantage to consumers just wasn't apparent.

TABLE 6-1 Six Factors Affecting Customer Purchase Decisions

1. Relative advantage	The benefits of adopting the new technology compared to the costs
2. Compatibility	The extent to which adopting and using the innovation is based on existing ways of doing things and standard cultural norms
3. Complexity	How difficult the new product is to use
4. Trialability	The extent to which a new product can be tried on a limited basis
5. Ability to Communicate Product Benefits	The ease and clarity with which the benefits of owning and using the new product can be communicated to prospective customers
6. Observability	How observable the benefits are to the consumer using the new product, and how easily other customers can observe the benefits being received by a customer who has already adopted the product

The same phenomenon is true for TiVo. Many potential buyers described it as a "super VCR." Does the enhanced ability to record and play back TV programs on a TiVo justify the additional cost? Based on TiVo's share of the TV watching households, only a small proportion of those households appeared to perceive an advantage.

2. Compatibility. Compatibility refers to the extent to which customers will have to learn new behaviors to adopt and use the innovation. Compatibility with existing ways of doing things, and with cultural norms, can hasten adoption and diffusion of innovation. Products that are incompatible with standard ways of doing things require more time in getting up to speed and require more education from the marketer. Especially in high-tech markets, issues of compatibility arise in terms of offering interfaces to legacy systems (e.g., between new desktop computers and older mainframes in which data are stored) and in terms of compatibility with complementary products (say, between HDTV and programming content or between TiVo and the videocassettes that many TV viewers already owned).

3. Complexity. Complexity refers to how difficult the new product is to use. Very complex products have slower adoption and diffusion rates compared to those that are less complex. Obviously, many new high-tech products are complex. Marketers should ask themselves how they can simplify their products and whether the level of complexity is absolutely necessary, in terms of customer requirements.

4. Trialability. Trialability is the extent to which a new product can be tried on a limited basis. Trialability reduces the risk that potential buyers perceive. This is a major issue since many new products or innovations are perceived as being complex and incompatible with older technologies. New products that can either be tried for a limited time without a commitment or that can be tried on a modular basis are generally adopted more rapidly than products that require irrevocable purchase or that are not divisible. TiVo could not be tried on a limited basis and thus was perceived by many as a risky purchase.

5. Ability to communicate product benefits. The likelihood of customer purchase is influenced by the ease with which the product benefits can be communicated to prospective customers. There are two issues pertinent to high-tech marketers here. First, for many high-tech products, the benefits are difficult to convey to customers. With the HDTV example, what does higher-quality resolution really mean in terms of benefit to the customer?

Second, many high-tech marketers tend to talk in technical terms when communicating about the product. Such communication typically focuses on product features and specifications, rather than the real benefit the customer will receive. For example, in 2003 new computer chips operated in the 2.4 gigahertz (GHz) range, operated with a bus interface (i.e., data transfer speed) in megabits per second (mbps), and offered exciting possibilities in terms of the operating capacity. But what did all this mean to a customer? Simply, it meant speed. However, from a customer's perspective, what did the increase from 1.6 GHz to 2.4 GHz provide? The answer for most customers' uses was "not much." The increase in speed provided only negligible improvements to most of their existing applications.[7]

6. Observability. Observability refers to, first, how observable the benefits are to the consumer using the new product and, second, how easily other customers can observe the benefits being received by a customer who has already adopted the product. Is the increased resolution of the HDTV really noticeable to the average TV viewer? Does the crisper picture really stand out compared to traditional sets? Moreover, can other customers observe the benefits that a user of the new technology gets? For products that are used in a *public* manner and for which the *benefits are clearly observable,* the likelihood of purchase is greater.

These factors must be assessed by inventors of new products in order to understand just how quickly their product might take off in the marketplace. Although the factors sound deceptively simple, they pose crucial barriers that high-tech marketers must overcome. They must educate buyers to overcome the "FUD" factor (fear, uncertainty, and doubt) and highlight benefits. Because breakthrough products don't connect easily with buyers' existing expectations, traditional approaches to marketing—which assume that customers understand the usefulness of the product and have the know—how to evaluate its features—are often insufficient.

Insight about the factors can be gained by involving customers in the new-product development process and by involving innovative customers who might be early adopters in evaluating new-product ideas. If a new idea does not fly well with innovators, it should raise a red flag. Even if the new idea does fly with innovators, it still doesn't guarantee success. However, without excited innovators, a new product rarely survives.

Purchase Decision

During the evaluation stage, the buyer forms opinions about the desirability of different alternatives. At the purchase stage, the buyer reaches agreement with the selected seller on the terms of purchase including: scope of the offering, price, terms of payment, and delivery.

Post-Purchase Evaluation

At this stage, the buyer assesses how well the product has lived up to its potential. Issues such as the following arise for the customer:

- Was I able to successfully learn how to use the new technology?
- Did the technology deliver the promised benefits?
- Were there hidden costs to using the new product?

These post-purchase issues (and potential buyer's remorse) loom large for techbuyers. For example, in the case of the ERP installations mentioned earlier, many expensive implementations have been scrapped, because after years of attempting to configure the organizational processes to reap the benefits, companies are frustrated. They see many hidden costs, such as training, customization, data integration between legacy and new systems, and so forth. In such situations, it is vital that the vendor be diligent in followup, in ensuring that the company and product deliver on its promises, and that this phase of the process be a positive one—particularly if the vendor hopes to rely on customer testimonials or word-of-mouth referrals.

Additional insights regarding the decision-making behaviors of organizational buyers are found in Box 6-1 on high-tech organizational buyer behavior. It is based on the notion that a particularly useful way to understand business buyer behavior is to understand the time sensitivity of information in making technology-related purchasing decisions. Rapid and uncertain changes in the environment make information time sensitive, which leads organizational buyers to use information differently than in more traditional contexts.

In assessing the rate of adoption and diffusion of innovation, inventors must take the perspective of the majority of the possible users of the product and not be biased by their own familiarity and ease with using technology. Nor should they be blindsided by the excitement and ease with which early adopters might use the product.

It is vital to understand which customers might be the first to embrace a new technology. In the case of the Internet, a company's best customers might be the last to embrace it if it poses a disruption in their current routines and procedures or because the Internet might not provide the service and performance such customers prefer. On the other hand, if a company ignores the people who embrace the Internet first, the innovation can creep up like a stealth attack.[8] Moreover, although an early market for a product might exist, innovators are usually not representative of "typical" customers. Therefore, it is vitally important to understand the different categories of consumers in terms of their likelihood of early adoption and what contributes to the disconnect between the early adopters of new technology and the reluctance of the mainstream market.

Categories of Adopters[9]

The categories of adopters discussed in traditional adoption and diffusion models include innovators, early adopters, early majority, late majority, and laggards. Innovators, early adopters and the early majority adopt an innovation *prior* to the average time of adoption, while the late majority and laggards adopt *after* the average time of adoption. Based on his extensive review of the research in this area, Rogers found that earlier adopters (innovators, early adopters, and early majority) tend to be younger and better educated; have greater upward social mobility; a greater capacity to cope with uncertainty and change; and have greater exposure to mass media and interpersonal communications than later adopters (late majority and laggards).[10]

Geoffrey Moore has adapted the theory of diffusion and adoption of innovations for the purchase of high-technology computing products in business markets. His adaptation is shown in Figure 6-3; a summary description of each of the categories is also shown.

Innovators

The early market for high-tech products is comprised of *technology enthusiasts,* people who appreciate technology for its own sake and are motivated by the idea of being a change agent in their reference group. Their interest in new ideas leads them out of narrow circles of peers into broaders circles of innovators. They are willing to tolerate initial glitches and problems that may accompany any innovation just coming to market and are willing to develop makeshift solutions to such problems. Geoffrey Moore believes that the enthusiasts want low pricing in return for alpha- and beta-testing new products. In the computer industry, often these technology enthusiasts work closely with the company's technical people to troubleshoot problems. Although not much revenue may come from this group, it is key to accessing the next group.

Early Adopters

The next category, the early adopters, are *visionaries* in their market. They are looking to adopt and use new technology to achieve a *revolutionary* breakthrough to gain dramatic competitive advantage in their industries. These people are attracted by high-risk, high-reward projects, and because they envision great gains in competitive advantage from adopting new technology, they are not very price sensitive. Customers in the early market typically demand personalized solutions and quick-response, highly qualified sales and support. Competition is typically between product categories (e.g., between DVDs and CDs) at the primary demand level. Communication between possible customer adopters cuts across industry and professional boundaries.

As an example of this early market in the electric car industry, the earliest adopters of electric cars in California paid nearly 25 percent more to lease their cars in 1998 than did people who leased a year later. These visionaries were willing to make do with the higher

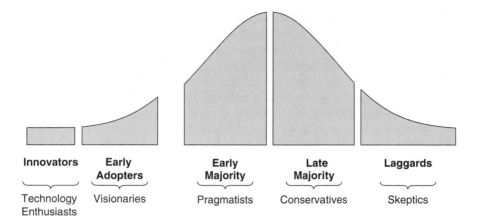

Innovators 〰️ Technology Enthusiasts	Early Adopters 〰️ Visionaries	Early Majority 〰️ Pragmatists	Late Majority 〰️ Conservatives	Laggards 〰️ Skeptics

Descriptions of Customer Categories[a]

Innovators	People who are fundamentally committed to new technology on the grounds that, sooner or later, it is bound to improve our lives. Moreover, they take pleasure in mastering its intricacies, just fiddling with it, and they love to get their hands on the latest and greatest innovations. Thus they are typically the first customers for anything that is truly brand-new.
Early Adopters	The first constituency who can and will bring real money to the table. They help to publicize the new innovations, which helps give them a necessary boost to succeed in the early market.
Early Majority	These people make the bulk of all technology infrastructure purchases. They do not love technology for its own sake, but rather, are looking for productivity enhancements. They believe in evolutionary not revolutionary, products and innovations.
Late Majority	These customers are pessimistic about their ability to gain any value from technology investments and undertake them only under duress, typically because the remaining alternative is to let the rest of the world pass them by. They are price sensitive, highly skeptical, and very demanding.
Laggards	Not so much potential customers as ever-present critics. As such, the goal of high-tech marketing is not to sell to them but, rather, to sell around them.

FIGURE 6-3   The Categories of Adopters

[a]Contributed by Jacob Hachmeister, University of Montana, Missoula, MT.
SOURCE: Adapted from Moore, Geoffrey A. (1991). The Product Adoption Curve in *Crossing the Chasm, Marketing and Selling Technology Products to Mainstream Customers,* New York: HarperCollins. Copyright © 1991 by Geoffrey A. Moore. Reprinted by permission of HarperCollins Publishers Inc.

price and hassles that accompanied being an early adopter. For early adopters of electric cars, the "hassle factors" came in the few stations that were capable of recharging batteries and the limited mileage range (90 miles between charges). The visionaries were willing to accept such inconveniences for the psychological and substantive benefits they received.[11]

Early Majority

The next group, moving into the mainstream market, are the *pragmatists* or the early majority. Rather than looking for revolutionary changes, this group is motivated by *evolutionary* changes to gain productivity enhancements in their firms. They are averse to disruptions in their operations and, as such, want proven applications, reliable service, and results.

Pragmatists generally want to reduce risk in the adoption of the new technology and therefore follow three principles:[12]

1. "When it is time to move, let us all move together." This principle is why adoption increases so rapidly at this point in the diffusion process, causing the landslide of demand.
2. "When we pick the vendor to lead us to the new paradigm, let us all pick the same one." This obviously determines which firm will become the market leader.
3. "Once the transition starts, the sooner we get it over with, the better." This is why this stage occurs very rapidly.

From a marketing perspective, these people are not likely to buy a new high-tech solution without a reference from a trusted colleague. A trusted colleague to a pragmatist is—who else?—another pragmatist, not a visionary or enthusiast who has a different view of technology. Obviously, this need for a reference from a pragmatist poses a real catch-22 to selling to this group: how to get just one pragmatist to buy, when the first won't buy without another pragmatist's reference. Yet, pragmatists are the bulwark of the mainstream market.

Late Majority

The late majority *conservatives* are risk averse and technology shy; they are very price sensitive and need completely preassembled, bulletproof solutions. They are motivated to buy technology just to stay even with the competition and often rely on a single, trusted adviser to help them make sense of technology.

Laggards

Finally, laggards are technology *skeptics* who want only to maintain the status quo. They tend not to believe that technology can enhance productivity and are likely to block new technology purchases. The only way they might buy is if they believe that all their other alternatives are worse and that the cost justification is absolutely solid.

These categories of adopters fall into a normal, bell-shaped curve. Although under most circumstances, a firm would likely target the innovators in a new-product launch, in some cases it might be more worthwhile to target the majority directly instead of the innovators. Firms will find it worthwhile to target the majority[13]

- When word-of-mouth effects are low.
- In consumer products industries (versus business-to-business situations).
- When there is a low ratio of innovators to majority users.
- When profit margins decline slowly with time.
- The longer the time period for market acceptance of a new products.

TECHNOLOGY EXPERT'S VIEW FROM THE TRENCHES

BUILDING A CUSTOMER CENTRIC APPROACH: A SOLUTION PROVIDER PERSPECTIVE

SUDIPTA BHATTACHARYA

Vice President
Manufacturing Solutions Management,
SAP, Dallas, Texas

New technologies, changing business processes, globalization, deregulation, government policies, new competitors, changed alliances, and changed customer behavior are dramatically affecting the business landscape. These changes are requiring businesses to continually and rapidly innovate to stay ahead of their competitors. One way in which customers are developing competitive advantage is through innovative uses of enterprise solutions software, with applications in areas of supply chain management, customer relationship management, enterprise resource planning, and product life cycle management.

Although most customers see enterprise solutions as an enabler of competitive advantage, the rate and timing of adoption vary. There is a set of customers (e.g., pragmatists) that follow the strategy of being followers when it comes to selecting a solution, preferring to go with tested solutions. On the other hand, there are customers (e.g., visionaries and innovators) that want to be first movers and implement the new solutions to create a competitive advantage by moving ahead on the learning curve. Justifications exist for both the scenarios and typically depend on the resources customers are willing to invest and risk, as well as the degree of competitive advantage that they specifically have to gain by being a first mover.

Irrespective of the philosophy adopted, *trust* between the customer and the provider of the enterprise application solution is what ultimately determines success and leads to the most innovative use of the technology.

TRUSTED ADVISOR ROLE

The ability to build trusting partnerships between the users of enterprise software applications and the providers of these applications frequently determines the rate of business innovation and, thus, business success. This requires a continual and close engagement among the customer, application solution provider, and other partners, not just until the sale and implementation are complete but throughout the relationship. There are two key phases in this relationship: initial engagement and recurring engagement.

Initial Engagement: The Engagement Lifecycle

Managing the engagement lifecycle enables the field organization to remain systematically engaged with the customer. It allows customers to gain solution competence earlier in the solution life cycle and to benefit from open account management and interaction with the solution provider team. There are four main phases in the engagement life cycle: Discovery, Evaluation,

Implementation, and Operations and Continuous Improvement.

The *Discovery* phase focuses on creating market awareness of the solutions among prospects and customers as they plan their business activities. The marketing and initial sales efforts aim at building awareness among potential customers, understanding their high-level needs, and making suggestions about areas where the solution can help them achieve their business goals.

In the *Evaluation* phase, cross-functional account teams are deployed to gain an in-depth understanding of the customer's operations, challenges, and opportunities. Only then is a customer-specific recommendation made. During this time, close interaction with the customer is essential to ensure that the proposed solution integrates with and enhances the customer's existing business and technology model and that best practices and key performance indicators are put in place.

In the *Implementation* phase, a detailed design of the customer's specific solution requirements, along with services tailored to deliver added value to the customer's business, are developed. During this process, proven tools and methods are used to increase the efficiency of the implementation and to minimize risk and total cost of ownership.

The *Operations* phase is initiated after the delivery of the business solution. In this phase the account team assists the customer in stabilizing its operations and focuses on initiating the process of measuring the actual performance relative to the targets identified in the earlier phases. The impact assessment of the project and learning for *continuous improvement* are key objectives here.

Recurring Engagement: The User Group Influence

Once customers are established and are familiar with the solution, they can provide important inputs into the product development process. This is typically done through the user groups that are comprised of customers who have implemented the solution. The user group model standardizes the process of communicating ideas and feedback about product, solution, service, and support issues directly to product managers and developers. The user group meetings allow members to provide inputs on: (1) continuous improvement, (2) product modifications in future releases, and (3) strategic product direction.

The *continuous improvement* level is an opportunity for members to fine tune existing solutions and submit development requests either individually or as a group of like-minded customers. These are typically minor modifications that may reflect industry-specific requirements or ease-of-use requirements.

In the case of *product modifications for future releases,* members are solicited for input prior to the development process. Usability testing, surveys, and polls are all done at this stage to determine modifications that satisfy the majority of customer needs. This is also an opportunity for customers to align their business requirements with products currently under development.

At the *strategic product direction* level, innovative solutions of the future are raised and discussed. This level draws heavily on emerging industry trends that customers identify as requiring enterprise solutions. These ideas are also discussed with other stakeholders, including the financial and academic communities, with their feedback being systematically incorporated into the decision process.

By following these processes in engaging with the customer, vendors of enterprise solutions will be viewed as trusted advisors by customers, and will be more likely to develop innovative solutions tailored to better meet customers' needs.

This chapter's first Technology Expert, Sudipta Bhattacharya, the Vice President of Manufacturing Solutions Management at SAP, explains how, regardless of whether customers are innovators or more mainstream customers, the role of the high-tech marketer as a trusted advisor is key.

Each category of adopters has unique characteristics. Moore characterizes the degree of these differences as gaps between each group in the marketplace.[14] These gaps represent the potential difficulty that any group will have in accepting a new product if it is presented in the same way as it was to the group to its immediate left. Each of the gaps represents an opportunity for marketing to lose momentum, to miss the transition to the next segment, and never to gain market leadership, which comes from selling to a mainstream market. The differences between the early market (innovators/early adopters) and the mainstream market (early majority) are more pronounced than differences between the other categories, and hence warrant special attention.

Crossing the Chasm[15]

The largest gap between categories of adopters is between the early market (innovators/early adopters) and the mainstream market (early majority, late majority, and laggards). This deep and dividing schism is the most formidable and unforgiving transition in the adoption and diffusion process. The **chasm** is the gulf between the visionaries (early adopters) and the pragmatists (early majority, mainstream market) and derives from critical differences between the two. Visionaries see pragmatists as pedestrian, whereas pragmatists think visionaries are dangerous. Visionaries will think and spend big, whereas pragmatists are prudent and want to stay within the confines of reasonable expectations and budgets. Visionaries want to be first in bringing new ideas to the market, but pragmatists want to go slow and steady. The chasm arises because the early market is saturated but the mainstream market is not yet ready to adopt. Hence, there is no one to sell to.

What contributes to this chasm, and how can it be overcome? The nature of a firm's marketing strategy in selling to visionaries is very different than the marketing that is required to be successful with pragmatists. Many firms do not understand this difference and are unable to make the necessary shift in strategies to be successful.

Early-Market Strategies: Marketing to the Visionaries

As mentioned previously, visionaries require customized products and technical support. Because such customization for several visionaries can pull a firm in multiple market directions, they can be a costly group of customers to support. However, for a new high-tech startup, sales to these visionaries represent the initial cash flows to the firm. Hence, given the demand from visionaries and need for cash flows, there is much pressure both to support their customization needs and to release products early to these customers. Just as customization can pull the firm in multiple market directions at a steep cost, early release of a product can backfire if it has not been adequately tested.

The goal of the marketer's firm at this point is to establish its reputation. In new high-tech startups, this time of selling to the early market is exciting and energizing. The product is often the focus; engineering and R&D folks play a critical role, and brilliance and vision are embraced. Firms try to develop the *best possible* technology for the market they pursue.

The Chasm

The bloom falls off the rose, however, when the firm takes on more visionaries than it can handle, given the high degree of customization and support they expect. No pragmatists are yet willing to buy, presumably because there is no credible reference for them. Hence, revenue growth tapers off or even declines. The goal of the high-tech marketer should be to minimize the time in the chasm. The longer the firm spends in the chasm, the more likely it is that it will never get out.

One implication of the chasm relates to relationships with venture capitalists and investors. Lack of knowledge of the existence of the chasm can create a crisis. Key personnel become disillusioned and management becomes discredited. Investors may pull out at the very time that more financing is necessary to get the product to the mainstream market. The ultimate demise of early-market success stories might be explained by the existence of this chasm.

The chasm in the Internet arena is often found after a company has made a decision to add an online sales channel—many times to the detriment of existing distribution channels. Due to this cannibalization, sales may actually decline in the short term after adding the online channel. As stated by Intel's VP of business development in discussing Internet investments, "In getting from point A to point C, B is hell." B is where revenues decline and profits go down, but there is simply no way for large companies to get to where they need to be in the future without going through it.[16]

Marketing to the Pragmatists

In contrast to marketing to the visionaries, who are willing to tolerate some incompleteness in the product and will fill in the missing pieces, marketing to the mainstream market requires that the vendor assume total responsibility for system integration. This need demands the development of a *complete, end-to-end solution* for the customer's needs, or the **whole product.** Identifying the whole product requires an exhaustive analysis of what it takes to fulfill the reasons the customer is buying. Asking what else the customer will need, from a systems perspective, makes apparent possible switching costs and exposure. For example, in the computer industry, the whole product includes hardware, software, peripherals, interfaces and connectivity, installation and training, and service and support. In conducting electronic business over the Internet, the whole product includes Web site design, hosting of the site on a server, connection to the Internet, security, financial transactions, and, depending upon the purpose, customer relationship management.

The job of the firm in the chasm is either to develop or to partner with firms to provide the whole solution to the initial customers in the mainstream market. Rather than developing the "best possible solution," the goal here is to develop the *best solution possible.* Yet, this is a very different skill set than that required to succeed in the early market. Now, the R&D team, rather than basing development on engineering solutions, must work closely with partners and allies on a project-oriented approach. For many, this is much less exciting than the pursuit of technological brilliance and requires painstaking work on compatibility, standards, and so forth. Moreover, this period may require that engineers go to customer sites to observe them in action. Customer service is a critical component of crossing the chasm. The vendor goal at this point is to bring in revenue.

Another crucial strategy in crossing the chasm and speaking to the needs of the mainstream market is simplifying, rather than adding additional features. Gadget makers tend to make new models bigger and more complicated—but not necessarily better—right at the time the mainstream market would buy if they were more user friendly. For example, Microsoft's Internet Explorer 5.0 did a good job of *not* focusing on dramatic breakthroughs. Rather, it focused on "thoughtful little advances that made using the Web simpler for average, nontechnical folks."[17] Similarly, the Palm III, introduced in 1998, recognized that the real growth in its sales would come from the explosion of third-party software and accessories that enhanced its abilities to link up with corporate networks.[18] And rather than clogging its new handheld device with many extra features, 3Com kept Palm III simple.[19] Because the developers took an evolutionary rather than revolutionary view, and because the Palm III was even easier to synchronize with a PC compared with the prior Palms, they followed the high-tech model perfectly.

Communication between pragmatic customers in the mainstream market tends to be vertical, or within industry and professional boundaries (rather than horizontal, across industry boundaries, as in the early market). Recent research shows that higher levels of communication between the early market adopters and the main market adopters is a key factor in mitigating the sales slump of the chasm.[20]

Competition is between vendors within a single category of solutions or product offerings. Indeed, pragmatists will want to see a competitor's proposal and product offering before making a decision. Competition, to the pragmatist, is actually a sign of legitimacy for the new technology. Customers typically demand some sort of industry standard to minimize their perceived risk.

Firms that succeed in the mainstream market have complemented their initially strong competencies in technological development with equally strong competencies in partnering and collaborative skills. Partners often drive further expansion, and so the firm's ability to interact with partners becomes a critical success factor. For example, although Intel's Pentium III chip offered "only a very modest speed improvement—at most 10%—over existing chips,"[21] to really take advantage of the improved speed, Intel had to work closely with software developers to get Pentium III–enhanced products to market quickly. Partnering requires critical skills that were discussed earlier in Chapter 3.

SAP, maker of enterprise resource planning (ERP) software, did an excellent job of identifying what a whole product solution would look like and creating a series of partnerships to develop the whole product. SAP developed a "solution map" for each of seventeen vertical markets, including automotive, media, oil and gas, and utilities, to name a few. The maps identified every function in each industry, specifying where SAP's software already offered a solution, where products from partner software developers were required, and where SAP would fill in gaps later. In essence, its solution map functioned as a technology map for SAP, corresponding closely to the steps and needs of high-tech product development (see Chapter 7). Moreover, in terms of partnerships, SAP readily recognized that it needed the smaller developers, but the smaller developers were somewhat hesitant to partner; they wondered how long their investment in developing their own company's products based around an SAP product would be viable, or whether SAP would ultimately enter that market too. So, to entice small developers to partner, to some extent SAP had to guarantee that it would not enter the partner's space in a two- to three-year time frame. (Despite this, SAP and its partners sometimes experienced conflict. For example, in working with i2, SAP announced it would enter i2's lucrative business of supply chain software, and the partnership died.) The challenge for the smaller developers was to partner with key market players (such as SAP, Oracle, or PeopleSoft) as a seal of approval in order even to be considered a viable option by large *Fortune* 500 customers (the pragmatists in the mainstream market). The smaller developers had to show *Fortune* 500 companies how well their products worked with the companies' existing ERP installations.

Partnerships generally revolve around the issues of power.[22] In the early market (enthusiasts and visionaries), power belongs to the technology providers and the systems integrators (firms that bring different suppliers' products together to create one integrated solution for a customer's needs). These market players (technology providers and systems integrators) make the decisions about whom to bring into the process as partners. In crossing the chasm and approaching the early mainstream pragmatists, the power is centralized in the hands of the company that has effectively picked the target customer, understands why they buy, and designed the whole product. In the pragmatist market, the market leader and its partners have the power. In later markets (mainstream conservatives), power is vested in the distribution channels or companies that provide superior distribution of product.

Until high-tech firms have established themselves in the mainstream market, they have not established staying power. How to succeed when the mainstream market takes off is addressed in Geoffrey Moore's second book, *Inside the Tornado,*[23] in which there are three distinct phases: the bowling alley, the tornado, and "Main Street."

The Bowling Alley.   The bowling alley is a period during which the new product gains acceptance in niche markets within the mainstream market, but has yet to achieve general, widespread adoption. During the bowling alley stage of market development, the market is typically not large enough to support multiple industry players. The successful firm will establish itself as the dominant market leader. One of the best ways to become the market leader is to follow a "whole product" strategy and partner extensively to create the *de facto* standard in the market.

The Tornado.   The *tornado* is a period when the general marketplace switches over to the new technology. It is driven by the development of a "killer app," or an application of the technology that is based on a universal infrastructure, is appealing to a mass market, and is commoditizable. For example, DVD players are the fastest growing consumer electronics product in history. In 2002, sales grew 39 percent and penetration levels grew to 35 percent of U.S. households. Similarly, the Internet exhibited a rapid takeoff. In three years' time, the Internet developed into a trading center of 90 million people. In contrast, it took radio thirty years to reach 60 million users, and TV fifteen years.[24]

The massive number of new customers entering the market in a rapid time period can swamp the existing system of supply. During this stage, companies have a huge opportunity to develop their distribution channels. In fact, when companies hit this stage of the cycle, they need to focus on operational excellence: getting their products out to the consumers through streamlining the creation, distribution, installation, and adoption of their whole product. This is typically best done by beefing up internal systems to handle the high-volume workload.

An important caveat during the tornado: Do not bet on preventing a tornado. If a market leader begins to develop, even if it is not your company, it is important to switch efforts to follow the emerging leader.

As noted previously, competing effectively in the tornado requires effective partnering skills. Moore looks at a number of the issues that must be addressed, one of the most important of which is: How do you dance with the market leader, "a gorilla," in a tornado and come away in one piece?[25] The answer is to sustain only enough ongoing innovation to stay out of the market leader's reach. In other words, don't try to pass the market leader by creating the next big thing. Do enough to remain stable, but not so much that the gorilla feels threatened—or it will crush you!

Main Street.   *Main street* refers to the period when the tremendous growth in the early majority/pragmatist market stabilizes. This period of after-market development is when the base infrastructure for the product's underlying technology has been deployed and the goal now is to flesh out its potential. Rather than focusing on generating sales from new customers, companies must sell extensions of their products to their current customer base to be competitive. Overall, it is important to emphasize operational excellence and customer intimacy, rather than product leadership.

Marketing to Conservatives

Finally, for continued success in the mainstream market, the high-tech firm will also need to reach out to the conservative market. This requires making the product even simpler, cheaper, and more reliable and convenient, and possibly splitting the product line into simpler components. From an engineering perspective, this is the anathema of good engineering work. Rather than adding more interesting features and cool "wow" factors, engineers should actually do the opposite. Because this is so foreign to product development in high-tech firms, many high-tech companies leave the conservatives' money on the table.

In summary, crossing the chasm requires a different type of marketing than the marketing strategies used in the early market. The whole product is the critical success factor in crossing the chasm and reaching out to the mainstream market, yet bringing the whole product together is expensive and time-consuming. Moore argues that the high-tech firm must make a decision about what one key market segment to put its resources behind and focus its efforts there. Going after too many market segments at once results in spreading its resources too thin and not building a strong reputation within that segment.

THE CHOICE OF CUSTOMER: SEGMENTING MARKETS AND SELECTING TARGET MARKETS

One of the most important issues with which high-tech firms wrestle is the choice of an initial target market to pursue with their promising new technologies. Yet, the choices that seem obvious in hindsight are rarely clear at the time of the decision. For example, Intel cofounder Gordon Moore rejected a proposal in the 1970s for a home computer built around an early microprocessor. He didn't see anything useful in it, so he never gave it another thought. In a list of possible uses for its 386 chip (written before the IBM PC), Intel omitted the personal computer, thinking instead of industrial automation, transaction processing, and telecommunications.[26]

The idea behind segmenting markets and selecting a target is to identify group(s) of customers who share similar needs and buyer behavior characteristics and who are responsive to the firm's offering. Directing marketing efforts toward a specific target is both more effective and more efficient than loosely attempting to reach as many customers as possible in the hopes that some of them might be interested and respond.

TABLE 6-2   Steps in the Segmentation Process

1. Divide the market into groups, based on variables that meaningfully distinguish between customers' needs, choices, and buying habits.
2. Describe the customer's profile within each segment.
3. Evaluate the attractiveness of the various segments and select a target market.
4. Position the product within the segment selected.

As shown in Table 6-2, market segmentation includes four steps.

1. ***Divide possible customers into groups,*** based on important characteristics that distinguish between customer groups in terms of the choices they make and the reasons they buy. For consumers, the traditional bases of segmentation include

   - Demographic variables, such as age, income, gender, occupation, and so forth.
   - Geographic variables, such as geographic location, rural versus urban, and so forth.
   - Psychographic variables, or consumers' values and beliefs that affect their lifestyles and, hence, purchasing behavior, such as an orientation toward a healthy lifestyle or being technologically current or environmentally friendly.
   - Behavioral variables related to the customer's behavior with respect to the specific product category, such as

     - Frequency/volume of usage of a product (i.e., heavy versus light users).
     - The benefits desired in a product (i.e., ease of use).
     - The usage occasion (such as use for work versus home).

   For businesses, traditional segmentation variables, such as SIC (Standard Industrial Classification) or NAICS (North American Industry Classification System) codes, firm size, and corporate culture, can be useful variables.

2. ***Profile the customers in each segment*** by describing a "typical" customer within each segment. A study compiled by technology consultant Forrester Research Inc., in conjunction with NPD Group, provides an example of steps 1 and 2.[27] This study identified ten key consumer segments based on traditional demographics combined with technology-related behaviors. Dubbed "Technographics," the study polled 131,000 consumers about their motivations, buying habits, and financial ability to purchase technology products. The ten segments are described in Figure 6-4.

   The insights offered by the Technographics segments influence the way technology companies make, sell, and deliver products. For example, consider two couples:

   - Cindy and Gary Williams, ages 46 and 44, respectively, from Tulsa, Oklahoma. Cindy is an administrative secretary for a health maintenance organization; her husband is a maintenance supervisor. They have two sons, ages 11 and 12. They have one PC they bought three years ago and have no Internet connections. They are considering an upgrade because their sons want speedier games than their sluggish machine can play.

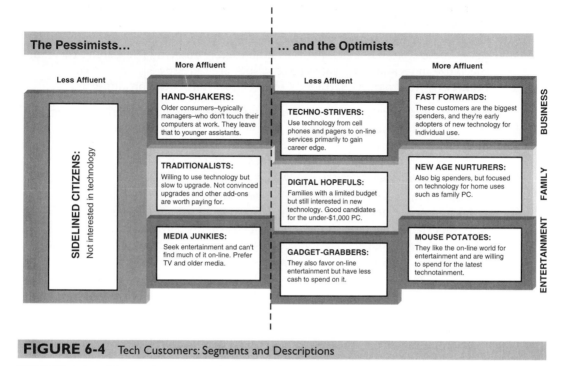

FIGURE 6-4   Tech Customers: Segments and Descriptions

SOURCE: Adapted from Judge, Paul (1998), "Are Tech Buyers Different?" *Business Week,* January 26, pp. 64–66.

Their family status and income are two traditional signposts that would high-light them as promising technology buyers. However, Forrester Research claims those factors are misleading and any technology company pitching products to the Williamses would likely be wasting its money. Its Technographics study pegs the Williams as *Traditionalists*—family-oriented buyers who are relatively well off but remain unconvinced that upgrades or other new technogadgets are worth buying. The key factor in their profile is the three-year age of their PC, making it ancient by tech standards. Traditionalists wait a long time before upgrading, mak-ing them not a very fertile part of the technology market.

- Carol and Robyn Linder, ages 46 and 53, respectively, from Milwaukee, Wisconsin. Carol is a customer service manager for Ameritech and her husband is a CPA; they have three school-age children, two pagers, and three PCs. Robyn spends time online for work.

Although similar to the Williams family in income and family status, the Linders are classified as *Fast Forwards,* using computers and other gadgets for job, family, and individual pursuits. So technology companies would find it desirable to target them.

Similarly, cable TV giant TCI knows that speed and performance were impor-tant to early users of cable modems, but those benefits are not necessarily the most compelling reasons for new types of buyers. Rather, kid-friendly Internet marketing targeted to family-oriented *New Age Nurturers*—the group focused on technology for home uses—or *Mouse Potatoes*—the group that likes entertainment-related technology—would be more successful.[28]

Research that focuses specifically on high-tech buyers can provide insights that more traditional segmentation variables may not. Other research firms that specialize in technology-related buyer issues are Odyssey Research (San Francisco), Yankelovich partners' Cyber Citizen, and SRI Consulting, Inc.

3. ***Evaluate and select a target market.*** After identifying meaningful segments in the market and understanding the customers within each of those segments, the third step in the segmentation process requires that the firm evaluate the attractiveness of the various segments in order to help narrow its choice of which to pursue. Four important criteria on which to evaluate each segment follow

- *Size.* Estimates of the potential sales volume within each segment are needed in order to identify segments that are large in size. Don't make the mistake of basing size estimates on the number of customers *per se,* because it is their purchase volume that counts. Segments with fewer people may have larger dollar purchases. For example, the 80–20 rule says that 80 percent of the sales in any one category are typically purchased by only 20 percent of the customers.

- *Growth.* Estimates of the growth rates of various segments are also needed to help evaluate possible attractiveness. Segments that are growing in size are attractive for at least two reasons. First, the growth means that the firm will be able to capitalize on customers' needs and grow with the market. Second, the growth means that, rather than stealing customers away from other firms, firms can capture new customers coming into the market.

- *Level of competition.* Estimates of the level of competitive intensity within each segment help a firm to identify how costly it may be to pursue that segment. High numbers of competitors or even a few powerfully entrenched competitors can pose formidable risks to a new firm. The issue is not so much "Is the new firm's technology better?" but how hard these competitors will work to defend their existing base of customers.

- *Capabilities of firm to serve the needs of that segment.* Finally, firms must take a good hard look at their core competencies and strengths to determine if they have the capabilities to serve the needs of a particular segment. Although partnering can augment some deficiencies, the reality is that customers will look to firms that offer the right set of skills to address their needs.

Geoffrey Moore refers to the selection of a target market as the identification of a **beachhead**, or a single target market from which to pursue the mainstream market.[29] A good beachhead requires that customers have a single, compelling, "must have" reason to buy that maps fairly closely on to the capabilities of the firm. A variety of compelling reasons to buy, and their relative attractiveness to different categories of adopters, include the following:

- Purchase of the new technology provides the customer a dramatic competitive advantage in a previously unavailable domain in a critical market. This reason-to-buy is difficult to quantify in terms of costs/benefits, which, although appealing to the visionary, is unpalatable to pragmatists and conservatives.

- Purchase of the new technology radically improves productivity on an already well-understood critical success factor, and there is no other alternative to achieving a comparable result. This reason has the greatest appeal to a pragmatist, because the cost savings in terms of a better return on resource expenditures can be quantified, typically in terms of incremental dollars.

- Purchase of the new technology visibly, verifiably, and significantly reduces current total overall operating costs. This reason will have the greatest appeal to a conservative because of the hard dollar savings. However, the risk factors surrounding the new technology are still too high for conservatives to take a chance, and the surrounding infrastructure for the whole product may not be sufficiently developed to convince these conservatives to take the purchase risk.

Moore believes that only the second reason represents a good choice for crossing the chasm, because it speaks directly to the pragmatist's concerns to generate incremental revenues from technology investments. The issue is whether the firm's capabilities offer this compelling reason for the customer.

For example, the digital networking firm Wam!Net, based in Minneapolis in 2000, provided a digital document delivery service to businesses. It identified the publishing industry as a strong beachhead. Customers such as Time Warner use thousands of suppliers to print their materials, and providing digital delivery improves both the quality and the speed of the process. Purchase of the new technology radically improved productivity on already well-understood critical success factors (quality and speed), and there were no other alternatives to achieving a comparable result. The cost savings could be quantified fairly easily. Wam!Net has identified adjacent segments with similar needs and positive spillover (word of mouth) effects in the entertainment industries.

The beachhead should provide clear opportunities to enter adjacent segments. If one uses a bowling pin analogy, the beachhead should be the lead pin; adjacent market opportunities would be the pins immediately behind the beachhead. The objective of the bowling pin model is to approach new market opportunities by leveraging knowledge about technologies or segments. As Figure 6-5 illustrates, adjacent pins are comprised either of (1) new segments into which the firm is selling existing applications or (2) developed segments into which the firm is selling new applications.

These adjacent segments can be "knocked off" more easily because of word-of-mouth relationships between customers in the two segments or similarities in whole product needs. Indeed, the definition of the whole product should be done within the confines of a single target market because needs will be relatively common within that target market but may differ significantly across segments. Moreover, the segment should be capturable in a short period of time.

Many high-tech firms make the mistake of attempting to pursue too many market segments at the outset. They are enthusiastic about the potential their innovations can offer to many different types of segments and are unwilling to limit their potential market opportunities. They also want to hedge their bets against selecting the wrong segment and so pursue several. However, most firms simply do not have the resources to be effective in multiple segments. They end up spreading their resources too thinly to be effective in any of the segments and, as

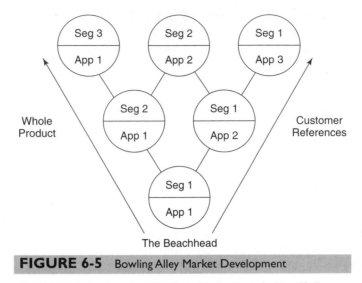

The Beachhead

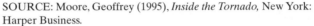

FIGURE 6-5 Bowling Alley Market Development

SOURCE: Moore, Geoffrey (1995), *Inside the Tornado,* New York: Harper Business.

a result, fail. A firm must be ruthless in its paring of its opportunities. Indeed, success in just one segment can be the catalyst required to succeed in other segments.

Some companies also may find it beneficial to penetrate more deeply into a segment by developing products or services that are related to the original application. The package of products and services comprises a total solution that addresses more of a customer's need. The classic example is Microsoft's development of its Office Suite of application programs that complements its operating system. Sprint PCS, one of the top six wireless telephone companies, also does a good job selling data services. Sprint now has over 5 million customers paying extra to use data services on its nationwide network. The popularity of emailing photos, which Sprint PCS has advertised heavily, is a major driver of data services. It should be noted that a market penetration strategy may keep a company focused within one segment, allowing competitors to outflank it.

The Technology Expert's View from the Trenches highlights one company's experience in selecting a beachhead.

4. *Position the product within the segment.* The fourth and final step of the segmentation process is to create a meaningful market position for the new technology. A market position is the *image* of the product in the *eyes of the customer, relative to competitors,* on critical attributes of importance. There are several important aspects of this definition.

First, a market position is based on customers' perceptions. After all, customers will be making the decision to purchase (or not purchase) the new technology, and what matters is what they believe about the new technology. Whether the firm thinks the customers' perceptions are wrong is totally irrelevant. The

---

TECHNOLOGY EXPERT'S VIEW FROM THE TRENCHES

A VERTICAL MARKETING SUCCESS, A BUSINESS FAILURE
CHRIS HALLIWELL,
Technology Marketing Consultant,
Los Altos, California

As a consultant in the area of strategic technology marketing, I've seen a fair number of success and failures over the years, particularly as companies make decisions about what their choice of an initial target market should be.

One Silicon Valley startup with which I've worked made network devices that allowed smaller companies to share a connection to the Internet among a few personal computer users. The company executed a successful positioning strategy based on ease of use and low price; it backed up that position with wide distribution through value added resellers (VARs).

The startup entered an emerging horizontal market, targeting small businesses who needed to give employees access to the Internet. This company's efforts and this target market were virtually unnoticed by the dominant player, Cisco Systems, who was focused on selling similar technology, but at higher prices for more complex solutions, directly to large enterprise accounts. However, the startup soon began receiving surprising multiple unit orders directly from large enterprises.

It is common for new market opportunities to present themselves as a confusing swarm of sales activity in unintended accounts. Although new sales often occur in a seemingly random pattern, the simple task of sorting will reveal areas of concentrated demand. In turn, based on this analysis, targeted sales programs and target segment solutions can be defined.

So, in order to make some sense of the initial sales activity, the startup's marketing department started sorting new sales invoices by order size and industry segment. In this case, the largest and most frequent orders were from consumer goods manufacturers (e.g., beverage firms) and from regional retail store chains. Similarities and differences in the two segments were identified as follows: Both segments were using private Internet networks to transmit financial information between a central corporate location and outlying locations (from the consumer manufacturer to its distributors, and from the corporate retail location to its retail outlets). Both segments valued the startup device's low cost and ease of use for remote locations with a few personal computers and little support staff. But there was one important difference that made one vertical market more attractive and less dangerous to the startup: Consumer goods distributors partnered with, but were not owned by, the manufacturer; as a result, they operated in a less controlled, heterogeneous computer and network environment. However, in the retail segment, the corporate parent typically owned local outlets, issuing corporate mandates for homogenous technology infrastructure. Because of standardization in this segment, it was clear that if Cisco were to sell a device at a low price, corporate retail offices would mandate a consistently all-Cisco network. In fact, retail segment customers were handing the startup's sales pitches to Cisco salespeople in order to convince Cisco that they needed to supply a low-cost solution!

The startup's marketing team argued to management that they should avoid the retail segment and quickly defined a "partner network" solution package to be sold through a new direct sales channel to consumer goods manufacturers. The challenge with this strategy, as with any vertical market strategy, is that it necessitates focus on the right segments and the ability to turn down business from the wrong segments. Sales in the consumer goods segment grew at a good pace, but retail segment customers continued to tempt management with the possibility of very large orders if only the startup could make its device more Cisco-like.

The startup had recently gone public, and pressure to grow from the financial community was intense. So, despite the marketing team's advice against pursuing the retail segment, the startup began complicating its network device designs to meet this segment's requests; perversely, this design change negated the company's core competitive advantages in simplicity, low cost, and easy maintenance. The company's existing value added resellers and small business users became confused and disappointed, company sales efforts became defocused and sluggish, new product launches multiplied and fell behind schedule, and the startup was eventually acquired at bargain prices by a Cisco competitor.

The lessons learned from this high-tech company's efforts are that:

1. Early adopters for new technology initially present themselves as in a horizontal array, with a few early adopters from each vertical market. Soon, one or two verticals, for whom the technology is strategic, will emerge as early adopting markets overall.

2. It is important to identify patterns of activity in the initial sales that a company gets. This sorting helps the company reveal areas of concentrated demand that will need similar "whole product" solutions and respond similarly to marketing messages.

3. Not all customer business is created equal! A dollar from some customers can actually cost a business money in the long run and even threaten the company's survival as resources become stretched thin and core advantages are compromised. A company must be willing to make tough calls regarding which customer dollars it wants to chase.

4. The exigencies of the financial pressures to perform can frequently pull a company in a direction that does not make sense. Discipline, courage, and vision can help to mitigate this unrelenting pressure—but it also takes experience and insight to know how to make that call.

issue is for the firm to use communications with customers to effectively create the market position the firm desires.

Second, a market position is always relative to competitors. Many new high-tech firms believe that they have no competition. Their innovation is so radical that no other firm provides anything remotely similar. Although this may be technically true, the customer's reality is that there are always other options (i.e., competition). The customer can choose to do things the old way or even do nothing. This is why positioning of the new product is generally achieved by focusing on how the new technology fits within existing market categories by referencing the older technology that is being displaced (i.e., product form competition).

It is tough to be successful marketing a product that violates the categorical scheme of the marketplace, which creates confusion both among consumers and in the retail channel.[30] If positioned as something that is totally new, retailers will not know (a) if it is something that their store should carry or (b) which department to put the product in. As a result, consumers will not know where to find the product. Moreover, if positioned as something totally new, pragmatist consumers will not be able to compare products because they don't know what to compare it to. Firms may need to collaborate with their rivals who also offer the new technology to overcome the prior technology successfully.

As the new technology is adopted by customers and begins to diffuse through the marketplace, competition based on the new technology will develop. Relative positioning strategies must now implicitly reference this brand competition (i.e., selective demand). Finally, in late stages of the adoption process, the market leader needs to create new products that will cannibalize its old products.[31] As a result, positioning may then reference the company's prior version of the product as the competition.

This chapter's Technology Tidbit focuses on a radical innovation that poses interesting issues for customer behavior/decision making.

TECHNOLOGY TIDBIT

SPACE TOURISM

According to preliminary market surveys, there are 10,000 would-be space tourists willing to spend $1 million each for a two-hour flight into space. Indeed, more than 130 people have already paid a deposit to Space Adventures in Arlington, VA to hold their spots, tentatively slated for flight in 2005.

The costs involved are astronomical: $10,000 per pound is what it costs to launch a simple satellite into orbit, with no oxygen, life support, or return trip necessary—not to mention the liability insurance!

Because NASA has no interest in this market, a host of smaller companies are working to design a reusable launch system that's inexpensive, safe, and reliable. Kelly Space (San Bernardino, CA) has a prototype that looks like a plane with rocket engines. Rotary Rocket in Redwood City,

Rotary Rocket

California, has a booster-with rotors to make a helicopter-style return to Earth. Kistler Aerospace in Kirkland, Washington has an elaborate parachute system.

Indeed, the market for space tourism is heating up. Hilton and Budget are planning to build space hotels, and some are talking about using Mir as a hotel.

The final topic in this chapter on the decision-making process and concerns of customers is the issue of customer strategies to avoid obsolescence. Customers' desires to avoid obsolescence require that high-tech marketers proactively manage upgrade options for customers.

CUSTOMER STRATEGIES TO AVOID OBSOLESCENCE

High-tech markets are blessed (cursed?) with fast and significant (revolutionary) improvements, which result in "inflection points" or technological discontinuities in the marketplace. The steady stream of improved and overlapping product generations typically makes the customers' investments in prior generations obsolete, even while those investments are still perfectly functional in use.[32] For example, computer microchips show a rapid pace of constantly improved generations available to the marketplace, even while customers are using prior generations. Indeed, as noted in Chapter 2's discussion of technology life cycles, successive generations tend to arrive when the current generation's sales curve is still rising and may continue to rise for some time.[33]

In high-tech markets, customers must make important decisions about if and when to adopt a new generation of technology. In the extreme, customers may "leapfrog," or pass entirely on purchasing, a current generation of technology in anticipation of a new, better innovation coming down the pike in the near future. This **leapfrogging** behavior, based on customer expectations of imminent improvement, can have a chilling effect on sales of current products.[34] The customer says, "If I wait to buy tomorrow, the product will not only be cheaper, it will also be better."

In essence, the realities of customer decision making create a tension for the firm between providing state-of-the-art technology through the introduction of new-product generations and customers' expectations and fears of obsolescence. Customer investments in equipment from a prior generation creates a "footprint of the past." Hence, high-tech marketers must manage interfaces between these **legacy systems** and newer generations. Moreover, marketers' decisions regarding both the launch of a new generation and the withdrawal of the old become equally significant with overlapping generations. Marketers must also consider how to help customers migrate from one generation to another.

Customer Migration Decisions

What affects a customer's decision to adopt the new generation of technology? *Customers' expectations about the pace and magnitude of improvements* in the marketplace play a big role in their adoption decisions. More specifically, customers form expectations of the pace and magnitude of performance improvements and price decreases. The customer must balance the value of the existing products against the value of new offerings and even future arrivals. When products improve rapidly and significantly, the old adage about starting with a high price (to "skim" those who buy early) and then lowering price to entice later purchasers may not hold.[35] In general, the greater the anticipated product improvement or expected price decline, the greater the customer's propensity to delay purchase. Moreover, if the customer has already purchased, then the greater the product improvements and price decrements of successive generations, the greater will be the customer's regret factor—especially for early adopters.

The need to manage upgrade options can be especially sensitive for business customers, who often face a gap between a product's useful technological life (shorter than three years, in many cases) and its "accounting" life (for depreciation purposes, usually five years for durable assets). The race to market accelerates customer demand for low-cost solutions that provide benefits of "new and improved" without scrapping the old version entirely. Upgrades allow business customers to protect their investments in technology in these circumstances.

Upgrades for businesses are particularly important in driving growth in tech spending. Spending on new technology and technology upgrades by businesses fell 6.2 percent between 2001 and 2003;[36] however, business outlays to capitalize on the automation of business processes via the Internet (known as "e-business," see Chapter 11) grew 4 percent in 2003. Indeed, e-business expenditures comprised 27 percent of total tech spending by businesses in 2003, representing an increase of 11 percent in e-business budgets. Tech spending by enterprise/business customers is vital for personal computer hardware and software makers, because these buyers in the corporate market allow hardware and software makers to survive the razor-thin margins in the high-volume home PC market. But some corporate buyers have decided that the price/performance calculations don't warrant buying next-generation computers just for fractions of increases in speed. Because most firms can make the case that the technology in place is pretty workable, what is needed to motivate another buying round for new technology is a real technological breakthrough or a "killer app" that convinces buyers their dollars are worth it.

For example, some argue that Wi-Fi technology (for Wireless Fidelity, otherwise known as 802.11b) that allows users to surf the Web wirelessly, is a killer app that is driving customers to adopt new Wi-Fi enabled computers (for example, with Intel's Centrino chip) that have Microsoft Windows XP operating system (which is adapted to handle Wi-Fi). Ownership of Wi-Fi enabled devices in turn will fuel demand for broadband connections, paving the way for the next generation of Internet services.[37]

Marketers' Migration Options

Marketers must help customers to manage the transitions between generations. One way they do this is to offer a **migration path**, or a series of upgrades to help the customer's transition between generations.[38] The various options along the migration path are based on the *degree to which the customer's options in the transition are more constrained versus enlarged.*

1. *Withdraw the older generation as soon as the new one is launched, with no assistance to the* **installed base** (the set of existing customers that adopted the firm's prior-generation technology). Lack of parts and service for these customers forces them into a decision about migration sooner than they might like.

2. *Withdraw the older generation when the new one is launched, but offer migration assistance,* which can be in the form of technical help, trade-ins, backward compatibility with gateways, and the like. Customers can upgrade and maintain the old version or move to the new version later.

3. *Sell old and new generations together for a period of time,* after which the old generation is withdrawn. Customers can continue with installed version A, migrate to next generation B, or skip B entirely (leapfrog) and go to C.

4. ***Sell both generations as long as the market desires them*** and provide migration assistance to the installed base.

Based on an options model developed by Grenadier and Weiss,[39] the firm's choice of which migration path to offer to customers is affected by the following factors, shown in Table 6-3.

Expectations of Pace of Advancements

When customers expect rapid advances, albeit small ones, it pays for a firm to increase options. Customers who anticipate a rapid pace of change tend to wait for price declines or bug fixes in the newly launched version. These customers can also bypass completely, waiting for some yet-to-be-launched future version.[40] Both stalling and leapfrogging are mitigated by migration assistance. Without such assistance, firms will find their revenues swinging wildly.

Expectations of Magnitude of Advancements

In contrast to the pace of advancements, customers who expect significant advances in technology recognize that smooth upgrades are simply not possible. In such a situation, few customers are willing to wait to purchase an older generation of the product at a reduced price, which will be made obsolete. In essence, there is less to be gained by keeping the customers' options open. As a result, where customers anticipate large discontinuities between generations, the firm may choose not to offer migration assistance. Even if the firm did, the reality is that the customers' existing investments have been destroyed.

Customer Uncertainty

When customers have fear, uncertainty, and doubt about their expectations, such a situation warrants migration assistance. A firm can choose to sell both the old and new versions to encourage customers with even older installed versions to migrate to the next step.

Hewlett-Packard is also using pricing strategies to manage customer migration decisions. Because demand drops for older Unix servers drop when new machines are nearly ready for the market, HP is using sophisticated dynamic pricing software to

TABLE 6-3   Migration Considerations and Options

Customer Perceptions	*Implication for Customer Behavior*	*Implication for Migration Path*
Customers expect rapid pace in technology advancements	Willing to wait for price declines	Marketer should provide migration assistance
Customers expect large magnitude of change in technology advancements	Recognize smooth upgrading is unlikely; therefore, waiting to purchase an older model at a lower price may result in obsolescence	Migration path less crucial, because the latest technology effectively obsoletes any path that was available
Customers have anxiety about making a decision	Need to feel that their decisions are safe	Marketer should provide migration path, possibly selling old and new models together for a period of time

analyze market trends. Rather than trying to slash prices on old machines too late in the process, HP calculates when to start discounting the old machines and by how much. So far, the markdowns made possible by this pricing strategy to manage customer migration decisions helped increase sales 1-1.5% in 2002.[41]

Any decision to offer upgrades must also account for complexities in managing relationships across the supply chain:

$$\text{Supplier} \rightarrow \text{Manufacturer/OEM} \rightarrow \text{Channel Members} \rightarrow \text{Customers}$$

Revenue from upgrades often flows to other members of the channel (e.g., the supplier in the case of chip upgrades, or manufacturers in the case of add-on components) and is smaller in size than that from selling entirely new units. Upgrades can also cannibalize sales of new products. So, the revenue implications and possible conflicts with other members of the supply chain must be monitored carefully with any decision.

SUMMARY

This chapter has provided an in-depth look at issues related to customer purchase decisions: what affects their purchase decisions, different categories of adopters, segmenting markets and selecting an attractive target, and customer strategies to avoid obsolescence. Marketers must understand where the market is in terms of its adoption of a new technology and the factors that have the greatest influence on the buying decision for the next wave of adopters. This can be accomplished only with an effective market research and segmentation process. The firm's marketing strategy should be designed to address the needs of buyers in the segment(s) it targets.

DISCUSSION QUESTIONS

1. What are the stages in the purchase process? What are the implications for marketing strategy in each stage?
2. What factors influence a customer's potential adoption of a new innovation? What are the implications of each of the factors for high-tech marketers?
3. What are the categories of adopters and their characteristics? What are the appropriate marketing strategies for each of the categories?
4. What is the chasm? Compare and contrast the marketing strategies that are necessary in the early market versus the mainstream market. For the conservatives?
5. What are the three phases of the pragmatist market (from *Inside the Tornado*)?
6. What are the four steps in segmenting markets?
7. What insights about market segmentation does the Technographics study offer?
8. What makes a good beachhead?
9. What is product positioning and how should a new innovation be positioned?
10. What are the issues in understanding customer strategies to avoid obsolescence? When should a migration path be offered?
11. How does knowledge of customer behavior in technology markets lead to insight in marketing? Provide examples.

GLOSSARY

Beachhead. A single target market from which to pursue the mainstream market. A good beachhead requires that customers have a single compelling, "must have" reason to buy.

Chasm. The large gap between the early market (innovators and early adopters) and the mainstream market (early majority, late majority, and laggards) in the adoption and diffusion process.

Installed base. Customers who have bought prior generations of a particular technology.

Leapfrogging. Passing entirely on purchasing a current generation of technology in anticipation of a new, better innovation coming down the pike in the near future. Leapfrogging

behavior, based on customer expectations of imminent improvement, can have a chilling effect on sales of current products.

Legacy systems. Investments in prior technology.

Migration path. Marketing tools (upgrades, pricing strategies, etc.) to help customers move from a prior generation of technology to a new generation.

Whole product. A *complete solution* of what it takes to fulfill the reasons the customer is buying. For example, in the computer industry, the whole product includes hardware, software, peripherals, interfaces and connectivity, installation and training, and service and support.

ENDNOTES

1. Judge, Paul (1998), "Are Tech Buyers Different?" *Business Week,* January 26, pp. 64–66.

2. Kirkpatrick, David (1998), "The E-Ware War," *Fortune,* December 7, pp. 102–112.

3. Judge, Paul (1998), op. cit.

4. Interested readers may also want to review the book by Allan Reddy (1997), *The Emerging High-Tech Consumer: A Market Profile and Marketing Strategy Implications,* Westport, CT: Quorum Books.

5. It is also important to note that some believe there is no such thing as a high-tech consumer and that consumers approach the buying decision for technology products in the same way they would any product. See Cahill, Dennis and Robert Warshawsky (1993), "The Marketing Concept: A Forgotten Aid for Marketing High-Technology Products," *Journal of Consumer Marketing,* 10 (Winter), pp. 17–22; and Cahill, Dennis, Sharon Thach, and Robert Warshawsky (1994), "The Marketing Concept and New High-Tech Products: Is There a Fit?" *Journal of Product Innovation Management,* 11 (September), pp. 336–343.

6. Rogers, Everett (1983), *Diffusion of Innovations,* New York: Free Press.

7. Wildstrom, Stephen (1999), "Pentium III: Enough Already?" *Business Week,* March 22, p. 23.

8. Byrnes, Nanette and Paul Judge (1999), "Internet Anxiety," *Business Week,* June 28, pp. 79–88.

9. Much of the material in this section is derived from Moore, Geoffrey (1991), *Crossing the Chasm, Marketing and Selling Technology Products to Mainstream Customers,* New York: HarperCollins, and Rogers, Everett (1995), *Diffusion of Innovations (4*th *ed.),* New York: Free Press.

10. Ibid.

11. McKim, Jenifer (1998), "Recharging Ahead," *Missoulian,* October 25, p. G1.

12. Moore, Geoffrey (1995), *Inside the Tornado,* New York: HarperBusiness.

13. Mahajan, Vijay and Etian Muller (1998), "When Is It Worthwhile Targeting the Majority Instead of the Innovators in a New Product Launch?" *Journal of Marketing Research,* 34 (February), pp. 488–495.

14. Moore, Geoffrey (1991), op. cit.

15. The majority of the information in this section is derived from Moore, Geoffrey (1991), op. cit.; revised edition (2002) by Harper Business.

16. Byrnes, Nanette and Paul Judge (1999), op. cit.

17. Mossbert, Walter (1999), "New Microsoft Browser Adds Some Nice Details for Simpler Use of Web," *Wall Street Journal,* March 18, p. B1.

18. Wildstrom, Stephen (1998), "The PalmPilot Flies Higher," *Business Week,* March 23, p. 20.

19. Himowitz, Michael (1998), "The Palm Pilot Sequel Is a Hit," *Fortune,* April 13, pp. 154–156.

20. Goldenberg, Jacob, Barak Libai, and Eitan Muller (2002), "Riding the Saddle: How Cross-Market Communications Can Create a Major Slump in Sales," *Journal of Marketing,* 66 (April), 1–16.

21. Wildstrom, Stephen (1999), op. cit.

22. Moore, Geoffrey (1995), op. cit.

23. Ibid.

24. Hof, Robert (1998), "The Click Here Economy," *Business Week,* June 22, pp. 122–128.

25. Readers interested in the idea of market gorillas might reference Moore, Geoffrey A. (1999), *The Gorilla Game: Picking Winners in High Technology,* HarperBusiness.

26. Gross, Neil and Peter Coy with Otis Port (1995), "The Technology Paradox," *Business Week,* March 6, pp. 76–84.

27. Judge, Paul (1998), op. cit.

28. Ibid.

29. Moore, Geoffrey (1991) op. cit.

30. Moore, Geoffrey (1995), op. cit.

31. Ibid.

32. John, George, Allen Weiss, and Shantanu Dutta (1999), "Marketing in Technology Intensive Markets: Towards a Conceptual Framework," *Journal of Marketing,* 63 (Special Issue), pp. 78–91.

33. Norton, John A. and Frank Bass (1987), "Diffusion Theory Model of Adoption and Substitution for Successive Generations of Technology Intensive Products," *Management Science,* 33 (September), pp. 1069–1086.

34. Weiss, Allen (1994), "The Effects of Expectations on Technology Adoption: Some Empirical Evidence," *Journal of Industrial Economics,* 42 (December), pp. 1–19.

35. Dhebar, Anirudh (1996), "Speeding High-Tech Producer, Meet the Balking Consumer," *Sloan Management Review,* Winter, pp. 37–49.

36. Mullaney, Timothy (et al.) (2003), "The E-Biz Surprise," *Business Week,* May 12, pp. 60–66.

37. Green, Heather (2003), Wi-Fi Means Business," *Business Week,* April 28, pp. 86–92.

38. John, George, Allen Weiss, and Shantanu Dutta (1999), op. cit.

39. Grenadier, S. and Allen Weiss (1997), "Investments in Technological Innovations: An Options Pricing Approach," *Journal of Financial Economics,* 44, pp. 397–416.

40. Weiss, Allen (1994), op. cit.

41. "The Price Is Really Right," *Business Week,* March 31, 2003, pp. 62–67.

CHAPTER 7

Product Development and Management Issues in High-Tech Markets

By the 21st century, scientists will be seriously considering mobile floating cities, which by 2000 will have been built in prototype. These will be used for space launchings, for plants, for processing food from the sea, and for special research.
—LLOYD V. STOVER, 1975

senior research scientist,
University of Miami
Institute of Marine Science

IBM's Strategic Shift Into Parts and Services

For much of its history, IBM was a vertically integrated hardware company, manufacturing everything from chips and disk drives to final assembled machines. Reflecting this history, its moniker originally stood for "International Business Machines." IBM came to dominate the computer industry in the 1960s and 1970s by making and selling "big iron" mainframe computers. In the early 1980s, it entered the personal computer market. Its initial success in PCs was short-lived as companies like Compaq, Dell, and myriad clone makers competed on price and pressured its margins. By the early 1990s, IBM was a company in disarray. The mainframe business was in decline, while the PC business was unprofitable. In 1992, under Chairman and CEO John Akers, IBM surprised analysts with a $5 billion loss instead of a projected $4 million profit. It became clear that IBM would have to cut tens of thousands of jobs for survival. Some industry analysts even thought IBM was headed for bankruptcy due to its size, bureaucracy, and insular corporate culture. John Akers resigned and the Board of Directors appointed Lou Gerstner, an outsider, as Chairman and CEO, with a mandate to turn around the company.

Gerstner observed that making and selling hardware "boxes" was becoming a commodity business. There were too many companies competing on the basis of price; margins were declining. He made two important product strategy decisions with regard to new growth opportunities for IBM. First, he decided to sell IBM's chips and technology to other companies, a major departure from the previous practice of exclusively using parts and technology in-house. Sony and Nintendo became important chip customers in the video game industry, as did Motorola in semiconductors and Dell in disk drives.

Second, Gerstner learned from large business customers the enormous difficulty they faced in managing complex information infrastructures requiring integration between different platforms across globally dispersed locations. So he made a major push into services by offering customers the option of having IBM become an outsource service provider, handling customers' computing needs in areas such as Web hosting and other information technology services.

As a result of this strategic shift, in the first quarter of 2001, revenue from parts and service exceeded 50 percent of IBM's total revenue for the first time in its history. In fiscal year 2000, IBM's global service business had revenue of $33 billion compared to number two Electronic Data Systems' revenue of $19 billion. During Gerstner's nine-year tenure, IBM's stock price grew from $13 to $120, an impressive contribution to shareholder value. IBM's turnaround is credited, in part, to Gertsner's bold move in shifting the company's product focus to parts and service. Product management, in high-tech companies, must include products, their underlying technology, and associated services in order to enhance profits.

SOURCES: Bulkeley, William F. (2001), "Gerstner's Legacy at IBM is Likely To Be Switch from Hardware to Help," *Wall Street Journal,* June 11; Gerstner, Louis V. (2002), *Who Says Elephants Can't Dance?* New York: Harper Collins Publishers.

The product development process in high-technology environments relies on many of the same concepts used in product planning in more traditional environments. For example, firms must be concerned with different ways of categorizing products. The most common classification for high-tech products (from Chapter 1) classifies innovations on a continuum ranging from incremental to radical innovations. Recall that incremental innovations are continuations of existing methods or practices and may involve extensions of products already on the market; they are evolutionary as opposed to revolutionary. Radical, or breakthrough, innovations employ new technologies to solve problems and, in so doing, often create totally new markets. Breakthrough innovations represent new ways of doing things. For success in commercializing innovations, the different types of innovations must be managed differently. Chapter 1 identified this notion as the contingency model for high-tech marketing.

Product managers also rely on the notion of product life cycles, which is modified in a high-tech environment. A technology life cycle (see Chapter 2) is the relationship between investments in improving the underlying technology of a product and its price–performance ratio. Typically, this relationship is S-shaped: Initial investments in a new technology may show modest improvements in the product's price–performance ratio, but after some threshold, additional investments show a drastic improvement in price–performance, which then levels off at some point. Importantly, technology life cycles help managers to understand that when a new technology enters the market, the new and the former technologies will compete with each other for a period of time, until the new technology eventually supercedes the former. Technology life cycles allow a product manager in a high-tech environment to anticipate when new technologies may supercede existing technologies and highlight the need to always be on the cutting edge. For example, as the life cycle of dial-up modem technology wanes, AOL Time Warner is trying to get its existing Internet customers to upgrade to a broadband DSL (digital subscriber line) or cable modem connection.

Classifications of innovations, technology life cycles, and adoption and diffusion of innovations are but three of the product management concepts on which high-tech marketers rely. With what other issues do product managers in high-tech environments need to be concerned? In this chapter, we delve much more specifically into other product development and management tools used in a high-tech environment. One important issue is to monitor technology trends and use these to guide technology development within the firm. Another is the decision about how to develop the technology. Firms may choose to develop new technology in-house, with their own resources and skills. Alternatively, high-tech firms may choose to partner with others in developing new technologies and products. Partnering allows firms access to others' skills and resources, which can both hasten development time and lower costs.

High-tech marketers must also consider how far to develop the technology prior to offering it for sale in the marketplace. For example, a firm may decide to market and sell basic know-how (say, through a technology transfer or licensing arrangement) or develop a whole product (from Chapter 6), marketing and selling a ready-to-use product. The decision about how close to final form the product should be developed in the product development process is not a clear-cut decision but one that warrants consideration. A related important issue is when to launch a new product. The timing of market entry is so important that it is usually made at the business unit or corporate level. The advantages and disadvantages of being a pioneer versus a follower have been discussed in Chapter 2.

Other product development/management issues in a high-tech environment that will be discussed in this chapter include the following:

- How should be a firm conceptualize its product architecture including its approach toward modularity, platforms, and derivatives?
- How should a firm use cross-functional teams in its new product development process?
- When or how should a firm halt investments and development in a new product whose success looks questionable?
- What are the unique issues involved in the development of technology-related services?
- How should intellectual property rights be controlled?

These and related topics are explored in this chapter. To get an idea of the product development issues confronting a small technology company, see the Technology Expert's View from the Trenches.

TECHNOLOGY DEVELOPMENT

A major technology trend over the past fifty years has been the "digital revolution" or the transition from analog to digital technology. Analog devices use as input continuously varying real-world signals (such as pressure, sound, or heat) and convert them into electrical signals like current and voltage. During this conversion process, there is some signal loss or distortion. Digital devices use as input electrical signals coded as zeros and ones, manipulate them, and transmit output as electrical signals, quickly and efficiently. Therefore, digital systems (based mainly on digital devices) have greater

CREATING NEW PRODUCTS
HANS WIJMANS
Director of Research and Development
Membrane Technology and Research (MTR), Inc.
Menlo Park, CA

If I were to ask you to give me some examples of new inventions, chances are you will mention high-tech products from the computer, communications or biotechnology industries. More easily overlooked are the less glamorous inventions that nevertheless have a major impact on our lives. For example, there are the tremendous innovations in the automotive industry which make your current car safer, more environmentally friendly and a better value all around than your previous car. Another less flashy category of innovations are the novel systems that are able to separate chemicals from each other and which reduce waste and increase energy efficiency in chemical processes, which is the business I am in.

The development of new products and/or technologies is a very costly process with no guarantee of success. So why do companies make the investment? Because their future depends on it! New products are a company's life-line and their successful development is essential to the long-term viability of the company. How does a company maximize its chances in this high-stakes game of poker? There are two different strategies to choose from.

The first strategy is *market-driven* in which a company conceives of a new product that it believes will fill a need in the marketplace. The company makes the product a reality using whatever technology is appropriate; if it does not have access to the technology required, the company develops the technology or acquires it. The risk of the market-driven approach is that one can be blind-sided by a radical new technology which changes the market as one knew it.

The second strategy is *technology-driven* in which a company creates and develops a new technology with the expectation that new products will be generated along the way. Most start-up companies are technology-driven and these companies dominate the development of truly new products which revolutionize the world. But it is equally true that the success rate of these companies is very low. The risk is that the technology does not work or that the company finds out that nobody has a need for it. So technology developers need to be mindful that success is not a nifty new technology, but rather, a useful product based on that technology. That means, among other things, that developers have to venture outside the laboratory and talk to potential users.

I believe that the two approaches are not exclusive and are best combined. The "best case" scenario that I have in mind starts with a technology-driven start-up which finds, through hard work and luck, the first winning application for its novel technology. The hurdle to market entry for a new technology is high, so this first product must be clearly superior and not just marginally better than the existing product. The company now has its first customer base and develops the infrastructure required to sell to these customers. Better understanding of the needs of these customers leads to product diversification, yet all products are still based on the original technology developed. Selling these additional products is somewhat easier, because the company and its technology are now known to the customers. The company is doing well and identifies potential applications for its technology in other industries. This leads to diversification of its customer base and to further growth of the company. The company realizes that other products, not based on its core technology but likely based on a related technology, could be sold through its existing marketing and sales organization. The company then decides to either develop or acquire the technology required to do so. The end result is a successful company, technology-driven and market-driven at the same time.

Not surprisingly, the above scenario is the one we at MTR are using. As a small R & D company, we pioneered a new way to separate organic vapors from gas streams based on a selective membrane permeation process. Membrane permeation is somewhat like filtration except that molecules rather than particles are separated. Based on this new technology, we have developed applications in the petrochemical industry and we sell to major petrochemical companies worldwide. The company has made the transition from generating ideas to selling products, and sales to the petrochemical industry have increased six-fold from 1996 to 2003.

Several years ago MTR identified natural gas processing as the next market for our technology. Using our existing marketing and sales effort, we made a few sales and became even more convinced that the opportunity for us in natural gas was real. However, it also became clear that, on our own, we would not be able to reach all potential customers. Additionally, natural gas project tend to be large and some of our customers have been reluctant to award such projects to a small business like MTR. Therefore, in 2002 we entered into a Marketing and Sales agreement with the natural gas equipment business of ABB Lummus, a leading engineering and construction firm. One year later, we are seeing the fruits of this collaboration with several orders placed and more in the pipeline.

MTR's future ambitions extend well beyond the petrochemical and natural gas industries. Our current Research and Development projects span a broad range of applications and the entire spectrum of membrane separation processes. These projects are expected to provide MTR with more products and more customers in the future.

---

speed and accuracy than analog systems. In industries as diverse as automobiles, computers, music, movies, television, and telecommunications, the move toward digital technology has produced better quality for consumers and tremendous business opportunities and challenges for technology companies. In mobile handsets, for example, Motorola lost its market leadership position to Nokia in the mid-1990s, in part, because it was slow in making the transition from analog to digital cell phones.

This transition from analog to digital devices highlights the need for high-tech companies like Motorola to monitor technology trends. An important tool used to monitor technology trends and systematically manage a company's resources is a **technology map.** A technology map defines the stream of new products, including both breakthrough and incremental products, that the company is committed to developing over some future time period. Companies that are most successful in defining next-generation products use the map to force decisions about new projects amid the technological and market uncertainty found in high-tech markets.[1] The use of a technology map can promote cohesion and commitment to new-product development plans and can be used to clarify possible sources of confusion, to allocate resources, and to make tradeoffs among various projects. Importantly, technology maps are not cast in stone but are updated and revisited regularly. Hence, rather than being a tool that inhibits innovation and creates blinders for a firm, a technology map should serve as a flexible blueprint for the future, updated and revised regularly.

Capon and Glazer offer the following steps in developing and managing technology resources[2] (see Figure 7-1).

1. Technology identification.   Technology identification requires taking an inventory of the firm's know-how to find those ideas having the most value. Technology know-how can be found in products, processes, and management practices. Although most firms can easily identify the technology that forms the basis of their products, it is more difficult to identify process technology that might have value outside the firm itself. And organizational and managerial know-how can be equally difficult to identify. However, as "best practices" routines are becoming more common—for example, superior skills in total quality management, electronic commerce, customer service, and the like—such organizational and managerial know-how has also become a revenue-generating asset. For example, IBM, under Gerstner, identified microprocessor chip manufacturing technology and consulting services as new sources of revenue (see vignette on page 200). At this stage, the firm is most concerned with its current technologies.

2. Make decisions about technology additions.   The technology identification from step 1 may highlight areas of weakness in the firm's strategy, identifying areas where it needs additional technologies in its platform and suggesting decisions about how to

---

FIGURE 7-1   Technology Map

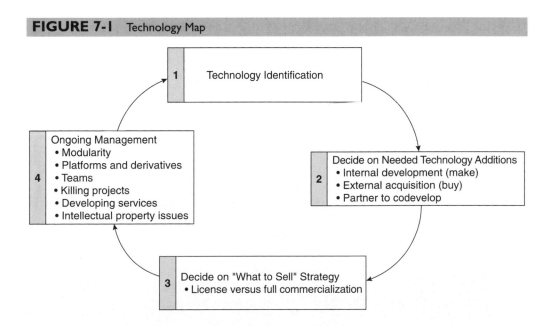

add them. Technology additions arise when the firm, through its technology identification process, has recognized technology arenas in which the firm would like to have skills or products to round out its offering. This should include explicit consideration of disruptive technologies outside the current portfolio so the firm can escape the tyranny of the served market (see Chapter 2).

It can choose to add these skills or products through internal development, external acquisition (buying another firm that has the requisite technology or licensing), or partnering. The decision about how to add needed skills and products is sometimes referred to as a "make versus buy" decision. "Make" refers to the decision to rely on internal development to develop new products, whereas "buy" refers to the decision to acquire externally the rights to a new product developed by another firm.

The key issue in the decision about how to add technology is *development risk*. It is sometimes best to pursue *internal development* (i.e., to make the product) if

a. The R&D area is close to current corporate skills.

b. The firm wishes to keep its technological thrust confidential.

c. The firm's culture fosters the belief that the only good technology is developed internally.

In some cases, internal R&D may be cheaper than external acquisition. Issues pertinent to internal development were discussed earlier in Chapters 2, 3, and 4, for example, the ideas of R&D–marketing interaction, product champions, and so forth.

On the other hand, *external acquisition* (i.e., to buy the product) makes sense if

a. Someone else has already developed the technology, and acquisition can save the firm time and effort.

b. The firm does not have all the necessary skills to develop the desired technology.

c. The firm wants to let others take big risks before participating in development.

d. The firm needs to keep up with a competitor whose new technology is potentially threatening.

e. The firm wants to obtain technology for products that can use present brand names, distribution channels, and so forth.

As a middle ground, a firm may choose to form a collaborative relationship for new-product development. Alliances between either competing firms or firms that provide complementary pieces to the product solution are frequently used. The unique issues associated with managing those alliances were discussed in Chapter 3.

In 1990, Boeing launched the development of the 777, its first major new-product platform after the 747 in 1969. As a result of step 1 (technology identification), it identified its need for new process technology, a software application for 3D design. With limited experience in software development, Boeing chose in step 2 (decisions about technology additions) to purchase a 3D design application that had been developed to design French fighter planes. Boeing partnered with IBM to enhance this software and ensure that it would be usable by nearly all of the thirty other companies that Boeing collaborated with to develop and supply structural components, systems, and equipment for the new jet.[3]

3. Make decisions about commercializing, licensing, and so forth.   After acquiring or developing the desired technological know-how, the firm faces the commercialization decision. Here, *marketing risk* is the critical issue. The firm must decide exactly how far in the development process to proceed before marketing and selling the product. This decision is explored in the next section as the "what to sell" decision. For example, Boeing sells aircraft, which are final products, ready for use. However, it does not compete in the business of passenger or cargo air travel. If a firm chooses to commercialize its technology, issues related to timing the entry into a new market (i.e., being a pioneer or a follower) also arise.

4. Ongoing management.   Finally, the firm will need to actively manage its technology asset base, including the development of product derivatives, whether to use product platforms in the strategy, when to "kill" new-product development projects, intellectual property management, and so forth. For Boeing's 777 project, a separate contract was negotiated with each major supplier to specify ownership of collaboratively developed innovations.[4] Boeing deliberately sought to share development costs and risks with suppliers, who had no other customers for their resulting specialized inputs. Thus, Boeing preserved intellectual property rights to the 777 as a whole, while allowing suppliers to retain ownership of the components they produced.

These issues are also discussed later in this chapter.

THE "WHAT TO SELL" DECISION

In a high-tech firm, technology itself either *is* the product (e.g., in a firm that licenses proprietary technology) or *gives rise to* the product (e.g., in a firm that chooses to commercialize products based on a new technology).[5] Firms that innovate technological know-how face a unique decision: Should they sell the knowledge itself or possibly license it? Should firms fully commercialize the idea—marketing, distributing, and selling a full solution including service and support? Or, given that final products can be "decomposed" into subsystems and components, should firms manufacture and sell some subsystem or component on an **original equipment manufacturer (OEM)** basis? Essentially, the decision about what to sell boils down to the basic issue of how to transform know-how into revenues.

Possible Options

As discussed next, firms can choose to sell know-how only; "proof-of-concept"; commercial-ready components to OEMs; ready-to-use final products or systems; or service bureaus that supply complete, end-to-end solutions for the customers' needs.

The decision is based on an underlying dimension: the required expenditures by customers to derive the intended benefits, above and beyond their acquisition costs of the focal purchase.[6] For example, funds can be spent for the core product, to purchase complementary items, services, and training, all of which can be necessary components to derive the intended benefit of the product. Customers who buy at the know-how end of the continuum must expend greater resources to realize the benefits of using the product, whereas customers who buy at the other end of the continuum (purchasing the complete product and all ancillary support services) can fully realize the benefits of the product with their purchase price.

Sell or license know-how only.   The sale of know-how requires the greatest additional expenditures of funds by the customer after the transaction to realize the intended benefit. For example, chemical firms may sell (or license) the rights to a specific molecule to downstream producers.

Sell "proof-of-concept."   The sale may include a prototype or pilot plant to establish that the know-how can indeed be made to work. Selling at this point on the continuum decreases the technological uncertainty for the buyer.

Sell commercial-grade components to OEMs.   Firms may manufacture and sell components that are ready to use in another firm's manufactured product. For example, Intel develops, makes, and sells computer chips to computer manufacturers (OEMs) that are ready to be inserted into the computer.

Sell final products or systems with all essential components, ready for use "out-of-the-box," to customers.   For example, computer manufacturers sell computers that are ready to be used by the customer.

Sell a complete, end-to-end solution.   This whole product solution delivers the intended benefits directly to customers with no need for them to incur additional expenditures on complementary items. For example, in addition to selling computer hardware and software, IBM has moved into the Web-hosting business, in which it will provide servers, software, and requisite services for any company that would like to have a Web site as part of its business solution.

What Decision Makes Sense?

The factors that affect a firm's decision about where along the continuum to generate revenues are shown in Figure 7-2 and discussed next.

In general, firms should lean toward the *selling know-how* end of the continuum when[7]

- The technology does not fit with the firm's corporate mission.
- The firm has insufficient financial resources to exploit the technology.
- The window of opportunity is tight and the firm cannot move quickly enough.
- The market potential is smaller than expected.
- The business cannot be made profitable by the firm.
- Allowing other firms access to the technology is the most appropriate action.
- The range of technologies in a market is very diverse.

In this latter case, it is difficult for firms to keep up with all the relevant technologies across all the components or subsystems of a final product. As a result, firms may find it desirable to compete closer to the know-how end of the continuum rather than the end-product level.[8] This means that high-tech industries experience a pull toward "componentization."

Allowing other firms access to technology (e.g., via licensing agreements) makes sense when the market is characterized by network externalities. Recall from Chapter 1 that many high-tech markets are characterized by a situation in which the more customers that adopt and use a particular innovation, the greater its value for all users. For example, the value of an Internet portal (to both customers and companies linked to the portal) increases as more and more customers use that portal. Known as "demand-side increasing returns," this type of network externality tends, on the one hand, to favor a position on the continuum closer to selling know-how.[9]

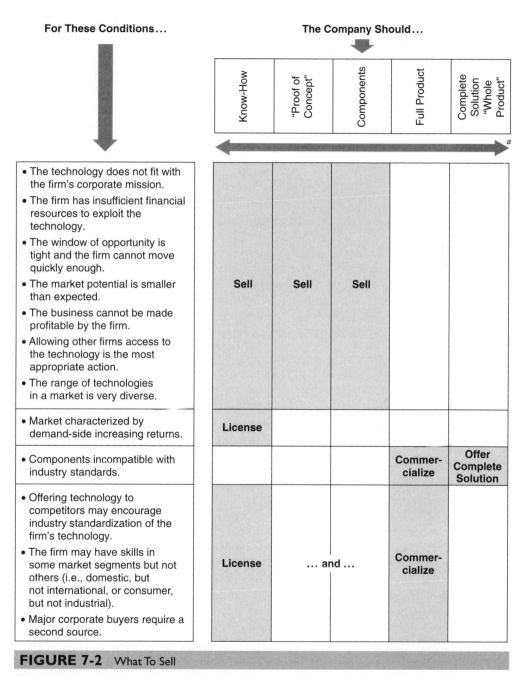

FIGURE 7-2   What To Sell

[a] Underlying continuum is based on customer expenditures needed beyond the initial acquisition cost to derive the benefits of their purchase.

But, on the other hand, know-how often involves tacit knowledge, making it difficult to value and trade. Hence, this latter tendency pushes firms to a position closer to the end product.

However, the best strategy may be to seek wide application and standardization of a technology (via licensing arrangements) in order to discourage other companies from producing substitute technologies. Sales of technology know-how are critical to such a strategy.[10]

Note that firms tend to *sell more toward the end-product level* when they offer components that are incompatible with general industry standards. In general, customers want to be able to mix and match compatible components. Poor compatibility across the range of relevant technologies raises the customer's costs of putting together an acceptable system. In such a situation, selling know-how or individual components is difficult, and selling new, improved components is even more difficult. For example, imagine a maker of computer disks with a new innovation in disks. If existing disk drives are incompatible with new disks, the innovator somehow has to persuade drive manufacturers to produce and market compatible drives. So firms that market incompatible products rather than industry standards may favor positions closer to the end user, in order to offer a complete package for customers.[11]

A firm that chooses to compete close to the end-user/system level of the continuum with a system or product that is incompatible with existing standards faces serious consequences.[12] Witness Apple Computer's travails in attempting to compete with the "Wintel" (Windows–Intel) duopoly. Apple's emphasis on its own proprietary standards over *de facto* open standards built around the Wintel platform made it extremely difficult to compete for third-party hardware and software developers. Sun Microsystems has faced similar issues as well.

Finally, firms may both commercialize and license technology when

- Offering technology to competitors may encourage industry standardization of the firm's technology.
- The firm may have skills in some market segments but not others (i.e., domestic but not international, or consumer but not industrial).
- Major corporate buyers require a second source.[13]

Rather than selling or licensing know-how, most firms have historically tended to sell close to the final user level. For example, royalty revenues (from licensing rights to know-how) have historically *not* been a major source of income compared to aggregate product sales revenue. Indeed, in many industries, royalty revenues are less than even the R&D expenditure itself.[14] But more and more high-tech firms are realizing revenues at multiple points along the what-to-sell continuum, say, by commercializing its technology in some markets and selling or licensing it in others.[15] One example of a company transforming itself to utilize a whole range of these options, of course, is IBM (see vignette on page 200).

One of the reasons high-tech firms are moving to sell more than just the products or components themselves is that, in a high-tech environment, sustainable long-term corporate growth depends on the continual development and leverage of a firm's technology. To maximize the rate of return on technology investment, a firm must plan for full market exploitation of all its technologies. These technologies may, but need not necessarily, be incorporated into that company's own products and services. Thus, a company's marketing strategy may—and probably should—provide for the sale of technologies for a lump sum or a royalty.[16]

For example, Texas Instruments, which previously relied on selling components and final products, has recently received more in licensing fees than its entire operating income.[17] As another example of a firm realizing revenues from different points on the continuum, Canon simultaneously sells printer subsystems on an OEM basis to both Hewlett-Packard and Apple, and ready-to-use laser printers to end customers.

A final issue related to this what-to-sell decision pertains to *international markets*. In many cases, technology transfers to other countries may take place on the basis of standard turnkey deals,[18] in which a company transfers rights (via licensing, for example) to use its technology to another company. Decisions to transfer technology to potentially low-cost producers must account for the effect on the company's own manufacturing plans. Many developing countries want to add value to their natural resources by buying sophisticated process technologies. Brazil, for example, wants to sell steel, not iron ore, and may be able to export steel relatively cheaply because it possesses key raw materials. The long-term effects of such a decision should not be ignored.

Technology Transfer Considerations[19]

Technology can be transferred from small companies/inventors to larger companies with the resources and expertise in the commercialization process. One specific form of technology transfer is the transfer of know-how developed with federal funds (say, at research universities and government labs) to the private sector. A key stimulus for technology transfer from public research institutions and universities to the private sector, with the purpose of commercializing promising technologies, was the Bayh-Dole Act of 1980, which allows transfer of technologies developed with federal funds (taxpayer dollars) to nonprofit institutions and small businesses the rights to inventions made with federal funds. Associated with this is a mechanism to share some of the profits from sales of the products or services with the inventor and the inventor's institution. Technology transfer has emerged as a specialized field, which requires legal, scientific, and business/marketing know-how. Its practice is proving to be one of the driving forces in economic development in the United States and other parts of the world.

One of the major hurdles in the technology transfer process is establishing a realistic and accurate value on the technology to be transferred. Many methods have been developed for this process, and they are referenced in publications from the Association of University Technology Managers (AUTM). Inventors often overvalue their innovations and undervalue the investment risk the purchaser of the invention takes on. Most inventors, for example, do not appreciate that over $200 million and seven to ten years may be spent in developing a drug or a vaccine with no assurance that it will get marketing approval or that it will be a commercial success if approved. Thus, one of the more difficult issues in technology transfer is the valuation of the invention. Participation in the AUTM organization provides a network of colleagues that may be consulted about their opinions on contentious terms in an agreement under negotiation.

A second key issue in the technology transfer process is the protection of intellectual property rights, both for the inventor and the licensing/purchasing company. A company is unlikely to make the enormous investments required to develop a product for market unless it has some period of exclusivity during which it can recoup its investment and earn a fair return. Not patenting an invention greatly reduces the incentive for a company to invest in its development and thus may delay or prevent application of the technology for public use. Note that if the inventor makes an "offer to sell" prior to filing for patent protection, it could bar an inventor from obtaining U.S. patent protection if the patent application is filed more than one year after the invention has been offered for sale.[20] In light of this, institutions advertising technologies available for licensing must exercise extreme caution that they do not destroy the "novelty" of an invention and, thus, its patentability. A low-cost way to prevent this is to file a "Provisional Patent Application" (discussed later in this chapter) before any attempt is made to attract licensees. This will protect U.S. and foreign patent rights for one year, during which the patent must be filed for continuing protection.

Regardless of where on the continuum a firm chooses to sell, the development of high-technology products and innovations necessitates enormous R&D investments. To better leverage these investments, technology companies need to carefully design their product architecture, dealing with the issues of product modularity, platform products and derivatives. This takes us to the fourth step in the technology development process, ongoing technology/product management, and the sub-issues of managing platforms and derivatives.

PRODUCT ARCHITECTURE: MODULARITY, PLATFORMS, AND DERIVATIVES

Modularity

Modularity is building a complex product from smaller subsystems that can be designed independently yet function together as a whole.[21] Modularity in design requires information to be partitioned into visible and hidden components. The visible information consists of design rules on how the subsystems should work together. The hidden information is about how to design each subsystem independently while following the visible design rules on working together with other subsystems. Different companies can take responsibility for each of the modules with the assurance that their collective efforts will create value for customers. Companies in diverse industries such as automobiles, computers, and software use modularity because it provides many benefits. Each supplier company is able to focus on a module and make it better, thus accelerating the rate of innovation in the industry. The company enforcing the visible design rules (the architect company) gets the best subsystems due to supplier competition. And customers are able to mix and match modules to suit their specific needs. However, modularity requires the architect company to have deep product knowledge on every subsystem so that the design rules can be specified in advance.

Modularity reduces the uncertainty in product design and makes things more predictable for a technology company. Accordingly, research shows that modularity results in product standardization, lower barriers to entry for competitors, and more incremental rather than breakthrough innovations.[22] For a significant breakthrough in product performance, individual components and subsystems have to be highly interdependent or integrated. But an integrated product strategy carries more design risk because a change in one component affects all other components and the overall product. It seems prudent to have an integrated product strategy in the early stage of the life cycle when performance is important to customers. Later, a modular strategy may be used when customer needs shift to convenience, customization, flexibility, and price.

The software industry offers a good example of modularity in the product development process.[23] The subsystems for most software programs include the file access, editing, graphics formatting, and printing subsystems, all with internal subsystem interfaces and a graphical user interface. In the software world, the interfaces are particularly important, and their design and evolution can lead to long-lived systems and market domination. For example, Microsoft effectively guides the innovation of thousands of independent software companies by having developed and promoted as a standard the interface mechanisms that allow different programs to communicate with one another in a distributed computing environment. The resulting compatibility, or *interoperability,* for customers reduces learning time from package to package and allows sharing of data. Moreover, by establishing the *de facto* standard for an industry, Microsoft enables other companies to build modules that operate on Windows. The development of these third-party products reinforces the standard, yet Microsoft need not share the costs of development or marketing. These independent third-party developers, in addition to being software producers, become advocates for Microsoft Windows.

Platforms and Derivatives

A **product platform** is a common architecture based on a single design and underlying technology. New product platforms have enhanced performance benefits and involve significant investments compared to existing platforms, hence they are also called "next-generation" products. A product platform can be shared by a set of **derivative** products that meet specialized needs of customers.[24] Derivative products include different models, brands, or versions of the platform product intended to fill performance gaps between the platform products. For example, each generation of Intel's microprocessor chips, such as the Pentium 4, is a platform product sharing the same underlying technology. However, the Pentium 4-M is a derivative of the Pentium 4 for mobile computing applications. Each company needs a technology map or product strategy encompassing its various platforms and derivatives.

Why Use a Platform and Derivative Strategy?

At least two underlying reasons exist for using a platform and derivative strategy in high-tech markets. First, recall one of the common characteristics underlying high-tech markets—unit-one costs. High-tech marketers typically face a situation in which the cost of producing the first unit is very high relative to the costs of reproduction.[25] For example, the cost of burning an additional DVD is trivial compared to the cost of hiring specialists to produce the content in the first place. This underlying feature of technology markets makes a product platform strategy very attractive. If the incremental costs of developing derivative products are relatively small, compared to the platform product, then proliferating versions of a common design to reach various segments adds incremental revenue.

As shown in Figure 7-3, a second reason for using a platform and derivative strategy is that, when a firm introduces a breakthrough product, it will inevitably create "gaps" in the marketplace.[26] These gaps, or holes, exist in the customer's migration path from the old technology to the new technology. Importantly, a firm should not overlook these gaps, which would, in essence, allow competitors to come in with gap-filling strategies and possibly even dislodge a firm from the very market it creates.

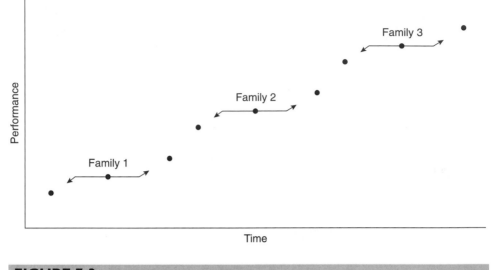

FIGURE 7-3   Product Platforms and Derivatives

It is vital that a firm understand who is buying current products and why they are buying, in order to make informed judgments about gaps in the market that a new product platform will create. Then, a firm must fill the holes with derivative products, which might include adding new features to the former model or scaling down versions of new products. The strategy addresses the needs of future customers while providing a migration path for current customers from the older to the newer product.[27]

Intel is a master at filling in the holes it creates by introducing new-platform products.[28] As shown in Figure 7-4, Intel has introduced derivatives to "gap fill" both the high and low end of its products. For example, with respect to the Pentium family only, Intel introduced the Pentium chip at 60 to 66 MHz speeds in March 1993; over time, it released successively faster versions that reached beyond 3.4 GHz. Each release involved a price cut that was made possible by cost reductions. The company also brought out compact versions of the chip for the laptop and notebook markets. Thus, Intel quickly filled all the performance, price, and applications gaps caused by the Pentium and so preempted the competition. Similarly, when the company introduced its new platform Pentium Pro chip, it filled the market niches created by that chip. And the Pentium Pro chip worked with software applications designed for previous Pentium chips, providing a migration path for existing customers. In 1997, the company plugged the gap in the multimedia market by introducing MMX technology for audio, video, and graphics.

A company's product strategy encompassing platforms and derivatives could be integrated or modular. If modular, the company can invite other players in the industry to develop derivatives. This may require the specification of interoperability design standards. Sometimes government or industry proposes standards specifications that are vital to the success of new technologies (Chapter 1).[29]

Making Decisions about Platforms and Derivatives

So how does one know what an appropriate common platform will be, so it can be "versioned" into multiple derivatives inexpensively and lucratively?[30] Rather than designing it to maximize its appeal to a specific segment, the platform should be

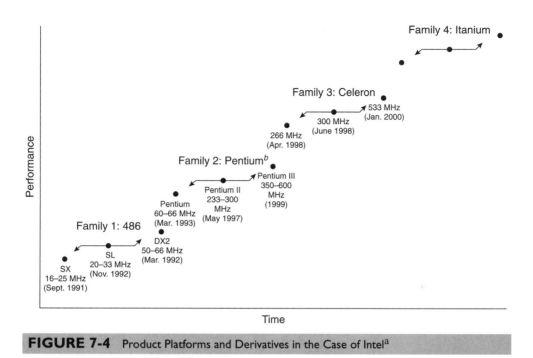

FIGURE 7-4   Product Platforms and Derivatives in the Case of Intel[a]

[a]Simplified summary of information from www.intel.com/pressroom/kits/processors/quickreffam.htm (2000).
[b] Not shown: MMX & XEON series.

designed for the high end of the user market and should incorporate as many of the desired features as needed for this segment.[31] Although the high end is not likely to be representative of the market as a whole, the large fixed development costs are more likely to be recovered from developing a design with the attributes desired by the highest willingness-to-pay segment. Then, subsequent versions can be sold at much lower prices with only modest incremental cost. It is the *subtraction* of features that is a lower incremental cost activity, rather than the addition of new, higher-end features.

This recommendation is consistent with both the notion of crossing the chasm and the lead user process. In crossing the chasm, products developed for innovators and enthusiasts will typically incorporate more technologically advanced features than those versions desired by conservatives. Similarly, because lead users tend to be more sophisticated users, developing products with input from lead users may result in more technologically-advanced products. Regardless, before delving into the development of these subsystems, designers must first engage in a careful study of the users' requirements and incorporate these discoveries into engineering initiatives.[32] Conjoint analysis can be a useful tool in assessing desired features in the platform (Chapter 5).[33]

The determination of how much better each new version should be, the time intervals between versions, and the positioning of versions relative to each other are complex issues that must be considered. Moreover, how to help OEMs and end users manage their migration choices (as covered in Chapter 6) must be part of these decisions.

NEW PRODUCT DEVELOPMENT TEAMS

Chapter 2 discussed the role of the product champion in jumpstarting the new product development process, especially for breakthrough innovations. However, in most situations, the product champion will need to recruit and nurture a team to bring the new product idea to fruition.

Most technology companies organize new product development through **cross-functional product development teams.** It is widely accepted that better product quality and market success occur when product development teams are comprised of individuals from different functional areas (such as marketing, research and development, manufacturing, engineering and purchasing).

A summary of recent research[34,35] on the performance of new product teams is presented in Figure 7-5. The outcomes studied have included product quality and product innovativeness of the output of new product teams. Product innovativeness is negatively related to product quality, reinforcing the common observation that some breakthrough products are "buggy" (have flaws or operating problems) in the beginning. Team performance is affected by team characteristics, customer characteristics, and the developer firm's characteristics. Among team characteristics, the team's ability

FIGURE 7-5 Performance of New Product Teams

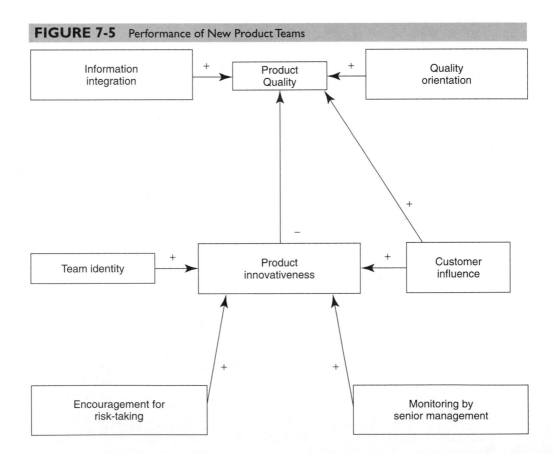

to integrate information across individual members results in higher product quality. Further, the development of an overall team identity and teams' encouragement of risk-taking results in greater product innovativeness. Customers' influence on the product development process results in higher product quality as does the developer firm's quality orientation. Finally, monitoring of the team progress by senior management results in greater product innovativeness.

Effective reward systems are used to motivate individual and group behavior, enhancing team performance. Two important managerial issues with regard to rewarding teams are (1) how to distribute rewards among team members and (2) on what criteria to base the team rewards. There are two approaches to distributing team rewards.

- Reward the team as a group. Here the team reward could be distributed to all team members *equally* or individuals could get different amounts based on their *position* or status in the organization. Research shows that when individual performance is easy to evaluate, position-based rewards are positively related to team member satisfaction. When individual performance is difficult to evaluate, equal rewards are negatively related to team member satisfaction.[36] Taken together, this means that when individual performance is difficult to evaluate, neither equal nor position-based rewards work satisfactorily. Thus, technology companies should invest in monitoring systems and procedures to better measure individual performance within teams.

- Reward individual performance of team members. Individual team members could be given *process-based* or *outcome-based* rewards. Process-based rewards are those tied to procedures or behaviors (e.g., completion of certain phases of the product development project). Outcome-based rewards are those tied to bottom-line profitability of the project. Team performance can be measured by *external* criteria (such as speed to market, adherence to budget and schedule, innovation, product quality, and market performance) or by *internal* criteria (such as self-rated team performance and team member satisfaction). For long and complex product development projects, outcome-based rewards are positively related to team performance, while process-based rewards are negatively related to team performance.[37] Employee stock option plans are used by technology companies as a specific way of providing outcome-based rewards for long-term individual and team performance.

Apple Computer has had many innovative hit products in recent years such as the iMac in 1998, the PowerBook G4 Titanium in 2001, and the iPod in 2003. It has many different teams working on new products. One of the most critical, early on, is the hardware engineering and design team. This team had two leaders in 2001, Jon Rubinstein, 46, an electrical engineer and Jonathan Ive, 36, an industrial designer from Britain, and a number of team members.[38] During the development of the PowerBook G4 Titanium, Mr. Ive and his product designers experimented with a number of metals to encase the laptop, from aluminum to stainless steel. Mr. Rubinstein and his product engineers rejected some suggestions as too expensive and others as too difficult to procure. Ultimately, both men and their team members decided that titanium was the best choice because it was thin, light, and durable. The team followed a five-phase development process including a rigorous conceptual stage where the goals of a product were hashed out. Steve Jobs, Apple co-founder and CEO, was a frequent attendee of these meetings, providing a clear product strategy and motivation. This streamlined process and the implementation of technology maps resulted in a reduction of time-to-market for Apple's innovative computers from 24 months to about 12 to 18 months.

A CAUTIONARY NOTE ON ISSUES RELATED TO "KILLING" NEW-PRODUCT DEVELOPMENT[39]

The decision about when to pull the plug on a new product that is not doing well in the market is extremely difficult. Managers often remain committed to a losing course of action in the context of new-product introductions; such a scenario is often referred to as "good money chasing after bad." Decision makers have strong biases that affect "stopping" decisions. Product champions and technology enthusiasts are perennial optimists about the future viability of their pet projects. Furthermore, because they have a personal stake in these projects, they tend to persevere at all costs. This escalation of commitment is a major problem in new-product introductions. Why does this happen?

Managers who tend to believe they can control uncertainties in their favor continue to escalate commitment to such projects. To justify their decisions, managers try to make data fit their desired expectations about the new product. So, if the decision maker is highly positive about the new product, then he or she will attempt to confirm this hypothesis by seeking out supporting data. For example, managers are much more likely to recall information that is consistent with prior (positive) beliefs, they tend to interpret neutral information as positive, and they will even ignore or distort negative information to support their desire to see their new pet project succeed. Indeed, they may interpret negative information as positive! In recent research, distorting negative information as positive occurs more than twice as often as distortions in the other direction (distorting positive information as negative).

These biases are one reason that improving the information affecting a decision to withdraw a new product from the market does not result in an actual withdrawal. Indeed, asking managers to set and commit to a stopping rule simply is ineffective. In light of these complications, managers should attempt to do the following.

1. It is clear that a change in direction will not be made unless managers are aware there is a problem in the first place. So, problem recognition requires that managers attend to possible negative feedback. However, in light of the confirmation bias noted previously, such recognition can be extremely difficult.[40]

2. Managers must also re-examine the previously chosen strategy, both clarifying the magnitude of the problem and possibly redefining the problem. This is equally difficult because of conflicting information and differences of opinion. Some stakeholders have a vested interest in maintaining the status quo, whereas others exert pressure to change direction.[41]

3. Managers should search for an alternative course of action, attempting to obtain independent evidence of problems and identifying new courses of action.

Creativity is vital in identifying a wide range of options, and the firm should foster a culture that encourages open questioning.[42] The enthusiasm of advocates may be contagious and could lead to collective blind faith, what social psychologists call "groupthink." Therefore, it may be a good idea, from the outset, to have naysayers or skeptics on the review commitee. Just as a new product development team needs a product champion, the team also needs an "exit champion," someone who can pull the plug based on hard objective evidence.[43]

4. Managers should prepare key stakeholders for an impending change and attempt to manage impressions. Active attempts to remove the project from the core of the firm will help in this step.[44]

5. Boulding, Morgan, and Staelin suggest that managers should attempt to decouple the withdrawal decision from previous investments in the project. Ideally, this decision decoupling requires the use of a different decision maker for the withdrawal decision.[45]

6. Alternatively, a firm can try to use a stopping rule that was developed early on, based upon information available at the time of the project's inception or product launch decision. However, if a firm attempts to update this stopping rule based on new, factual information available at the time the withdrawal decision is to be made, this decision will likely not be effective because it will suffer from the biases already noted. A decision procedure that is most effective at reducing the escalation of commitment takes "out of play" information closely tied to those with a vested interest in the project and renders a decision based on benchmarks established prior to incremental commitments to the project.[46]

DEVELOPING SERVICES AS PART OF THE HIGH-TECHNOLOGY PRODUCT STRATEGY

The process of developing services offers different challenges from developing physical products. Many firms involved in developing, manufacturing, and distributing high-tech products have turned to services as a way to augment their revenue streams. For example, IBM spent $100 million in 2001 to develop training, consulting, and support services for Linux operating systems (note that the operating system itself, based on open-source computing, is free). Similarly, resellers and distributors of high-tech equipment have turned to training and other services to augment their shrinking margins on hardware.

The range of services related to high-technology products is wide. For example, the Bureau of Labor Statistics (BLS), in its review of employment opportunities in R&D-intensive industries, identifies a category called "R&D-intensive services." This category includes services such as management and public relations, computer and data processing, engineering and architectural, and research and testing services. Recall that the BLS's definition of R&D-intensive are industries with 50 percent higher R&D employment than the average proportion for all industries surveyed. In other words, these service categories have a higher degree of technical workers employed than other service industries. These examples show the range of ways that technology and services potentially interact, as exhibited in Figure 7-6.

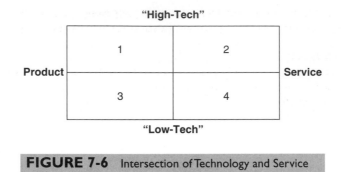

FIGURE 7-6   Intersection of Technology and Service

In cell 1, companies that sell a tangible high-tech product such as hardware or consumer electronics can augment their product sales with services revenue, as in the IBM example. There are several ways to do so. First, a company might offer consulting services in conjunction with its product sales. This is one of the key reasons that high-technology companies have purchased (or attempted to purchase) some of the major consulting firms. For example, IBM purchased the consulting service business of Price Waterhouse Coopers and Lybrand in order to beef up the technology consulting services it can offer. Of course, like the issues the accounting industry faced in the early 2000s, it is imperative that the consulting arm of the company be viewed as credible and not merely a quasi-sales force attempting to find ways to sell more of the company's own products. As another example of this strategy, a company that offers a product comprised of, say, a handheld computing device coupled with a Global Positioning System and a digital camera (used for a variety of field-based applications, such as emergency rescue response, forest fire fighting, and even retailer merchandising monitoring for manufacturers offering cooperative advertising monies) can also offer the ability to manage and analyze the data that the device collects on an ongoing basis. Second, a company that sells high-tech products might offer training, repair, or maintenance contracts to supplement its revenues from product sales.

Both of these strategies can offer a company an ongoing revenue stream after the customer has purchased the product. Indeed, the focus on long-term customer relationships suggests that augmenting product sales with service revenues might be the best strategy for many high-tech companies. Critical issues to be addressed by the firm in this intersection of services and technology are:

- Does the company have sufficient service personnel, and are they appropriately trained, to effectively manage the customer's service needs and questions?
- Can the company develop proficiency in services, without losing its core competencies in product innovation and development?

In cell 2, companies that offer high-tech services might include contract research firms, consulting companies that advise customers on implementation of technology solutions, as well as service providers that offer outsourced information technology services to corporate customers (as in on-demand computing, for example). Moreover, some of these "high-tech" services companies must rely on high-technology products to deliver their services. An example of this would include the use of CAD/CAM workstations by an engineering or architectural firm to provide design services for its clients. Issues that must be addressed are:

- Can the technical service personnel communicate in user-friendly ways with customers?
- Are the underlying technologies used to deliver the service reliable in terms of up-time? (Consider that Webhosting services for e-tailers must ensure 24/7 access to shoppers.)
- Does the company invest continuously in upgrades to maintain cutting-edge technologies?
- Does the company invest adequately in training service personnel to be at the leading edge of practice?

In cell 3, for companies that offer more traditional products, issues related to the intersection of services and technology include the adoption of technologies to improve customer service or make the supply chain more efficient. For example, the use of retail self-scanning devices by consumers to check out their own goods and pay by credit card and the use of the Internet to check order and delivery status help improve customer service and reduce costs. As another example in cell 3, some companies that offer traditional products (such as Nike shoes) must rely on technology solutions within their companies in order to stay abreast of cutting-edge business practice. For example, Nike recently spent $500 million on a supply chain management system from SAP in order to automate its supply chain.

Finally, in cell 4, companies that offer "low-tech" services also must use technology as part of their service-delivery strategy. Many traditional service industries—not R&D-intensive services—are affected by new technological innovations. The movement to self-service technologies (SSTs) has allowed many industries to automate previously labor-intensive operations.[47] The use of automated teller machines (ATMs) and online banking are examples in the banking industry. Another example is the adoption of DNA testing to apprehend and convict criminals, an improvement over traditional gumshoe detective work à la Sherlock Holmes.

These examples from cells 3 and 4 highlight another key dimension in the intersection of technology and services: Is it the customer who uses the technology (i.e., a "customer-facing" external technology solution), or is it the employees of the company who use the technology to gain business process efficiencies (i.e., an internal technology solution)?

In the case of customer-facing external technology solutions, critical issues are:

- Are customers comfortable using the self-service technology? Have they been trained in how to use it effectively? Does it provide them with real value, or do they feel that the company is merely pushing indirect service costs onto them, forcing them to do themselves what they had expected the company to do for them? Do customers miss the human touch in their interactions with the company?
- Do customers have access to the technology necessary to engage in self-service (i.e., do they have Internet access, for example)?
- What are the other elements of the whole product (or end-to-end solution) that the company must offer in order to make the value proposition to the customer compelling?

In the case of internal technology solutions, critical issues are:

- Are internal employees willing to embrace the new technology? Have they been appropriately trained?
- Have internal business processes been redesigned in such a way that the efficiencies the technology solution can offer can be realized?

Unique Characteristics of Services: Implications for High-Tech Marketing

Regardless of the specific intersection of services and technology, by their very nature, services have some critical characteristics that make the marketing of them particularly challenging.

Intangibility

First, services are **intangible,** meaning that they can be neither touched nor examined prior to the customers' purchase decision and consumption. Moreover, the customer of a service does not take home a tangible good. In other words, the intangibility exacerbates the anxieties a customer may experience in a high-tech purchase decision. The marketers' role here is to reduce the perceived risk of the high-tech offering for the customer through product demonstrations, free trials, money-back guarantees, extended warranty, training, or technical support.

Inseparability

Second, the production of a service (when the service provider is providing the service) cannot be separated from the consumption of a service (when the customer actually gets the delivery of the service). For example, when a high-tech company is providing training for a customer on its new technology implementation, the customer is simultaneously receiving the service experience. In this sense, the **inseparability** means that the company cannot produce/inventory the good in advance of customer demand. Thus there may be peaks and valleys in demand. One way to handle this is to configure service operations to be flexible or scalable. Additional service providers can be hired during peak periods and let go during lean periods. For example, online retailers may add additional customer service representatives to meet the increased workload at call centers during the holiday shopping season. Another way to manage the perishable nature of services is to price such that customers have an incentive to utilize services during low-demand periods. This can be found, for example, among cell phone service providers who provide a fixed number of weekday calling minutes at a monthly subscription fee and unlimited night and weekend minutes.

Inseparability also means that the quality with which each service experience is provided might vary considerably, from occasion to occasion, and service employee to service employee. For example, the SAP personnel assigned to manage Nike's supply chain management application may vary in expertise and customer communication skills from those assigned to manage Intel's SAP application of supply chain management software. This potential for inconsistency in the service-delivery experience means that companies that have service as part of their high-tech marketing strategy absolutely MUST train their customer-facing personnel so that there is a level of consistency and reliability, regardless of which employee team is assigned to which customer.

PINING FOR A PET?

Have you thought about getting a pet for your home? Are you daunted by the daily chores involved, such as feeding, grooming, and walking the pet? Has that discouraged you from taking the plunge and getting one? If so, technology has a solution for you: the Sony AIBO entertainment robot. The AIBO can be trained to recognize and respond to owner voice commands. For example, it can wave its front paw upon hearing its name. AIBO can walk, dance, wag its tail, chase a ball, and be taught to heel. You never have to take it to a vet. And it doesn't even need any pet food. It can walk back to its charging station when it runs low on battery power.

The question is often asked, what kind of an animal is AIBO? A dog? A cat? Sony is careful not to answer this question. This is a clever marketing strategy because Sony wants consumers to define this for themselves. AIBO is whatever your favorite animal is. Consumers have begun to treat these robots as real pets, with a number of Web sites dedicated to AIBO. For example, at www.jp.aibo.com, AIBO owners can communicate with each other, and they can post pictures and video of their pets. There are also offline AIBO fairs where owners and their pets can interact with each other in a fun setting, including games, music, and food.

When the first generation of AIBOs went on sale in Japan in 1999, all 3,000 of them sold out on the Internet within 20 minutes. The 2,000 units made available for the U.S. market sold out on the Internet within four days. The new Sony AIBO ERS-7 is available in the United States for a list price of $1,599 in 2003:[a] a small price to pay for the love of a pet—without any inconvenience.

SOURCE: Phelan, David (2003), "Fast Forward: Full of artificial goodness; Remember Sony's AIBO, the robot dog who captured everyone's heart a decade ago? Now there's a new kid on the block, and, says David Phelan, he's even better," *The Independent*, September 20, pg. 49.

[a] If you have broadband Internet access, you can download a 5-minute video clip of the second-generation AIBO from Stephen Bell's Web site at http://www.stephenbell.net/archive/2002_08-01_archive.htm

---

As this section highlights, understanding the relationship between high-technology environments and services is an important piece of the high-tech marketing strategy and necessitates attention to the unique characteristics of services.

PROTECTION OF INTELLECTUAL PROPERTY

High-technology firms face an environment characterized by frequent innovation, high mortality rates, a high priority on research and development, stiff competition in a race to the marketplace, and partnerships with firms that may be potential competitors. In such an environment, the management of sensitive information is particularly critical.[48] One manager at a high-technology firm summarized the issue:

Product life cycles are very short and development times are very long. That sets up a situation where having knowledge is extremely valuable. If one of our competitors were to acquire information, it would give them an edge. They could respond with their offerings in a stronger fashion than they might have.[49]

Information is vital to the success and vitality of an organization, but the potential of seepage from strategic alliances, employee turnover, and poor information-security procedures poses a real threat. Concerns about intellectual property and protection of trade secrets are important in many industries, but they are paramount in high-technology industries, in which the basis for competitive advantage is most likely to be superior technological know-how. The Software Information Industry Association reports that **software piracy** in 2002 was valued at $13 billion worldwide and $2 billion in the United States.[50] Thefts of intellectual property are especially common in today's high-tech industries, in which the most prized assets can be stored on a disk or shared over the Internet.

Information presents something of a double-edged sword to high-technology firms. On the one hand, as noted in previous chapters, high-tech companies want to be skilled in gathering and utilizing information—which includes information about their competitors—to gain advantage. On the other hand, each company wants its information to remain proprietary and its firm's boundaries to be impenetrable to the competitive intelligence-gathering efforts of its competitors. Protection of information can be particularly crucial when strategic alliances fail.[51] So, how is a company to manage this situation, which requires that it be both open and restrictive in information sharing at the same time? Knowledge of the various strategies to protect intellectual property is vital.

The high-tech industry is littered with the sagas of many companies who have sued for infringement of intellectual property rights. In 1997, Digital and Intel filed a series of suits against each other. Digital first sued Intel, accusing Intel of violating patents on a Digital chip called Alpha. Intel filed a countersuit demanding that Digital return confidential information about upcoming chips, such as its Merced chip being jointly developed with Hewlett-Packard, that it had received as a major customer. Intel alleged that Digital was misusing confidential information.[52] Then Digital filed yet another suit, accusing Intel of using monopoly power to harm Digital by demanding the return of its technical documents. Digital argued that Intel's recall of the confidential data about its chips could leave the company stranded in the technology race in its ability to develop products for the next-generation chips.[53] Digital did return the Merced information, because it decided not to pursue products for that chip's design, but said it wouldn't return other documents that Intel also supplied to other computer makers.

Intellectual property refers to original works that are essentially creations of the mind. The U.S. legal system has protected a creator's right to enjoy due economic returns from his or her creations since the inception of the Constitution in 1789 (Article 1, Section 8). Intellectual property is an expansive term, referring broadly to a collection of rights for many types of information, including inventions, designs, material, and so forth. In the business arena, intellectual property may include not only creations from research and development but also general business information that a firm has developed in the course of running its business and that it needs to maintain as proprietary in order to remain competitive.

Companies have several options available to them in protecting the proprietary information that forms the basis of their competitive advantage: patents, copyrights, trademarks, and trade secrets.[54] Each of these is briefly reviewed in turn.

Patents

Grant of a **patent** confers to the owner(s) the right to exclude others from making, using, offering for sale, or selling the invention claimed in the patent in the United States, or from importing the invention into the United States, for a specific time period, usually twenty years from the date of filing an application for patent with the U.S. Patent and Trademark Office (PTO). Patentable subject matter includes any "new and useful process, machine, article of manufacture, composition of matter, or any new and useful improvement thereof."[55,56]

As noted above, a patent gives the inventor(s) the *right to exclude*. It is a common misunderstanding that grant of a patent confers the positive right to make, use, or sell the invention, when in fact the patent provides merely a right to *exclude*, typically by bringing legal action. It is also a common misconception that the grant of a patent for an invention limits further use of the technology to benefit humankind. For example, the statement is often heard that "I will not patent this because I want to make it freely available to everyone." To the contrary, there are classic examples wherein a decision not to seek patent protection delayed introduction of a new technology that could have benefited society, for example, when lack of patent protection prevents firms from licensing technological know-how because they have no guarantee of the rights to the revenue stream from that know-how.

An invention is patentable only if it meets three requirements: It must be useful, novel, and nonobvious. How are these three requirements met? To meet the *utility* requirement, the invention must be useful. That is, the invention must function as intended and provide some benefit to society. Examples of inventions lacking utility include those that conflict with known scientific principles, such as a perpetual motion machine, or those that are unreasonably dangerous, such as nuclear weapons. Examples of inventions for which the utility requirement is usually not an obstacle are those in the mechanical and electrical arena. For inventions in the chemical and biotechnology field, the utility requirement poses some concern, and an applicant for patent must specify a meaningful use for any claimed chemical compound, nucleotide sequence, or protein sequence.

The *novelty* requirement is met when no single prior public disclosure or literature document—for example, a published patent application, patent, journal article, conference publication, or meeting abstract—describes, explicitly or inherently, each aspect of the claimed invention. If even a single published writing is found that describes the invention, or if it can be shown that the invention was known or used by others in the United States, the invention is not novel.

The *nonobviousness* requirement means that there is no implicit or explicit suggestion in prior literature of the subject matter claimed as the invention. In determining if an invention meets the nonobviousness requirement, the teachings in multiple literature references may be combined. If the teachings, when collectively considered, show or suggest the invention, then the invention is not patentable.

Patent procurement is expensive and procedurally complex. Often a preliminary assessement of patentability is done prior to preparing or filing an application (by either the inventor or a patent attorney on behalf of the inventor) in order to determine—to the extent possible—that the idea is new and nonobvious. An assessment of patentability is made by searching in a variety of databases, including patent databases such as those available at www.uspto.gov/patft/index.html and at http://espacenet.com, and literature databases relevant to the field of the invention. Two useful search sites are www.ncbi.nlm.nih.gov/pubmed and www.scirus.com. An assessment of patentability can be made by determining if there is an absence of any reference or combination of references that teaches or suggests the invention. It is also important, prior to preparing or filing an application, to confirm with the inventors that the idea has not been disclosed to anyone. If the inventor has previously disclosed the idea, either orally or in writing, then it might no longer be considered novel, and as a result, a patent may not be granted.

Most inventors use a patent agent or attorney to prepare and file the patent application and to act as their representative before the PTO during the examination process. Representative costs for filing are shown in Table 7-1 on patenting fees. Appendix 7A to this chapter details the steps in obtaining patent protection for international patents. The Technology Expert's View from the Trenches provides the perspective of a patent attorney and details hidden pitfalls in the patenting process.

TABLE 7-1 Patenting Fees

U.S. Patent Costs		*Foreign Patent Costs[1] in Selected Countries*	
Application preparation	$8,000–$16,000	PCT filing fee[1,2]:	$2,000–3,000
Searching/patentability	$2,000–$4,000	Costs[1,3] associated with entry of the PCT application into selected national countries:	
Drafting application	$5,000–$11,000		
Filing fees	$750	Australia	$2,500
Examination/prosecution		Canada	$2,000
Attorney's time (per response)	$2,000–$4,000	China	$3,500
PTO fees for "late" responses	$110–$1,970	Europe	$5,200
		Japan	$6,500
Issue fee	$1,300	Costs[1,4] associated with examination of an application before selected national patent offices:	
Typical Total Range	**$12,000–$25,000**		
Maintenance fees		Australia	$1,500–2,500
3 1/2 years	$ 890	Canada	$3,000–5,000
7 1/2years	$2,050	China	$4,000–6,000
11 1/2 years	$3,150	Europe	$3,000–6,000
		Japan	$3,500–8,000
		Costs[1,5] associated with validation and grant of a European Patent in five selected European countries:	
		Germany	$4,000
		France	$3,500
		Italy	$3,500
		Spain	$4,000
		United Kingdom	$1,500

[1]*Costs are estimated using Global IP Estimator® from Global I.P Net (www.globalip.com) and are based on an application with 30 pages of text, 2 pages of claims with 20 claims, and 3 pages of drawings. The costs are based on minimum fee schedules supplied by foreign associates and are for processing of straight-forward applications. All costs are subject to currency fluctuations.*

[2]*PCT: Patent Cooperation Treaty. See Appendix A.*

[3]*National filing fee, U.S. attorney time, foreign attorney time, translation costs for China and Japan*

[4]*U.S. attorney time, foreign attorney time, translation of written responses filed with Chinese Patent Office and Japanese Patent Office*

[5]*Granting fees and translation into national language*

ISSUES IN THE PATENTING PROCESS

JUDY MOHR J.D., PH.D. (CHEMICAL ENGINEERING)
Perkins Coie LLP, Menlo Park, California

Patent attorneys or agents function at the crossroad between a company's research and development effort and its business and marketing goals. Patent agents operate from a very privileged vantage point, hearing first-hand from scientists about their budding inventions and early results. From computer disk drives with more storage capacity to gene sequences that express a protein for treatment of multiple sclerosis, patent attorneys are part of the forefront of science, working with corporations to develop a strategic patent portfolio. And for many products—for example, therapeutic products derived from gene sequences discovered from the ever-expanding research and development effort in biotechnology—early patent protection is crucial due to the competitive nature of the field.

For any given company, its employee-scientists find numerous potentially patentable inventions each year. Because of the expense involved in procuring patent protection, careful thought must be given to discern which of those inventions best fit with the business strategy. Knowing with certainty what will be the best fit, however, is a bit like predicting the future. It is entirely possible that a seemingly unimportant invention becomes a cornerstone

product for a company. Further complicating the decision is the fact that in today's highly competitive, fast-paced world, filing for patent protection before a competitor is crucial. This means filing an application as soon as possible, and the luxury of doing a few more experiments to resolve one or two questions is often not possible. Decisions on whether to file for patent protection are often made prior to a full understanding of the commercial viability of the invention.

With the right decisions and a bit of luck, the result is amazing. For example, one small company started with a handful of people working on administering drugs for the treatment of various cancers. They worked with their patent agent to file patent applications on the scientists' findings. With time, the early findings grew into prototype products and, ultimately, into marketed products for cancer therapy in humans. That small handful of people grew to over 200 people—research scientists, product development people, and marketing representatives working together. The core patent position of the company effectively excluded the larger corporations from producing copycat products. With a patent-protected product on the market and several more in the pipeline, this young company was purchased for over $500 million by a major pharmaceutical firm. All that from just a seed of an idea described in an early patent. Amazing.

What are some common trouble spots with obtaining a patent? One, typically uncovered early in the preparation process, is that the inventor has previously "disclosed" (either orally or in writing) the invention prior to filing the patent application. For example, it's not uncommon for an inventor to give a talk at a conference or to publish an abstract or even a paper on the research before the patent application is filed. This is a fatal mistake for obtaining patent protection outside of the United States, since disclosing the invention to the public prior to filing for a patent is an absolute bar to obtaining a patent in most countries. The United States allows for a one-year grace period, in which the disclosure is not considered "prior art" if the application for patent is filed within one year of the disclosure. Be sure your company has a disclosure control program that watches over the scientists' presentations at meetings and publications.

A second common trouble spot is the misconception that, upon issuance of a patent, the patent owner is free to practice the invention. Grant of a patent confers to the patent owner the exclusive right to prevent others from making, using, offering for sale, or selling the claimed invention for twenty years from its filing date. However, there are instances where a patent owner is unable to practice the invention because of a patent granted to another. That is, the

patent owner is free to exercise his or her right to exclude others from practicing his or her invention; however, the patent owner is not able to practice the invention because doing so would infringe another patent. For example, suppose Company A has a patent that sets forth a claim to the chemotherapeutic drug paclitaxel and derivatives of paclitaxel for treatment of cancer. Company B has found that paclitaxel, when derivatized in a particular way, is significantly more effective in fighting colon cancer. So, Company B applies for and obtains a patent that claims this new derivatized paclitaxel for treating colon cancer. In this case, the patent issued to Company B falls within the broad claim to paclitaxel obtained by Company A, and hence, if Company B practices the invention, they would infringe Company A's patent. In this case, Company B can practice its invention only if a license to Company A's patent is obtained. Likewise, Company A is not free to practice the invention claimed in the patent to Company B. Although Company A has a broad claim to paclitaxel and its derivatives, Company A is excluded from using the paclitaxel derivative of Company B for treatment of colon cancer. Company A would need a license from Company B to practice the invention claimed in Company B's patent.

Types of Patent Applications

In the United States, there are two types of patent applications: the provisional application and the utility application. A *provisional application* is filed with the United States Patent and Trademark Office (PTO) with a filing fee[57] of $160 and by definition has a lifetime of only one year. During that one-year life, the application remains in the PTO unexamined by a patent examiner. One year after filing, the application, to avoid abandonment, must be refiled as a utility application (see next type of patent application) for examination. The benefit of the provisional application is that the inventor has that one-year period to do further research on the invention or to obtain investor monies to develop the idea. Filing of a provisional application secures a "filing date," also referred to as a "priority date."

Alternatively, if an inventor does not need time to do further research, time to obtain monies for development, or is anxious to obtain a patent, the application can be filed with the PTO as a *utility application,* along with the filing fee of $750. (Note that certain fees charged by the U.S. Patent and Trademark Office are reduced by 50 percent for so-called "small entities"—any person, nonprofit organization, or small business concern that has not assigned, granted, conveyed, or licensed, and is not under an obligation to do so, any rights in the invention to an entity having more than 500 persons.)

The application will be assigned to an examiner, who will render an initial report regarding patentability. If that initial decision is unfavorable, the applicant can amend the claims or submit remarks in writing to the examiner addressing the unfavorable report. The examiner will then issue a final decision as to patentability of the claims. If the decision is favorable, a patent is granted, and the PTO will publish a patent (Figure 7-7). If the examiner's final decision is unfavorable and the patent application is denied, the applicant may appeal the examiner's decision to a three-person panel at the PTO, the Board of Patent Appeals and Interferences. Alternatively, the application can be refiled (for another $750 fee) with amended claims.

Disadvantages in Using Patents

Although a patent provides the obvious benefit of giving the patent owner, or any licensee, the right to exclude others from making and using, i.e., practicing, the invention, there are some disadvantages in using patents as a means to protect intellectual property. One potential disadvantage is that, in order to obtain a patent, a full description of the invention must be provided in the application at the time of filing with the United States Patent and Trademark Office. In the United States, as well as in foreign countries, patent applications are published eighteen months from their filing date, making the proprietary information available to the public prior to grant of the patent. This provides a way for competitors to know about a company's intellectual property prior to grant of a patent. Hence, because patents are public information, risks exist with this form of protection. One study found that 60 percent of patented innovations were "invented around" by other firms within four years.[58] Competitors may be able to design around the claimed invention by modifying a minor aspect of the invention in order to avoid infringing the patent.

Another disadvantage is that it is the patent owner's burden to enforce the patent—that is, to keep watch that competitors and others are not "infringing" the patent. Enforcement can and often does lead to lawsuits, an expensive undertaking. Moreover, a patent in such a lawsuit may be found to be invalid by the court, because grant of a patent confers a presumption, but not a guarantee, of validity.

Changes in Patent Law

Patent law, like copyright and trademark law, is concerned with fostering creativity by securing for the inventors, or authors in the case of copyright law, the exclusive right to their discovery. The law seeks to balance the right of the general public to useful information in a new discovery with the right of the inventors to benefit from their work. Recently, the balance has been shifting in favor of property owners, in part because of pressure from companies with valuable intellectual property portfolios.[59]

Until the late 1990s, business information and other business know-how were not patentable: Patents for business methods were prohibited. However, on July 23, 1998, the Federal Circuit Court of Appeals held that a business method that uses a mathematical

The Commissioner of Patents and Trademarks

Has received an application for a patent for a new and useful invention. The title and description of the invention are enclosed. The requirements of law have been complied with, and it has been determined that a patent on the invention shall be granted under the law.

Therefore, this

United States Patent

Grants to the person(s) having title to this patent the right to exclude others from making, using, offering for sale, or selling the invention throughout the United States of America or importing the invention into the United States of America for the term set forth below, subject to the payment of maintenance fees as provided by law.

If this application was filed prior to June 8, 1995, the term of this patent is the longer of seventeen years from the date of grant of this patent or twenty years from the earliest effective U.S. filing date of the application, subject to any statutory extension.

If this application was filed on or after June 8, 1995, the term of this patent is twenty years from the U.S. filing date, subject to any statutory extension. If the application contains a specific reference to an earlier filed application or applications under 35 U.S.C. 120, 121 or 365(c), the term of the patent is twenty years from the date on which the earliest application was filed, subject to any statutory extension.

Bruce Lehman
Commissioner of Patents and Trademarks

Arnita Manley
Attest

FIGURE 7-7   Sample of Patent

US005770789A

United States Patent [19]

Mitchell-Olds et al.

[11] **Patent Number:** **5,770,789**

[45] **Date of Patent:** **Jun. 23, 1998**

[54] **HERITABLE REDUCTION IN INSECT FEEDING ON BRASSICACEAE PLANTS**

[75] Inventors: **Storrs Thomas Mitchell-Olds; David Henry Siemens**, both of Missoula, Mont.

[73] Assignee: **University of Montana**, Missoula, Mont.

[21] Appl. No.: **496,016**

[22] Filed: **Jun. 28, 1995**

[51] Int. Cl.^{6} **A01H 5/00**; A01H 1/04; A01H 1/06; C12N 15/01

[52] U.S. Cl. **800/200**; 800/230; 800/DIG. 15; 800/DIG. 17; 435/6; 435/172.1; 47/58; 47/DIG. 1

[58] **Field of Search** .... 435/172.1, 6; 47/58, 47/DIG. 1; 800/200, 230, DIG. 17, DIG. 15

[56] **References Cited**

PUBLICATIONS

Berenbaum et al., *Plant Resistance to Herbivores and Pathogens Ecology, Evolution, and Genetics*, Chapter 4:69–87 (1987), Univ. Chicago Press, Chicago.
James et al., *Physiologia Plantarium*, 82:163–170, (1991).
Koritsas et al., *Ann. Appl. Biol.*, 118:209–221, (1991).
Lister et al., *Plant J.*, 4:745–750, (1993) No. 4.
Magrath et al., *Heredity*, 72:290–299, (1994) No. 3 Mar.
Magrath et al., *Plant Breeding*, 111:55–72, (1993).
Richard Mithen, *Euphytica*, 63:71–83, (1992).
Mithen et al., *Plant Breeding*, 108:60–68, (1992) No. 1.
Parkin et al., *Heredity*, 72:594–598, (1994).
Thangstad et al., *Plant Molecular Biol.*, 23:511–524, (1993).
Giamoustaris et al., *Ann. appl. Biol.* 126:347–363, 1995.
F.S. Chew, "Searching for Defensive Chemistry in the Cruciferae, or, do Glucosinolates Always Control with Their Potential Herbivores and Symbionts? No!," *Chemical Mediation and Coevolution*, Ed., Kevin C. Spencer, Academic Press, Inc., Ch. 4, pp. 81–112 (1988).

Lenman et al., "Differential Expression of Myrosinase Gene Families," *Plant Physiol.*, 103:703–7111 (1993).
Xue et al., "The glucosinolate–degrading enzyme myrosinase in Brassicaceae is encoded by a gene family," *Plant Molecular Biology*, 18: 387–398 (1992).
Höglund et al., "Distribution of Myrosinase in Rapeseed Tissues," *Plant Physiol.*, 95: 213–221 (1991).
Ibrahim et al., "Engineering Altered Glucosinolate Biosynthesis by Two Alternative Strategies," *Genetic Enginering of Plant Secondary Metabolism*, Ed., Ellis et al., Plenum Press, New York, pp. 125–152 (1994).
Hicks, "Mustard Oil Glucosides: Feeding Stimulants for Adult Cabbage Flea Bettles, *Phyllotreta cruciferae* (Coleoptera: Chrysomelidae)," *Ann. Ent. Soc. Am.* 67: 261–264 (1974).
Reed et al. 1989 Entomol. exp. appl. 53(3):277–286.
Bartlet et al. 1994. Entomol. exp. appl. 73(1):77–83.
Butts et al. 1990. J. Econ. Entomol. 83(6):2258–2262.
Bodnaryk et al. 1990. J. Chem. Ecol. 16(9):2735–46.
Haughn et al. 1991. Plant Physiol. 97(1):217–226.
Mithen et al. 1987. Phytochemistry 26(7): 1969–1973.
Jarvis et al. 1994. Plant Mol. Biol. 24(4):685–687.

Primary Examiner—David T. Fox
Attorney, Agent, or Firm—Fish & Richardson, P.C.

[57] **ABSTRACT**

A method for producing plants of the Brassicaceae family that have reduced feeding by cruciferous insects is disclosed. The method comprises selecting for the heritable trait of altered total non-seed glucosinolate levels or for the heritable trait of increased myrosinase activity. Selection may be performed on Brassicaceae cultivars, mutagenized populations or wild populations, including the species *Brassica napus*, *B. campestris* and *Arabidopsis thaliana*. Plants having such altered levels show reduced feeding by cruciferous insects, including flea beetle, diamond back moth and cabbage butterfly. Plants selected for altered levels of both glucosinolates and myrosinase also show reduced feeding by cruciferous insects.

25 Claims, 5 Drawing Sheets

FIGURE 7-7 *(Continued)*

---

formula can be patented, as long as it meets the three traditional criteria for legal protection (useful, novel, and nonobvious).[60] In particular, the Boston-based Signature Financial Group was granted a patent for a unique data-processing software program that cut down on the cost of crunching numbers.[61]

This decision reflected an important shift in legal thinking and was quite controversial. Critics contend that those few who hold a patent to a business method will slow the spread of valuable commercial innovations, to the detriment of society. Although processes are patentable, ideas—for the most part—are not. It is the fine line between processes and ideas that has allowed such business methods to be patented.

With the change in legal precedent that allowed business methods to be patentable, the PTO was flooded with applications for business method patents on e-commerce methods with the hope that such patents could lock in competitive advantage for a few firms. E-commerce patents included that of CyberGold Inc. for a system of giving awards to people who click on an advertising message. Another was Priceline's patent for being the first to innovate a buyer-driven e-commerce system. It was granted a broad patent on a method of auctioning goods and services on the Internet.[62] Amazon.com was granted a patent for its one-click shopping cart, and Barnes & Noble was excluded from using a one-click shopping cart on its Web site. Some suspect that the field of medicine may see a surge of new patents, not only in methods of treatment but also in the management of patient and claims processing.

Critics say that because these methods are so common, they should not be eligible for patents. Patented processes are required to be novel, but some believe that the recently awarded Internet patents are not novel. These critics argue that merely transferring a marketing technique to the Web does not necessarily constitute novelty.[63] Patents are also supposed to be nonobvious, but with so many people using similar strategies on the Internet, these critics also say they cannot be considered nonobvious.

Moreover, patents have historically been granted to protect the common good by providing an incentive to innovate. However, some believe that patents granted for broadly defined Internet business models do not protect the common good; rather, they create protected profits for the few to the detriment of many,[64] resulting in burdening with inefficiency an area with so much potential to realize economic efficiencies. Do these patents act as an incentive to innovate? In many industries, the costs to develop and bring a new product to market are enormous and warrant protection. However, Internet venture often do not have the same startup costs as, say, a new drug or manufacturing process.

The biggest issue in deciding whether to patent business methods on the Internet may be the financial value of such protection. With technology changing so quickly, some patents may be obsolete by the time they are awarded (usually two to three years after application).

Other Considerations Regarding Patents and Competition

Outside the Internet domain, although some recent court decisions suggest that big businesses are getting the benefit of the doubt when it comes to intellectual property, other decisions suggest that consumer welfare and fair trade practices are more important.[65] For example, Intel's settlement with the Federal Trade Commission in the spring of 1999 forced Intel to make a key concession: It will no longer withhold vital data about its products from customers with whom it has patent disputes. Rulings with respect to both Intel and Microsoft indicate that the government can set conditions on how a company can use its intellectual property. To defuse government scrutiny, Intel has now licensed its interface technology to other companies. Microsoft has allowed Dell Computer to delete the icon for Microsoft's browser from some PCs.

Some experts warn that if patent holders can't fully control their property, it might chill innovation. However, others believe that more open information about intellectual property and preventing monopolists from abusing their power arising from intellectual property will actually enable inventors to experiment and innovate even better inventions.

Copyrights

Copyrights are similar to patents in granting protection of one's creations and in granting the owner exclusive rights to reproduce and distribute the copyrighted work. However, copyrights protect the form or manner in which the idea is expressed, not the idea itself. Copyrights are particularly useful not only for artistic creations (music, literature) but also for products such as software that are mass marketed (which makes the protection of the information difficult). For these types of products, although it is the underlying concept or generic capability that gives rise to its expression in the product, only the tangible representation of the idea is copyrightable.

Hence, because copyright laws do not offer protection of the idea itself, the ideas embodied in the creative work can be freely used by others. For example, the idea of a computer program that manages finances is not subject to copyright protection; however, the program code of Quicken, a particular software program, is protected by copyright. Therefore, although copying the Quicken program without permission constitutes copyright infringement, prior law suggests that writing another program that accomplishes all the same tasks would not be an infringement[66]—unless the business method is patented.

Copyright infringement occurs when someone other than the copyright owner performs one of the exclusive rights afforded the copyright owner, most typically unauthorized reproduction or distribution of the copyrighted work. The term for copyrighted works, for individual authors, is the life of the author plus 70 years. For works created by employees for their employers—i.e., corporate authorship—the copyright term lasts for 95 years from the date of publication or 120 years from the date of creation of the work, whichever occurs first.[67]

Obtaining copyright protection is really quite easy. A copyright exists from the moment a work is created, without any formal action on the part of the author. Although there is a copyright registration procedure, registration is not required to obtain copyright protection. However, registration with the U.S. Copyright Office in Washington, DC, provides benefits should ownership of the copyright ever be challenged. Unless a copyright is registered, a copyright infringement suit cannot be filed should a firm need to file suit against an infringer.

Although information on the Internet is in the public domain, owners have not given up copyrights. Copyright owners should identify their materials subject to copyright protection by using a copyright notice with either the word copyright or © followed by the year in which the work was first published, followed by the name of the copyright owner. In addition, an explicit warning not to copy, or explaining the limits of permissible copying, is advisable.[68] There are digital systems to help owners find stolen material in use on the Net. Fingerprinting Binary Images (or FBI for short) embeds unique identifiers in online images. The Web can be searched for the use of these identifiers, and violators identified. Net crawlers can also search for protected material on the Internet.[69]

Trademarks

Often serving as an index of quality, **trademarks** are words, names, symbols, or devices used by manufacturers to identify their goods and to distinguish them from others. Trademarks provide protection against unscrupulous competitors who would attempt to trade on the firm's previously established goodwill and reputation by engaging in actions that confuse, deceive, or mislead the customer as to the identity of the producer of the goods.[70] If a product has not been patented, another firm can still copy and sell a similar product as long as consumers are not confused about who is producing the product. Hence, trademarks do not offer protection against such situations.

Trade Secrets

Trade secret protection in the United States evolved as common law in each of the fifty states and was standardized in 1979 in the Uniform Trade Secrets Act. This law establishes rules for fair competition among businesses with respect to proprietary information. **Trade secrets** are generally defined as any concrete information that

- Is useful in the company's business (i.e., provides an economic advantage or has commercial value);
- Is generally unknown (secret);
- Is not easily ascertainable by proper means; and
- Provides an advantage over competitors who do not know or use that information.

The company must have a precise description of its trade secrets in order for the courts to recognize them as valid. The definition of a trade secret does not mean that other people or competitors do not possess the information or have not "discovered" the same information independently, but that such information is not commonly known. In addition, to be protected as a trade secret, the information must have been developed at some expense by the owner, and the company must maintain a rigorous program to protect the information that forms the basis of its success. Issues related to such a program are covered in Appendix 7B on **proprietary information programs.**

Hence, trade secrets are broadly defined and mean all forms and types of financial, business, scientific, technical, economic, or engineering information including patterns, plans, compilations, program devices, formulas, designs, prototypes, methods, techniques, procedures, programs, or codes; whether tangible or intangible; and whether stored physically, electronically, photographically, or in writing.[71] Examples of trade secrets might include product formulas, designs, manufacturing processes, customer lists, new-product plans, advertising plans, cost and pricing data, financial statements, employee information, and analyses of competitors.

Trade secret protection is premised on the notion that proprietary information is generally shared only by parties who have been held in confidence by one another. Violations of that trusting relationship would indicate a breach of, and the basis for intervention by, the law. Hence, many companies protect themselves and signal to the courts their confidential relationships through a series of contractual obligations with business partners.

Contractual Obligations with Respect to Proprietary Information

Nondisclosure agreements (also referred to as confidentiality or proprietary rights agreements) set forth the nature of the secret and indicate that the firm both owns the information and expects the signer not to use or disclose the information described without permission. Employees, customers, and suppliers may be asked to sign such agreements that define ownership of the information and responsibilities to protect it.

Noncompete agreements specify the rights of an employee who leaves the firm's employ with respect to future employment opportunities. Such contracts often specify time and territorial limits that prohibit the employee from joining or establishing a firm that competes directly with the former employer. Hence, noncompete agreements prevent the signer from competing with the firm for a given period of time in a given territory. (The courts have found many of these agreements to be restrictive; hence, care must be taken in their wording.)

Invention assignment clauses (and ownership of copyright notices) are signed by employees and assign to the company rights to all inventions that the employee makes during the term of employment and, sometimes, for a period thereafter. Again, these agreements specify the employee's role and rights to information relative to the employer. It is important that the firm realize that the employee can keep and freely use general knowledge or skills acquired on the job.

These three types of legal agreements are designed to protect against the loss of trade secrets. Indeed, in this information age, companies are trying harder to tie down their most important assets: knowledge in the employees' brains. The issue of who owns the knowledge in a person's head is an increasingly important one. As knowledge and intellectual property become more important than physical capital, businesses feel compelled to protect that intellectual capital by taking extraordinary steps.

Who Owns the Knowledge: The Employer or the Employee?[72]

Almost anything a person creates, develops, or builds while on the employer's payroll can be considered the employer's property. Unless a person can prove the idea or list was developed personally and not as part of the company's work product, it belongs to the company. Because it can include important information on customers, even an employee's Rolodex may belong as much to the employer as to the person.

The *doctrine of inevitable disclosure* recognizes that people who possess sensitive competitive information may, in the course of doing a new job, use information from their former employers. Hence, courts have been willing to ask a person to sit out of the industry for a time until the information they know is no longer as sensitive. For example, the president of DoubleClick, a company that has been selling advertising space on the Internet since 1996, found that two of his key employees were planning to set up their own company, Alliance Interactive Network. Because they possessed highly sensitive information about pricing and product strategies, databases, and plans for future projects, he was able to obtain an injunction that prohibited the two employees from selling or placing advertising on the Internet in any capacity for six months.

Companies can protect against departing employees taking trade information with them by following these steps:

- Have the employee sign a noncompete agreement and a nonsolicitation statement (agreeing not to solicit customers of the firm during or after exit). These must be narrowly drawn with respect to geography and length of time.
- Have the departing employee sign a release document protecting the employer from postemployment lawsuits about alleged mistreatment during the course of his employment.
- Pay severance in installments, which can keep former employees on a tighter rein.
- Insist employees return all company documents and disks before leaving.

- Conduct a thorough exit interview to ask the employee if he or she learned or developed trade secrets at the company. Have the employee confirm the obligations of confidentiality. There should be another person to attend the interview and serve as a witness.

The employee's perspective is covered in Box 7-1, "Employee Considerations in Signing Noncompete Agreements."

Patents or Trade Secrets?

Many types of business information do not qualify for patent protection (i.e., because they are not within the definition of patentable subject matter, are not novel, or are obvious). Even when such information does qualify for patent or copyright protection, trade secret status may be preferred. For example, a trade secret can be protected as long as the company successfully prevents the secret from becoming widely known. In contrast, a patent has a lifetime of twenty years from the date of filing the application,

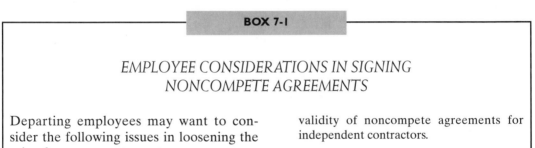

BOX 7-1

EMPLOYEE CONSIDERATIONS IN SIGNING NONCOMPETE AGREEMENTS

Departing employees may want to consider the following issues in loosening the grip of a noncompete pact.

1. It is important to negotiate acceptable terms before ever signing one, say, by shortening the duration of the agreement or the territory covered. It is also possible to negotiate "carve outs," or specific jobs and places for which the clause doesn't apply.

2. It is important to manage the departure so that the former employer is less likely to accuse the departing employee of unfair competition. Even if the departing employee is innocent, it might not seem that way for people who have signaled their intent to depart to be in the office after hours or to take files home.

3. Both employers and employees should be aware that some states may void such clauses. For example, Florida didn't recognize the validity of noncompete agreements for independent contractors.

Even if a person cannot loosen the noncompete, there are things to do in the interim. Finding employment in a related field can provide new and useful knowledge, keep skill levels current, and add to personal networks. For example, one person left a company and faced a year in which he couldn't work for a competitor and would have to relinquish any software he invented in the meantime. Violating the agreement would have cost him his severance package (several million dollars). So, in the year he faced out of work, he contacted possible customers for his new business idea and asked them to share their requirements and future needs. When his noncompete expired, these contacts helped fund his firm's new product.

SOURCE: Lancaster, Hal (1998), "How to Loosen Grip of a Noncompete Pact After Breakup," *Wall Street Journal,* February 17, p. B1. WALL STREET JOURNAL. EASTERN EDITION [STAFF PRODUCED COPY ONLY] by HAL LANCASTER . Copyright 1998 by DOW JONES & CO INC. Reproduced with permission of DOW JONES & CO INC in the format Textbook via Copyright Clearance Center.

TABLE 7-2 Trade Secret versus Patent Protection

	Trade Secret	*Patent*
Protects	A secret	A public disclosure
Lifetime	As long as information is secret	20 years from date of filing
Property Rights	Prevents unauthorized use by a person who acquired secret improperly	Excludes others from making, using, offering for sale, or selling
Scope	No protection for reverse engineering or someone else having same idea	Protects invention and equivalents thereof

Makes Sense When	Secret is not eligible for patent protection
	  e.g., a way of doing business
	Product life cycle is short
	  e.g., a computer chip with a life of $1\frac{1}{2}$ to 2 years
	Patent protection difficult to enforce or would be narrow
	  e.g., a process or method of making a computer chip
	Secret is not detectable in product
	  e.g., Coca-Cola composition
		Product has long market lifetime
		  e.g., a drug or pharmaceutical composition
		Product can be reverse engineered
		Corporate policy is to "patent it" for reasons including:
		  Importance of patent portfolio to financial backers
		  Employee mobility
		  Professional growth of employees via publication, conference presentation of research
		Patent protection is enforceable

after which time the information is available for public use. Given this, for many companies, classifying and treating intellectual property as trade secrets are the preferred ways to manage these assets. The comparison between trade secret and patent law is shown in Table 7-2 and summarized next.

When is trade secret protection preferred over patent protection?

- When the secret is not eligible for patent protection—for example, a way of doing business (an exception is the business method patent discussed in July 1998 ruling by the Court of Appeals for the Federal Circuit).
- When the product life cycle is short—say, a computer chip with a life of one to two years.
- When the patent would be hard to enforce or would offer only narrow protection, for example, a process or method of making a computer chip—can it be known from inspection of the end product (chip) that a particular method was used for its preparation?
- When the trade secret is not detectable in the product (i.e., a secret component, ingredient, or process of making cannot be discerned via reverse engineering of the end product, the notorious example being Coca-Cola).

On the other hand, there are times when patents are preferred over trade secrets:

- When the product will have a long market lifetime, such as a drug or pharmaceutical composition.
- When the product can be reverse engineered.

- When it makes sense as a matter of corporate policy (i.e., as an indicator of financial viability, to enhance employee mobility, or for the professional growth of employees via publication of research).
- When protection is enforceable.

Finally, because of the value of trade secrets, they are often targeted for theft by other countries and other companies. The Economic Espionage Act of 1996 makes stealing of trade secrets a federal criminal offense. The law bars stealing and using, in any fashion, another's trade secrets and makes it illegal to receive or possess trade secrets with the knowledge that they were stolen or misappropriated. The Act applies to conduct occurring in the United States by a U.S. citizen or company or by foreigners. The Act also applies to conduct occurring outside of the United States if the offender is a U.S. citizen, resident alien, or organization substantially owned or controlled by a U.S. citizen or company. Anyone caught stealing business secrets for a foreign government, company, or agent could get 25 years in prison and a $250,000 to $10 million fine. In addition to foreign economic espionage, the bill also targeted Americans who steal secret information from one business for the benefit of another. The maximum punishment for an individual would be 15 years in prison, or a fine of $250,000, and for an organization, a fine of $5 million.

Managing Intellectual Property[73]

In today's successful high-tech companies, the management of intellectual property has become a core competence. Because intellectual assets rather than physical assets are the principal source of competitive advantage, unlocking the hidden power of these assets is often a key source of success. One study reported that 67 percent of U.S. companies failed to exploit technology assets, and these companies let more than 35 percent of their patented technologies go to waste because they had no immediate use in products. Yet, active management of intellectual property assets is vital because

- Patents can be tapped as a revenue source (via licensing, for example).
- Costs can be reduced by cutting maintenance fees on unneeded patents (that could be donated to universities or nonprofits for a tax write-off).
- Patents can be repackaged to attract new capital and communicate an asset picture in a more attractive way to investors.

In addition to helping companies in the market by protecting their core technologies and business methods, patents can help a firm manage its product line. The potential strength of patents can help a firm establish R&D priorities. For example, Hitachi tries to develop only those products for which patents can help it establish market dominance. Similarly, in the biotech arena, Genetics Institutes says the strength of the potential patent position is a leading factor in deciding which research to pursue. Moreover, a patent strategy can help companies respond to shifts in the marketplace in an effective manner, by acquiring or partnering with firms that own patent rights to important developments.

Each of these issues points to the reality that intellectual property rights must be considered strategically in the product management process and not relegated solely to the realm of corporate attorneys.

SUMMARY

This chapter has covered a wide range of topics that must be understood in order to effectively develop and manage high-tech products and services. Organized around the framework of steps in the technology development/management process, the chapter has addressed ways to manage a firm's products to maximize success. The first step is to take an inventory of all of a firm's know-how (product, process, management) and figure out how to use this know-how to create and deliver value to customers. If there are gaps in technology identified in step 1, the next step is to figure out how best to add these technologies, whether by making them in-house, buying them on the market, or partnering to co-develop these. The next step is to decide on the firm's revenue-generating product offerings, whether to just license its technology or go the whole way into commercializing end-user products. The final step is the ongoing management of the new product development process, including decisions on the product architecture (modularity, platforms and derivatives), the management of cross-functional product development teams, the criteria for discontinuing projects, a services strategy to complement product offerings, and management of intellectual property issues.

DISCUSSION QUESTIONS

1. What is a technology map? What four steps does managing technology resources involve?
2. What are the various routes to product development that a firm can take? Why would a firm choose to pursue internal development or external acquisition?
3. What is the underlying dimension of the what-to-sell continuum? What factors affect a firm's decision about what to sell?
4. Is the mobile handset industry going to evolve along a path similar to the PC industry? Which companies are likely to make money in the future?
5. What is modularity? What are the pros and cons of following a modular approach to product design?
6. What is a product platform? What are the advantages to developers? To users?
7. Select and research a technology company of your choice. Assess this company's product strategy in terms of its technology map, including platforms and derivatives. Do you see any opportunities here for new product development?
8. Why do technology companies use cross-functional teams for new product development? What rewards or incentives may be used to enhance team performance?
9. What issues does a high-tech firm face in the decision to stop or kill a particular project?
10. Take the organization where you work for or study. Which cell of Figure 7-6 would you place it in with regard to the intersection of technology and service? What are the roles of technology and service in the product offering? Is technology being used for external or internal purposes? What are the issues the organization faces for increasing supply chain efficiency or enhancing customer value?
11. What are the three criteria for a patentable innovation? What are the steps in the patenting process?
12. What are your thoughts on the use of patents in highly innovative markets? Do they encourage or stifle innovation?
13. What are the three ways to signal a confidential business relationship?
14. What steps can companies take to protect against departing employees taking trade information with them?
15. What are the pros and cons of using patents versus trade secrets to protect intellectual property?

GLOSSARY

Copyright. Copyrights protect the *form or manner* in which the idea is expressed, *not the idea itself;* they are similar to patents in granting protection of one's creations and in granting the owner exclusive rights to reproduce and distribute the copyrighted work.

Cross-functional product development team. A project team charged with the responsibility of developing a new product with members representing different organizational functions or departments.

Derivative. Spin-off product from a common underlying technology platform that includes either fewer or additional features to appeal to different market segments.

Inseparability of services. The production of a service (when the service provider is providing the service) cannot be separated from the consumption of a service (when the customer actually gets the delivery of the service).

Intangible services. Services cannot touched or examined prior to consumption.

Intellectual property. An expansive term, referring broadly to a collection of rights for many types of information, including inventions, designs, material, and so forth. In the business arena, intellectual property may include not only creations from research and development but also general business information that a firm has developed in the course of running its business and that it needs to maintain as proprietary in order to remain competitive. Simply, intellectual property is original works that are essentially creations of the mind.

Invention assignment clause. An agreement signed by an employee assigning to the employer rights to all inventions that the employee makes during the term of employment and, sometimes, for a period thereafter.

Modularity. Building a complex product from smaller subsystems that can be designed independently yet function together as a whole.

Noncompete agreement. A contract that prohibits an employee from joining or establishing a firm that competes directly with an employer for a specified time period after leaving.

Nondisclosure agreement. An agreement that sets forth some trade secret owned by a firm and prohibits the signer from using or disclosing the information described without permission.

Original equipment manufacturer (OEM). Companies that buy components (such as disk drives) from suppliers that they integrate in a manufacturing process into a finished product such as a computer.

Patent. A form of protection for intellectual property; grant of a patent confers to the owner(s) the right to exclude others from making, using, offering for sale, or selling the product or process described and claimed in the patent for a specific time period, usually twenty years from the date of filing an application for patent with the U.S. Patent and Trademark Office (PTO). Patentable subject matter includes any new and useful article of manufacture, composition of matter, or process. An invention is patentable only if it meets three requirements: It must be useful, novel, and nonobvious.

Product platform. A common architecture based on a single design and underlying technology from which a stream of derivative products can be efficiently developed and produced.

Proprietary information program. A company's policies and procedures regarding steps to protect proprietary information, including marking of sensitive documents, copying, distributing, securing, mailing, and storing. Must be in place in order to prove trade secret protection.

Software piracy. Software piracy is the copying and/or distributing of copyrighted software without the permission of the copyright holder.

Technology map. Defines the stream of new products, including both breakthroughs and derivatives (incremental improvements based on the new technology) that the company is committed to developing over some future time period. Companies that are most successful in defining next-generation products use the map to force decisions about new projects amid the technological and market uncertainty found in high-tech markets.

Trade secret.   Any information that is useful in the company's business (i.e., provides an economic advantage), is generally unknown, is not easily ascertainable by proper means, and provides an advantage over competitors who do not know or use that information. Trade secrets are broadly defined and mean all forms and types of financial, business, scientific, technical, economic, or engineering information including patterns, plans, compilations, program devices, formulas, designs, prototypes, methods, techniques, procedures, programs, or codes; whether tangible or intangible; and whether stored physically, electronically, photographically, or in writing.

Trademarks.   Words, names, symbols, or devices used by manufacturers to identify their goods and to distinguish them from others; often serve as an index of quality; provide protection against unscrupulous competitors who would attempt to trade on the firm's previously established goodwill and reputation by engaging in actions that confuse, deceive, or mislead the customer as to the identity of the producer of the goods. Trademarks do not offer protection against copying and selling a product that has not been patented, as long as consumers are not confused about who is producing the product.

ENDNOTES

1. Tabrizi, Behnam and Rick Walleigh (1997), "Defining Next-Generation Products: An Inside Look," *Harvard Business Review,* November–December, pp. 116–124.
2. Capon, Noel and Rashi Glazer (1987), "Marketing and Technology: A Strategic Coalignment," *Journal of Marketing,* 51 (July), pp. 1–14.
3. Cohen, Isaac (2000), "Philip Condit and the Boeing 777: From Design and Development to Production and Sales," *North American Case Research Association,* October.
4. Ibid.
5. Capon and Glazer (1987) op. cit.
6. John, George, Allen Weiss, and Shantanu Dutta (1999), "Marketing in Technology Intensive Markets: Towards a Conceptual Framework," *Journal of Marketing,* 63 (Special Issue), pp. 78–91.
7. Capon and Glazer (1987) op. cit.
8. John, Weiss, and Dutta (1999) op. cit.
9. Ibid.
10. Ford, David, and Chris Ryan (1981), "Taking Technology to Market," *Harvard Business Review,* 59 (March–April), pp. 117–126.
11. John, Weiss, and Dutta (1999) op. cit.
12. Ibid.
13. Capon and Glazer (1987), op. cit.
14. Thurow, Lester (1997), "Needed: A New System of Intellectual Property Rights," *Harvard Business Review,* September–October, pp. 95–103.
15. Capon and Glazer (1987) op. cit.
16. Ford and Ryan (1981) op. cit.
17. Capon and Glazer (1987) op. cit.
18. Ford and Ryan (1981) op. cit.
19. Our thanks to Jon A. (Tony) Rudbach, Director of Technology Transfer, University of Montana, Missoula, Montana, for his assistance in the first edition with the material in this section. For additional information, see Steele, Thomas, W. Lee Schwendig, and George Johnson (1990), "The Technology Innovation Act of 1980, Ancillary Legislation, Public Policy, and Marketing: The Interfaces, " *Journal of Public Policy and Marketing,* 9, pp. 167–182.
20. *Pfaff v. Wells Electronics, Inc.*, 525 U.S. 55 (1998), rehearing den., 525 U.S. 1094 (1999).
21. Baldwin, Carliss Y. and Kim B. Clark (1997), "Managing in an Age of Modularity," *Harvard Business Review* (September–October), 84–93.
22. Fleming, Lee and Olav Sorensen (2001), "The Dangers of Modularity," *Harvard Business Review* (September), 20–21.
23. Meyer, Marc and Robert Seliger (1998), "Product Platforms in Software Development," *Sloan Management Review,* 40 (1), pp. 61–75.
24. Tabrizi and Walleigh (1997) op. cit.

25. John, Weiss, and Dutta (1999) op. cit.

26. Tabrizi and Walleigh (1997) op. cit.

27. Meyer and Seliger (1998) op. cit.

28. Tabrizi and Walleigh (1997) op. cit.

29. Ford and Ryan (1981) op. cit.

30. Shapiro, Carl and Hal Varian (1998), "Versioning: The Smart Way to Sell Information," *Harvard Business Review,* November–December, pp. 106–114.

31. John, Weiss, and Dutta (1999) op. cit.

32. Meyer and Seliger (1998) op. cit.

33. Moore, William L., Jordan J. Louviere, and Rohit Verma (1999), "Using Conjoint Analysis to Help Design Product Platforms," *Journal of Product Innovation Management,* 16 (January), pp. 27–39.

34. Sethi, Rajesh (2000), "New Product Quality and Product Development Teams," *Journal of Marketing,* 64 (2), 1–14.

35. Sethi, Rajesh (2001), "Cross-Functional Product Development Teams, Creativity, and the Innovativeness of New Consumer Products," *Journal of Marketing Research,* 38 (1), 73–86.

36. Sarin, Shikhar and Vijay Mahajan (2001), "The Effect of Reward Structures on Cross-Functional Product Development Teams," *Journal of Marketing,* 65 (2), 35–54.

37. Ibid.

38. Tam, Pui-Wing (2001), "Designing Team Helps Shape Apple Computer's Fortunes," *Wall Street Journal Online,* July 18.

39. Except where noted, this section is drawn from Boulding, William, Ruskin Morgan, and Richard Staelin (1997), "Pulling the Plug to Stop the New Product Drain," *Journal of Marketing Research,* 34 (February), pp. 164–176.

40. Keil, Mark and Ramiro Montealegre (2000), "Cutting Your Losses: Extricating Your Organization When a Big Project Goes Awry," *Sloan Management Review,* Spring, pp. 55–68.

41. Ibid.

42. Ibid.

43. Rover, Isabelle (2003), "Why Bad Projects Are So Hard to Kill," *Harvard Business Review,* February, 48–56,

44. Keil and Montealegre (2000) op. cit.

45. Boulding, Morgan, and Staelin (1997), op. cit.

46. Ibid.

47. See, for example, Meuter, Matthew L., Amy L. Ostrom, Robert I. Roundtree, and Mary Jo Bitner (2000), "Self-Service Technologies: Understanding Customer Satisfaction with Technology-Based Service Encounters," *Journal of Marketing,* 64 (July), 50; and Bitner, Mary Jo, Amy L. Ostrom, and Matthew L. Meuter (2002), "Implementing Successful Self-Service Technologies," *The Academy of Management Executive,* 16 (November), 96–109.

48. Mohr, Jakki (1996), "The Management and Control of Information in High-Technology Firms," *Journal of High-Technology Management Research,* 7 (Fall), pp. 245–268.

49. Ibid.

50. "Software Industry Statistics Page," Software Information Industry Association, downloaded from http://www.siaa.net.

51. MacDonald, Elizabeth and Joann Lublin (1998), "In the Debris of a Failed Merger: Trade Secrets," *Wall Street Journal,* March 10, p. B1.

52. "You Sank My Battle Chip!" (1997), *Time,* June 9, p. 47.

53. Takahashi, Dean and Jon Auerbach (1997), "Digital Files Antitrust Suit Against Intel," *Wall Street Journal,* July 24, p. B5.

54. Our sincere thanks to the efforts and insights of Judy Mohr, J.D., Ph.D., Perkins Coie LLP, Menlo Park, California, for her technical advice on the intellectual property sections and Appendix 7A in this chapter.

55. Prior to June 8, 1995 (the effective date of the GATT-TRIPS legislation), the term of a United States patent was 17 years from the date the patent issued. The U.S. participation in the Uruguay Round Agreements included an Agreement on Trade-Related Aspects of Intellectual Property (TRIPS) that harmonized the United States patent term with the rest of the world, by changing the term from 17 years from date of issuance to 20 years from the filing date. As a result, the present patent term for patent applications filed before June 8, 1995 is the greater of (1) 17 years from the date of issuance; or (2) 20 years measured from the filing date of the earliest referenced application. For applications filed on or after June 8, 1995, the patent term is 20 years measured from the earliest claimed application filing date.

56. 35 U.S.C. § 101.

57. The dollar amounts indicated reflect the fees that took effect November 1, 2003. Fees are

57. increased periodically; a current fee schedule can be found at www.uspto.gov/main/howtofees.htm.

58. Mansfield, E., M. Schwartz, and S. Wagner (1981), "Imitation Costs and Patents: An Empirical Study," *Economic Journal,* 91, pp. 907–918.

59. Updike, Edith (1998), "What's Next—A Patent for the 401(k)?" *Business Week,* October 26, pp. 104–106.

60. *State Street Bank & Trust Co. v. Signature Financial Group, Inc.,* 149 F.3d 1368, (CAFE 1998).

61. Updike (1998) op. cit.

62. Ibid.

63. France, Mike (1999), "A Net Monopoly No Longer?" *Business Week,* September 27, p. 47.

64. Gurley, J. William (1999), "The Trouble with Internet Patents," *Fortune,* July 19, pp. 118–119.

65. Garland, Susan and Andy Reinhardt (1999), "Uncle Sam's Balancing Act: Patent Rights vs. Competition, *Business Week,* March 22, pp. 34–35.

66. Stavish, Sabrina (1997), "Copyrights on the Internet . . . Protecting Yourself," *Advertising and Marketing Review,* January, p. 6.

67. Sonny Bono Copyright Term Extension Act, October 1998.

68. Stavish (1997) op. cit.

69. Morris, Glen Emerson (1997), "Protecting Intellectual Property on the Internet," *Advertising and Marketing Review,* January, p. 24.

70. Stern, Louis and Thomas Eovaldi (1984), *Legal Aspects of Marketing Strategy,* Upper Saddle River, NJ: Prentice Hall.

71. Shapiro, Barry (1998), "Economic Espionage," *Marketing Management,* Spring, pp. 56–58.

72. The information in this section is drawn from Lenzner, Robert and Carrie Shook (1998), "Whose Rolodex Is It, Anyway?" *Forbes,* February 23, pp. 100–104.

73. Rivette, Kevin and David Kline (2000), "Discovering New Value in Intellectual Property," *Harvard Business Review,* January–February, pp. 54–66.

APPENDIX 7A

Steps in Obtaining International Patent Protection[a]

Under the auspices of the World Intellectual Property Organization (WIPO), an international application can be filed under the Patent Cooperation Treaty (PCT). Although WIPO had 179 member countries, as of Fall 2003, not all members have ratified the PCT. Member countries of the PCT are shown in Table 7A-1. Under the PCT, an applicant can file an international application within one year of the filing date of a national application for patent (e.g., a U.S. provisional or utility application) and claim priority back to the filing date of the national application. The PCT enables an applicant to file one application in his or her home language and have that application considered for patent in any or all of the member states of the PCT. At the time of filing the International PCT application, the applicant has auto-matic and all-inclusive designation of all states that are bound by the Treaty, thus preserving the right to seek patent protection in any or all of the member states.

Like U.S. applications, an application filed under the PCT is published eighteen months from the priority date. For example, suppose an application, either a provisional application or a utility application, is filed in the United States, establishing a "priority date." Within one year from the filing date of that U.S. application, a PCT application must be filed in order to preserve the right to claim back to priority date. The PCT application is then published eighteen months from the priority date. Since patent applications are published prior to examination, a company must be mindful that even if granting of a

TABLE 7A-1 PCT Contracting States (January 2004)

Name of State followed by the two-letter code

Albania AL
Algeria DZ[1]
Antigua and Barbuda AG
Armenia AM[1]
Australia AU
Austria AT
Azerbaijan AZ
Barbados BB
Belarus BY[1]
Belgium BE
Belize BZ
Benin BJ
Bosnia and Herzegovina BA
Botswana BW
Brazil BR
Bulgaria BG
Burkina Faso BF
Cameroon CM
Canada CA
Central African Republic CF
Chad TD
China CN
Colombia CO
Congo CG
Costa Rica CR
Côte d'Ivoire CI
Croatia HR
Cuba CU[1]
Cyprus CY
Czech Republic CZ
Democratic People's
    Republic of Korea KP
Denmark DK
Dominica DM
Ecuador EC
Egypt EG
Equatorial Guinea GQ
Estonia EE
Finland FI[2]

France FR[1,3]
Gabon GA
Gambia GM
Georgia GE[1]
Germany DE
Ghana GH
Greece GR
Grenada GD
Guinea GN
Guinea-Bissau GW
Hungary HU[1]
Iceland IS
India IN[1]
Indonesia ID[1]
Ireland IE
Israel IL
Italy IT
Japan JP
Kazakhstan KZ[1]
Kenya KE
Kyrgyzstan KG[1]
Latvia LV
Lesotho LS
Liberia LR
Liechtenstein LI
Lithuania LT
Luxembourg LU
Madagascar MG
Malawi MW
Mali ML
Mauritania MR
Mexico MX.
Monaco MC
Mongolia MN
Morocco MA
Mozambique MZ[1]
Netherlands NL[4]
Namibia NA
New Zealand NZ

Nicaragua NI
Niger NE
Norway NO[2]
Oman OM[1]
Papua New Guinea PG
Philippines PH
Poland PL[2]
Portugal PT
Republic of Korea KR
Republic of Moldova MD[1]
Romania RO[1]
Russian Federation RU[1]
Saint Lucia LC[1]
Saint Vincent and
    the Grenadines VC[1]
Senegal SN
Serbia and Montenegro YU
Seychelles SC
Sierra Leone SL
Singapore SG
Slovakia SK
Slovenia SI
South Africa ZA[1]
Spain ES
Sri Lanka LK
Sudan SD
Swaziland SZ
Sweden SE[2]
Switzerland CH
Syrian Arab Republic SY
Tajikistan TJ[1]
The former Yugoslav
Republic
    of Macedonia MK
Togo TG
Trinidad and Tobago TT
Tunisia TN[1]
Turkey TR
Turkmenistan TM[1]
Uganda UG

(*Continued*)

TABLE 7A-I *Continued*

Ukraine UA[1]	Tanzania TZ	Viet Nam VN
United Arab Emirates AE	United States of America	Zambia ZM
United Kingdom GB[5]	US[6,7]	Zimbabwe ZW
United Republic of	Uzbekistan UZ[1]	

* All PCT Contracting States are bound by Chapter II of the PCT relating to the international preliminary examination.

[1] With the declaration provided for in Article 64(5).

[2] With the declaration provided for in Article 64(2)(a)(ii).

[3] Including all Overseas Departments and Territories.

[4] Ratification for the Kingdom in Europe, the Netherlands Antilles and Aruba.

[5] Extends to the Isle of Man.

[6] With the declarations provided for in Articles 64(3)(a) and 64(4)(a).

[7] Extends to all areas for which the United States of America has international responsibility.

patent on the application is denied, others still have access to the information in the application.

The fees for filing the PCT application depend on the number of pages of the application and the country selected to conduct the prior art search and initial assessment of patentability. A typical range for filing fees for a PCT application having 25–35 pages is about $3000–$5000 (see Table 7-1, p. 226).

Shortly after filing the PCT application, an examiner in Europe or in the United States will review the application and conduct a search of the prior art and issue an international search report that lists documents the examiner believes to be the closest prior art. The examiner will also prepare a preliminary, non-binding written opinion on whether the claimed invention appears to meet the requirements for patentability in light of the closest prior art. Copies of the search report, the cited art, and the written opinion on patentability are sent to the applicant. The applicant can respond in writing to the examiner's opinion and amend the claims, if desired.

These activities provide the applicant an early evaluation of patentability before incurring the major costs involved in bringing the application into individual member states to obtain national patent protection. The PCT application is brought before any or all of the PCT member states in which the applicant desires a national patent. Specifically, the application is reviewed by an examiner in each national country. To facilitate this, a translation of the application into the home language of each country is required, along with the national filing fee for each country. Costs to prepare multiple translations and to pay for the national filing fees, as well as the attorney fees for handling the application before each national patent office, quickly escalate to a range of $1,500 to $10,000 per country. The examiner in each country will consider the preliminary examination performed during the PCT phase; in some countries the PCT examiner's recommendation is merely "rubber-stamped," whereas in others the examiner's review yields a different outcome.

International patent applications can also be filed directly in the patent office of those countries where patent protection is desired. This is the only way to obtain patent protection in countries that are not members of the PCT, such as Taiwan or Argentina.

Considerations in filing for international patents include the differing standards of enforceability in each country and the expensive translation costs. For example, it may be difficult to enforce patent rights in China or elsewhere. Translation costs, particularly for Japan, South Korea, and so forth, account for the rapid escalation in international patent costs for a U.S. company. In contrast, Canada and Australia are relatively inexpensive because no translation is needed.

Online Resources for Intellectual Property

IP Worldwide, the Magazine of Law and Policy for High Tech (www.ipmag.com)
PIPERS Virtual Intellectual Property Library (www.piperpat.co.nz)
U.S. Patent and Trademark Office (www.uspto.gov)
Patent Office Sites Around the World (www.pcug.org.au/~arhen/)
Free copies of U.S. patents (www. patent-fetcher.com/FetchPatent.php) www.espacenet.com www.wipo.int

The following documents can be found free at http://www1.oecd.org/dsti/sti/index. htm
Facilitating International Technology Cooperation
Fiscal Measures to Promote R&D and Innovation
Foreign Access to Technology Programmes
Government Venture Capital for Technology-based Firms
Industry Productivity: International Comparison and Measurement Issues
The Knowledge-based Economy
National Innovation Systems
Patents and Innovation in the International Context
Policy Evaluation in Innovation and Technology
Regulatory Reform and Innovation
Technology Diffusion Policies and Programmes
Technology and Environment: Towards Policy Integration
Technology Foresight and Sustainable Development
Technology Incubators: Nurturing Small Firms
Venture Capital and Innovation

[a]Source: Judy Mohr J.D., Ph.D. (Chemical Engineering), Perkins Coie LLP, Menlo Park, California.

APPENDIX 7B

Proprietary Information Programs[a]

What factors affect a firm's ability to manage and control the proprietary information that forms the basis of its competitive advantage?
1. Employees. Possibly the single biggest danger to effective information management is a firm's own employees. Close to two-thirds of all U.S. intellectual property losses can be traced to insiders.[b] Unique management practices in high-technology industries, including efforts at team building (which include open information sharing and decentralized decision making), professional associations and ties, handling of new employees (many of whom may have been hired from a competitor), potential job dislocations, and so forth, greatly complicate the firm's ability to manage and control employees. Having a healthy organizational climate, with positive

employee morale, results in employees motivated to see their organization succeed. Disgruntled employees are more likely to divulge information recklessly than are more satisfied employees. Moreover, a corporate culture that values information and tacitly understands the need to be circumspect in sharing information can be a strong tool in managing information.

2. Senior managers. Senior managers must send appropriate signals regarding the control of information. Information security can be viewed as just another nuisance, unless corporate officers stand behind the security program.

3. Team-building efforts. The use of team-building efforts and open information sharing can make it difficult to signal the sensitivity of information. Indeed, as access to information is more diffused, information is more likely to seep out of the organization. And generalized access to sensitive information can jeopardize the company's ability to successfully prosecute or defend a firm's claims to trade secrets. As a result, many firms attempt to share information on a *need-to-know* basis only.

4. Internal politics. All organizations are subject to the dynamics of power and politics. Access to information and knowledge is a signal used to convey power. As people try to gain access to information to boost their sense of power in the organization or to use it politically, they may ferret out information that is considered sensitive, but may not understand that it is.

5. Proprietary information policy. Most firms have a policy on handling of proprietary information (for one such policy, see the example in Table 7B-1). Typically, the security department identifies breaches of security. Monitoring can help find leaks of information and may deter information abuses. And some level of monitoring is required to gain trade secret status in legal proceedings. Excessive monitoring can demotivate

employees and cause them to sabotage or circumvent the system.

6. Geographic, professional, and friendship ties. It can be difficult to manage information in high-tech industries given the close geographic (say, of Silicon Valley or the Boston Corridor), professional, and friendship ties that develop. People who switch employers may often maintain friendships with those at the former employer, and communication with these people may be viewed as less risky because of the nature of the relationship. Moreover, engineers traditionally have had a culture of trading trade secrets on a reciprocity basis.[c] Engineers may view information sharing as either their right (because they "invented" the information) or their obligation (to disseminate research results at a professional meeting).

7. Partnering relationships. Partnering relationships with other firms also affect the ability to manage proprietary information. Preferred customers may be beta-test sites for new products. Suppliers may be provided with proprietary information to aid them in tailoring their processes to better meet the buyer's needs. Codevelopers also need access to product plans. Although partners may be asked to sign nondisclosure agreements, the extent to which they observe such agreements is questionable. In fact, some companies actually rely on the fact that customers will *not* observe nondisclosure agreements, scheduling customer visits so that their firm is the last one that a customer will visit in order to hear customers' comments regarding competitors' strategies.

> A lot of our customers [on customer visits] are under nondisclosure, but they'll make comments in these briefings that just blow your mind. [You think,] "I don't know why you said that...." They violate nondisclosures constantly—unwittingly, I might add.[d]

With respect to suppliers, similar issues may arise.

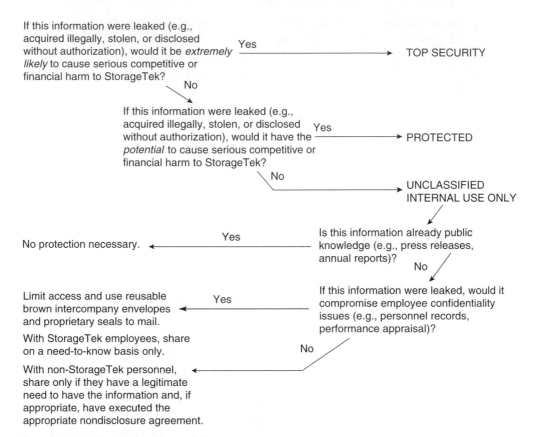

If this information were leaked (e.g., acquired illegally, stolen, or disclosed without authorization), would it be *extremely likely* to cause serious competitive or financial harm to StorageTek?

Yes → TOP SECURITY

No ↓

If this information were leaked (e.g., acquired illegally, stolen, or disclosed without authorization), would it have the *potential* to cause serious competitive or financial harm to StorageTek?

Yes → PROTECTED

No ↓

UNCLASSIFIED INTERNAL USE ONLY

Is this information already public knowledge (e.g., press releases, annual reports)?

Yes → No protection necessary.

No ↓

If this information were leaked, would it compromise employee confidentiality issues (e.g., personnel records, performance appraisal)?

Yes → Limit access and use reusable brown intercompany envelopes and proprietary seals to mail.

With StorageTek employees, share on a need-to-know basis only.

No → With non-StorageTek personnel, share only if they have a legitimate need to have the information and, if appropriate, have executed the appropriate nondisclosure agreement.

Proprietary Information Pyramid

The percentages listed below are current industry standards. As you move higher in the pyramid, fewer documents are affected, but more protection is required.

It is undesirable—and unnecessary—to classify too much information at the highest level of protection.

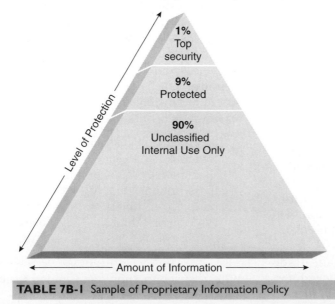

- 1% Top security
- 9% Protected
- 90% Unclassified Internal Use Only

Level of Protection

Amount of Information

TABLE 7B-1 Sample of Proprietary Information Policy

TABLE 7B-1 *Continued*

Handling Strategies	Top Security	Protected	Unclassified—Internal Use Only
Marking	A Top Security cover sheet must be used on appropriate documents. (Copies of the cover sheet may be used.) Every document containing Top Security information must have every page marked Top Security. (Stamps are available for this purpose. Headers or footers may also be used.) The automatic Declassification Date/Event must be marked on the document.	A Protected cover sheet must be used with appropriate documents. (Copies of the cover sheet may be used.) Internal page markings are optional, but recommended. The automatic Declassification Date/Event must be marked on the document.	No markings are required.
Disclosure	It is the responsibility of the person making the disclosure to determine the need to know and to make the recipient aware of the information's classification.	Same as Top Security.	Take reasonable precautions against disclosure to outside parties or employees without a need to know.
Distribution List	Must be maintained. Serialization is recommended.	Employees are strongly encouraged to maintain a distribution list. At management's discretion, distribution lists may be required.	Distribution list is not required to be maintained.
Storage	When not in use or under the direct supervision of authorized personnel, all Top Security documents must be locked in a desk, file cabinet, or office.	Protected documents must be placed out of sight in a desk or file cabinet when not in use.	No special requirements. Judgment should be used to prevent casual disclosure when it is appropriate.
Reproduction	When copies are made of Top Security documentation and distributed to those with a need to know, a distribution list must be developed, accurately maintained, and periodically reviewed to reconfirm need to know. Cover sheets must be affixed to the copies. When, in the judgment of the originator, a Top Security document should not be copied, it is recommended that every page be printed on "DO NOT COPY" background paper before it is distributed.	Employees should use discretion and minimize the number of Protected documents that are reproduced. Cover sheets must be affixed to the copies. Employees are strongly encouraged to maintain a distribution list. At management's discretion, distribution lists may be required.	This material has no special restrictions; however, unnecessary copies should not be made.

TABLE 7B-I   *Continued*

	Top Security	*Protected*	*Unclassified—Internal Use Only*
Disposal/Destruction	Holders of Top Security documents must either destroy them or dispose of them in a proprietary information container. Contact the manager, corporate information security, for guidance on purchasing destruction devices (e.g., shredder) or proprietary information containers.	Same as Top Security.	Recycle or place in regular trash. Place confidential employee information in proprietary information containers.
Mailing	**Interoffice Mail–Pouch** Hand deliver when possible. When transmitted within the company, place in a red proprietary envelope. The red security envelope is for one time use only and should be opened only by the addressee. Pouch mail service is not recommended for Top Security material.	**Interoffice Mail–Pouch** Use brown interoffice envelope and proprietary information seal.	**Interoffice Mail–Pouch** Use regular interoffice mail envelopes. For confidential employee information, use brown envelope and proprietary information seal.
	External Postal–Delivery Services Must have appropriate markings (e.g., cover sheet or tag) and must be enclosed in an opaque, security StorageTek letterhead envelope. Certified mail must be used. Return receipt is required.	**External Postal–Delivery Services** Same as Top Security, except that pouch or regular mail may be used.	**External Postal–Delivery Services** Use pouch or regular mail.

Source: Storage Technology Corporation, Louisville, Colorado. Reprinted with permission.

8. Competitive intelligence gathering efforts by other firms. Further, control of proprietary information is affected by the efforts of other firms to gain access to the information. For example, although many tools can be used legitimately to gather competitive intelligence, as advocated by the Society for Competitive Intelligence Professionals, many more nefarious and insidious tactics can be used as well. Industrial espionage by foreign governments has been on the rise and warrants careful security procedures, both at home and when employees travel abroad. Even hiring away employees with the express aim to access strategic information is common in high-tech fields, in which relationships sometimes characterized as "incestuous" are commonly used.

Awareness of these factors can help a firm to design proprietary information programs for success.

NOTES

a. Excerpted from Mohr, Jakki (1996), "The Management and Control of Information in High-Technology Firms," *Journal of High-Technology Management Research,* 7 (Fall), pp. 245–268.

b. "Eyeing the Competition" (1999), *Time,* March 22, pp. 58–60.

c. von Hippel, Eric (1988), "Trading Trade Secrets," *Technology Review,* 91 (February–March), pp. 58–64.

d. Mohr, Jakki (1996), op. cit.

CHAPTER 8

Distribution Channels and Supply Chain Management in High-Tech Markets

Isaac Asimov predicted in 1981 that, in the 21st century, "under global sponsorship, the construction of solar power stations in orbit about the Earth will have begun."
—ISAAC ASIMOV,

science fiction writer and biochemist,
quoted in The Book of Predictions

Cisco's Reseller Network: Moving up the Value Chain

Starting as a manufacturer of Internet routers, Cisco has become the leading supplier of Internet infrastructure, with an extensive global network of resellers. These resellers allow the company to serve a large, geographically dispersed customer base.

Anticipating the commoditization of hardware, Cisco made a major strategic shift to providing both hardware and software in integrated end-to-end solutions. This business redefinition, from selling physical products to selling network-related solutions, placed new demands on the corporation's distribution system. Many of Cisco's 36,000 resellers were ill-equipped for this change, lacking the technical knowledge and resources to service customers in such a highly skilled fashion. They functioned well in their previous capacity, but now a move up the value chain was required of them, from value-added resellers (VARs) to systems integrators.

As VARs, resellers bundled Cisco's hardware with other hardware and software to provide customers with a system of working components. However, as systems integrators, they needed to manage much larger, more complex, and customized projects, consulting directly with senior managers at client companies to provide end-to-end solutions.

Cisco invested substantial resources to enhance its resellers' skills, designing and implemented a multistep process with each reseller. Cisco provided customer-focused business models and helped each reseller to determine current and ideal positions in the value chain. They were assisted in identification of existing strengths and in matching those strengths to an appropriate market niche. Next, Cisco offered rigorous technical training and certification programs, including annual audits to ensure quality standards. This approach created collaborative relationships between Cisco and its smaller, autonomous partners.

By anticipating and seizing the opportunity to redefine its existing reseller network and augment its skills, Cisco has been able to raise both prices and margins at a time of increasing industry commoditization. Managers in high-technology companies need to recognize and manage distribution channels as a valuable asset, critical to their marketing efforts.

SOURCE: Mitchell, Tom. "Cisco Resellers Add Value," *Industrial Marketing Management,* 30 (2), 2001, 115–118.

R epresenting the various firms and players involved in the flow of product from producer to customer, distribution channels are an important tool in high-tech markets. Manufacturers must manage both the flow of product between production and consumption and the different relationships between firms at the various stages. Figure 8-1 shows a sampling of the various supply chain and distribution options that might be used for high-tech products.

Distribution channel activities include traditional *logistics and physical distribution functions*—such as inventory, transportation, order processing, warehousing, and materials handling decisions—as well as the *activities used to structure and manage distribution channel relationships* (e.g., the selection and management of distribution channel structure and players).

Distribution channels can be inefficient because suppliers and manufacturers, as well as manufacturers and distributors, often work at odds with each other; they may have conflicting goals and objectives and often don't think in terms of solving joint problems. Effective distribution channels allow a firm to identify redundancies and

FIGURE 8-1   High Tech Supply Chain Options

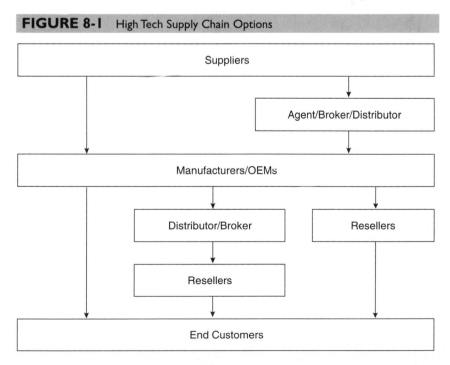

inefficiencies in the system, to develop relationships and alliances with key players, and to achieve both cost advantages and improved customer satisfaction. Some members of the channel may perform the functions more efficiently than others; a good distribution channel meets customer needs for channel functions in the most effective and efficient mode possible. The goal of channel management is to manage all the various logistics and distribution processes to provide value to the end customer effectively and efficiently.

New technologies are a major impetus today in redesigning supply chains and distribution channels. For example, in a logistical partnership with UPS based on proprietary software, Motorola radically redesigned its distribution process, eliminating multiple handoffs and cutting delivery time by 75 percent. It consolidated products from eight Asian semiconductor plants and delivered directly to its U.S. customers' doors in a processing time of just seven days from order placement to delivery. The logistical partnership allowed Motorola's original equipment manufacturer (OEM) customers to receive a higher level of quality, cost controls, and delivery.[1]

Channel members may also be used as marketing partners, playing a role in branding the product for the user and providing services to the customer. A study by the Gartner Group[2] found that when customers go to dealers for assistance, 77 percent do not have a specific brand in mind. Of those customers who are not "prebranded," resellers report that nearly 90 percent of the customers purchase a brand recommended by the reseller. And, even if the customer is prebranded, resellers report that they switch the customer to an alternative brand 53 percent of the time. Hence, distribution channels are not only an order fulfillment mechanism, they are also an important tool used to establish brand identity and preference in the marketplace. Channel partners are used to assess customer needs, develop solutions, consult with the customer, and provide service.

To a large extent, high-tech marketers face distribution issues similar to those faced by marketers in more conventional contexts. Although this chapter will provide a brief review of basic channels strategy, it will focus primarily on managing some of the complexities in high-tech distribution channels. These complexities arise from the high value of many technologically sophisticated products, the rapid pace of market evolution, the need to maintain sales and service support, and the ease with which some high-tech products can be pirated. Other complexities arise from the addition of the Internet as a new distribution channel.

This chapter begins with a brief review of the basic issues in distribution channel design and management. It then addresses issues specific to high-tech marketplaces. The third section of the chapter addresses the addition of the Internet as a new distribution channel. The chapter concludes with an expanded view of distribution to include supply chain management issues and resources.

ISSUES IN DISTRIBUTION CHANNEL DESIGN AND MANAGEMENT

Distribution channels exist to perform vital functions in completing marketing exchanges. From providing assortments for customers in the amount and variety they desire, providing service and other facilitating functions (credit terms, training, installation, etc.), to communicating with end users, and so forth, these functions are necessary for a successful exchange between buyer and seller. In designing a distribution channel, firms face the following decisions or issues, shown in Table 8-1:[3]

1. Consideration of Channel Objectives, Constraints, and External Environment. Firms have a variety of needs to consider in structuring and designing a distribution channel. Customer needs and buying habits are some of the most important considerations. Purchase quantity, location convenience, delivery speed, product variety, and service needs must be considered. The channel structure used by competitors and product characteristics must also be examined.

2. Choice of Channel Structure: Direct versus Indirect.   A *direct* channel structure is one in which a manufacturer sells directly to the customer, say with its own sales force, through company-owned stores or via the Internet. Of course, the Internet is a major force leading to more direct channels across a variety of industries, and its role is examined in detail later in this chapter. An *indirect* channel is one in which a manufacturer uses some type of intermediary(ies) to market, sell, and deliver products to customers. Direct and indirect channels are not mutually exclusive, as a firm may use some combination of them to get its products to customers. The combination of direct and indirect channels is known as a **hybrid channel**, or a **dual channel**.

The reality is that most firms typically juggle both direct and indirect channels in reaching their customer base. They may use a direct sales force for some customers, use channel intermediaries for others, and have an e-commerce Web site, too. Especially when a firm uses a hybrid channel, channel management is quite complex. One complexity is that indirect channels are subject to less management authority than are direct channels, posing a control issue. Further, if different types of channels compete for customers and revenue, conflict between various members in the channel can increase. These complexities are addressed in the section on adding new channels.

Some people argue that by using a direct channel, the intermediary can be eliminated and the price of the product lowered. However, a careful examination of the underlying role of a distribution channel shows the fallacy of this thought. Although intermediaries can be eliminated, the functions they perform—providing assortments for customers in the amount and variety they desire, providing service and other facilitating

---

TABLE 8-1   Issues in Distribution Channel Design and Management

1. Consideration of channel objectives, constraints, and external environment
2. Choice of channel structure: direct versus indirect
3. Choice of type of intermediary
4. Penetration/coverage: number of intermediaries
5. Channel management
    a. Selection and recruitment of channel intermediaries
    b. Control and coordination
    c. Consideration of legal issues
6. Evaluation of performance

functions, communicating with end users, and so forth—cannot. If a firm were to use a direct channel, either the manufacturer or the customer must assume responsibility for those functions. And, in either case, although the price of the product may be lower, the costs for one or the other channel members have increased because of the additional functions performed.

3. Choice of Type of Intermediary. If a firm chooses an indirect channel, it can use different types of intermediaries. Distributors usually buy directly from the manufacturer and sell to other intermediaries such as resellers or retailers who, in turn, sell to end users. For example, Tech Data Corporation based in Clearwater, Florida, distributes a variety of information technology products and services to customers ranging from small retailers to large corporations. Resellers, who typically operate locally, have a closer relationship with end users by providing products and services matched to their needs.[4] Many resellers of high-tech products are referred to as **value-added resellers (VARs)**, who purchase products from one or several high-tech companies, add value through their own expertise, and usually market bundled solutions to particular vertical (i.e., industry-specific) markets. For example, Meridian IT Solutions based in Schaumberg, Illinois, is a VAR for Cisco Systems (see vignette on page 252) and other manufacturers. **Systems integrators** are specialized resellers who typically manage very large or complex projects involving hardware, software, or services from different vendors and take responsibility for customized implementation for specific customers. For example, Science Applications International Corporation (SAIC), a Fortune 500 company based in San Diego, California, provides information technology systems integration services to many government customers.

Inbound versus outbound dealer organizations are other terms used to identify channel members in the high-tech setting. **Inbound dealers** typically have a retail storefront, and their primary customers are walk-in traffic generated through traditional advertising and promotion means. **Outbound dealers** have a sales force that makes calls on customers, typically at the customer location (the dealer may or may not have a storefront).

Depending upon the type of high-tech product being sold, many traditional channel intermediaries found in the retail sector may also play a channel role. Mass merchandisers, "category killers," small mom-and-pop stores, and franchises (such as ComputerLand or Radio Shack) may all be members of a firm's distribution strategy.

4. Penetration/Coverage: Number of Intermediaries. If a firm uses an indirect channel, it must decide how many intermediaries to use within each region or territory. An important tradeoff is that between the *degree of coverage* and the *degree of intrabrand competition*. Firms typically want as much market coverage as possible and, as a result, may sell through as many intermediaries as they can. Although such a decision may provide additional market penetration, this penetration comes at a cost. When a firm has many dealers in any area, each dealer competes with other dealers in its territory. To the extent that such competition occurs between different brands or products in the marketplace (referred to as "**interbrand competition**"), it can be healthy. However, when such competition occurs between dealers of the same manufacturer's brand, or **intrabrand competition**, it can cause problems.

Dealers competing against each other to sell the same brand often rely on price competition. Not only can this be damaging to the manufacturer's reputation and perceived quality in the market, but the dealers themselves often end up making a lower margin on those manufacturer's sales. As a result, it is difficult for them to support the level of service and training that the high-tech product often requires. So, too much penetration can actually lead to problems over the long term, in which the product is neither supported nor valued by channel intermediaries and end-user customers in the way a firm might desire. Hence, in deciding on the degree of coverage in a market, a firm must maintain a balance between too limited coverage and too much coverage (which invites intrabrand competition). Vertical, or territorial restrictions, in which a select distributor is granted exclusive rights to a particular territory, can be used to inhibit intrabrand competition.

5. Channel Management.   Channel management includes the ongoing activities that a firm uses to maintain channel relationships and effective channel performance over time.

a. ***Selection and recruitment of channel intermediaries.*** Once a channel structure is decided upon, a firm must attract and recruit intermediaries for the firm's product. Attending trade shows can be a useful strategy for doing so, as can using a targeted direct-mail campaign, effective publicity and public relations, or personal selling.

b. ***Control and coordination.*** Depending upon the particular type of intermediary, many channel members are interested in creating an identity and position for their store in a local market. Manufacturers generally have less interest in which particular store a customer buys its goods, so long as the customer chooses the manufacturer's brand over competing brands. Given this discrepancy between manufacturers' and channel members' goals and objectives, manufacturers must use coordination mechanisms to manage, guide, and monitor their resellers' activities. Recall from Chapter 3 the discussion of unilateral and bilateral governance structures, or the terms, conditions, systems, and processes used to manage the ongoing interactions between two partners. Similar tools can be used to guide and manage the behavior of channel intermediaries.[5]

Authoritative (unilateral) control tools reside in one channel member's ability to develop rules, give instructions, and, in effect, impose decisions on the other. Such control might arise from ownership (via vertical integration) of the channel member or the authority arising from formal, centralized decision making (as in franchising). Formal controls and monitoring focus attention on desired behaviors and outcomes and, where coupled with a basis in authority, realign a partner's interests and activities to ensure desired ends. Authoritative control might also arise from one party's power over another. Importantly, power arising from unilateral authority does not necessarily imply an exploitative relationship; power can be used in a benevolent fashion as well. As a governance tool, power provides a basis for administering and managing exchange relationships.

Bilateral control mechanisms originate in the activities, interests, or joint input of *both* channel members. An important form of bilateral control involves relational norms or shared expectations concerning channel members' attitudes and behaviors in working together to achieve mutual goals. The spirit of such sentiments is reflected by a commitment to flexibility and adaptation to market uncertainty, mutual sharing of benefits and burdens, information sharing, collaborative communication, and so forth. These norms establish a social environment in which individually oriented attitudes and behaviors are discouraged in favor of mutual interest seeking.

Moreover (recall from Chapter 3), joint interdependence and commitment also can serve as an effective basis for bilateral channel control. Interdependence creates incentives that regulate and motivate each party's conduct. Jointly shared (high symmetric) dependence is conducive to more flexible, long-term channel relationships. In such relationships, the need for (and power over) one another tempers inclinations toward self-interest seeking and motivates mutually beneficial behaviors.

Trust, or the extent to which partners jointly believe that each will act in the best interest of the partnership, can also provide effective bilateral coordination. Mutual trust can provide the basis for conflict reduction, higher performance, and satisfaction in ongoing exchange. Mutual trust alleviates the fear that either party will act opportunistically toward the other. If both parties trust one another, each possesses confidence in the other's integrity and reliability and expects the other to act responsibly toward it in the furtherance of their exchange relationship.

Legal tools can also be used to control the relationship. Vertical restrictions, exclusive distribution, and similar tools can be used to motivate and align channel members' behavior.

  c. ***Consideration of legal issues.*** Two legal issues have recently received increased attention in high-tech distribution channels. **Tying** occurs when a manufacturer makes the sale of a product in high demand conditional on the purchase of a second product. Among other things, the Justice Department accused Microsoft of tying its operating system to its Internet browser in 1997. Tying may also come in the form of bundled rebates that give buyers discounts for hot products if they also purchase the company's other products.[6]

  Exclusive dealing arrangements restrict a dealer to carrying only one manufacturer's product. Ostensibly, such arrangements are put into place to ensure adequate service, but antitrust implications arise if large companies use their dominance to restrict customers' access to competitors' products.

6. Evaluation of Performance.   It is important to assess the performance of the various channels and channel members. Questions about efficiency typically guide the selection of one type of channel structure over another. According to Dataquest,[7] the cost of selling through channels ranges from 10 percent of a firm's selling, general, and administrative expenses for a VAR channel to 11.5 percent for a VAD (value-added dealer) channel to 18 percent for a retail channel (which requires market development funds for advertising, returns, etc.).

In terms of selection and evaluation of particular intermediaries, both quantitative and qualitative performance indicators are pertinent.[8] *Quantitative* indicators might include sales volume moving through a particular intermediary or the manufacturer's market share in the intermediary's relevant territory. *Qualitative* indicators might include the dealer's satisfaction with and commitment to the manufacturer and the dealer's willing coordination of activities with the manufacturer's national programs. Table 8-2 identifies a variety of channel performance indicators.

TABLE 8-2   Channel Performance Indicators

Reseller's contribution to supplier profits
Reseller's contribution to supplier sales
Reseller's contribution to growth
Reseller's competence
Reseller's compliance
Reseller's adaptability
Reseller's loyalty
Customer satisfaction with reseller

SOURCE: Kumar, Nirmalaya, Louis Stern, and Ravi Achrol (1992), "Assessing Reseller Performance from the Perspective of the Supplier," *Journal of Marketing Research,* 29 (May). pp. 238–253.

CHANNEL CONSIDERATIONS IN HIGH-TECH MARKETS

How do these issues in channel design and management play themselves out in high-tech markets? Because of the high value of many technologically sophisticated products, and because of the rapid pace of market evolution, high-tech marketers face serious incentives to minimize the number of products held in inventory in the channel. The average selling price for information technology products has declined, placing pressure on vendor profitability. So, manufacturers look for the most cost-efficient channel to take their products to market. As mentioned previously, different types of indirect channels have different costs. Moreover, the advent of the Internet as a distribution channel for high-tech products has changed the nature of the relationship between channel members and manufacturers. Channel design and management must address the characteristics in the external environment for high-tech products shown in Figure 8-2. The implications of these characteristics for distribution channels are addressed here.

Blurring of Distinctions between Members in the Supply Chain

In high-tech environments, the line between suppliers and channel members is blurring. Dell Computer used to be a reseller of Hewlett-Packard printers, scanners, handheld computers, and digital cameras. When Dell announced it would begin selling its own brand of printers to its customers, HP perceived this as a major competitive threat. As a result, HP terminated its eight-year relationship with Dell, an important customer.[9]

At a different level in the supply chain (that between a supplier and its OEM customers), a similar phenomenon is occurring. Relationships between Intel and computer OEMs have been somewhat fractious due to OEMs' legitimate concerns that Intel would one day no longer be content with being a chip supplier and enter the computer market. The fact that Intel is making and selling complete motherboards (a complete subassembly of integrated circuits that can be "dropped into" the casing of the computer and sold as a finished product) gives some credence to this fear. Customers and suppliers now have the potential to become competitors.[10]

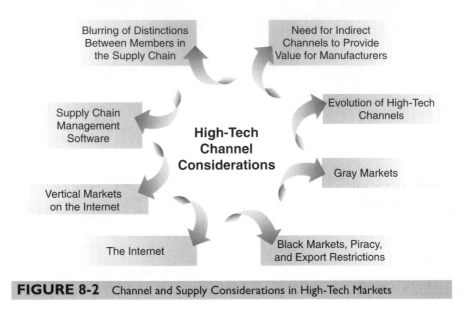

FIGURE 8-2   Channel and Supply Considerations in High-Tech Markets

Need for Indirect Channels to Provide Value for Manufacturers

Traditionally, costly and complicated high-tech products were sold from manufacturer to distributor to reseller to end user. But now, many firms are moving to a direct model. Prior to its merger with HP in 2002, Compaq had shifted from a ratio of traditional channel sales to direct sales of 98–2 in 1998 to 80–20 in 1999.[11] With the direct-sales model becoming increasingly popular for computer and network products, distribution partners who played a key role in the early stages of the industry are scrambling for ways to add value to their position in the supply chain.[12] As noted previously, some channel members are reaching into functions traditionally provided by other channel partners, such as manufacturers and resellers. Two new strategies are channel assembly and colocation.

In **channel assembly**, manufacturers ship semifinished products to distributors that configure them to client specifications and complete production before shipping the products. The advantages are customization and speedy turnaround for customers. Although resellers also do assembly, the function is increasingly shifting to distributors who have warehouse space to tailor products to order more efficiently, cutting down on time and expense.

> Rather than trying to anticipate demand by stocking huge amounts of inventory that can become obsolete, strategies are evolving to building products once true demand is known. Channel partners now deliver exactly what is needed at the time it's needed.[13]

As another example, Pinacor Inc., a distributor in Arizona, partnered with Lucent Technologies to perform final configuration, testing, and distribution of a telecommunications system. The partnership reduced lead time from thirty to forty-five days down to ten days, and the enhanced speed increased the odds the distributor competed successfully for the sale. Some believe that channel assembly is still too complex to have multiple partners assemble products in multiple locations and that, ultimately, the direct model will be more efficient, with one assembly point for each region.

In another strategy to bolster the value of indirect channels, **colocation** has the distributor's employees work from a vendor's manufacturing site to ship completed, high-demand models to either resellers or end users. Colocation can shave seven to ten days from the delivery cycle. Major distributors, such as Ingram Micro, Merisel, and Tech Data, partnered with IBM, have begun colocation efforts. Both colocation and channel assembly put customization all under one roof.

Although buying, bundling, and selling hardware and software are important for resellers, the shrinking profit margins on product sales are forcing them to *shift into services* (see IBM vignette, page 200). Hardware margins average only 1 to 8 percent, but service margins are 17 to 35 percent.[14] Services to help clients keep up with rapidly changing technology is vital. Customers want solutions to their problems, and as a result, more and more resellers are moving into territory typically held by consultants. Resellers are providing more client services, such as technical support, product maintenance, designing and installing networks, and implementing complex software solutions.

Evolution of High-Tech Channels

The type of channel a firm uses typically changes over the course of a technology's life cycle,[15] shown in Figure 8-3. When new technolozgies first appear in the market, sales strategies are naturally focused on original equipment manufacturers, independent software vendors, and integrators as the products struggle to gain support and a presence in the industry. Once the technology gains a toehold in the market and the target moves to the early adopters, technically astute value-added resellers join integrators as a key sales channel. VARs and early adopters determine whether a new technology establishes a presence in the market. And to leverage full efficiencies of the indirect channel, vendors work with distributors to expand their reach and grow their base of VARs.

As the technology approaches critical mass and enters a high-growth phase, a fairly traditional dealer channel is necessary. National distributors, value-added dealers, and traditional retailers (such as computer superstores) each serve to add coverage in a growing market. At this point in the technology life cycle, the earlier channel members (such as the value-added resellers) may transition to newer technologies and opportunities, even while maintaining their presence in the current technology as a convenience to their customer base.

Once technology reaches maturity as a standardized technology, the distribution cycle shifts again and mass-market channels become increasingly important. As the technology reaches maturity, increased use of mass merchants, consumer electronics, and office products stores may be seen.

Of course, there are exceptions to the model depicted here. Some products' first success might be with a retail channel. For example, if the product has an application in a home environment, then vendors must sell where this customer buys, which is typically at a retail store or via the Internet. Or, if the market is comprised of business customers with an installed base of existing technology, customers face headaches in migrating to a new technology and so may be more conservative and require more personal selling.

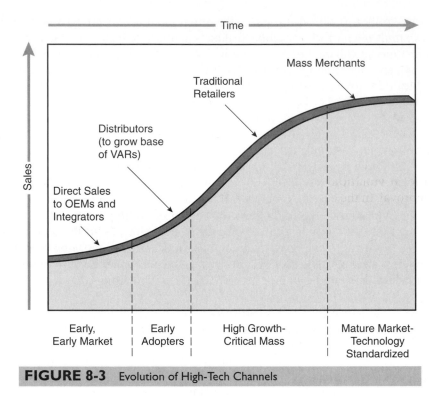

FIGURE 8-3 Evolution of High-Tech Channels

Geoffrey Moore believes that a direct sales channel is the most effective one to create demand for a new product and to cross the chasm.[16] But he argues that volume and predictability of revenues determine whether a direct-sales model is even viable. To support a single consultative salesperson requires a revenue stream of anywhere from $500,000 to several million dollars, depending on the amount of pre- and post-sales support needed. For a quota of $1.2 million per year, the salesperson must close $100,000 per month. If the sales cycle is six to nine months, and if the close rate is one of every two opportunities, then twelve to eighteen $100,000 prospects must be in the pipeline at all times, or some smaller number of significantly larger deals.[17]

A retail channel model may also be successful for a mainstream market, but it is not ideally suited for crossing the chasm. A retail channel does not create demand (rather, it is suited for situations in which customers are looking for a channel to fulfill demand), and it does not help develop the whole product.[18]

Understanding Gray Markets

In managing and controlling distribution channels, manufacturers want to ensure that the channels they use reach the appropriate customer segments with the combination of products and services each segment desires. In some cases, however, some channel members, rather than selling and marketing the firm's products to legitimate customers, may sell to unauthorized distributors or markets. For example, if a manufacturer offers large volume discounts, a distributor may feel compelled to buy a large quantity to take advantage of the discount; however, rather than inventorying those goods to be sold at a later date to legitimate customers and resellers, the distributors may "divert" them to unauthorized distributors, exporters, or other markets. Alternatively, if a large price differential exists between export markets (say, due to tariffs or other conditions), channel intermediaries in one market may try to take advantage of the price differential and sell to unauthorized distributors in those markets.

Known as the **gray market**, such unauthorized distribution refers to the selling of goods at discounted prices through resellers not sanctioned by the producer[19] and results in a lack of control over the product, distribution, and services. It is considered a legal violation for firms to sell trademarked goods without the manufacturer's approval. In the case of international markets, the appropriate customs service can levy cease and desist orders or prevent the import of offending goods.[20]

Although some may believe that the gray market offers another way to move product through another channel—and indeed, it may be only a small amount of total sales that is being diverted—it can cause serious problems. Legitimate channel members and resellers may become confused and angry over the unfair advantage the unauthorized distributors gain in buying and selling at a lower price point. Both the legitimate outlets and the firm's own sales force may lose business to the gray marketers. Gray marketers typically don't provide the sales and service that authorized distributors do. In turn, legitimate channel members become less motivated to push the firm's products. In some cases, the company's products may wind up competing across different channels, lowering the price point. Ultimately, the negative backlash can end up hurting the firm's brand and position in the marketplace.

Causes

In order to address the issue of gray markets, a firm must have a solid understanding of its root causes. As shown in Table 8-3, the most commonly cited cause of gray markets is a *firm's pricing policies*.[21] Manufacturers tend to structure discount schedules in favor of large orders, which causes distributors and other customers to buy more than they can sell or use (known as "forward buying") and then to resell the rest to unauthorized resellers. This problem can be exacerbated if distributors and customers must commit to purchases far in advance, with penalty clauses for canceling orders. Although manufacturers may have strong production reasons for such conditions, it contributes to the gray market problem.

TABLE 8-3 Causes and Solutions for Gray Markets

Causes	*Solutions*
Volume discount price policies	Eliminate sales to the source of the gray market
Differentials in exchange rates	Eliminate the arbitrage problem: one-price policy
Different resellers' cost structures	Increase market penetration
Highly selective distribution	Gather information on gray market problem
Producers performing many marketing functions	Institute consistent performance measures internally
Inconsistent internal policies	

A second cause is the price arbitrage opportunity that arises from *differentials in exchange rates in international markets.* Termed "parallel importing,"[22] this type of gray market activity occurs when goods intended for one country are diverted into an unauthorized distribution network, which then imports the goods into another country. Globalization has increased this type of gray market opportunity.

A third cause arises from the *cost differences between different resellers' cost structures.*[23] For example, full-service resellers have a higher cost structure than do discounters. Full-service resellers tend to provide functions including advertising, product demonstrations, post-sales service support, and so forth. Gray marketers can free ride on these functions. Resellers may also use some products as loss leaders, selling products at or below cost in order to attract traffic. Loss-leader tactics are common among gray marketers of popular brand-name products. Full-service resellers look to brand-name products to perform a very different role in their businesses—the product is expected to generate sufficient profit to cover the costs associated with providing full service.

Fourth, gray markets may be fostered *when suppliers practice highly selective distribution.* Territorial restrictions, exclusive dealing, and so forth may lessen intrabrand competition, but can draw unauthorized dealers into the market if demand is strong.[24]

Fifth, gray markets can also develop *when producers assume a range of marketing functions* that might otherwise be provided by full-service dealers. When such services are available from the manufacturer, the buyer's risk in buying from minimal-service dealers is reduced. Moreover, heavy advertising by the producers may build total demand and brand image and decrease buyers' dependency on resellers' reputations.[25]

Sixth, *inconsistent and incompatible policies* regarding a manufacturer's own departments can contribute to the gray market problem.[26] Plant managers may view gray markets positively when they contribute to the ability to operate plants at full capacity. Sales personnel may overlook gray market activity if the volume contributes to the quota in their territory. Such problems make it difficult to establish a clear solution to the gray market problem.

Solutions

In light of the causes, the solutions are many and varied. First, a firm may *use serial numbers on products* to track the source of units sold to gray marketers. Then, the firm can cut off the offender and is legally justified in doing so. Although the firm may lose some sales in the short term, it may sell more units through remaining authorized distributors and mitigate price erosion. Such a move sends a strong signal of commitment to the authorized distribution channel. However, this solution can be costly in terms of the time and administrative burden of identifying the offending dealers. Moreover, cutting off a distribution network may be satisfying to the authorized dealers but may actually cut off a market where the firm may have some competitive advantage.[27]

A second solution is to eliminate the source of the arbitrage and *offer a one-price policy with no quantity discounts.*[28] This strategy, although useful in eliminating one of the root causes of the problem, forecloses valid price–differentiation opportunities among different types of customers that have different transaction costs and receive different benefits from the product. It doesn't reward the larger, full-service dealers in the network, which may have other options available to them.

A third solution is to *move toward increased penetration in the market,*[29] balancing the potential of attracting unauthorized distributors with restrictive distribution against the increased intrabrand competition from too intensive distribution. Above all, it is important to have *information on the extent to which gray markets exist* in the distribution system, coordinated pricing brackets, and *consistent performance measures.*

Black Markets, Piracy, and Restricted Exports

Counterfeit, high-quality knockoffs pose yet another problem that is endemic to high-tech industries. Given the unit-one cost structure, in which the cost of producing the very first unit is high (due to R&D investments) relative to the reproduction costs of subsequent units, pirated copies of software and related items can be made relatively easily and inexpensively. In 2002, the global software piracy rate was 39 percent and in the United States the piracy rate was 23 percent. This amounted to revenue losses of $13 billion worldwide and $2 billion in the U.S.[30] A study on the economic impact of software piracy in 57 countries found that a 10 percent reduction in the global piracy rate by 2006 could boost the growth in the global information technology sector by an additional 15 percent. Specifically, China could see its IT sector grow five times while Russia could double its IT sector.[31]

In addition to being aware of potential counterfeit problems, firms selling strategic high-tech products, such as satellites, must also be aware of **export restrictions**. For example, chip makers must submit applications to sell microprocessors to certain countries, such as the former Soviet Union states or China. Sales of "dual use" products—nonmilitary items with military applications—are restricted to some countries. Even when restricted items are sold legally, problems can arise.

In 1994, McDonnell Douglas sold China machine tools for a civilian machine center in Beijing and subsequently found the tools had been diverted to a military complex. The Commerce Department imposed a $2.12 million civil penalty on McDonnell Douglas after a six-year investigation for violation of federal export laws.[32] Other items sold to China legally have included computers sold to the Chinese Academy of Sciences that could be used in nuclear-fusion projects.

The U.S. satellite company Loral hired China to propel its communications satellite into orbit using the Chinese-made Long March rocket. However, upon liftoff in the Sichuan province in February 1996, the rocket exploded. A committee of Western aerospace experts investigated the explosion and faxed its report to the Chinese government. A subsequent U.S. federal investigation concluded that this technical feedback may have helped China improve the accuracy of its rocket and missile program and that Loral had engaged in serious export control violations. Loral agreed to pay a fine of $20 million after a four-year investigation, the largest fine imposed on a US company under the Arms Export Control Act.[33]

The purpose of rules against technology sales is to protect U.S. security interests abroad. But the policy is fraught with problems. Some industry experts say that it is impossible to prevent the goods sold to friendly countries from ending up in restricted-access countries. Intel spokesperson Bill Calder says, "We ship chips to thousands of distributors all over the world, who aren't prevented from selling to these countries. There's a disconnect there."[34] Moreover, some believe that the controls actually undermine the United States' position as a technology leader. For example, the U.S. share of the international satellite market dropped from 73 to 53 percent during the 1999 to 2000 time frame; and French, Canadian, and German firms picked up the satellite contracts that the U.S. firms couldn't fill due to export restrictions.[35] Others believe that, rather than restricting access to such products, it may actually serve U.S. strategic interests to have countries such as China use U.S. technology—the United States will understand the technology being used. Moreover, strict restrictions could drive these countries to other suppliers, which might put even more information in the hands of other countries.

Manufacturers of high-tech goods must be aware of issues surrounding export controls in order to ensure that their distribution channel operates legally. And manufacturers must take proactive steps to protect their goods against counterfeiting.

ADDING NEW CHANNELS: THE INTERNET

Probably one of the most compelling challenges firms face in their distribution channels is managing the changes brought about with the Internet. Hewlett-Packard's medical-products unit provides a case in point. This division used 500 sales representatives and dozens of distributors to sell more than $1 billion of equipment per year worldwide.[36] The sales model used by these people was face-to-face interactions to build customer relationships and demonstrate products. However, changes in the healthcare field forced hospital chains to gain efficiencies. One way they did so was to demand one-stop shopping on the Internet, allowing the chains to buy ultrasound machines to electrodes without ever seeing a salesperson. So, HP was in a quandary: Should it offer customers the sales channel that they wanted and, in doing so, risk mutiny from the traditional sales force and distributors? Or should it keep its existing channel members happy but risk losing customers to competitors who delivered what they wanted? A middle ground option could use some combination of traditional channels and Web-based sales. But this also invites conflict between the manufacturers and traditional retailers.

Companies adding an Internet sales channel to their existing distribution channels have found that, inevitably, with multiple channels pursuing customers, conflicts arise as different channels simultaneously pursue the same customers. In addition, customers may become confused and angry as both the manufacturer and its channel members pitch the same account, often with wildly different terms and conditions. The situation is another example of coopetition: Manufacturers want to cooperate with their channel partners in reaching the segment(s) of the market the partners serve, but they also will compete with those partners when they go directly online. Powerful retailers can strong-arm their suppliers to avoid the Web, letting them know that a supplier who chooses to sell products is ultimately viewed as a competitor rather than a partner. Even in cases when a firm agrees to avoid the Web, its products may still wind up there if gray market activity exists.

Disintermediation refers to the situation in which a company adds an online distribution channel that bypasses existing intermediaries in favor of a direct-sales model. There are three strategic factors to consider in such a situation.

- Will the addition of an Internet channel generate incremental revenue or cannibalize existing sales?

- How can the conflict that the addition of an Internet channel generates be effectively managed?
- How can a hybrid channel model be used to strategically add an Internet channel?

Compaq provides a case in point for illustrating these strategic factors.

Compaq's Experience[37]

In 1998, prior to its merger with HP, Compaq decided it had to launch a Web site to compete with Dell Computer. Compaq attempted to balance the needs of traditional computer dealers with the competitive urgency to provide products to Internet buyers. To avoid conflict with its distribution channel, it created a unique set of business-oriented Prosignia computers for Internet-only sales. It targeted only small and medium-size businesses, which weren't the dealers' primary sales focus. And it created a way for dealers to profit from Internet referrals. So, what went wrong?

Resellers saw Compaq's embrace of the Internet not as a way to take business from rival Dell, but as a sign of Compaq's indifference toward their role. The fact that Compaq also slashed the number of distributors in North America from thirty-nine to four contributed to this perception. Its reason given for doing so was to cut the costs of maintaining inventories at resellers—costs that direct-model competitors such as Dell never incur. The fact that Compaq also started to deal in a more rigid fashion with its online resellers was little consolation to its traditional dealers. For example, Compaq's Internet-only resellers that sold the home-PC line had to adhere to minimum-advertised-price rules *and* provide repair services. And its five corporate dealers who were approved for online sales also had to follow specified rules when selling PCs from their own Internet sites. Regardless, dealers and other resellers felt left out by the direct-sales Internet strategy and competing business line, and they shunned selling *any* Compaq PCs. So, even though its traditional personal computer dealers were very unhappy about this new channel, Compaq argued that it had to roll out its own Web site, selling computers at low prices directly to small business customers and individuals, in order to stay competitive.

Generating Incremental Revenue or Cannibalizing Existing Sales

One important question for any company adding an online channel is whether the new Internet channel creates a new value proposition for end users, or whether it merely creates a more efficient distribution structure online.[38] Channels with clearly delineated value propositions are likely to attract new segments of customers and, hence, will be less likely to cannibalize existing revenues and more likely to generate incremental sales. For example, some firms are finding that an Internet presence extends the brand to shoppers they were not reaching with traditional retail stores. Yet, in other cases, the Internet channel merely cannibalizes an existing sales channel. This is why, for example, the Internet has displaced sales of airline tickets from traditional travel agents to the Internet: Merely buying over the Internet is not going to increase the number of vacations or trips purchased. On the other hand, the Internet has empowered and encouraged more individual investors to buy and sell stocks and mutual funds than was ever done through investment houses or brokers. The ease of access and ability to transact online created a new value proposition. This is why both the frequency and number of transactions have increased.

TECHNOLOGY TIDBIT

SMART FUEL PUMPS

The term *disintermediation* may soon take on new meaning in the realm of gas stations, where filling up your gas tank is becoming more and more like an episode of the Jetsons. Motorists in Indianapolis can fill up from the comfort of their driver's seat while a robotic gas pump does the work.

The SmartPump is a robotic mechanism that first reads a transponder placed on the windshield of the car to identify its make and model. The arm has a suction cup that opens the fuel tank door, and the unit engages the nozzle through a spring-loaded gas cap. The process works on nearly any car made since 1987 (unless the gas tank is underneath the license plate). The technology was developed by a combination of companies including H. R. Textron, which manufactures the service robot; a pump manufacturer; a fuel-cap manufacturer; Texas Instruments, which developed the windshield transponder; and Canada's International Submarine Engineering, which helped with motion sensor technology.

A similar technology, Autofill, has been used in Sweden for several years already.

Photograph reprinted with the permission of Shell Oil Company, Houston, Texas.

Shell conducted a test market in Sacramento, California, with 500 customers in 1997, and the project has gone live in a suburb of Indianapolis. Other companies are also developing different versions of the robotic system.

What are reactions in the United States so far? Some drivers are worried about damage to their vehicles. Others don't like the $1 transaction fee that accompanies each fill-up. Ultimately, the convenience could be worth it, though.

SOURCE: Rowell, Erica (2000), "Pumping It Up with Robotics," ABC News.com (www.abcnews.go.com/sections/tech/DailyNews/smartpump000310.html), March 10.

What factors predict whether an Internet channel will lead to incremental sales or instead cannibalize or displace existing sales? The flowchart in Figure 8-4 provides an assessment of critical factors that must be considered.

In Compaq's case, customers clearly were willing to take advantage of the efficiencies online ordering allowed; in that sense, they wanted an Internet channel. Compaq thought it would attract new customers by offering a unique line of computers targeted only to small and medium-sized businesses (not the dealers' primary focus). What appeared to be missing from Compaq's consideration of its channel strategy was the backdrop of the tenor of its relationships with existing dealers. As noted by Jap and Mohr,[39] effective use of the Internet is a function of whether the business partners

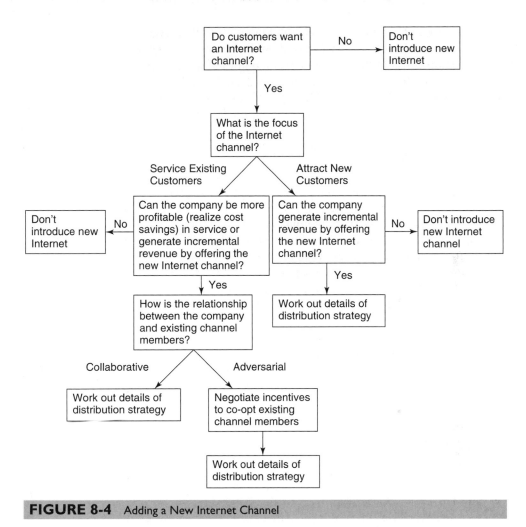

FIGURE 8-4   Adding a New Internet Channel

involved have a more adversarial relationship or more collaborative relationship. When manufacturers add a new online channel with the goal of reaching new customer segments, such efforts often backfire when the tenor of their relationships with existing channel members is more adversarial in nature. In Compaq's case, it had a fairly long history of adversarial relationships with existing dealers. In neglecting to consider the tenor of its relationships, Compaq did not anticipate the backlash from its existing channel, despite its careful planning. Clearly, in adding an Internet channel, a company must factor in the tenor of its existing business relationships.

Additional Considerations in Adding an Online Channel

The additional details of a distribution strategy (see boxes in Figure 8-4) include the following:[40]

- Does the company currently sell products through a catalog?
- Are the company's products simple in nature (no configuration required, not integrated with products from other manufacturers)?

- Is the sales process clear-cut and nonconsultative?
- Are the products easy to install and maintain?
- Does the company have an existing infrastructure to support direct sales (order fulfillment, returns, customer service, etc.)?
- Do customers usually know what they want when they are ready to buy, or do they need information about competitors' products and product benefits from a third party?
- Is the company willing to promote the Web site enough to attract sufficient prospects?

If a company can answer "yes" to all seven of these questions, then a Web-direct sales model does make sense. If, however, the answer to any of the questions is "no," then a Web-direct sales model may be problematic. (Note, however, that some companies, such as Dell Computer, have been successful, despite the product complexity, need for bundling, and a moderate level of installation complexity.)

In many cases, when a company chooses to add an Internet channel, conflict with existing channels is likely to result. Table 8-4 shows ways companies may avoid or manage conflict when faced with this situation.

Avoiding Conflict

1. *Use Web sites to disseminate product information only.* For example, 3M lists hundreds of products on its Web site but generally doesn't provide any way to order them directly. This is deliberately done out of concern for channel members. This solution uses the Internet to perform an information dissemination function only and relies on other channels for the sales functions.

2. *Use the Web only to generate leads; direct potential buyers to the nearest dealer or dealer's Web site.* U.S. automakers are prohibited by law from selling cars directly to consumers. On Ford's Web site, for example, a consumer can enter his or her zip code and find the nearest dealer from whom they can get the product. Unfortunately, this approach doesn't take full advantage of the Internet's ability to wring costs out of the system.

TABLE 8-4 Managing the Transition to the Internet

Avoid Conflict with Existing Channel

1. Use Web sites to disseminate product information only.
2. Use the Web only to generate leads; direct potential buyers to the nearest dealers.
3. Use Web sites only for limited merchandise offerings.
4. Take online orders for small customers only; direct larger sales to dealers.
5. Launch a Web site with no publicity.

Manage Conflict with Existing Channel

1. Keep prices on the manufacturer's Web site aligned with traditional channels.
2. Give a cut of each Internet sale to dealers.
3. Improve the flow of information.
4. Follow steps to manage hybrid channel.

3. *Use Web sites only for limited merchandise offerings.* For example, the Sharper Image tends to sell merchandise on its Web site that is excess or out-of-season inventory. Similarly, Compaq attempted to sell only one line of products dedicated solely for Internet sales.

4. *Take online orders for small customers only; direct larger sales to dealers.* For example, Jackson Products, a St. Louis company that makes safety goggles and welding gear, sells products over the Web, but online shoppers who need to purchase over $1,000 in merchandise are directed to a distributor.[41] Another variation of this strategy is to pursue online sales in geographic areas retailers don't cover.

5. *Launch a Web site with no publicity.* Many companies that have e-commerce Web sites do not promote them separately for two reasons. First, they want to avoid conflict with their existing offline dealers. Second, these companies may want the majority of their customers to come through the offline dealers for a better service experience.

Managing Conflict

Capitalizing more fully on the distribution efficiencies the Web has to offer will invite conflict, but it may be the only viable option. Not to capitalize on the Web in order to avoid conflict with distribution channels may threaten long-term business survival. Actively cannibalizing one's own distribution channel, although traditionally considered an evil to be avoided, is consistent with the notion of creative destruction.[42] Indeed, the new capabilities that the Internet has enabled threaten the very foundation of many businesses—and not just in distribution. For example, for Eastman Kodak, embracing digital imaging means undermining its specialized investments in production, processing, and distribution of silver-halide film. General Electric's Internet business units were referred to as "destroy-your-business.com."[43] Creative destruction in the online world essentially means survival by suicide. In order to compete with the pure-play Internet startups, established firms must turn from brick-and-mortar to **bricks-and-clicks** strategies, offering an online channel ("clicks") to augment their existing offline ("bricks") business. Companies that make decisions based on keeping the peace with their existing distribution channel likely will not succeed in the new world. As the experts say, cannibalize before there is nothing of value left to cannibalize.[44]

In this vein, some companies are willing to be more aggressive in using the Web as a distribution channel, but adopt one or more of the following strategies as a way to more actively manage channel conflict (rather than merely attempting to avoid it):

1. *Keep online prices equivalent to offline (dealer) pricing.* For example, Hewlett-Packard rolled out a Web site to let major hospitals buy online. Online prices were carefully aligned with those in other sales channels. Intuit offers Quicken and TurboTax on its Web site only at list price. (Firms that have significantly lower online prices are inviting a gray market situation.)

2. *Give a cut of each Internet sale to dealers and salespeople,* regardless of whether they played a role in generating it. At Hewlett-Packard, online orders generate commissions for the sales reps who typically handled the account. Compaq paid resellers an "agent fee" for steering business online. Although this strategy may

seem counterintuitive, it actually weights the compensation structure in favor of the success of the new channel.[45] Although new channels are added to gain cost efficiencies in coverage, the new channel is unlikely to contribute a large percentage of revenues in the short term. Rather, as customers are migrated to the new channel, it is important to alleviate concerns of channel members whose customers are migrating. Without concern for these issues, channel members may sabotage the success of the new channel.

3. *Improve the flow of information with resellers.* Channel management software products are available that allow customized information to specific resellers. In such a way, pricing, leads, and promotions can be more effectively and efficiently communicated. Moreover, forecast accuracy can be improved by improving communication flows from the channel members to the manufacturer.

Regardless, rather than adding new channels incrementally, without a clear vision of an ultimate go-to-market architecture—which can create conflict and morale problems internally and confuse customers externally—managers must design and manage channel systems strategically to achieve competitive advantage.[46] Ultimately, the synergies that can be found in offering products simultaneously through both traditional channels and on the Internet may be superior to either pure brick-and-mortar channels or Internet "pure plays" (companies that have only a Web presence and no physical stores, such as Amazon.com). These "bricks-and-clicks" channels are integrated (or harmonized) in the sense that customers can order products online and use physical stores for return centers, for example. Recall that a combination of a direct (such as the Internet) and indirect (such as VARs or other intermediaries) channel is referred to as a hybrid channel.

Steps in Managing Hybrid Channels

The objectives of a hybrid channel are to (1) increase market coverage while (2) maintaining cost efficiency. Increased coverage and lower costs can create a competitive advantage for firms that understand how to implement and manage a hybrid channel effectively. The key steps in effectively implementing and managing hybrid channels follow.[47]

1. *Identify customer target segments.* Customers can be targeted on the basis of size of customer, geographic region, products purchased, or buying behavior/needs.

2. *Delineate the tasks or functions that must be performed in selling to those segments.* Tasks include activities such as lead generation, sales prospect qualification, presales activities, closing the sale activities, post-sales service and support, and ongoing account management.

3. *Allocate the best (i.e., efficient and effective) channels to those tasks.* The various channels/tools/methods that can be used include national account management, direct sales, telemarketing, direct mail, retail stores, distributors, dealers/VARs, Internet, and so on. Not all channels must perform all tasks. Rather, channels should be combined to optimize costs and coverage relative to the tasks they are performing for various customer segments.

This model is the familiar contingency model, depicted in Figure 8-5. The idea is that no one channel can be used for optimum performance. Rather, the type of channel

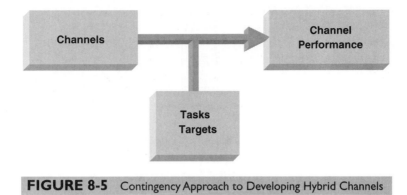

FIGURE 8-5   Contingency Approach to Developing Hybrid Channels

used must be matched to the tasks needed by particular customer segments (or target markets). Only by effectively matching the appropriate channel tasks to the channel used to perform the tasks will optimum channel performance occur.

Another way to look at this model is as a grid that aligns the various tasks performed by the different channels across customer segments.[48] For example, the grid in Figure 8-6 shows that for one particular company, direct mail was desirable to generate leads for three customer segments based on size (small, medium, and big). Once leads are generated, telemarketers perform the lead qualification task for medium and big customers, but smaller customers are directed to a retail channel for the remainder of the tasks. Presales and sales closing activities for large accounts are then performed by a direct sales force, whereas telemarketers perform those tasks for medium accounts. Ongoing account management is similarly allocated based on customer size.

As an example of this system at work, a major manufacturer of computer-aided manufacturing systems (CAD/CAM) sells its offerings in the United States and Europe through a direct sales force; in Japan, it uses an exclusive distributor. Because the channels and customers are physically separated, little conflict occurs (except in global accounts). Similarly, Xerox used product boundaries when it entered the personal copier market. It sold mid-range and high-end machines through a combination of direct sales and dealer distribution and low-end machines exclusively through retail channels (electronics and appliance stores, mass merchants, etc.).

Effective management of a hybrid channel requires three additional considerations:

- Managers must *assess the magnitude of potential conflict.* With no revenue in conflict, channel members may become complacent and not pursue their tasks effectively enough. On the other hand, with too much conflict, dysfunctional effects distract channel members from performing their jobs with clarity and enthusiasm. One guideline is that having more than roughly 10 to 30 percent of a firm's revenue in conflict between multiple channels provides a dysfunctional amount of conflict and can result in angry feedback from both customers and marketing personnel.[49]

Tasks \ Channels	Lead Generation	Qualify Sales	Presales	Close Sales	Postsales Service	Account Management
National Account Management						B
Direct Sales			B	B		B
Telemarketing		B M	M	M	B M	M
Direct Mail	B M S					S
Retail Sales		S	S	S		
Distributors						
Dealers/VARS						

(S) Small customers   (M) Medium customers   (B) Big customers

FIGURE 8-6   Grid: Allocating Tasks to Channels

SOURCE: Reprinted by permission of Harvard Business Review. "Grid Allocating Tasks to Channels" from "Managing Hybrid Marketing Systems" by Rowland, Morarity and Ursula Moran, November-December 1990. Copyright © 1990 by the Harvard Business School Publishing Corporation; all rights reserved.

- After assessing the magnitude of conflict, *clear boundaries and guidelines for managing the conflict,* based on who owns which customer, must be established.[50] As indicated in the hybrid channel model, customers can be delineated on the basis of size, order size, decision process, industry, or geography; alternatively, firms can bound the conflict based on product line (low end, mid-range, or high end). Using such heuristics to identify which channels can pursue which customers will keep the level of conflict in a manageable, functional range.
- Finally, compensation and communication issues are vital. Existing channels must be compensated for the potential loss of income that flows to the new channel, or the new channel may be sabotaged. Communication with all channel members (new and old) about the logic behind the strategy can go a long way towards keeping the focus on ultimate objectives

As with most marketing issues, it is vitally important to keep the customer in mind. What benefits does the channel deliver to customers? How can the Web be used to enhance customer value, either through the channel or with a direct-sales model? Going direct on the Web may work for some companies, but the inherent complexities can mean that a middle-ground approach may also be part of the answer. One company's struggle with these issues is detailed in the Technology Expert's View from the Trenches.

DISTRIBUTION CHANNEL STRATEGY AND MANAGEMENT

MICHAEL MCDONOUGH

*Consultant to telecommunications
companies; formerly Senior Vice
President—Marketing and Sales, GTE Wireless
(now a part of Verizon Communications)*

Distribution channel productivity is often a critical success factor for many businesses. The importance of distribution channel strategy development and effective management of the channels is growing, as many businesses move to a more complex channel design with multiple direct and indirect traditional channels as well as rapid expansion of electronic channels. For example, GTE Wireless had a direct marketing channel with direct mail and telemarketing, an Internet-direct channel, and a sales force for business accounts, as well as retail, wholesale, and mass merchandisers and traditional agents. How well a company manages the change to more complex channels can be the difference between winning and losing in the marketplace.

GTE Wireless used the following steps to improve the probability of success and to minimize the likelihood of having to deal with the questions: "Why aren't we getting the results?" "Was the strategy flawed or was it poor execution?"

First, the channel strategy must be integrated with the total company strategy of what products go to what markets. Even the best distribution strategy will fail without well-founded product and target decisions.

Second, analytical rigor is important. The assessment of market opportunity, coverage models, channel-specific benefit/cost analysis, etc. is needed to determine channel resource allocation. But, more importantly, logic and quantification are needed to communicate to the organization what you are doing, why, and what you are asking them to do. It is the numbers and facts behind the logic that make the changes compelling and understandable.

Third, the channel strategy should include not only the "front-end" of sales and sales support, but also the "back-end" of order processing, inventory fulfillment, and customer care. One cannot make channel decisions in isolation of these back-office functions.

Finally, the changes must be led by sales management, including first-line sales managers. This is the key leverage point. If the salespeople do not buy into the changes, they may sabotage or undermine the changes. One of the key ways to get buy-in is consistent reinforcement of the strategy delivered by sales leadership and a compensation structure supportive of that strategy.

GTE Wireless found this four-pronged approach to making channel decisions and channel changes enhanced the odds of success.

EXPANDING THE VIEW: FROM DISTRIBUTION CHANNELS TO SUPPLY CHAINS

Distribution channels trace the flow of product after it leaves the manufacturer, through the various intermediaries and institutions that add value along the way, to the end user. Supply chain considerations broaden the focus to include not only the distribution of the product from the manufacturer to downstream customers, but also the logistical management of all of the incoming components and pieces used in the manufacturing process of a particular product. Whereas distribution channels are concerned about inventory levels and customer service in the channel, **supply chain management** is concerned about matching the inflow of supplies and other materials used at every stage of the manufacturing process to the actual demand exhibited by customers in the marketplace.

Again, given the characteristics of high-tech products, firms are very concerned about efficient supply chain management. For example, when Intel saw the life of its microprocessors going from about eight years for the 386 series in 1985 to less than a year and a half for its Pentium II Xeon processor in 1998, it realized it needed smaller inventory.

> When the product is being devalued so fast, I want to ship it at today's price, not tomorrow's, and I want to know the right amount of new stuff to build on the right day. Our product, ounce for ounce, is more valuable than platinum, and the cost of inventory is a number beyond anything you can imagine.[51]

So, Intel adopted SAP's enterprise resource planning (ERP) software to handle what had previously been managed by four separate systems. The new system handled orders from the United States, Japan, Asia, and Europe, and tracked them all together. The system allowed Intel to understand where its inventory was on a given day and to track orders and shipping costs. The system was integrated to handle parts and materials coming in from suppliers as well as goods outbound on the way to customers. It reshuffled Intel's global logistics, focusing on three new warehouses next to critical airports. Routing international shipping through these three points allowed Intel to guarantee that contractors' plants operated at full capacity, or close to it.

The new supply chain reduced finished goods inventory from eight weeks' worth in 1995 to four weeks' worth in 1998. Now Intel can see its inventory anywhere in the world, commit it to a customer, and deliver it in three days. It reduced the chip-making cycle time and used electronic links to customers (personal computer manufacturers) to reduce their chip inventories. For example, the system is hooked electronically to five leading personal computer makers. The software automatically triggers a new shipment when customers' inventory falls to about four days' worth. The customer's inventory might get as low as one day's supply before a new shipment arrives. This halves the time customers hold valuable inventory. In essence, the system has traded information for inventory. And links to the retail chains that are the PC makers' biggest customers get the system one layer closer to the end customer in its market-sensing capacity.[52] This section's Technology Expert speaks to Intel's continuous improvement in managing its complex supply chain for the Intel Centrino chip, used to support Wi-Fi-enabled devices.

HIGH TECH SUPPLY CHAIN DESIGN
WADE SIKKINK

Outsourcing Program Manager
Intel Corporation, Hillsboro, Oregon

Managing supply chains for new, rapidly growing high-tech products is challenging. When done in an outsourced environment, where the company outsources the production/manufacturing of the high-tech product—the reality of high-tech manufacturing today—it's even more challenging. This is the situation we found ourselves in at Intel when ramping up production for the wireless module used in the Intel® Centrino™ Mobile Technology platform.

High-tech product life cycles are getting shorter and shorter, which means the production cycles ramp steeper and steeper, both during new product introduction and at the end of the product's life. Also, the days of steady, reliable forecasts from large OEM customers are gone. Demand is constantly changing, causing tremendous pressure on material supply chains to support the customer's needs without taking unwarranted risk or incurring unnecessary liability. The goal of the supply chain manager is to partner with other divisions within the company to design an entire supply chain from raw material to end customer that meets all the goals of service levels, availability, flexibility, and cost.

Initially, the Intel Centrino launch contained a great deal of uncertainty for the supply chain. As a completely new product concept, no one was quite sure how well the product would be accepted by the market. Our standard approach to supply chain management was untested in this highly uncertain environment and would need improvement. So, our team engaged across functional areas to determine the goals of the supply chain, enlisting the efforts of procurement, planning, operations, outsourcing, and marketing.

The Intel Centrino Mobile Technology platform consists of three pieces: a Mobile Intel Pentium 4 Processor-M, Intel 855 Chipset, and Intel Pro/Wireless LAN MiniPCI Adapter. Our specific area of responsibility was for the manufacture of the Intel Pro/Wireless LAN MiniPCI Adapter, so we set a top-level goal of ensuring that our piece of the three pieces never prevented a shipment of the platform to a customer. In other words, availability was our primary goal. We then developed secondary goals on standard supply chain concerns like flexibility and cost.

A goal of high availability can be achieved by building a very large amount of product, far above what a company expects to sell. However, in doing so, the likelihood that a company will have to scrap a large amount of inventory once the product reaches end of life is high. This is neither efficient nor cost effective. Our challenge was to achieve high availability while at the same time minimizing exposure to inventory obsolescence.

In designing the supply chain, our team decided to start at the end of the channel (customer) and work backward. We could accomplish the most immediate customer needs with finished goods inventory. However, finished goods represent the highest level of inventory cost exposure. To limit the overall exposure we conducted a sku-level analysis[a] of demand levels and probable lead times. We set finished goods inventory levels and channel locations to best support the highest demand skus with the most inventory closest to the customer. For the remaining skus, we used lower inventory levels and supported a higher percentage of product requests with newly built product out of the factory. Next, we looked at component inventories both in the factory and in the pipeline. By considering cost, lead time, number of suppliers, and overall industry conditions, we determined the inventory level for each component, where in the pipeline that inventory would be held, and when it should move from one stage to the next.

Ultimately, through intelligent supply chain design, we were able to produce all forecasted products, as well as handle special requests from the field, and double our output every quarter without missing a single customer shipment. The team felt our success was a result of staying focused on the customer and great collaboration across intra-organizational boundaries.

[a]A sku (shop keeping unit) level analysis essentially means analyzing demand by the individual products bought by customers. For example, in the cereal industry, this means analyzing sales for each of the various types of Cheerios, and the various size boxes of Cheerios.

Intel, Intel Centrino and Intel Pentium are trademarks or registered trademarks of Intel Corporation or its subsidiaries in the United States and other countries.

Effective Supply Chain Management[53]

Again, following the contingency model for effective high-tech marketing, an effective supply chain strategy matches the type of innovation (incremental versus breakthrough) to the type of supply chain needed. A supply chain can serve two functions:

- *Physical functions* convert raw materials, transport goods, and so forth.
- *Market mediation functions* ensure the variety of products reaching the marketplace matches what customers want.

Costs arise when supply exceeds demand or when supply is less than demand (the opportunity cost of lost sales or unhappy customers). As the matrix in Figure 8-7 shows, the appropriate match between the supply chain focus and the type of innovation is imperative.

For *incremental,* more functional products—or those with stable, predictable attributes, long life cycles, and fairly low margins—market mediation is fairly easy. Because the products have a long history and don't change significantly, matching products to customer needs can be straightforward. Instead of focusing efforts on flexibility and reading market signals, companies can focus on managing physical costs. They can manage the flow of information within the chain between players to coordinate activities. Close coordination between suppliers and distributors offers the greatest opportunities to slash lead times and inventory.

For *innovative* products (unpredictable demand, short life cycle), however, firms must read early-market signals to react quickly and get crucial information from the market to the players in the chain. A firm must know where to position inventory and production capacity in order to hedge against uncertain demand. For these innovative products, critical issues are speed and flexibility more than low cost. The primary challenge is

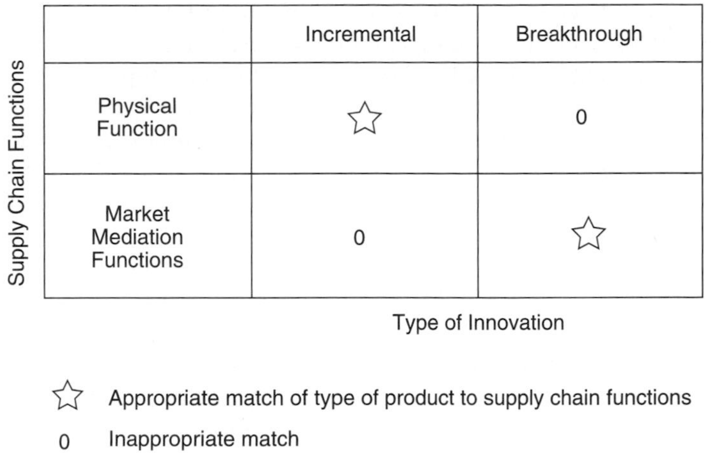

	Incremental	Breakthrough
Physical Function	☆	0
Market Mediation Functions	0	☆

Supply Chain Functions *(vertical axis label)*

Type of Innovation

☆   Appropriate match of type of product to supply chain functions

0   Inappropriate match

FIGURE 8-7   Matching Type of Innovation to Supply Chain Functions

SOURCE: Reprinted by permission of Harvard Business Review. "Matching Type of Innovation to Supply Chain Functions" from "What is the Right Supply Chain for your Product" by Marshall Fisher, March–April 1997. Copyright © 1997 by the Harvard Business School Publishing Corporation; all rights reserved.

responding to uncertainty. Uncertainty can be reduced with data and information systems, with a build-to-order model, or by manufacturing closer in time to when demand materializes. *For innovative products, rewards from investments in supply chain responsiveness are greater than rewards from efficiency investments.* Improvements in responsiveness result in savings from fewer stock-outs and mark-downs.

For example, consider a typical innovative product with a contribution margin of 40 percent and an average stock-out rate of 25 percent. The lost contribution to profit and overhead resulting from stock-outs alone is huge: 40% * 25% = 10% of sales—an amount that usually exceeds profits before taxes. Consequently, the economic gain from reducing stock-outs and excess inventory is so great that intelligent investments in supply chain responsiveness will nearly always pay for themselves. This is why, for example, Compaq chose to produce high-variety, short life cycle circuits in-house rather than outsource them to a low-cost Asian producer; local production gave the company increased flexibility and lead times. Note that this logic doesn't apply to functional products. With a contribution margin of 10 percent and an average stock-out rate of 1 percent, the lost contribution to profit and overhead amounts to only 0.1 percent.

The changes in the distribution channel for computer products, including channel assembly and colocation, provide other good examples of changing from a physically efficient supply chain to one more usefully focused on market mediation. Although inventory turns are a familiar measure, the market increasingly demands responsiveness.

Dell is a sophisticated practitioner of supply chain management. Employees at its flagship desktop computer factory in Austin, Texas, sort through orders placed online or by phone. They decide which computers should be built in the next two hours, pass those orders on to the factory and order materials from suppliers, which are delivered within 90 minutes. Factory workers unpack supplies as they arrive and set up building kits for every computer. Conveyer belts move each kit to a build cell, where a worker builds each computer, often within three minutes. The computer is then tested and loaded with software, ready for shipment. In this way, the factory builds 650 computers per hour on each of its production lines. While a single desktop computer can be built, tested, and shipped within two hours after the order is placed, any order can be delivered usually within five days of being placed.[54] This phenomenal productivity comes with the lowest costs in the industry. Dell's overhead costs consume 11.5 cents of every sales dollar compared to 16 cents for Gateway and 22.5 cents for HP.[55]

Dell's supply chain management system gives it tremendous pricing flexibility in the marketplace. It is not uncommon to find different prices for the same Dell laptop on a given day by logging into the different sections of its Web site, e.g. for small business, healthcare, or government. Prices also vary day by day. Dell's more than 5,000 salespeople continuously collect intelligence from customers on their purchase plans and competitive offers. Suppliers also update Dell continuously on their costs. The actual price for a given segment takes into account customers' price sensitivity, supplier costs, and competitive considerations.

By selling computers directly over the Internet and building machines to order, Dell is in closer touch with its customers and better able to gauge price sensitivity. Dell's twenty-five main suppliers provide regular updates on costs and prices that enable Dell to forecast the cost of each component several months into the future. Suppliers are expected to participate in price reductions initiated by Dell. Through Internet portals, each supplier is also able to see Dell's orders and inventory position. All of this information enables Dell and its suppliers to avoid overproduction and potential material shortages.[56]

Trends in Supply Chain Management

The Internet has had a major impact on improving supply chain management. It enables three kinds of efficiencies to be realized in the supply chain.[57] First, information sharing between suppliers and customers helps in better matching supply and demand conditions. Second, the Internet enables suppliers to increase their reach and find new customers at lower cost. Finally, dynamic pricing, as in auctions, helps suppliers clear unsold inventory and helps customers procure goods and services at lower cost. Three trends that high-tech firms have relied on to gain such efficiencies are participation in vertical markets, the use of supply chain management software, and outsourcing.

Vertical Markets on the Internet

Vertical electronic markets, *exchanges,* or *e-marketplaces* are industry-specific websites that provide electronic business solutions ranging from e-commerce (sales of products) to document sharing, bill presentment, order tracking, contract negotiations, and credit services. These hubs use e-business solutions at many points along the supply chain to build links between suppliers, customers, manufacturers, and even competitors. For example, Covisint is a vertical market or exchange bringing together suppliers and customers in the auto manufacturing industry with a view to increasing the efficiency of the supply chain. The business solutions offered can add or create value by streamlining and eliminating redundancy of current or traditional processes being used within an industry.

However, the performance track record of B2B vertical markets or exchanges has been equivocal at best. The first versions of these were *public exchanges,* open for participation to any company (buyer or seller) that wanted to join and was willing to pay a membership fee. The number of these public B2B exchanges went from a peak of about 1,500 in the year 2000 to about 700 in 2002 and were expected to dwindle to about 180 in 2003.[58] A major reason was the exchanges' lack of offline relationships with buyers and sellers necessary to generate critical mass. The second iteration of online B2B exchanges were *consortia* sponsored by some of the leading players in the industry. For example, Covisint was co-founded by Ford, General Motors, and Daimler Chrysler. However, such industry consortia are also facing a critical mass problem due to lack of trust by other industry players who are not affiliated with the exchange. These exchanges tend to favor buyers who are able to compare prices across different suppliers easily. The third stage in the evolution of these exchanges is the emergence of *private exchanges,* and they seem to hold the most promise for the future.[59] Companies like Dell invite their trusted partners to join their private online network so that all parties benefit from supply chain efficiencies.

Because public, open exchanges easily allow for price-based competition (using reverse auctions and dynamic pricing, for example), they are well-suited for short-term transactions of commodity products between businesses. (Such tools can seriously undermine the tenor of more collaborative, trusting relationships where parties may have worked together over a longer-time horizon to jointly develop innovations or open new markets.) Conversely, because private exchanges allow trusted business partners to share sensitive information and open new lines of business (say, by reaching new customer segments), they are more suited to collaborative relationships; in more collaborative relationships (unlike in more short-term, transaction-oriented situations), firms can be more confident that their information and knowledge will not be unfairly used by their business partners. Because awareness of the use of online exchanges for supply chain relationships is vital for high-tech marketers, they are explored further in the next chapter.

Supply Chain Management Software

With the advent of ever-increasingly sophisticated information technology, companies are using new software programs to redesign their business processes. Known as **enterprise resource planning (ERP)**, enterprise software refers to the subindustry of software for companies looking to use technology to improve virtually every key corporate function, including manufacturing, finance, sales, marketing, human resources, supply chain management, and the data juggling behind the e-commerce scenes. The market for ERP software is anticipated to grow from $20.6 billion in 2002 to $21.6 billion in 2004.[61] ERP programs were originally focused on automating business processes, rather than on the strategic use of information. Now, they function as an electronic nervous system for the business operation, tracking orders, receivables, products, warehousing, inventory, accounting, suppliers, customers—anything of interest to the increased efficiency and effectiveness of the business process.

A subset of ERP software, supply chain management programs, bring data from manufacturing, inventory, and suppliers to create a unified picture of the elements that go into building a product. For example, one vendor's products, i2 Technologies, can help a firm decide if it makes more sense to build just one factory or three separate ones that are closer to its customers. Tracking the manufacturing of products and the various business processes as the product moves from order entry to order fulfillment can help in managing facilities, vendors, subcontractors, and so forth. Other back-office programs help companies analyze their operations to unearth ways to cut costs in areas such as order processing, customer service and returns, and order fulfillment. The market for supply chain management software is expected to grow from $5.6 billion in 2003 to $6 billion in 2004.[62]

Supply chain management used to be handled with proprietary EDI links between suppliers and OEM customers. Now, cheaper, Internet-based software is more popular. With the increased velocity and unpredictability in the high-tech environment, accurate predictions, available through the use of this software, have increased in importance.[63]

Although supply chain management software that forecasts and mediates supply and demand provides important efficiencies, it could wreak havoc on traditional supply chain relationships. For example, traditional parts brokers may become obsolete in an era of online virtual communities and vertical networks. Former single-source suppliers may be pitted against each other in auction bids. These issues have yet to be reconciled in the still-emerging world of online supply chain management.

Outsourcing

Another trend is the outsourcing of the entire manufacturing process to suppliers at a higher level in the supply chain. In recent years, many high-tech companies have outsourced production to contract manufacturers in low-cost countries. For example, Singapore-based Flextronics International manufactures products such as Nortel routers, Microsoft Xbox consoles, and Ericsson cell phones in Mexico, China, and other countries.[64] Indeed, at the extreme, Hewlett Packard has completely outsourced its production and no longer makes any of its own computer products.

The trend toward outsourcing is particularly pronounced in technology-related industries because of shorter product life cycles and increased global competition. Intel was able to effectively manage its supply chain for its Centrino family of products in an outsourced environment.

Although a key trend in supply chain management is the outsourcing of production, other functions are also affected by the outsourcing phenomenon. U.S. companies are relocating a greater number and range of functions to foreign countries, including call centers, stock analysis, accounting, tax return and insurance claims processing—and even software programming and other tech jobs. Clearly, this trend is fueled by the need to lower costs in a competitive, slow global economy. But there are downsides as well. When there was a labor slowdown at the port of Oakland, California, in 2002, the supply chain of many U.S. companies on the West Coast was disrupted because products shipped from China and Taiwan did not reach their intended destinations in time. Similarly, the SARS epidemic in many Asian countries in 2003 prevented business executives from traveling to meet suppliers and customers, causing delays in the supply chain. U.S. companies also have to deal with the political backlash from labor unions and state legislatures protesting the migration of U.S. jobs to other countries.[65] Clearly the benefits of lower costs through overseas outsourcing have to be weighed against the loss of control over the supply chain.

SUMMARY

This chapter has provided an overview and understanding of some of the issues pertinent to designing and managing distribution and supply chain relationships in the high-tech environment. Given the characteristics of the high-tech environment, firms are seeing a blurring of distinctions between members in the supply chain, and indirect channels are changing to provide more value. This chapter also addressed the way in which high-tech channels evolve as an innovation becomes more mainstream, the causes and solutions to gray market problems, piracy concerns, and export restrictions.

The Internet is having a significant impact on distribution channels; therefore, the chapter provided a decision framework to integrate the Internet into existing distribution channels. In addition to distribution channels, the Internet is having a significant impact on optimizing inventory in the entire supply chain, using sophisticated software programs such as Enterprise Resource Planning (ERP). Other trends in supply chain management helping to make supply chains more efficient are the use of vertical electronic markets or exchanges for procurement of goods and services, and the outsourcing of many business functions including manufacturing, software development and customer service.

DISCUSSION QUESTIONS

1. What are the six basic issues in designing and managing a distribution channel?
2. What are the various types of intermediaries commonly used in high-tech channels?
3. What is the nature of the trade-off between the *degree of coverage* and the *degree of intra-brand competition?*
4. What tools can be used for control and coordination of channel members?
5. What is tying? What are its legal implications?
6. What is an exclusive dealing arrangement? What are its legal implications?
7. How can channel performance be measured?
8. How and why is the line between suppliers and channel members blurring?
9. What are channel assembly and colocation? Why are they being used in high-tech channels?
10. What is the typical evolution pattern for channels in high-tech markets?
11. What is the gray market? Why is it a problem? What are its causes and solutions?
12. What is the purpose of export restrictions?
13. What factors are used to assess whether an Internet channel will cannibalize existing sales or lead to incremental sales?
14. What are some of the intermediate steps in managing the transition to an Internet channel?
15. What is the source of conflict in a hybrid channel? How can it best be managed?
16. What are the objectives of a hybrid channel? What are the key steps in effectively implementing and managing hybrid channels?
17. What is supply chain management? How should supply chain functions be matched to the type of innovation?
18. What are vertical electronic markets? How are these used in supply chain management?
19. What is ERP? How do supply chain management programs work?
20. What are the pros and of outsourcing in the supply chain?

GLOSSARY

Authoritative controls.   Reside in one channel member's ability to develop rules, give instructions, and in effect, impose decisions on the other; include ownership (via vertical integration), formal, centralized decision making (say, as in franchising), formal controls and monitoring, or one party's power over another.

Bilateral controls.   Originate in the activities, interests, or joint input of *both* channel members and include relational norms (shared expectations concerning channel members' attitudes and behaviors in working together to achieve mutual goals) such as flexibility, mutual sharing of benefits and burdens, information sharing, and so forth; joint interdependence, commitment, and mutual trust.

Bricks-and-clicks.   A business that offers both a physical market presence via retail stores ("bricks") and an online channel ("clicks") to its customers.

Channel assembly.   Manufacturers ship semifinished products to distributors that configure them to client specifications and complete production before shipping the products. The advantages are customization and speedy turnaround for customers.

Colocation. When the distributor's employees work from a vendor's manufacturing sites to ship completed, high-demand models to either resellers or end users.

Disintermediation. The adding of an online distribution channel that bypasses existing intermediaries in favor of a direct-sales model.

Enterprise resource planning (ERP). The redesign of business processes and functions, including manufacturing, finance, sales, marketing, human resources, supply chain management, and the data analysis used to support e-commerce, to improve effectiveness and efficiency; typically based on ERP software and the Internet.

Exclusive dealing arrangements. Restricting a dealer to carrying only one manufacturer's product; antitrust implications arise if large companies use their dominance to restrict customers' access to competitors' products.

Export restrictions. Some governments prohibit companies from selling certain products to specified countries in the national interest.

Gray market. Unauthorized distribution of goods at discounted prices through resellers not franchised or sanctioned by the producer; results in a lack of control over the product, distribution, and services.

Hybrid channel or **dual channel.** When a firm uses a combination of direct and indirect channels.

Inbound dealers. Typically have a retail storefront, and their primary customers are walk-in traffic generated through traditional advertising and promotion means.

Interbrand competition. Competition between different brands or makes in the marketplace.

Intrabrand competition. Competition between dealers on the same manufacturer's brand.

Outbound dealers. Have a sales force that makes calls on customers, typically at the customer location (the dealer may or may not have a storefront).

Supply chain management. The management of processes that make available materials, components and services required for the operations of an organization in order to be able to deliver products and satisfaction to customers. It includes management of inventory, warehousing, distribution and transportation with suppliers, inside the organization, and out to customers.

Systems integrator. A type of dealer who manages very large or complex computer projects, typically creating customized solutions for customers by bundling and reselling different brands of equipment.

Tying. When a manufacturer makes the sale of a product in high demand conditional to the purchase of a second product.

Value-added reseller (VAR). Smaller channel intermediaries that purchase products from one or several high-tech companies and add value to them, via integrating with proprietary expertise, to meet their target market's (often a vertical market) needs.

Vertical electronic markets. Industry-specific websites that provide electronic business solutions ranging from e-commerce (sales of products) to document sharing, bill presentment, order tracking, contract negotiations, and credit services. These hubs use e-business solutions at many points along the supply chain to build links between suppliers, customers, manufacturers, and even competitors.

ENDNOTES

1. "Managing Customers As Assets" (1995), *Fortune,* May 29, Special Advertising Section, p. S4.

2. "How Technology Sells" (1997), Dataquest, Gartner Group, and CMP Channel Group, CMP Publications, Jericho, NY.

3. Stern, Louis and Frederick Sturdivant (1987), "Customer-Driven Distribution Systems," *Harvard Business Review,* July–August, pp. 34–41. This section introduces the basic elements of channel design and management; the intent is not to provide all the decision

tools and tradeoffs faced in designing a distribution channel, but rather to provide a foundation for the subsequent development of particular issues faced in distribution channels for high-tech products. More detail is available in Coughlan, Anne, Erin Anderson, and Louis Stern (2001), *Marketing Channels,* 2nd ed., Upper Saddle River, NJ: Prentice Hall.

4. Briones, Maricris (1999), "Resellers Hike Profits Through Service," *Marketing News,* 33 (February 15), pp. 1, 14.

5. Mohr, Jakki, Christine Page, and Greg Gundlach (1999), "The Governance of Inter-Organizational Exchange Relationships: Review and State-of-the-Art Assessment," Working paper, University of Montana, Missoula, MT.

6. France, Mike (1998), "Are Corporate Predators on the Loose?" *Business Week,* February 23, pp. 124–126; Stremersch, Stefan and Gerrard J. Tellis (2002), "Strategic Bundling of Products and Prices: A New Synthesis for Marketing," *Journal of Marketing,* 66(1) (January), 55–72

7. "How Technology Sells" (1997), op. cit.

8. Kumar, Nirmalya, Louis Stern, and Ravi Achrol (1992), "Assessing Reseller Performance from the Perspective of the Supplier," *Journal of Marketing Research,* 29 (May), pp. 238–253.

9. Tam, Pui-Wang (2002), "Hewlett Packard Discontinues Printer-Sales Deal With Dell," *Wall Street Journal Online,* July 24.

10. "Line Blurs Between Supplier and Supplied" (1997), *Computer Retail Week,* October 6.

11. Briones, Maricris (1999), "What Technology Wrought: Distribution Channel in Flux," *Marketing News,* 33 (February 1), pp. 1, 15.

12. Ibid.

13. Ibid.

14. Briones, Maricris (1999), op. cit.

15. "How Technology Sells" (1997), op. cit.

16. Moore, Geoffrey (1991), *Crossing the Chasm, Marketing and Selling Technology Products to Mainstream Customers,* New York: HarperCollins, chapter 7; revised edition (2002).

17. Ibid., p. 173.

18. Moore, Geoffrey (1991), op. cit.

19. Duhan, Dale and Mary Jane Sheffet (1988), "Gray Markets and the Legal Status of Parallel Importation," *Journal of Marketing,* 52 (July), pp. 75–83; Corey, E. Raymond, Frank V. Cespedes, and V. Kasturi Rangan (1989), *Going to Market: Distribution Systems for Industrial Products,* Boston: Harvard Business School Press, chapter 9 (The Gray Market Dilemma).

20. Myers, Matthew and David Griffith (1999), "Strategies for Combating Gray Market Activity," *Business Horizons,* November–December, pp. 2–8.

21. Corey, E. Raymond, Frank V. Cespedes, and V. Kasturi Rangan (1989), op. cit.

22. Duhan, Dale and Mary Jane Sheffet (1988), op. cit.

23. Corey, E. Raymond, Frank V. Cespedes, and V. Kasturi Rangan (1989), op. cit.

24. Ibid.

25. Ibid.

26. Ibid.

27. Ibid.

28. Ibid.

29. Ibid.

30. *Eighth Annual BSA Global Software Piracy Study* (2003), International Planning and Research Corporation, June.

31. *Expanding Global Economies: The Benefits of Reducing Software Piracy* (2003), IDC, April 23.

32. "Commerce Department Imposes $2.12 million Civil Penalty on McDonnell Douglas for Alleged Export Control Violations" (2001), press release, US Department of Commerce, http://usinfo.state.gov/regional/ea/uschina/charms.htm, November 14.

33. "U.S. Customs Probes Into Satellite Manufacturer Results in Record $20 Million," (2002), press release, U.S. Customs Service, http://usinfo.state.gov/regional/ea/uschina/charms.htm, January 10.

34. Cohen, Adam (1999), "When Companies Leak," *Time,* June 7, p. 44.

35. Gay, Lance (2000), "U.S. Satellite Controls Have Backfired," *Missoulian,* June 1, p. A7.

36. Anders, George (1998), "Some Big Companies Long to Embrace Web, but Settle for Flirtation," *Wall Street Journal,* November 4, pp. A1, A13.

37. McWilliams, Gary (1999), "Dealer Loses?" *Wall Street Journal,* July 12, p. R20; Useem, Jerry (1999), "Internet Defense Strategy:

Cannibalize Yourself," *Fortune,* September 6, pp. 121–134.

38. Kumar, Nirmalya (1999), "Internet Distribution Strategies: Dilemmas for the Incumbent," *Mastering Information Management,* March 15, pp. 6–7.

39. Jap, Sandy and Jakki J. Mohr (2002), "Leveraging Internet Technologies in B2B Relationships," *California Management Review,* 44 (4), 24–38.

40. Kirsner, Scott (1998), "Channel Concord: The Web Isn't Just for Alienating Partners Anymore," *CIO Web Business,* November 1, pp. 32–34.

41. Ibid.

42. Useem, Jerry (1999), "Internet Defense Strategy: Cannibalize Yourself," *Fortune*, September 6, pp. 121–134.

43. Ibid.

44. Ibid.

45. Moriarty, Rowland and Ursula Moran (1990), "Managing Hybrid Marketing Systems," *Harvard Business Review,* November–December, pp. 146–155.

46. Ibid.

47. Ibid.

48. Ibid.

49. Ibid.

50. Ibid.

51. Brown, Stuart (1998), "Wresting New Wealth from the Supply Chain," *Fortune,* November 9, p. 204X.

52. Ibid.

53. Except where noted, this section is drawn from Fisher, Marshall (1997), "What Is the Right Supply Chain for Your Product?"
Harvard Business Review, March–April, pp. 105–116.

54. Pimantal, Benjamin (2003), "Dell's Big Sell," *San Francisco Chronicle* Online, April 13.

55. McWilliams, Gary (2001), "Dell Fine-Tunes Its PC Pricing To Gain an Edge in Slow Market," *Wall Street Journal Online,* June 8.

56. Ibid.

57. Jap and Mohr (2002), op. cit.

58. Day, George S., Adam J. Fein and Gregg Ruppesberger (2003), "Shakeouts in Digital Markets: Lessons from B2B Exchanges, *California Management Review,* 45 (2), 131–150.

59. Harris, Nicole (2001), "Private Exchanges May Now Allow B2B Commerce to Thrive After All," *Wall Street Journal Online,* March 16.

60. Jap and Mohr (2002), op. cit.

61. "AMR Research Releases Enterprise Applications Market Projections," (2003), press release, AMR Research, http://www.amrresearch.com/Content/viewpress.asp?id=16218&docid=10622, June 5.

62. Ibid.

63. Gross, Neil (1998), "Leapfrogging a Few Links," *Business Week,* June 22, pp. 140–142.

64. Engardio, Pete (2003), "Weathering the Tech Storm," interview with Flextronics CEO Michael E. Marks, *BusinessWeek Online,* May 5.

65. Schroeder, Michael (2003), "Unions, States Seek to Block Outsourcing of Jobs Overseas," *Wall Street Journal Online,* June 3.

CHAPTER 9

Pricing Considerations in High-Tech Markets

*In 1955, the U.S. Atomic Energy Commission projected the nation
would have 1,000 nuclear power plants by the year 2000.
Instead, the number was 63
(Rachel Emma Silverman, Wall Street Journal, 1/1/2000).*

Ups and Downs of Pricing Services

In December 1996, AOL switched to monthly flat rate pricing from per-hour usage pricing for its dial-up Internet access service. The demand for dial-up access at $21.95 per month was so high that AOL had to invest in a major expansion of its capacity and infrastructure at that time. Monthly subscription prices would cover the acquisition costs of each new subscriber, and additional advertising revenue would contribute to profits. The promotion techniques included direct mail and retail distribution of AOL software on a CD, backed by 24-hour telephone technical support, free usage for a 30-day trial period, after which the monthly subscription would kick in. These promotions were highly successful at attracting "newbies," people new to PCs and the Internet.

In May 2001, after merging with Time Warner, AOL increased the monthly subscription for existing and new users to $23.90 per month. The price increase did not seem to have any negative impact on AOL's subscriber base, despite Microsoft's decision to keep its price at $21.95 per month and hard-hitting Microsoft ads urging AOL users to switch to MSN. By March 2002, AOL had 34 million paying members.

Around this time, subscriber growth slowed because two-thirds of the market already had Internet access. The remaining one-third, laggards, would be difficult and expensive to convert into users. AOL started bundling its dial-up software and service with PC manufacturers like Compaq, Dell, and Gateway. AOL paid these manufacturers for the privilege of bundling and the manufacturers paid AOL a discounted rate for Internet access. Frequently, these deals were not profitable for AOL, but it tolerated them in the hope that consumers would pay full price after the free trial period from the PC manufacturer ended.

A concurrent technology trend that took AOL by surprise was the introduction of broadband Internet access via DSL and cable modem. Subscribers started switching to broadband Internet service providers, and AOL had no offering to compete with them. Further, in dial-up service, competitors started offering free or discounted service. AOL, with its $23.90 per month dial-up service, was getting squeezed from both sides. At the high end, broadband service providers like Verizon began offering DSL service at monthly rates of $29.95. At the low end, companies like United Online offered no-frills dial-up service at $9.95 per month.

In 2003, AOL began taking aggressive steps to defend its position. It started offering unlimited broadband Internet service through bundled deals with cable and DSL service providers at $54.95 per month, including multiple simultaneous logins and a broadband modem. For consumers who wanted to have some other broadband service provider, AOL offered unlimited access to its exclusive content (much of it from Time Warner) for $14.95 per month. AOL announced that its Netscape brand of discount dial-up service would be available in early 2004 for $9.95 per month. AOL hoped that these pricing moves would stabilize its subscriber base in the short term and restore the business to growth and profitability in the longer term.

SOURCES: Angwin, Julia (2002), "AOL Finds More Subscribers Don't Pay to Use Online Service," *Wall Street Journal,* March 19.
Bilstein, Frank F. and Frank Luby (2002), "Manager's Journal: Casing AOL's Flat-Price Model," *Wall Street Journal,* December 10, page B2
Angwin, Julia (2003), "AOL to Launch Dial-Up Service Called Netscape at a Discount," *Wall Street Journal,* October 14.

THE HIGH-TECH PRICING ENVIRONMENT

What forces impinge on high-tech pricing decisions? As shown in Figure 9-1, the forces are varied and strong. Many high-tech firms might find it desirable to price at a high level, in order to recoup investments in R&D and to signal high product quality. However, many factors conspire to push prices down.

High-tech firms face an environment characterized by ever-shortening product life cycles, with the inevitable rapid pace of change and potential obsolescence of products. Moore's Law[1] operates unforgivingly: Every eighteen months or so, improvements in technology double product performance at no increase in price. Stated a different way,

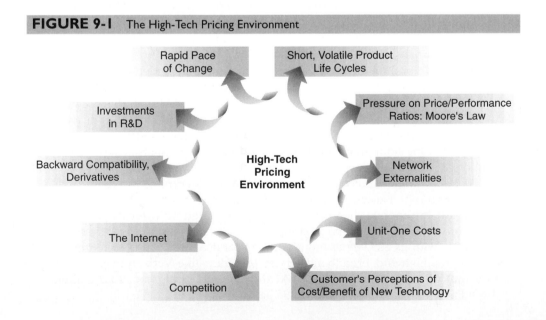

FIGURE 9-1 The High-Tech Pricing Environment

every eighteen months or so, improvements in technology cut price in half for the same level of performance. So introductions of product versions with better price/performance ratios are a given, creating downward pressure on prices.

Moreover, as identified in Chapter 1, network externalities and unit-one costs operate in the market. Recall that network externalities exist when the value of the product increases as more users adopt it; examples include the telephone, portals on the Internet, and so forth. Unit-one costs refer to the situation in which the cost of producing the first unit is very high relative to the costs of reproduction for subsequent units. For example, the costs of pressing and distributing a CD-ROM are trivial compared to the cost of hiring programmers and specialists to develop the content recorded on it. Both of these factors create pressure to acquire a critical mass of users through lower price structures.[2]

Furthermore, customer perceptions of the cost/benefit of the new technology affect pricing strategy. Customer anxiety may cause delays in adoption. For example, as firms introduce one new-and-improved version after another, consumers may postpone purchases in the hope (fear?) that prices eventually will come down and performance will improve substantially.[3] For example, the initially slow adoption of broadband Internet service in the United States was explained, in part, by its high price. In such a situation, marketers may need to lower the prices of newer technologies aggressively to reduce possible switching costs, to offer special deals for upgrades, or to entice customers switching from a competing application.[4]

The effect of customer anxiety on purchase is further complicated by the upgrading considerations covered in Chapter 6: The customer's perceptions of the performance gap between the old and new generations relative to the cost to upgrade have a strong influence on purchase behavior. Because of this anxiety, marketers may have to lower prices for future generations to encourage upgrades.

Other complicating factors include the fact that high-tech firms must ward off competitors. Moreover, the Internet has allowed both consumers and organizational customers the ability to compare prices and negotiate for lower prices to a much greater degree than in the past. Issues of backward compatibility (with older version of the product), support for existing products, changing operating standards, pricing for product derivatives, and so forth, all must be considered in pricing strategy.

Pricing even in conventional marketing contexts is a very complex decision; this overview of the high-tech pricing environment shows that it is doubly complex for high-tech products and services. As in prior chapters, this chapter does not go over pricing basics (such as calculating payback periods, return on investment, break-even points, experience curves, leasing, competitive bidding, price elasticity, penetration versus skimming strategies, etc.); interested readers should consult other resources to learn the basics. Rather, this chapter addresses how to make pricing decisions that incorporate and address many of the complications in the high-technology context, mentioned previously. We start by examining the three major factors that all marketers must systematically consider when setting prices: costs, competition, and customers.

THE THREE Cs OF PRICING

The three Cs of pricing—costs, competition, and customers—are analogous to a three-legged stool, shown in Figure 9-2. Stools with only two legs are unbalanced and likely to topple over. Similarly, setting price on the basis of considering only one or two of the

Competition

Costs   Customers

FIGURE 9-2 The 3 Cs of Pricing

three Cs results in an unstable situation. Solid pricing strategy must be based on a systematic consideration of all three factors.

Costs

Costs provide a floor, generally below which marketers ought not to price. Companies that position on a low-price basis should not do so unless they have a strong, nonimitable cost advantage in the industry that is unlikely to disappear with future generations of technology. For example, a cost advantage based on economies of scale arising from large volume sales based on existing technology may not translate to a cost advantage when a new generation of technology comes down the pike.

A firm that bases prices primarily on its own cost structure (i.e., on a markup, cost-plus, or target-return basis) often fails to recognize the impact that market factors have on profitability. Overlooking the impact of the market on pricing and profitability can be a fatal mistake in high-tech markets, in which market considerations are so important.

Competition

Competition provides a benchmark against which to evaluate prices. A firm might let competitors set prices and then establish its price below, equal to, or above those of competitors, depending upon its position in the market. While Dell tries to position itself as the low price leader, Apple tries to differentiate itself with innovative products and premium pricing.

In the high-tech arena, a firm that introduces a radical innovation to the marketplace often (wrongly) believes that, because its innovation is so new, there is "no" competition. However, this belief is not necessarily the case from the customer's perspective. A customer can always choose not to adopt new technology, but to solve problems based on former solutions (which provide the competitive benchmark for radical innovations). Indeed, one executive from Motorola said: "Our biggest competitor isn't IBM or Sony. It's the way in which people currently do things."[5]

Customers

Customers' perceptions of value provide a ceiling above which marketers should not price. Simply, customers balance the benefits of a purchase against its costs. High-tech marketers often find it difficult to understand fully the customer's perceptions of benefits and costs. The innovating firm may find the new technology so compelling, so sophisticated, or so "innovative," that it assumes the benefits are obvious to users. Similarly, the innovating firm may not fully appreciate the customer's perceptions of costs.

Product benefits might include the following:[6]

- *Functional benefits.* The utilitarian aspects that might be attractive to engineers or technology enthusiasts.
- *Operational benefits.* The product's reliability and durability, and the product's ability to increase efficiency for the customer.
- *Financial benefits.* Credit terms, leasing options, and so on. For business customers, does the product help boost revenues or profits?
- *Personal benefits.* The psychosocial satisfaction from being an early adopter, purchasing a well-known brand to avoid risk, and being professionally rewarded for making good organizational buying decisions.

The costs a customer perceives are similarly diverse and might include the following:[7]

- *Monetary costs.* The price paid, transportation and installation, and so forth.
- *Nonmonetary costs.* The risk of product failure, risk of obsolescence, obsoleting of a prior piece of equipment or related products (e.g., VHS tapes that won't play on a new DVD player), risk of late delivery, switching costs and the like. For business-to-business goods, nonmonetary costs might include factory downtime for repair and maintenance of machinery.

The **total cost of ownership** (or **life cycle costing**) is one way to look at customers' costs; it reflects the total amount a customer expends in order to own and use a product or service. Total cost of ownership includes the price paid for the good (including financing fees), as well as delivery or installation costs, service costs to maintain and repair the good, power costs to run the equipment, supplies, and other operating costs *over the life of the equipment.* In 2002, the total cost of ownership of a corporate personal computer was estimated to be $6,400 per year including hardware, software, installation, training, maintenance, infrastructure, and support.[8] Using the total cost of ownership in pricing strategy can help a firm position its products relative to those of competitors. Showing that the total cost of ownership of a product is lower than the competitor's can be a compelling benefit to a customer—despite an initially higher outlay for the product.

Firms such as Microsoft use this approach when selling to corporate customers. Microsoft's Systems Management Server 2003 (SMS) leverages the Internet to deliver business services to computers and its Operations Manager product is used to improve operations, enhance performance, and maximize system and application availability. The company claims that network managers who use SMS and Operations Manager experience reductions in operational costs of up to 33 percent over the cost of previous solutions, and improved manageability for Windows environments. The message to corporate purchasers is that even though some of Microsoft's network solutions may have a higher initial outlay (in terms of the purchase price, or site license fee on a per-user basis), users can still see significant cost savings in the total costs of operating and maintaining the network.

The Linux operating system offers similar total-cost-of-ownership savings for customers. For example, Jeffrey Birnbaum, the managing director for computing at Morgan Stanley's Institutional Securities Division decided in 2003 to replace 4,000 high-powered servers running traditional software from Microsoft or Sun, with much cheaper machines running Linux (a free, open-source operating system). His projected five-year savings from the switch: $100 million![9] This value proposition is part of the reason that the market share of Linux (in the market of server operating systems) has grown from 0 percent in 2000, to 13.7 percent in 2003, and is expected to jump to 25.2 percent in 2006.

In summary, solid consideration of costs, competitors, and customers is vital in establishing a successful pricing strategy. Focusing on costs alone can be myopic and can cause problems. Similarly, focusing on competition can be hard in high-tech markets, when the competition for a radical innovation might be the customer's current behavior pattern. Both of the drawbacks in focusing solely on costs or competition point to the value in taking a customer perspective in pricing. Taking a customer perspective in pricing forces the marketer to realize that the firm's costs to manufacture a product and its investments in R&D are relatively unimportant to the customer's perceived value. Moreover, the customer tends not to care about the firm's costs so much as his or her own costs in buying and using the product. Iridium, the satellite-based international wireless telephone service launched by Motorola and its partners in 1998 failed, in part, due to the telephone handset's price of $3,000 each and a per minute charge in excess of $3. At these prices, only 15,000 users signed on instead of the expected 500,000.[10] The service later filed for bankruptcy.

Because of the importance of a customer orientation in pricing, and because of the benefits to high-tech marketing of being customer-focused, this leg of the stool deserves additional consideration.

TECHNOLOGY TIDBIT

WIRED HOME OF THE FUTURE

SOURCE: Courtesy LG Electronics

You've finished a tiring day at work and have a long commute ahead of you to get back home. If all your home appliances were connected to the Internet, you could program your sauna to be ready at the right temperature when you get home, have your favorite dinner entrée ready to eat after that, and watch your favorite TV show that you missed after dinner (thanks to TiVo)—all with a few commands issued from the computer in your car. Sounds far-fetched? Not really. Most of the technology pieces are already in place.

The centerpiece of LG Electronics' Home Networking vision is the 26 cubic foot, Titanium Internet Refrigerator. It has two ports at the back, one for cable TV and another for broadband Ethernet. There is a built-in Windows PC with a 20 gigabyte hard drive on the top and a 15-inch flat-panel color screen in the door, on which you could watch cable TV or surf the Internet. The refrigerator also comes equipped with a digital music player and four speakers, a digital still and movie camera, email, address book, and calendar. There is a built-in cookbook with 100 recipes organized by course, ingredient, and country. A "stored food" program brings up a diagram of the inside of the refrigerator and lets you manually record the different foods stored in various compartments. The program knows the normal shelf life of the different foods and sends alerts to the screen for the owner to check and dispose of foods beyond their normal shelf life.

The LG Internet Refrigerator was launched with a retail price of $8,000 in 2002. At this price, it may be of value only to affluent technology innovators, who want to show off their "kitchen of the future" to their friends. Eventually, at a price point of $1,500, it may actually become the centerpiece of the wired home of the future.

SOURCE: Mossberg, Walter (2002), "Internet Refrigerator Should Be Kept On Ice," *Chicago Sun–Times,* November 16, pg. 34.

CUSTOMER-ORIENTED PRICING

Steps in Customer-Oriented Pricing[11]

In order to price products based on the value that customers perceive, marketers can use the steps shown in Table 9-1.

1. A firm must understand exactly how the customer will use its products. Customer-oriented pricing requires that a marketer completely understand how customers apply and use the products they buy from the firm. Each end use of a product may have a different cost/benefit analysis. For example, a customer who purchases a Quicken tax program to run a small business doing tax preparation and consulting would place a different value on the product than a person who purchases the same program to do his or her individual taxes. Because of the varying ways in which customers use products, marketers may need to segment on an end-use (usage occasion) basis.

In the business market, such end-use segments are referred to as **vertical markets**. The idea is to examine how different customer segments in different industries use a product and to price accordingly. Because of the different requirements in their end-to-end (or total) solution, customers in different vertical markets evaluate costs and benefits of a specific product in terms of a complete usage system, and not just in terms of an isolated part of that system. For example, if a small business decides to use a Web-based solution for its business processes (e.g., customer relationship management, supply chain management, customer service and billing), it must also have

TABLE 9-1   Steps in Customer-Oriented Pricing

1. Understand exactly how the customer will use a firm's products.
2. Focus on the benefits customers receive from using the products.
3. Calculate all relevant customer costs, and understand how a customer trades off costs versus benefits in the purchase decision.

an Internet service provider, a Web-hosting service, and technical support (whether in-house or outsourced). Evaluating the costs/benefits of, say, the Web-hosting service, really cannot be considered in isolation of the total value to be gained from the Web-based business process in the digital arena. (Recall the discussion of the whole product from Chapter 6.) And, obviously, for critical applications, customers will perceive greater value. Therefore, companies like IBM could charge higher fees for its Linux services being used in a corporate e-business environment, than in a public university setting.

2. A firm must focus on the benefits customers receive from using its products. The various types of benefits a customer can obtain were previously discussed and include functional, operational, financial, and personal benefits. In analyzing benefits, firms must not fall into the trap of focusing on product features at the expense of benefits. A familiar example is that the person who buys the quarter-inch drill bit does not want the drill bit, but wants the *capability* to drill quarter-inch holes. Customers buy benefits, not features. High-tech firms often mistakenly stress the cool technical wizardry of their inventions and are hard pressed to identify the real benefits customers receive. Additionally, the benefits that the technical/development personnel think are compelling are often confusing or not clearly important to the customers. Focusing on customer needs is a good way to overcome this problem.

For example, in marketing computers, advertisements frequently discuss terms such as *megahertz, megabytes, pixel resolution,* and so forth. Although customers might know that greater numbers on each of these categories are presumably better, they might not know what the "improved performance" really delivers. Speaking in terms of processing speed (less wait time for functions to be performed), greater storage capacity (for the ever-increasing size of software programs), and greater clarity of the screen can help customers understand what they are getting.

Another example of understanding customer benefits is exhibited in the refinement of the pricing strategy of RightNow Technologies. RightNow Technologies offers a software solution that automates the delivery of customer service over the Web. Customer support employees can cost companies on average $100,000 to $150,000 in salary, benefits, and equipment. The question was: How much money per month does RightNow's product save companies because fewer phone calls come into the customer support center? Based on calculations for a typical customer's call volume, the company estimated that customers were saving thousands to hundreds of thousands of dollars. Rather than licensing the product for $15,000 per year (the initial price point), this company raised the price to roughly $35,000 to $50,000 per year. Quantifying the value proposition for the customers allowed the pricing strategy to be based on sound logic—for both the customers and the company.[12]

3. A firm must calculate customer costs, including product purchase, and other relevant costs (discussed previously) including transportation, installation, maintenance, training, and nonmonetary costs, and *understand how a customer trades off costs versus benefits in the purchase decision.*

For example, in considering the purchase of a high-definition TV, typically priced upward of a couple thousand dollars, marketers have focused quite heavily on the aspect ratio and greater resolution of the picture. A customer-oriented perspective on pricing would ask:

- How or why will customers be using the product?
- What are the tangible benefits a customer receives from the features of aspect ratio and greater resolution?
- What are the costs that a customer perceives, in addition to the purchase price?

Customers who buy the product to watch TV at home for personal enjoyment will likely assign different value to the attributes than will sports bars and other businesses whose competitive advantage is wrapped around viewing programs. For at-home customers, the tangible benefits of greater resolution might not be all that clear (pun intended). In addition to the purchase-price outlay, customers might have to consider the costs of obsolescence of their existing TV sets, and the "cost" that not many programs are broadcast in digital format in the early stages of this product's life cycle. Hence, in terms of a tradeoff of costs/benefits, many typical at-home customers may find it difficult to justify the high price tag.

An example of step 3 in the business-to-business arena is exhibited in corporate installations of Wi-Fi. In 2003, the cost of installing a Wi-Fi hotspot in a corporation ran $1,000; maintenance and installation costs added another $3,000. However, the leaps in productivity that come from being able to share data instantaneously anytime/anywhere is why United Parcel Service is equipping its worldwide distribution centers with wireless networks at a cost of $120 million—it expects a 35 percent gain in productivity from its loaders and packers as they scan packages and the information instantaneously is shared to the UPS network. And, CareGroup Inc. hospitals in Massachusetts installed wireless systems to connect more than 2,000 doctors and nurses to the corporate system. This way, they could access patient records, add observations to the database, and check on medicines whether they were in the emergency room or intensive care. The system has reduced costly medical errors by 50%.[13]

A customer-oriented perspective on pricing is provided in the Technology Expert's View from the Trenches by Daria Schuster at IBM.

Implications of Customer-Oriented Pricing

The implications of these steps in customer-oriented pricing should help marketers in the following ways. First, this analysis helps marketers to realize that *pricing considerations should not be made* after *a product is developed and ready for commercialization, but* early *in the design process*. Treating price as a design variable helps the firm to understand the relevant cost/benefit trade-offs involved for the customer.[14] Recall that conjoint analysis (Chapter 5) is a useful tool in this regard. Many firms take a customer-oriented perspective on pricing early in the design process, and then develop the product around the relevant price point. For example, Hewlett-Packard, in its initial foray into the digital photography market in the mid-1990s, had research showing that a $1,000 price point was the maximum a consumer would be willing to pay for a scanner and printer for digital photography needs. As a result, HP worked its price analysis backward from the customer value point, through the retail channel, subtracting out the margin that retailers would take, ending with a **target cost** figure that HP had to meet in product design and manufacturing. It then did the sourcing and manufacturing

TECHNOLOGY EXPERT'S VIEW FROM THE TRENCHES

IBM E-BUSINESS ON DEMAND
DARIA SCHUSTER
Director of Pricing Strategy, Sales and Distribution,
IBM Corporation, White Plains, New York

An on demand business is *responsive, resilient, variable,* and *focused.* As the advertisement for IBM's approach to on demand business shows, an on demand business has the ability to respond rapidly to unpredictable changes in market conditions as well as the needs of its constituents. It operates with consistent availability, privacy and security. It can adapt its cost structure and business processes to be more flexible, thereby reducing risk, increasing productivity, and controlling costs. It has greater capital efficiency and financial predictability. Most importantly, such a business concentrates on its core competencies and areas of differentiation and employs tightly integrated strategic partners to manage its non-core activities.

So, in reality, on demand is more than just an information technology (IT) offering—it's about managing business, our business and our clients' businesses. This initiative was created to address clients' points of pain with respect to their IT investments. Our clients don't always just need a new server or a new application. Our clients need IT solutions that are responsive and resilient, that increase the variability of their cost structures, and that enable them to focus on their core business.

Competition is more fierce than ever. Our clients need to leverage their technology to drive growth in their business as well as improve the efficiency and effectiveness of their processes. In addition, clients want to transfer the ownership of risk to their IT providers. These are the criteria used for successful development, delivery, and pricing of on demand offerings; our focus is on the client.

Pricing for this new IT marketplace requires a deep understanding of client segments. Once we understand our clients and their specific needs, we develop offerings to address those needs at price points which enable customers to reduce their total cost of ownership and increase their return on investment.

Our on demand portfolio provides clients more flexibility in their IT bills and increases the variability of their cost structures. Increased "flexibility" means something different to each client. In its broadest sense, this flexibility can mean paying an IT bill that is similar to a phone bill, where a standard rate will be charged for the service with usage-based pricing for the units required. This is the underlying objective of our utility pricing strategy: we charge our clients based on increments of server processing capacity, number of purchase orders, or other combinations of infrastructure and business metrics so the customer pays only for what they need when they need it. As a result, our utility pricing enables flexibility through the use of fixed and variable pricing with quantity based discounts billed on metrics which more clearly align with our clients' needs.

Endless possibilities will drive growth opportunities in the IT industry for IBM and its partners. For example, a client can pay by the month for a financial accounting system, rather than building and maintaining the necessary technology resources in-house. The value of this decision for our client goes beyond the outlay of cash to pay the IT bill. The additional value is that this client now has highly responsive computing capabilities that it can leverage to drive its top line revenue growth by focusing on its core business, or to reduce expense to increase profits.

As the service provider, we at IBM invest in technologies that reduce our own cost structure. In order to accomplish this effectively, we apply the on demand principles within our business. These principles include standardized and secure technologies that are scalable, as well as more efficient technologies that are more "virtually" dedicated, rather than "physically" dedicated. Naturally, as standardization and virtualization increase, savings increase for both the client and the service provider.

Utility pricing based on the idea of technology as a service is more affordable to the client because fewer resources and investments are needed to manage for peak IT requirements. In the past, a client would purchase assets to manage its peak IT requirements. Utility pricing enables a client to avoid unnecessary outlay of capital. In addition, money is saved because they can spread out their payments.

We are in the process of a major transformation. Our clients need to drive growth of their businesses. And, their technology and business processes can leverage new technology advances where physical IT resources are consolidated and scaled, and also provide interoperability of different varieties of hardware and software. When you add to this environment the different possibilities of pricing permutations, the role of pricing in technology is one of the most exciting areas in business.

Because of this evolution as well as client-focused development, pricing strategies can no longer belong solely to the Finance function; they need to also be a priority for the Marketing function. The Finance perspective is critical, but as important is optimizing value for the client and the service provider.

around this target cost. Similarly, in the high-definition TV example, working more diligently with other industry players early in the design process on the "whole product," which would include programming considerations, might help tip the balance more positively toward higher benefits versus costs.

Second, this analysis shows that *different customers in different segments will value the same product differently.* Prices must account for both the perceived value of the product to customers and the cost to serve a particular customer account. Understanding that different customers value the product differently, and that different customers require distinct levels of service, means that the profitability of different customer accounts can vary widely—and differentially affect the profitability to the firm. Customer-oriented pricing requires that companies manage their customers based on profits, not just sales.[15] High-tech firms must be attuned to the costs of serving customers and filling orders, which can vary significantly by customer, depending upon the sales support, design or applications engineering, and systems integration required. Costs to serve customers can include pre-sales costs (e.g., sales calls, applications engineering), production costs, distribution costs, and postsales service costs. Unfortunately, the price paid by a particular customer often does not correlate with the costs to serve that customer.

With the adoption of activity-based costing practices[16] and customer relationship management (CRM) software (from companies like Siebel, SAP, Oracle, or PeopleSoft, to name a few), it is now possible for businesses to track profitability at the level of each individual customer. This can provide more useful insights for pricing policy than segment-level profitability analysis. For example, based on a study of customers of a U.S. high-tech corporate services provider, a U.S. mail-order company, a French retail food business, and a German direct brokerage house, Reinartz and Kumar found that loyal customers can cost more to serve and pay lower prices than newer customers.[17] If loyal customers turn out to be unprofitable, prices may need to be revised upward.

This implication of a customer-oriented view of pricing (focusing not just on sales, but on profits) is consistent with and reinforces the customer relationship management strategies identified in Chapter 3. As noted in that chapter, and a key implication for pricing: *Firms should track the profitability of different customer accounts.*

In analyzing the profitability of customer accounts, one implication that can arise is that *companies may actually decide* not *to serve some customers*[18]—unless there are mitigating reasons for doing so (e.g., the lifetime value of a particular account is likely to be positive, or ancillary products and services might be sold at a profitable level). At Tech Data, a distributor of computer-related components and accessories, customer profitability is calculated with activity-based costing and analysis of 150 costs, including freight charges, average order size, and account gross margin. "We've learned how to calculate the cost of serving every customer," says CEO Steve Raymund, "and that has enabled us to make money in bad times." Unprofitable customers are identified and encouraged to order more efficiently or take their business elsewhere. For example, rather than placing ten orders for $100,000 each, a smaller client might be encouraged to make one order of $1 million instead. After Tech Data began using this approach, the company reduced expenses to 3.5 percent of sales, compared to the industry average of 5 percent.[19]

Similarly, using vast data warehouses of customer information and sophisticated hardware and software, FedEx has divided its customers into the "good," the "bad," and the "ugly" with respect to profitability.[20] The good (profitable) customers are carefully monitored and receive regular customer service followup to prevent defections. The bad customers who spend much but are expensive to serve may be charged higher prices. And the ugly customers, who don't spend much and are expensive to serve, are not targeted with marketing communication at all.

PRICING OF AFTER-SALES SERVICE

Many manufacturers of durable high-tech products earn significant revenue from after-sales service. As discussed in Chapter 7, services have the potential to provide higher margins and competitive differentiation to sellers. Pricing of services poses a unique challenge because the benefits are often intangible to customers and companies lack data on unit production costs. As a result, many companies default to pricing service contracts by intuition. Some use uniform pricing based on a fixed percentage of the sales price of the equipment. This technique is too simplistic, since service costs can vary by accessibility of the customer, age of equipment, usage, and operating conditions. At the other extreme, some companies have a bewildering array of special contract terms negotiated with each customer. These may be costly to negotiate and be perceived as unfair to customers. Technology companies, thus, can end up losing money on services.

A better approach, and one that is consistent with the steps in customer-oriented pricing, is to price services based on careful segmentation of customer requirements. Customer needs for service usually include one or more of the following: technical support, training, maintenance, response times, parts coverage, after-hours availability, and add-on services. The McKinsey Consulting Company has found that most companies' service customers can be segmented into three categories:[21]

- "Basic needs customers" want a standard level of service with basic inspections and periodic maintenance.
- "Risk avoiders" want to avoid big bills but don't care as much about response times.
- "Hand-holders" need high levels of service, often with quick and reliable response times and are willing to pay for the privilege.

The three types of service pricing approaches, which will vary for the three categories of customers, are fixed price contract, time and materials, and full coverage. On the basis of this segmentation, it makes sense to offer the basic needs customers a fixed-price, well-defined, limited service contract, while the hand-holders should be happy to invest in a full-coverage contract. The risk avoiders' needs may be met best with a combination of fixed price plus time and materials add-on option. This type of service pricing strategy, based on customer needs and provider costs, has a better chance of profitability than either the more simplistic or complicated strategies.

THE TECHNOLOGY PARADOX

Probably one of the most significant factors high-tech marketers face is the rapid pace of price declines. Competition is forcing down prices in products ranging from semiconductor chips to finished personal computers; the pace of declines has reached 20 percent or more annually.[22] This situation requires huge gains in volume if a firm is to maintain sales revenues, let alone profitability. Falling prices can help a firm or an industry sell more units—some believe that the demand for digital resources is almost infinitely elastic[23]—and increasing volumes can allow for more price cuts. But the cycle is spinning ever faster, and companies have to scramble to keep up.[24]

Known as the **technology paradox,**[25] businesses can thrive at the very moment when their prices are falling the fastest—if they understand how to thrive in such an environment. At a minimum, the situation requires exponential growth in the marketplace, such that volume grows faster than prices decline. However, at its extreme, technology is virtually free, and companies cannot count on volume to provide profits when they are literally giving the products away. Extremely low-price or even free offers are attractive to the late majority adopters who can be difficult to acquire. For example, for those who are still not online and don't have a PC, it will take a really good offer to get them to adopt. But the cost to obtain these sales has a serious effect on the bottom line.

What can companies do to thrive when prices are falling so quickly?

Solutions to the Technology (Pricing) Paradox

Obviously, one implication of the technology paradox is that high-tech companies must know how to *keep costs falling faster than prices.* Moreover, the question of how to be competitive when technology is free requires a whole new paradigm for profitability. Companies must redefine value in an economy driven by unit-one costs. In such an environment, there is no single set of rules, as value can be found in several solutions.[26] For example, some companies will thrive by charging a premium for their products (e.g., Intel and Microsoft). Others can make money in selling products like commodities (e.g., disk drives). But, in the middle, companies must be inventive with their pricing strategies, as the solutions in Table 9-2 suggest. As so eloquently stated in *Fortune* magazine, "as lower prices undermine already crummy margins, anyone who wants to be top dog in computers must master some new tricks—and the initiatives have little to do with selling PCs."[27]

Second, *technology companies must make every effort to avoid getting stuck making commodity goods.* Commodity markets compel companies to follow supply/demand dynamics, and pricing power dissolves altogether. For example, Lucent Technologies no longer makes telephone handsets, which had become a commodity; rather, it sells network solutions.[28] When products become near-commodities, firms must focus on giving customers something that provides value above and beyond the competition's offerings. This might include customization (e.g., the Dell model), 24-hour technical support or maintenance agreements, or a strong brand name (Chapter 10). **Mass customization,** or serving mass markets with products that are tailored to individual customers, can be a compelling source of competitive advantage and provides knowledge of individual customer tastes and preferences. Amazon.com has taken this strategy into the Internet world.

Third, *firms must have agility and speed.* If a firm can't get to market on time, it might have missed its chance for profitability, because the price point will have moved.[29] Relatedly, engineers must focus less on the *best possible solution* and more on the *best solution possible* in the fastest time frame.[30] Efficient design and systems are probably less important in a market in which prices decline rapidly than getting the product to market quickly. As noted in Chapter 5, Guy Kawasaki refers to this as rule number 2 for revolutionaries: "Don't worry, be crappy."[31] His inflammatory rhetoric means that it is sometimes acceptable to strive not for perfection, but for the minimum level of market acceptability with the first generation of a radical new product.

Fourth, *companies can strive to find new uses for their products.* For example, Intel has actively been cultivating partnerships with a wide variety of companies, including toy companies, car companies, appliance manufacturers, and so forth, to expand the markets and uses for its chips.

Fifth, rather than being found in selling hardware or software, *a real source of value is found in developing long-term relationships with customers.* When the cost of manufacturing one more unit is negligible (unit-one costs), the goal of the firm changes from

---

TABLE 9-2 Solutions to the Technology Paradox

1. Squeeze out cost inefficiencies.
2. Avoid commodity markets.
3. Have agility and speed in getting products to market.
4. Find new uses for products.
5. Develop long-term relationships with customers.

making a high margin on each product sold to building relationships with customers. The telecommunications companies are recognizing this as they use sophisticated database marketing to sell customers their whole range of telecommunications services in a one-stop shopping model including local, long distance, Internet service, wireless and mobile commerce ("m-commerce") solutions. Other companies are recognizing this as they move away from focusing on sales of hardware or software to providing ongoing services that are a sustainable source of revenue—and competitive advantage. For example, IBM has steadily moved away from being a provider of computer equipment and software to a provider of information-technology–related services (see vignette on page 200).

Companies can justify extremely low product pricing, or at the extreme, giving away products for free, when it allows them to build strong customer relationships that establish the following:

- *A market hold.* Establishing a market hold with a large volume of customers is a viable strategy when customer attention is the most valuable commodity. Grabs for "mind share" are part of a high-tech, attention-driven economy, based on the scarcity of customer time.[32] In an **attention-based economy,** the consumer's attention is considered to be more valuable than the money paid for the product. Getting big fast, gathering enough consumer eyeballs, and acquiring knowledge about those consumers' shopping habits are the goals. Because customer time is scarce, and because technology keeps getting more costly in terms of the time required to master it, firms can grab attention by making products easy to use, exciting, or both.

  Establishing a market hold with a large volume of customers was one justification Amazon gave for its customer acquisition strategies. It strives to develop personalized knowledge of each individual's tastes and preferences and then capitalize on that knowledge by being the provider of choice for related products and services.

- *An installed customer base that will buy additional products and services.* One form of establishing an installed customer base is known in traditional marketing as **captive product pricing**. The basic idea is that a firm can be highly profitable by giving away the base or foundation product and making money on the complementary goods required to make the product useful. For example, Nintendo charted a business model in which the game consoles would be given away to consumers at or below cost, in order to boost sales of its game software. Virtually all of Nintendo's profits flow from sales and license fees, on the game software.[33] Cell-phone companies subsidize the price of handsets and make money with monthly service bills.

  Another form of this strategy (establishing a customer base that will buy additional products) is to *focus on the whole product,* or the entire set of items needed by a customer for a smoothly functioning system. For example, Gateway Computers recognized that the personal computer is only the "enabler" of all the activities that go on around the box itself.[34] And a typical 5 percent margin on a $1,500 PC yields only $75 in profit. So, Gateway made a major move to expand into marketing a whole product; it bundles software, maintenance, services, peripherals (printers and scanners), and Internet service. Customers can pay for this package over time with credit and can trade in for a new machine in the future.

From Free to Fee

One drawback to giving away products for free is that the strategy can create the perception that the product being given away for free may not be worth much. Further, once customers get used to something being free, it is very difficult to get them to pay later on. This can be a problem for high-tech marketers to attain breakeven and become profitable. During the Internet boom, many companies like FreePC gave away free PCs to customers who agreed to watch advertising. They all closed down. Online merchants like Webvan promised free delivery to customers if they ordered more than $50 in groceries. Webvan had to revise this minimum order upward to $75 and then to $100 to accelerate the attainment of breakeven, but went out of business anyway. Portals like Yahoo and MSN attracted users with free email, then sold these "eyeballs" to advertisers. With the benefit of hindsight, we now know that such price policies may shortchange companies on their path to profitability. The challenge for many high-tech businesses is how to wean customers away from "free" and make them pay "fees" for content, software and services.

A case in point is Yahoo. Its online advertising revenue took a nosedive along with the Internet bust. In 2001, under new CEO Terry Semel, Yahoo embarked on an ambitious strategy to convert users into paying subscribers.[35] Its Hotjobs service started charging employers for job listings. Under a partnership with Overture, it started charging advertisers for paid search listings. And it got a boost in e-commerce revenue from merchants selling products and services on its Web site. After several unprofitable quarters, Yahoo turned a strong profit in the fourth quarter of 2002. In 2003, about 40 percent of the company's revenue came from non-advertising sources. But much remains to be done in the transition from free to fee. For example, in 2003, only about 2 million users (out of its 200 million-plus user base) pay Yahoo fees for things like personal ads or premium email services.[36]

Antitrust Considerations in Free Pricing

In addition to being aware of the negative impact on perceived value, firms must also be aware of potential antitrust considerations in low-price or free offers. Given the existence of network externalities, there is a tremendous incentive to give away high-tech products, such as software, to build an installed base. And, because there is a strong tendency for consumers to band around one standard, once a company gains a decisive lead in an industry such as computing, it is almost impossible for rivals to unseat it. In such a situation, there is likely to be a high payoff from predatory pricing. Moreover, by scaring off potential competitors, firms with a strong reputation for predatory pricing find it cheaper to conquer new markets.[37] This is particularly effective in high-tech markets, because there's a continuous stream of new products. Because of these issues, some experts believe that software is the perfect market for predatory pricing.[38] Where is the line drawn between effective high-tech pricing strategy and predatory pricing?

The Microsoft Case

Large firms must be aware that when they give products or services away for free, they may be scrutinized more closely for anticompetitive practices than smaller players. For example, like others, Microsoft has frequently used the strategy of giving away its products in order to steal market share away from competitors. It has bundled free giveaways, such as its Internet browser and many parts of its lucrative corporate software (disk compression, firewalls, Web site management, and database analysis), into existing products.[39] Is Microsoft merely responding to customer needs? And is the software industry so dynamic that any dominant position is inherently short lived?

One test that is used in an assessment of the anticompetitive impact of business practices is the effect on prices. In the Microsoft case, one issue was whether Microsoft ended up raising prices to above-market levels to reap the benefits of being a predator. Some believe the answer to that question was "yes": The price it charged PC manufacturers for its Windows operating system doubled from 1991 to 1998. And Microsoft also tightened the terms of its user licenses for corporate clients.[40]

The U.S. Justice Department found in November 1999 that Microsoft used its dominant position in the market to "bludgeon the competition."[41] The court found that it bundled its Internet Explorer browser into Windows just to beat out Netscape and, in bullying its competition, caused the demise of innovations that would have "truly benefited consumers." And, its bundling actually came at the expense of its own product's performance; bundling a Web browser into Windows 98 actually slowed down the operating system, increased the likelihood of a crash, and made it easier for viruses to find their way from the Internet onto computers.[42]

The lesson from the Microsoft case is that firms must consider their market position (in terms of potential monopoly power) prior to using any marketing strategy, and particularly pricing.

THE EFFECT OF THE INTERNET ON PRICING DECISIONS[43]

Another factor that is exerting downward pressure on prices is the Internet. The Internet creates **cost transparency**, which allows buyers to more easily find information about manufacturers' costs and prices, providing them more leverage in making product choices. For example, through the Internet, customers are better armed with information about features and benefits. More knowledgeable customers know more about how to gauge value. The Internet makes a buyer's search more efficient. **Reverse auctions**, in which suppliers make lower and lower bids in order to "earn" the right to sell a manufacturer supplies for its business, allow customers to identify suppliers' price floors, or the lowest price at which they are willing to sell a product or service. Moreover, due to the transparency of pricing information online, some believe that the Internet makes it more difficult for a firm to engage in different pricing strategies in different markets—something that was commonly done in international markets in the past. And the increasing frequency of low-priced or free offers on the Net makes customers more sensitive to prices.

In light of these challenges, how can firms work to overcome these downward pressures? **Price lining**, or versioning, follows the practice of offering derivative products and services (Chapter 7), at various price points to meet different customers' needs. For example, broadband Internet service is available at a lower speed and price for residential customers and a higher speed and price for commercial customers. Alternatively, **price bundling** strategies, where a company offers two or more goods as a package at one price, can make it more difficult for buyers to discern a manufacturer's costs. Again, the bundling of broadband Internet service with cable TV or telephone service is a way to mitigate cost transparency.

Probably, the optimal way to mitigate the Internet's downward pressure on price is through maintaining a steady stream of *innovations* that allow a firm to avoid price competition. For example, eBay's ongoing innovation in services such as alerts, fixed price purchase, and convenient electronic payments, gives customers a reason to pay higher prices.

In contrast to the downward pressure on prices that the Internet exerts, some argue that the Internet and online marketing strategies can actually afford companies the opportunity to charge higher prices.[44] Smart pricing (or **dynamic pricing**) uses data on customer shopping habits to adjust prices real-time on the Internet. This allows a company to identify customer preferences and gauge how sensitive certain customers might be to price differentials.[45] Web pricing systems (based on sophisticated software and data analysis) go hand-in-hand with datamining and one-to-one marketing techniques, that allow marketers to target individual customers geared to their profitability and volume (see also Chapter 10).

The final section of this chapter explores other pricing issues germane to high-tech markets.

ADDITIONAL PRICING CONSIDERATIONS

The role of pricing in any market is to transfer rights of the product to the buyer, in exchange for some form of payment. High-tech products are valuable because of the know-how embedded in them. Recall from Chapter 7 that revenues can be generated by selling the know-how in multiple forms; firms can sell the know-how itself, they can sell components to OEMs, they can sell complete systems in a ready-to-use form, or they can operate a service bureau, providing complete, hassle-free solutions to customers (as IBM does with its e-business and Web-hosting solutions).

Because of the embedded nature of know-how, it can be difficult to price high-tech products at different levels on the "what to sell" continuum. A firm can price for a complete transfer of rights, whereby the buyer owns the product and related know-how completely and operates without restrictions, or it can use highly restrictive licensing arrangements that specify volume, timing, and purpose of usage.[46] The issue for a firm is how to maximize profits by choosing the "right" amount of property rights to transfer. In this section, the following options are briefly examined: outright sale versus licensing agreements, licensing restrictions for a single use versus multiple users, pay-per-use versus subscription pricing, and leasing.

Outright Sale of Know-How versus Licensing Agreements[47]

An outright sale of know-how assumes that the net present value of the technology over the relevant time horizon be estimated. However, with technological uncertainty, it is hard to assess the value of the technology at the time of transfer, so outright sales of know-how can be difficult to consummate. On the other hand, short-term licenses require an estimation of value over specific fields of use, which can be more readily estimated. When compared to an outright transfer of rights—for which buyers will presumably pay more—short-term licenses may reduce the revenue stream. However, given the high levels of technological uncertainty in high-tech markets which make it difficult to valuate know-how, firms might be more willing to use short-term licenses. Although short-term licenses generally yield lower revenue than outright sales of know-how, they are easier to valuate and execute. Rather than undervaluing the know-how, they yield a guaranteed stream of revenue for a specified time and use.

Licensing Restrictions

For many technology products, especially software, individuals are given a license to use the product. Sometimes the license is given free for one-time use or for a specified time period. For instance, in 2003, the RealOne audio and video subscription service offers individuals free use of their downloadable software for a fourteen-day trial period. After that, there is a monthly license or subscription fee of $9.95. When a fee is charged for a license, additional conditions may apply, including restrictions on the transferability of the license, time period of use, number of users permitted or number of physical products on which the software may be used. For example, in 2003, the terms of the license to use the Apple iTunes Music Store for music downloads at $0.99 per title permit export, burn, and copy of music for personal noncommercial use only, and the use of the music on three Apple-authorized computers at any time.

When licensing technology products, especially software, to enterprise (or corporate) customers, a major pricing issue is, should a company continue its licensing policy for individual users or offer a site license for multiple users. In general, a discounted site license for multiple users provides more value to enterprise customers than the individual license policy.

The standard pricing policy for an enterprise software company is to charge a one-time site license fee for a particular version of the software. Microsoft used to follow this policy, which gave its corporate business customers the right to choose if and when they wanted to upgrade to a new version. If they decided to upgrade, they would have to pay for a new site license. In October 2001, Microsoft moved to a new program, Licensing 6.0. This required enterprise customers to sign up for a two- or three-year subscription plan where, in addition to the one-time site license fee, enterprise customers would be required to pay additional annual subscription fees of 29 percent of the site license fee and would get free upgrades and support in return (this is an example of bundling, see section below). This strategy evoked a negative reaction among Chief Information Officers of Microsoft's enterprise customers, who are seriously considering migrating to the Open Source Linux operating system in order to avoid these hefty fees.[48]

Pay-per-Use versus Subscription Pricing[49]

An additional consideration for pricing of technology-based services is whether customers should pay per use or a subscription fee. *Subscription plans* charge one fee, regardless of usage, for a time period—e.g., monthly or annually. If the technology is new and unfamiliar to consumers, *pay-per-use* (also called *micropayment*) encourages trial at lower risk. However, as consumers become familiar with the technology, beyond a certain usage volume, the subscription plan will offer more value to consumers. The huge success of the iMode wireless cell phone service in Japan[50] and the accelerated diffusion of SMS text messaging in northern Europe is credited, in large part, to the pay-per-use pricing of the service providers.

In setting subscription prices, one has to consider the impact on consumption. There is some evidence that a large upfront fee such as an annual subscription paid in advance may have a negative impact on consumption and result in lower usage.[51] Thus, monthly subscription fees may serve as a better reminder for technology utilization than annual subscription fees.

Price Bundling

As noted previously, *price bundling* is the sale of two or more separate products in a package at a discount, without any integration of the products.[52] Customer value is created through the discount compared to the sum of the prices of the separate products. A software suite like Microsoft Office is a price bundle that offers more value to consumers than buying Word, PowerPoint, Excel, and Outlook separately. There are two types of price bundling, pure and mixed. Microsoft's pricing strategy for Office is an example of *mixed price bundling;* it continues to sell the bundle as well as the individual products separately. Mixed price bundling enables the charging of different prices across different market segments. *Pure price bundling* is when the company sells only the bundle at a fixed price and not the individual products separately. This makes sense when the products in the bundle are complementary or when the company wants to build volume or share of one or more products in the bundle. Pure price bundling, also called "tying" by economists and attorneys, requires caution as its use may be considered illegal for a company that has high market power and when a substantial amount of commerce is at stake. Courts can rule "tying" as illegal if the effect of the bundling hurts buyers or lessens competition in the market. This was the main charge against Microsoft in its antitrust case with the U.S. Justice Department (see The Microsoft Case on page 303).

Leasing

Another option for pricing in high-tech markets is to offer leasing to customers. This issue is explored by an expert at Babcock and Brown, one of the premier lease brokers, in the Technology Expert's View from the Trenches.

TECHNOLOGY EXPERT'S VIEW FROM THE TRENCHES

LEASING CONSIDERATIONS IN THE HIGH-TECH ARENA
LEONARD SHAVEL

Partner, Babcock and Brown,
Greenwich, Connecticut

My firm is an investment bank specializing in long-term financing of large-scale capital assets. In our typical role, we advise the purchasers of major capital assets, such as telecommunications networks and satellites, in the arrangement of leases to finance these acquisitions. Less frequently, we are hired by the equipment vendors looking to use attractive financing as an inducement to purchase their equipment; these vendors may also hire us simply to assist their customers so that they, as vendors, will not have to supply their own funding.

Whether hired by the buyer or the vendor, we:

- Identify and evaluate for the client the various financing alternatives that may apply to their type of equipment and its usage.

- Execute the transaction by placing the requisite debt and/or equity components with investors.

- Facilitate the entire process through negotiation of terms and closing of documentation.

As a financial adviser, our participation in the purchase decisions of our clients as it relates to choosing a given technology extends only to the question of how a certain type of equipment may be financed. But technology does play a role. An example would be the build-out of PCS networks in the 1990s by several of the major "baby bell" telecommunications companies. In two separate series of transactions, we were hired to arrange low-cost lease financing of the major network components (specifically, the mobile switching centers, base switching centers, and base transceiver stations). Although the form of financing (cross-border leases) was the same, the clients had chosen different technologies for their networks (technically: TDMA, or time division multiplexing, and CDMA, code division multiple access) and needed to be taken into consideration. As it turned out, both technologies lent themselves to an attractive leasing program, but investors needed to be convinced of the future value and utility of these systems before committing to finance them.

In most cases our involvement begins after the selection of the vendor has occurred. Exceptions would be: (a) where the client's decision to purchase a given asset hinges on the availability of attractive financing, or (b) where we are representing a bidding vendor who is seeking to create a competitive advantage by means of the financing element. These situations arise from the fact that an equipment user often can extract favorable financing from its vendor as an inducement to make a large or precedent-setting order. For a vendor, financing can be used as a means to differentiate itself from other manufacturers of similar equipment, or as a way to maintain its desired price level in the face of competition, or both.

As a proxy for the high tech sector, our experience with telecommunications companies may be instructive. Given the pace of modernization and expansion in the telecommunications industries (media, voice, data), there has been tremendous growth in the purchase of new digital technology by these companies, precipitating in turn increasing demands on the financial sector—and our firm—to provide innovative funding. In general, one can divide the types of telecommunications companies we work with into two categories:

1. Well-established companies with strong balance sheets and high credit ratings who use their newly acquired, high-valued assets

to get the most advantageous financing (in terms of cost, efficient use of tax benefits, and/or favorable accounting treatment).

2. Startup companies, who need to raise money any way they can and for whom an asset-based structure might be a means for attracting investors who otherwise would not want to take their company's business risk.

Examples of our clients that fit into the first group are companies like Inmarsat (satellites), Telenor (telephone switching equipment), and TCI (digital set top boxes), all investment-grade rated operators that could have financed their equipment purchases in a number of different ways (including out of operating cashflow) but were able to lower the cost of acquiring and owning the respective assets by means of domestic or cross-border leases. Such well-heeled companies often share the following objectives:

- To generate savings through the use of tax benefits.
- To minimize balance sheet impact and indemnification risks associated with the investor's participation.
- To maintain operating flexibility, including the ability to substitute and/or replace equipment over time.

These companies also view their asset-based financings as a means to develop relationships with new capital sources.

The second category (startups in need of financing) grew exponentially in the 1990s, given the creation of new telecommunications technologies and business opportunities. In the United States alone, there are many examples of early stage projects that were the first to commercialize new technologies: Echostar and USSB (direct broadcast satellite television), Sprint PCS and PrimeCo. (wireless telephony), Sirius and XM Radio (satellite radio). By using structured financing, companies like these have been able to obtain earlier stage funding (in most cases, prior to the start of commercial operations) than they would have otherwise. The typical form has been either long-term debt secured by their "crown jewels" equipment, or "venture leases," wherein a lessor obtains warrants in the lessee company in exchange for assuming an unrated credit exposure.

The key consideration for the investors in such "asset-based" financings is: Will I have a valuable asset to sell, and by doing so, hopefully recover a good share of my investment even if the management of this company does not succeed? Satellites and wireless telephone networks, while not as fungible as aircraft and railcars, became recognized to have such value, independent of the specific business plan. With the perspective of the last three years, we now know that this assumption was not always a good one. Many investors in satellites and new technology telecommunications equipment ended up with virtually worthless assets due to failed business plans and overcapacity.

Perhaps the greatest risk associated with arranging structured financings is failure to close. These are complex transactions involving two or more parties, sometimes in different jurisdictions, each seeking to achieve results that may be in direct conflict with the other's. A lessee (i.e., the equipment purchaser) may want to obtain a very low lease rate as the result of the lessor's (investor's) making certain favorable assumptions. The allocation of risks associated with these assumptions can be the most time-consuming aspect of a transaction, sometimes leading to impasse. With this in mind, we advise companies to approach these transactions on an opportunistic basis, wherever possible having alternatives "just in case."

SUMMARY

This chapter has addressed salient issues in pricing in high-tech environments. After examining the many factors that create a complex pricing environment, the chapter presented an overview of the 3Cs of pricing (cost, competition, and customers), a framework for the issues that must be simultaneously considered prior to setting prices. Because of the vital importance of a customer orientation in the high-tech arena, the chapter delved more deeply into customer-oriented pricing concerns. The case is made for analyzing profitability of individual customers and revising pricing as a result of this analysis. High-tech marketers need to pay special attention to the pricing of after-sales service.

One of the most significant factors high-tech marketers face is the inexorable decline in prices over time; therefore, special attention was given to strategies to generate revenue in light of price declines. Known as the "technology paradox," businesses who understand these strategies can be profitable despite falling prices. The strategies and solutions offered for the technology paradox are not without their disadvantages, however, and astute marketers balance antitrust concerns and brand reputation with their pricing strategies. The practice of giving away free content and services in the early days of the Internet created consumer expectations that cannot be fulfilled any more by high-tech marketers. The trend now is to start charging for content and services, to move from "free" to "fee."

As with other chapters in this book, focused attention was also given to the effect of the Internet on pricing strategies. Because the Internet provides what is known as "cost transparency" for buyers, buyers have a better understanding of manufacturers' costs of doing business. Tools to handle pricing strategies in light of cost transparency were addressed. The use of dynamic pricing strategies were also explored.

Finally, special pricing considerations, such as sales of know-how, licensing, pay-per-use, subscription pricing, price bundling, and leasing, were addressed. For the many reasons outlined, pricing decisions are very difficult; despite this difficulty, marketers must systematically address the issues presented here in order to minimize the odds of making a mistake. Importantly, success is difficult to guarantee.

DISCUSSION QUESTIONS

1. What lessons on pricing can be learned from the AOL vignette in the opening box of this chapter?
2. What are some of the complicating factors in the high-tech pricing environment? What is the impact of these factors on price?
3. What are the 3 Cs of pricing strategy? Describe the importance of each.
4. What are the relative costs and benefits of the purchase of a high-tech product from the customer's perspective?
5. What is the total cost of ownership? What is its pricing implication?
6. What is customer-oriented pricing? What are the steps in customer-oriented pricing? What are the implications of understanding this approach to pricing?
7. What are the various options for a company in pricing after-sales service?
8. What is the technology paradox in pricing? What five strategies can a firm use to stay profitable, despite the downward pressures on price, or even free products?
9. How can a company justify giving away products for free? What are the various strategies to make a profit in such a situation?
10. What are the potential drawbacks of the solution to price low to gain customer relationships? What is the lesson to be learned from the Microsoft case on low pricing?
11. What is cost transparency? How can firms address it in their pricing strategies?
12. What is dynamic pricing?
13. When should a firm use the following pricing strategies?
    - Outright sale of know-how versus licensing
    - Single-use licenses versus multiple-use licenses
    - Pay-per-use versus subscription pricing
    - Price bundling
    - Leasing
14. What strategies from the chapter can be seen in the views of the pricing expert?

GLOSSARY

Attention-based economy. The consumer's attention is often more valuable than the money paid for the product.

Captive product pricing. A strategy of giving away a base or foundation product that is required to use a corollary product and making money on the complementary goods required to make the base product useful.

Cost transparency. Buyers have information about manufacturers' costs and prices, providing leverage in purchase decisions.

Dynamic (smart) pricing. The use of sophisticated software which allows analysis of data on customer shopping habits to adjust prices in real-time.

Mass customization. Use of technology (both for ordering the goods and producing the goods) to allow tailoring of goods typically produced in mass quantities for an individual customer's tastes and preferences.

Price bundling. The offering for sale of two products packaged together at a single price point.

Price lining. Offering different versions of a base product or service at various price points to meet different customers' needs.

Reverse auction. On-line bidding format in which suppliers make lower and lower bids in order to "earn" the right to supply a business customer.

Target cost.   When the market price is taken as given, and desired profit margins are deducted from the price that customers are willing to pay, a company arrives at the target cost. Then engineering, product development, and operations must make the product or service available at this target cost.

Technology paradox.   Businesses can thrive at the very moment their prices are falling the fastest.

Total cost of ownership (or life cycle costing).   The total amount of money expended by a customer in order to own a product or use a service; includes the price paid for the good (including financing fees), as well as delivery or installation costs, service costs to maintain and repair the good, power costs to run the equipment, supplies, and other operating costs over the life of the equipment.

Vertical markets.   Customer segments in the business market with each segment having a unique end-use application for a product or service.

ENDNOTES

1. This phrase was coined by Gordon Moore, cofounder of Intel, in the semiconductor industry.
2. Smith, Michael F., Indrajit Sinha, Richard Lancioni, and Howard Forman (1999), "Role of Market Turbulence in Shaping Pricing Strategy," *Industrial Marketing Management,* 28 (November), pp. 637–649.
3. Dhebar, Anirudh (1996), "Speeding High-Tech Producer, Meet the Balking Consumer," *Sloan Management Review,* 37 (2), pp. 37–49.
4. Smith, Michael F., Indrajit Sinha, Richard Lancioni, and Howard Forman (1999), op. cit.
5. Martin, Justin (1995), "Ignore Your Customer," *Fortune,* May 1, p. 122.
6. Shapiro, Benson and Barbara Jackson (1978), "Industrial Pricing to Meet Customer Needs," *Harvard Business Review,* 56 (November–December), pp. 119–127.
7. Ibid.
8. Orr, Tim (2002), "Reducing Total Cost of PC Ownership," White Paper, http://www.nextbend.com/TCO.htm, July 11.
9. Greene, Jay (2003), "The Linux Uprising," *Business Week,* March 3, 78–86.
10. Quentin, Hardy (1999), "Iridium Plans to Cut Prices, Alter Marketing Strategy," *The Wall Street Journal,* June 22, page B9.
11. Shapiro and Jackson (1978), op. cit.
12. Ryan, Rob (2001), *Entrepreneur America: Lessons from Inside Rob Ryan's High-Tech Start-up Boot Camp,* New York: Harper Business.
13. Green, Heather (2003), "Wi-Fi Means Business," *Business Week,* April 28, 86–92.
14. Shapiro and Jackson (1978), op. cit.
15. Shapiro, Benson, V. Rangan, R. Moriarty, and Elliot Ross (1987), "Manage Customers for Profits (Not Just Sales)," *Harvard Business Review,* 65 (September–October), pp. 101–108; Myer, Randy (1989), "Suppliers—Manage Your Customers," *Harvard Business Review,* (November–December), pp. 160–168.
16. Cooper, Robin and Robert S. Kaplan (1991), "Profit Priorities from Activity Based Costing," *Harvard Business Review,* 69 (3), (May-June), pp. 130–136.
17. Reinartz, Werner and V. Kumar (2002), "The Mismanagement of Customer Loyalty," *Harvard Business Review,* 80 (7), (July), pp. 86–94.
18. Bishop, Susan (1999), "The Strategic Power of Saying No," *Harvard Business Review,* (November–December), pp. 50–61.
19. Cruz, Mike (2001), "Tech Data Adds Pricing Tiers," *Computer Reseller News,* May 7.
20. Judge, Paul C. (1998), "Do You Know Who Your Most Profitable Customers Are?" *Business Week* (September 14), http://www.businessweek.com/1998/37/b3595144.htm
21. Bundschuh, Russell G. and Theodore M. Devzane (2003), "How to Make After-Sales Service Pay Off," *The McKinsey Quarterly,* 4, http://www.mckinseyquarterly.com/article_print.asp?ar=1343&L2=16&L3=19&srid=17&gp=0
22. Wysocki, Bernard (1998), "Even High-Tech Faces Problems with Pricing," *Wall Street Journal,* April 13, p. A1.
23. Gross, Neil and Peter Coy with Otis Port (1995), "The Technology Paradox," *Business Week,* March 6, pp. 76–84.

24. McDermott, Darren (1999), "Cost-Consciousness Beats Pricing Power," *Wall Street Journal,* May 3, p. A1.

25. Gross and Coy (1995), op. cit.

26. Ibid.

27. Kirkpatrick, David (1998), "Old PC Dogs Try New Tricks," *Fortune,* July 6, pp. 186–187.

28. McDermott (1999), op. cit.

29. Kirkpatrick (1998), op. cit.

30. Ibid.

31. Kawasaki, Guy and Michele Moreno (1999), *Rules for Revolutionaries,* New York: Harper Business.

32. Gross and Coy (1995), op. cit.

33. Ibid.

34. Kirkpatrick (1998), op. cit.

35. Mangalindan, Mylene (2003), "Yahoo Enjoys a Rebound That Rivals Haven't Seen," *Wall Street Journal Online* (January 15).

36. Mangalindan, Mylene (2003), "Yahoo Posts Solid Profit, Sales With Rise in Premium Services," *Wall Street Journal Online* (January 16).

37. France, Mike and Steve Hamm (1998), "Does Predatory Pricing Make Microsoft a Predator?" *Business Week,* November 23, pp. 130, 132.

38. Ibid.

39. Ibid.

40. Ibid.

41. Cohen, Adam (1999), "'Microsoft Enjoys Monopoly Power . . . ,'" *Business Week,* November 15, pp. 61–69.

42. France and Hamm (1998), op. cit.

43. This section is drawn from Sinha, Indrajit (2000), "Cost Transparency: The Net's Real Threat to Prices and Brands," *Harvard Business Review,* March–April, pp. 3–8.

44. Koch, James (2003), "Are Prices Lower on the Internet? Not Always!" *Business Horizons,* January–February, 47–52.

45. Keenan, Faith (2003), "The Price Is Really Right," *Business Week,* March 31, 62–67.

46. John, George, Allen Weiss, and Shantanu Dutta (1999), "Marketing in Technology Intensive Markets: Towards a Conceptual Framework," *Journal of Marketing,* 63 (Special Issue), pp. 78–91.

47. Ibid.

48. Koch, Christopher (2003), "Showdown at 6.0 Corral," *CIO,* (March 15), p. 1.

49. John, Weiss, and Dutta (1999), op. cit.

50. Moon, Youngme (2002), "NTT DoCoMo: Marketing i-mode," Case # 9-502-031, Harvard Business School Publishing.

51. Gourville, John and Dilip Soman (2002), "Pricing and the Psychology of Consumption," *Harvard Business Review,* (September), pp. 90–96.

52. Stremersch, Stefan and Gerrard J. Tellis (2002), "Strategic Bundling of Products and Prices: A New Synthesis for Marketing," *Journal of Marketing,* 66 (1) (January), 55–72.

RECOMMENDED READING

Nagle, Thomas T. and Reed K. Holden (1994), *The Strategy and Tactics of Pricing: A Guide* *to Profitable Decision Making,* London: Pearson.

CHAPTER 10

Advertising and Promotion in High-Tech Markets: Tools to Build and Maintain Customer Relationships

If operating systems ran the airlines, Linux Air would have been started by a group of visionaries who built the planes, ticket counters, and paved the runways themselves. They would charge a small fee to cover the cost of printing the ticket, but you could also download and print the ticket yourself. When customers board the plane, each is given a seat, four bolts, a wrench, and link to a manual on the Web. Once the seat is installed, it is quite comfortable. The plane leaves and arrives on time without a single problem, and the in-flight meal is wonderful. In trying to tell customers of other airlines about the great trip, they can't seem to get past the seat: "You had to do what with the seat?"
—FROM THE INTERNET

Intel Centrino: Unwire Your Life

In March 2003, Intel launched a $300 million integrated marketing communications campaign in eleven countries for its Centrino mobile technology. This technology included a mobile processor, related chipsets and 802.11 wireless networking capability (also called Wireless Fidelity, or Wi-Fi) optimized to work together. The benefits to end users of notebook computers include seamless wireless capability, extended battery life, and thinner, lighter notebooks. The "Unwire" campaign kicked off March 3 with teaser ads on TV, saying, "On March 12, Intel will change not only how you work but where you work." Then, on March 12, Intel launched TV spots targeted at business and mobile computer users with taglines such as "Introducing Intel Centrino Mobile technology, the new wireless laptop technology designed to help you unwire your life." Print ads, including eight-page inserts, had copy describing technology features and continued the "unwired" theme: "Intel has an urgent message for the wired world: Unwire." The campaign also included outdoor advertising and sponsorships.

On September 25, 2003, Intel sponsored "One Unwired Day," an event that provided mobile PC users with free wireless access at thousands of locations all over the United States. The event included live concerts, product demonstrations, and prize packages at locations such as South Street Seaport in New York, North Riverside Plaza in Chicago, and Justin Herman Plaza in San Francisco. Intel worked with *New Yorker* magazine and Zagat to produce the Zagat Survey mini-guide on "2003 Wi-Fi Hotspots," which was published in the *New Yorker*, September 22 issue. A new blitz of global TV advertising reinforced the benefits of the Intel Centrino Mobile technology as bringing greater freedom, flexibility, and convenience to people's personal and professional lives. Print and outdoor ads featured the Intel Centrino's distinctive blue and magenta logo with catchy benefit statements like "No excess baggage," "Make all your connections," "Travel at the speed of really light," and "Need more battery life?" An online Web site (www.intel.com/unwire) provided more information on Wi-Fi. Intel also worked with *PC Magazine* and Yahoo! to create print and online resources for wireless users to find product information, places to buy, and practical trips on how to unwire. Intel expected 130 notebook designs with Centrino mobile technology to be available by year-end 2003.

The agency Euro RSCG MVBMS in New York created this integrated marketing campaign encompassing television, print, online, outdoor, and promotional advertising that was expected to run at least through the end of 2003. The thrust of the campaign was to increase primary demand for wireless mobility: As more users demand this capability, PC manufacturers will be motivated to include Intel Centrino in their notebook computers. With its consistently delivered "Unwired" message, Intel hopes this campaign will be as successful as its trail-blazing "Intel Inside" branding campaign of the 1990s.

SOURCES: "Intel Centrino Mobile Technology Advertising Begins," wifiareas.com, March 3, 2003; Black, Jane (2003), "The Magic of Wi-Fi," *Business Week* (March 18); "Intel's New Advertising Campaign Shows Unwired Moments," Intel Press Release, Santa Clara, California, September 22, 2003.

A solid advertising and promotion mix is as important in high-tech markets as in traditional markets. Some of the key tools that can be relied upon include traditional advertising (in both mass media as well as trade journals), trade shows, sales promotions (contests, incentives, etc.), public relations (event sponsorships, etc.), publicity (articles in the news media), the Internet, direct marketing (mail, telemarketing), and personal selling. High-tech marketers should avail themselves of a useful handbook that covers the basic advertising and promotion (A&P) tools and issues. Rather than covering A&P basics, this chapter delves into issues to which high-tech marketers must pay particular attention—tools that are often overlooked or not used to the extent that would be useful.

For example, engineers and technical personnel often disparage the important role advertising can play in developing brand awareness and brand equity. Yet fear, uncertainty, and doubt often plague the customer's buying decision. In such a situation, customers rely on heuristics to help them make safer, easier decisions, and a solid brand name is one such heuristic. Many high-tech companies, as well as Web portals and online communities, have realized this and rely heavily on traditional advertising to develop, reinforce, and sustain brand equity.

Moreover, the timing of new-product announcements can be vitally important in high-tech markets. Preannouncements help customers know what new products are coming down the pike and can delay them from buying a competitor's product in anticipation of another one coming in the near future. However, the pros and cons of preannouncing new products must be considered carefully. Finally, high-tech marketers need to understand how to use marketing communication to build and sustain relationships with customers.

Before delving into the specifics of this chapter, a useful device for planning and coordinating advertising and promotion tools is presented, the advertising and promotion pyramid.

ADVERTISING AND PROMOTION MIX: AN OVERVIEW

The advertising and promotion (A&P) pyramid,[1] shown in Figure 10-1, positions advertising and promotion tools based on two dimensions:

- The degree of coverage, or reach, of the target audience.
- Cost efficiency. One useful way to compute cost efficiency is based on **cost per thousand (CPM,** *M* is the Roman numeral for 1,000).

$$CPM = \frac{\text{\$ Cost of the advertising and promotion tool}}{\text{Number of people the tool reaches}} * 1,000$$

$$\text{(say, an ad in a particular trade journal)}$$

FIGURE 10-1 Advertising and Promotion Pyramid.

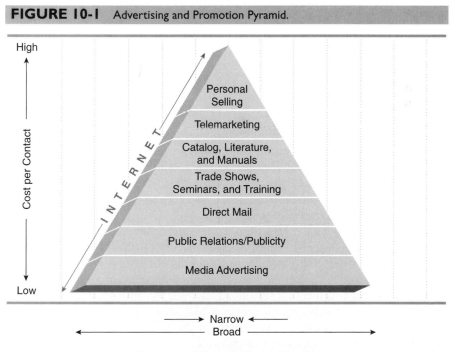

SOURCE: Adapted from Ames, B. Charles and James D. Hlavacek (1984), *Managerial Marketing for Industrial Firms*, New York: Random House, p. 253.

At the base of the pyramid are tools that have wider coverage of the target audience and lower cost on a per-contact basis. Tools at the top of the pyramid generally have narrower coverage and higher cost on a per-contact basis. While the pyramid implies a tradeoff between coverage and cost per contact using traditional communication tools, the Internet as a new communication tool is scalable and extremely cost efficient and can be used at all levels in the pyramid.

The idea behind using the pyramid as a coordinating device for the advertising and promotional tools is that a firm should not use the tools at the highest level of the pyramid in isolation of the tools at lower levels. The role of the tools at the lower levels of the pyramid is to create product and brand awareness, and to "warm up" prospects prior to using the more expensive, narrower tools.

For example, a new product may be announced with free product reviews in trade journals. Leads from that announcement that come in (either via a Web site or an 800 number) can be sent a direct-mail piece, possibly consisting of a brochure or other collateral material. As prospects continue to express their interest, they can be contacted more personally via either telephone, email, or a salesperson's visit. Importantly, the use of higher-level tools is leveraged on the efficiency and effectiveness of the lower-level tools and is geared to the prospect's continuing interest. Using all the promotional tools together in an orchestrated campaign to deliver a consistent message to the target audience constitutes what is known as **integrated marketing communications** (see vignette on Intel's Unwired campaign on page 313).

Matching the advertising and promotion tools to the appropriate task is based on the strengths and weaknesses of each tool, relative to the objectives each is to perform. The tool used must be matched (based on effectiveness and efficiency considerations) to the task at hand. A brief discussion of pertinent issues for each of the elements in the advertising and promotion mix follows.

Brief Overview of A&P Tools

Media Advertising

Advertising is a paid, non-personal form of communication using mass-media channels such as print, radio, television, billboards, and so forth. Starting at the base of the pyramid, high-tech companies can use media advertising in both mainstream media and trade journals. Critical issues are the degree to which the audience of the media vehicle overlaps with the firm's target market, the cost efficiency (CPM) of the vehicle, the fit of the editorial content of the media vehicle with the brand message, and the size and frequency of the ads that will be run.

How can high-tech firms identify appropriate media outlets? An advertising agency is a good start. For example, Godfrey Q and Partners in San Francisco is an example of an advertising agency that focuses exclusively on high-tech clients.[2] If a firm does not utilize the services of an ad agency, the Standard Rate and Data Service (www.SRDS.com) is a useful resource to search by industry, audience, and so forth for appropriate media vehicles. Moreover, CMP Media Inc., a high-tech media company, prints trade publications serving a wide technology spectrum including the builders, sellers, and users of technology worldwide. All of its publications are available from www.cmp.com/totallist. Finally, it is possible to get a media kit either online or from the advertising sales representative of any media outlet, which will provide useful information about not only the specific media vehicle but also its audience, ad rates, editorial calendar with upcoming special issues, other publications in the industry, and pertinent industry statistics.

Identifying appropriate media options is only half of the equation in using media advertising. A firm also must decide on an effective message that breaks through the competitive clutter of the many ads vying for the viewer's attention, while simultaneously reinforcing a brand message in a quick and easy manner. Striking a balance between gaining attention and reinforcing a brand message can be tough. Some firms err on the "gaining attention" side, using cute techniques (humor, sex, babies, etc.) to gain attention in a way that actually has little to do with the brand message, and so may even distract the viewer from the brand message. Other firms err on the side of providing so much technical detail—in jargon, which is, at best, uninteresting to the audience, and, at worst, unintelligible—that attention is quickly lost.[3] The best ads will both break through the clutter and use an attention-getting device that quickly and effectively delivers the key benefit without being lost in details—details that would be better provided in the next contact with a customer. For example, Cisco's "Empowering the Internet Generation" ad campaign has succeeded in increasing unaided brand awareness 80 percent since 1999 with simple messages emphasizing reliability, precision, innovation, globalization, and human interaction.[4]

Public Relations/Publicity

Public relations (PR) includes the activities a firm undertakes to develop goodwill with its customers, the community in which it does business, stockholders, and other key stakeholders (e.g., government regulators, industry trade associations). These activities might include sponsoring events, such as sporting events or charitable causes, cause-related marketing (aligning with a nonprofit organization), corporate advertising regarding the firm's position on critical issues or its philosophy of doing business, or other outreach activities (speeches by company executives, tours, etc.). *Publicity* refers to any coverage the company receives in the news media (print or broadcast) regarding its products or activities. Firms can attempt to gain favorable news coverage by holding press conferences, sending press releases, or staging events.

High-tech firms should not ignore the value that comes from maintaining a positive public image. Maintaining good relations with the media is extremely important in nurturing a positive image. When bugs are discovered in new high-tech products after launch, it is up to the media either to make a big issue of it or to relegate it to several paragraphs on an inside page of a newspaper or magazine. In 1994, Intel had a major public relations problem when customers discovered a bug in the first Intel Pentium microprocessor that introduced computational errors in complex calculations. That problem caused Intel to take a $475 million write-off for a product recall. Having learned from that negative experience, Intel is now more proactive in its relations with the media and the public. In April 2003, when Intel engineers discovered a glitch in a small number of Pentium 4 microprocessors, the company halted shipments immediately until they found a fix. This time around, the problem did not have a major impact on Intel's image.[5]

One of the most important types of publicity for smaller high-tech firms is the use of company or product announcements that many trade journals feature. By sending information about a firm or its products to the appropriate contact, information will be printed, typically for free, by that trade journal. This initial publicity can prove to be a valuable source of sales leads. For example, Channelweb (www.channelweb.com) has an online directory that permits searching for technology vendors. Any technology company may submit its listing free to the directory. Similar free submissions of company and product information could be made to online directories like Yahoo.

Although PR and publicity are shown at a higher level in the pyramid in Figure 10-1 than is media advertising, it is important to note that publicity can often be less expensive (as well as more credible) than media advertising. Furthermore, media advertising can also reach a narrower audience than can PR efforts, and so, the relative position of these two tools in the pyramid may be juxtaposed depending upon the specific media vehicle or public relations tool.

Direct Mail

Because direct mail can generally be more precisely targeted than can advertising or public relations, it is placed higher in the pyramid than those two tools. Lists can be obtained from a number of list brokers, including the mailing lists of many trade publications or Dun and Bradstreet, to name just two sources. Additional information on possible lists can also be found in the Standard Rate and Data Service. Costs are usually determined on a per-name basis, with more targeted lists that are frequently updated costing more. Firms must decide how many pieces to mail, as a function of the potential size of the target market, as well as the frequency of mailings. One mailing is typically insufficient to achieve results.

Trade Shows, Seminars, and Training

Trade shows, seminars, and training reach yet a narrower group of customers at a proportionately higher cost than tools lower in the pyramid. In the computer industry, Comdex–Fall in November in Las Vegas remains the industry launchpad for new products and innovative technologies, as well as a key environment to compare, contrast, and test-drive them. At Comdex–Fall 2003, 500 vendors showed thousands of new products to about 50,000 attendees, including possible business customers, trade channel members, and original equipment manufacturer (OEM) partners.[6] Individual registration packages were priced between $995 and $2,495.[7]

Trade shows can be quite expensive for exhibitors, including expenses for exhibitor fees, design and setup of a booth, personnel to staff the show (with the attendant travel, meals, and entertainment costs), and so forth. Estimates of trade show costs vary widely by show, industry, and location. On the low end, *Business Week Frontier* cites one example of a small cost-conscious exhibitor who spent roughly $10,000 for a three-day show with two other people to staff the booth.[8] At Comdex–Fall 2003, exhibitor space ranged in price from about $2,000 for a 10×10 booth to about $13,000 for a 20×20 booth—not including carpeting, lighting, personnel, or any other costs.[9]

Especially in such a large show as Comdex, one must trade off the large exposure potential from exhibiting at the show against the investment needed to break through the competitive clutter. Large players tend to dominate the space, and unless attendees have a compelling reason to visit a particular booth, a firm may not realize a return on its costs of attending the show. For example, Parker Manufacturing attended the A/E/C Systems show, a show for technology companies that serve the architecture, engineering, and construction industries, in Los Angeles in May 1999, hoping to talk to structural engineers about their pneumatic button punch, a tool used to crimp steel building seams tight. Linda McPherson and Jim Parker set up a

10-by-10-foot booth, and $8,000 later, languished with little traffic and no sales to show for their investment. They had no attention-getting device to compete with larger machines and had poor booth position (people were focused on bypassing them in favor of the food court).[10]

At a minimum, a company must absolutely plan on an accompanying advertising and promotion campaign just to generate traffic at its booth. Additionally, some attention-getting device can be helpful, to the extent it ties in with the product's message. Follow-up on the leads generated and a post-show evaluation are also critical. Despite the costs and risks of attending trade shows, trade show statistics are compelling. A national survey done by Data & Strategies Group Inc. in Framingham, Massachusetts, showed that closing a sale from a lead generated at a trade show costs an average of $625 and takes 1.3 followup calls, compared to the average $1,117 and 3.7 calls for other leads.[11]

Catalogs, Literature, and Manuals

Companies must have brochures or other collateral material of some sort to provide customers with additional information. The material must build on earlier communications with customers and showcase the key benefits of the products in terms customers can understand. Supporting technical detail is appropriate in these later follow-ups with customers. Issues such as relative advantage (costs/benefits), compatibility/interoperability, scalability, service, and warranties should be addressed. Especially for new technologies that may cost more than existing solutions, communicating benefits in terms of total cost of ownership can be effective.

Telemarketing

Telemarketing can be done on an outbound or inbound basis. *Outbound* calls are made by a company's personnel to existing customer accounts or on a cold-call basis. *Inbound* calls are made by customers or prospects calling into the company's call center, typically via a toll-free number. Having the opportunity to maintain customer relationships with a person dedicated to particular accounts can be an efficient use of resources. Telemarketers can help provide support to the field salespeople, answering questions, maintaining contact with accounts, and staying abreast of account changes in between visits from field sales personnel. In October 2003, Congress passed legislation creating a national "Do Not Call" list for those who don't want to be bothered by telemarketers' calls. However, as this book goes to press, the law is being challenged in court by the telemarketing industry. If the law, or some version of it, is allowed to stand, it will reduce the effectiveness of telemarketing as a sales promotion tool as some 50 million U.S. consumers have already signed up to be on the "Do Not Call" list.[12]

Personal Selling

At the highest level in the pyramid is the use of personal selling, where each salesperson can generally reach just one customer at a time. One implication of the use of the advertising and promotion pyramid is that small high-tech companies that are resource constrained generally should *not* use their company founders and salespeople to call on prospects *unless the prospects have first been contacted using less expensive, broader-reach tools*. Although small firms may say that they lack the resources to fund public relations or a direct-mail campaign, the real issue is whether they can afford to use their existing resources in an inefficient manner. By leveraging the tools at lower levels in the pyramid, firms ensure that the value of their high-expense tools is maximized.

Do not misunderstand this message. It is *not* meant to imply that personal selling and wooing customer accounts with top executives is unimportant. For example, EMC Corporation, maker of computer storage devices, hears from its customers that sending its top sales executive to its customers' headquarters has been a key tactic in winning customers' business away from companies such as IBM.[13] The lesson to be learned is that the value of such resource-intensive tools is maximized by ensuring that customers are appropriately primed to receive the company's message.

Internet Advertising and Promotion

The Internet can be used to complement many traditional advertising and promotion tools. Banner ads, sponsored web pages, search engine optimization, paid search, paid inclusion, contextual ads, targeted permission-based email advertising, and personalized customer contact are all cost efficient tools that the Internet offers. Following the dotcom bust, advertising spending on the Web declined steadily from 2000 to 2002. In 2003, the Internet's share of advertising spending was expected to remain flat at 2.5 percent[14] with a total value of $6.3 billion.[15] However, growth in online advertising is expected to rise again in the future, at least through 2008. This section covers the different types of Internet ads, their relative effectiveness, and associated costs, as depicted in Figure 10-2. It is important to delineate the role of each of these tools in the overall communications mix so as to create an effective integrated marketing communications campaign with a consistent message.

Types of online advertising

Banner ads remained a popular form of online advertising in 2003, accounting for 58 percent of all Internet ads, but their use has been declining over time.[16] The effectiveness of banner ads is often questioned. In a study of 2000 companies that do Internet advertising, only 20 percent expressed satisfaction with the results from banner ads.[17] Banners are easy to ignore, and most surfers may not want to leave the site they are viewing to check out an ad. Indeed, the click-through rates on banner ads have been decreasing over time. However, banner ads are cheap in delivering impressions. With enough creativity and repetition, an impact can be made. The current thinking in the industry seems to be that banner ads, like traditional advertising, are better at long-term brand building rather than generating short-term sales.

Innovations in banner ads designed to address their drawbacks are *live banners* that use video and sound. Such forms of rich media move, talk, flash, or play music to engage the Web surfer. The click-through rates of rich media ads are higher than that for non–rich-media ads (5.41 percent versus 1.57 percent).[18] Rich media ads work because they allow an advertiser to introduce drama, humor, or other emotions into their message. However, such ads are expensive and require high-speed broadband access to be viewed properly. The growth in the use of rich media ads is shown in Table 10-1.

One kind of rich media ad is the *interstitial ad,* or pop-up ad, which opens up a separate window when users click through to another site. A variant of this is the *superstitial ad,* a patented method from Unicast that grabs the full screen of the user for 15 seconds between web page transitions. Some surfers find such ads annoying because one must actively click to remove them from the screen. Another rich media

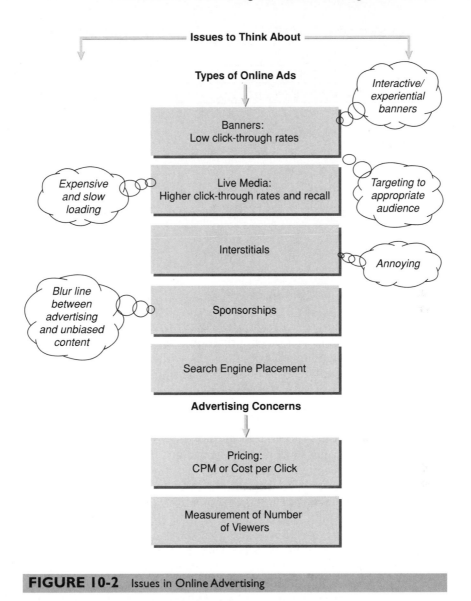

FIGURE 10-2   Issues in Online Advertising

ad without a pop-up window is the *shoskeles* from United Virtualities, which has animated ads floating on top of a Web page.

The degree to which the ad is appropriately targeted to a receptive audience has a
positive effect on its impact. Using cookie data, or information stored on a computer's
hard drive about which sites users have visited, users can be sorted into different categories, say, those interested in entertainment, business, or sports. Then an advertiser can
direct its ads to users in a particular category. What matters is not how many people the
ad reaches, but how many of the right people. A special case of targeting is *sponsored*
Web pages where an advertiser seeks to associate its brand with popular Web sites
using all kinds of large-format, rich-media ads. For example, Saturn sponsored The Red
Room on Launch Music on Yahoo!

TABLE 10-1   Rich Media as a Percentage of Total Ad Impressions Served

Quarter	Percentage
Q1 2002	17.3
Q2 2002	19.3
Q3 2002	23.2
Q4 2002	24.9
Q1 2003	27.8
Q2 2003	31.7
Q3 2003	36.6

SOURCE: Doubleclick

Directories and Search Engines

Directories and search engines are the yellow pages of the Internet. Web sites submit their pages to a directory like Yahoo, which employs human editors to review, classify, and list the pages under various information categories. A search engine like Google operates in a fundamentally different way. It sends out automated software "spiders" or "crawlers" that explore the billions of pages on the Web, index their content, and include them as listings in their database. A search box on Yahoo or Google enables users to enter keywords for the information they are looking for. These keywords are matched against the content of the pages in the database and a listing of the pages is produced in order of decreasing relevance. So any Web site that wants to promote itself needs to submit its pages to all the major directories and search engines. It also needs to design its pages with certain search keywords in mind that are most relevant to the content of its pages. This practice of designing Web pages so that they rank high in relevance when users search with certain keywords is called **search engine optimization**.[19]

Historically, most of the submissions to directories and search engines were free, and the results of user keyword searches were ranked on the basis of objective criteria such as relevance or popularity of the Web site. That changed in 1998 when Overture introduced the concept of *paid search*. Overture invited advertisers to bid for certain search keywords in real time. For example, Miller Brewing could buy the word *beer* on Yahoo for a defined time period. Every time someone conducted a Yahoo search using that term, a Miller Web page would come up as the first sponsor result or sponsored link. Those who bid less for the same keyword would be ranked in decreasing order of their bids. The advertiser would be charged the bid price only if a user actually clicked on their link. So paid search is also called *pay for clicks, pay for performance,* or *pay for placement.* Paid search includes boxed text ads on the top right of the user screen identified as "sponsored links" or "advertisements" in response to specific keyword searches. Unpaid listings follow below the sponsored links and advertisements. Paid search has emerged as the most effective of all online advertising tools in recent years with the highest return on investment.[20]

A search engine can take weeks or months to get to new Web pages with its spiders or crawlers. Further, spiders and crawlers sample only some of the pages on a Web site. For faster processing and inclusion in its database, some search engines like Yahoo charge a fee per page called *paid inclusion.* This is another promotional tool available to Web sites. However, the paid inclusion fee does not promise any position in the ranking of search listings beyond the objective criteria of relevance.

Contextual ads combine elements of online ads and paid search. Google, with its AdSense service, acts like an ad agency on behalf of smaller advertisers. Web sites that register with AdSense allow Google's search technology to monitor the content of their Web site. If Google finds editorial content on a registered Web site that is of interest to one of its advertisers, it will serve up a text ad on that Web site to match the content.[21] For example, WiFinder (www.wifinder.com), a directory of Wi-Fi hotspots around the world, is registered with Google AdSense. Google regularly serves up ads on behalf of many Internet Service Providers (ISPs) on this site. The idea is that visitors to this Web site are good targets for ISP service. Each time a visitor clicks on these ads, WiFinder gets compensated by Google, which gets compensated by the ISPs. So this is a potential win-win-win for the advertiser, Google, and the Web site publishing the ads. We will probably see more of this type of online advertising in the future.

The highly specialized field of search engine positioning is one that not only is very technical, but also changes rapidly. Interested readers are referred to the technical appendix to this chapter.

Pricing of Online Advertising: CPM or Cost per Click?

As with traditional media, one approach is to price online advertising as a function of the number of viewers and the desirability of and ability to focus on specific demographic profiles. The more targeted an audience a site offers, and the more desirable its demographics, the higher the rates that can be charged. Cost per thousand (CPM) is the cost to display a particular ad one thousand times. As one would expect, the CPM is higher for Web sites that offer a tighter, more targeted audience. For example, the Wall Street Journal Online's CPM is upward of $35[22] compared to the average CPM for Internet advertising of $30.

Cost per click refers to the ad rate charged only if the surfer clicks on a displayed ad as in paid search. This requires measuring how many surfers click on the ad. Whether an advertiser is willing to pay on a CPM or cost-per-click basis likely depends upon its online advertising goals. If the advertising is designed to support broad-based awareness and brand familiarity, a CPM pricing model might make sense. On the other hand, if the goals of the campaign are to develop a database of possible customers and move toward a direct sale, then paying on a cost-per-click basis might make more sense.

The ability to charge for online advertising rests on industry adoption of agreed-upon metrics. The Internet Advertising Bureau is a trade association of online media companies. It has adopted standard definitions of five key online metrics.[23] An *ad impression* is the measurement of a response of an ad delivery system to an ad request from a user browser. A *click-through* is the measurement of a user-initiated action of clicking on an ad element, causing a redirect to another Web location. *Visit* is one or more page downloads from a site, without 30 consecutive minutes of inactivity, attributed to a single browser for a single session. The number of *unique visitors* is the number of actual individual people, within a designated reporting timeframe, with one or more visits to a Web site. A *page impression* is the measurement of a response from a Web server to a page request from a user browser. While ad impressions are the units for buying CPM, click-throughs are the units for buying paid search.

Use of Affiliates

Affiliates are Web sites that provide sales to another Web site with whom they have a marketing agreement. For example, Amazon.com enlists Web site owners that offer Amazon's books, music, and videos on their own sites. The company pays affiliates a commission of 5 to 15 percent for any sales on their sites, but pays nothing for the added exposure.

Viral Marketing

Viral marketing, or referral marketing, refers to making offers so compelling that people voluntarily pass them around to their friends. It takes advantage of the power of contacts and shared interests to stimulate word of mouth via email. For example, Nike and Absolut vodka both allowed visitors to their sites to create videos and music messages to email to their friends. Email is the primary tool in this technique. When people respond to the offer via email, it enables the company to capture names and email addresses of possible customers.

Viral marketing is sometimes thought to be a subset of what marketers refer to as "buzz" marketing.[24] **Buzz marketing** is based on the idea of harnessing word-of-mouth communications to generate interest and excitement in a company's product. This technique is often done in a subtle (even covert) fashion by marketers, who seek out trendsetters in particular areas, who are then compensated in some fashion to "talk up" the product to friends. Buzz strategies can be so effective that they result in overwhelming product demand, sometimes known as "the tipping point"[25] in becoming a market success story.

Permission Marketing[26]

The idea behind **permission marketing** is to ask customers or prospects to *opt in* to receive marketing messages based on their interests in a particular topic. This is very different from nearly all conventional advertising and promotion, which might be characterized as "interruption marketing." Furthermore, permission marketing takes a very direct approach in marketing to customers, rather than the more veiled persuasion inherent in most traditional marketing attempts. The Internet is a natural environment for permission marketing because of email communication and its growing ability to close sales and, in the case of purchasing informational or digital products, to fulfill sales. In a comparative study of the ROI of email marketing versus direct response TV and direct mail conducted by the Direct Marketing Association, email marketing came out on top.[27] Among email marketing campaigns, those targeted to existing customers through inhouse databases did dramatically better than those intended to prospect for new customers.

Once a company has received the customer's permission to receive email information and product offers, a business should follow *relationship marketing* strategies. Rather than pouring marketing dollars into prospecting for new customers, a business should try to up-sell and cross-sell to existing customers. Customers who trust the vendor to give them reliable products and services at a fair price are receptive to such marketing. In relationship marketing, the true value of any customer is based on the lifetime value of his or her future purchases across all the product lines, brands, and services offered by a firm. So, although companies spend anywhere from $10 to $200 to acquire a new customer, their belief is that the lifetime value of a customer is so great that all-out marketing and low initial prices may pay off.

Importantly, permission-based marketing avoids the use of its antithesis, spamming. *Spamming* is sending unsolicited, undesired email to people. Its real-world analogy is junk mail and cold-call telemarketing. "Netiquette," or the informal rules of the Internet in terms of etiquette and decorum, is fiercely opposed to the use of the Internet in such a manner. A host of unscrupulous Web sites can inform possible Web marketers about how to engage in spamming, but doing so invites the ire of the very clientele a firm is trying to reach. Avoid spamming at all costs.

Mobile Advertising

A new platform for advertising, a mobile (cell) phone handset, is fast catching on in Europe and Asia. The value of this new medium to advertisers is that it lets them target their messages to individuals who carry mobile devices with them most of the time. Further, it can be an interactive medium by requiring message recipients to participate in some kind of game or competition. Rather than voice messages, this medium makes greater use of SMS text messages to target a youth audience. With opt-in permission, Radiolinja Oy conducted a promotional campaign with its mobile phone customers in Finland and got response rates of 30 percent. While novelty may partly explain the high response rate, it is widely believed that this method of advertising will grow in the future. The London-based research firm, Ovum Limited has estimated that the mobile advertising market will grow globally from $48 million in 2002 to $12 billion in 2006.[28] The advertising opportunities will expand even more when embedded chips using global positioning satellite (GPS) technology make possible the customization of location-based advertising services to consumers. For example, while walking past a pizza parlor, a consumer could get an SMS text message on the cell phone to come in for some fresh hand-tossed pizza. This chapter's Technology Tidbit focuses on these innovative uses of mobile commerce.

TECHNOLOGY TIDBIT

MASTERING MOBILE MARKETING

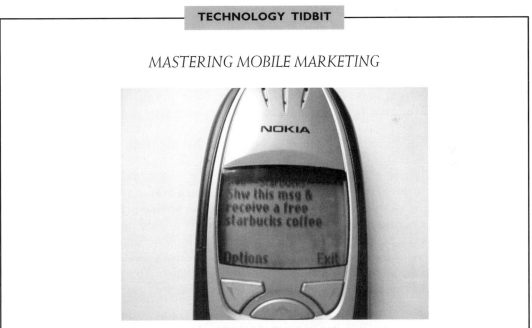

SOURCE: Courtesy G02Mobile Solutions, Ltd.

We know that certain countries in Asia and Europe are ahead of the United States in the adoption of wireless mobile technology. But which country would you guess has a promising future in mobile marketing? The future has already arrived in Ireland. Seventy-seven percent of the Irish own a mobile phone and this figure goes up to 100 percent for the 15- to 24-year-old age group. The Irish send out, on average, 31 SMS text messages per week. Over half of mobile phone users are interested in receiving commercial text messages on phones. One in five has already received advertising on mobile phones. While response rates to SMS campaigns in

Europe have averaged around 11 percent, Go2Mobile, an Irish mobile messaging services company, reports response rates up to 70 percent within 24 hours for some of its clients. Most of these campaigns have been successful at targeting the youth market with products and services, including consumer packaged goods, travel, recruitment, property, hospitality, and entertainment (wireless ticketing). Successful mobile messaging campaigns have been those that were integrated with a brand's overall marketing campaign, those that reinforced brands rather than tried to create awareness.

Here's an example of a successful SMS text messaging campaign. On your birthday, you are wondering how to celebrate it with your friends. Your mobile phone beeps to alert you to an incoming text message. You view the message to learn that it's your favorite local pub inviting you to come down with your friends. They promise to treat you and your friends to a free first drink to help you celebrate. Is that too good an offer to refuse? Irish coffee anyone?

SOURCE: Hickey, Gerreth J. (2003), "The Future of Mobile Marketing," TIU Management Consultants, Dublin, Ireland, http://www.techwatch.ie/files/6_3/The%20future%20of%20mobile%20marketing.htm.

THE IMPORTANCE OF BRANDING IN HIGH-TECH MARKETS

The importance of strong brands, built by efforts to establish and nurture a corporate identity, can be seen in many examples. Microsoft, Intel, Hewlett-Packard, Apple, Cisco, IBM, Dell, AOL, and Yahoo! are names that come to mind. These companies use mass media TV, print, radio, and billboards to reach a large audience to define their brands.

Business Week ranks the world's top 100 brands every year based on brand value or projected future earnings. In 2003, the top ten global brands, in order of decreasing brand value, were Coca-Cola, Microsoft, IBM, GE, Intel, Nokia, Disney, McDonald's, Marlboro, and Mercedes.[29] It is interesting to note that five of the top ten global brands could be classified as high-tech brands. Sony was ranked number 20 on this list in 2003 while Samsung was 25. Samsung's objective is to close the gap with Sony in brand value. Their strategy for accomplishing this is discussed in Box 10-1.

Clearly, in addition to being a marketing concern, **branding,** is also a financial concern. There is evidence that **brand equity** is positively related to firm financial performance and stock market valuation. In a study of firms in the computer industry, Aaker and Jacobson documented that brand attitude was positively related to return on equity, and changes in brand attitude were positively related to abnormal stock returns.[30] Further, they identified the following five possible drivers of changes in brand attitude:

- Major new product introductions (dramatic and visible) with strong advertising support, like the IBM ThinkPad or the Apple iMac, were associated with positive changes in brand attitude.
- Reports of product problems, like Intel's "floating point flaw" in its Pentium processor in 1994, were associated with declines in brand attitude.
- Changes in top management, like IBM's induction of Lou Gerstner in 1993 or the reinvolvement of Steve Jobs with Apple in 1997, were associated with improving brand attitude.
- Competitor actions, like hard-hitting direct comparison advertising, had a negative impact on brand attitude.

BOX 10-1

SAMSUNG'S GLOBAL BRANDING STRATEGY

For almost three decades Samsung Electronics used to be a low-priced manufacturer of commodity Dynamic Random Access Memory (DRAM) chips to OEM manufacturers like Dell and Nokia. In 1997 Samsung was in deep trouble following the Asian financial crisis, with a huge debt and a brand associated with cheap, me-too TVs and microwave ovens. A new CEO, Yun Jong Yong, started out by cutting 24,000 jobs and selling $2 billion in noncore businesses. He sensed the opportunity in electronics that the technology shift from analog to digital would create and decreed that Samsung would sell only high-end products. Yun hired new blood from outside the company. Eric Kim, who moved to the United States from Korea at age 12 and had worked in many U.S. tech companies, was hired in 1999 as executive vice president of global marketing, based in Seoul. Together, Yun and Kim developed and executed the global branding strategy.

For a start, Mr. Kim severed the unwieldy relationships Samsung had with 55 advertising agencies and consolidated these into one, awarding Madison Avenue's Foote, Cone and Belding a $400 million contract to build a global brand to rival Sony. The global advertising campaign featured ethereal models equipped with Samsung gadgets. They advertised heavily in the 2002 Winter Olympics in Salt Lake City and sponsored the World Cup Soccer tournament in Seoul in 2002. An electronic billboard in Times Square prominently featured Samsung advertising and the tagline "Samsung DigitAll" is becoming quite familiar to U.S. consumers.

In the face of competition from Nokia and Ericsson, Samsung forged a deal to supply co-branded mobile phone handsets to Sprint PCS. Samsung was able to supply 1.8 million handsets to Sprint PCS in 18 months, half the contracted time. Now Samsung is Sprint PCS's largest supplier of sophisticated handsets.

Mr. Kim successfully courted Best Buy and convinced the retailer to carry Samsung products. Best Buy shares information with Samsung on what features consumers prefer, and Samsung incorporates

these into their gadgets. Two of Best Buy's best selling products in 2001 were a DVD/VCR combo player and a cellphone/ PDA hybrid product, both from Samsung.

Today, Samsung is well positioned as a global, high-end maker of digital television sets, mobile handsets, DVD players, and MP3 players. It has the best-selling brand in TVs priced above $3000, the number 2 position in DVD players and the number 3 position in mobile phone handsets and MP3 players. It is well on its way to catching up with Sony's consumer brand value. But this required significant investment in advertising, product development, manufacturing, and retail distribution channels.

SOURCES: Solomon, Jay (2003), "Samsung Vies for Starring Role in Upscale Electronics Market," *Wall Street Journal*, June 14, page 1; Edwards, Cliff, Moon, Ilwahn, and Pete Engardio (2003), "The Samsung Way," *Business Week*, June16.

- Finally, legal actions were also associated with changes in brand attitude. For example, as the Justice Department case against Microsoft gathered momentum in the fourth quarter of 1997, its brand attitude declined noticeably.

In 1998, AOL's brand-name recognition dwarfed that of Yahoo!, Netscape, or even the Microsoft Network. As the AOL brand built strength, it was able to spend less on marketing. In a two-year time frame (1996 to 1998), AOL's cost of acquiring a new subscriber dropped from $375 to $90.[31] AOL's brand equity and stock market capitalization rose to such an extent that it was able to pull off the biggest media merger in history, with Time Warner in 2001. Soon after that, things began to unravel at AOL Time Warner. The recession lead to a sudden decline in AOL's advertising revenue. Content synergies expected before the merger did not materialize due to infighting between Time Warner and AOL business units. Accounting irregularities were discovered at AOL, and several top executives left. Ultimately, in October 2003, Time Warner decided to drop the AOL moniker from its corporate name.[32] The decline of the AOL brand illustrates the myriad factors that impact branding and highlights the need for high-tech companies to continuously monitor, nurture, and invest in maintaining their brands.

Strong brands possess important advantages in a competitive marketplace. Well-known brands generally are priced at a premium, resulting in higher margins to the companies that sell them. Strong brands are used as a badge or emblem that bestows credibility and attracts attention in new markets, be it a new country, a new category, or a new industry.[33] Hence, a strong brand can reduce risks a company faces in introducing new products, because customers may be less vigilant about examining the specifics.

From a customer's perspective, strong brands stand out as a beacon to the harried customer, a safe haven from the daily cacophony of technologies, new products, and media clutter around them.[34] Strong brands help customers simplify their choices, providing a safe shortcut in their decision making. A sizable segment of customers tends to migrate to the familiar in a cluttered and confusing world. Some believe that in high-tech markets, in which products change rapidly, a strong brand name is even more important than in the consumer packaged goods industry[35]—an industry that "wrote the book" on developing strong brands. Because customers may lack the ability to judge the quality of high-tech products, they may use brand reputation as a means to reduce risk.

For example, Cisco Systems makes routers, computers that direct the streams of information packets that travel the Internet. In the $6.6 billion worldwide market for routers, Cisco's 73 percent market share in 2003 dwarfed competitors like Juniper Networks.[36] About 90 percent of Cisco's sales in 2003 came over the Web. Its early advertising was aimed at creating a brand that business customers could trust for reliability and performance, and in 2000 Cisco spent $60 million airing its "Are You Ready?" television commercials. In 2003, Cisco turned its focus to the consumer or home networking market, expected to grow from $3.7 billion in 2002 to $7.5 billion in 2006. Home networks enable consumers to share broadband Internet connections, printers, files, digital music, photographs, and games over a wired or wireless local area network. That is why Cisco acquired Linksys in 2003, the leading provider of home networking products.[37] After a one-year hiatus during 2002 due to the technology slump, Cisco embarked on a new $100 to $150 million advertising campaign in 2003 to jump-start sales with a new theme, "This is the Power of the Network. Now."[38]

Kevin Keller, one of the premier experts on brand equity, notes that the short product life cycle for high-tech products has several significant branding implications.[39] First, it puts a premium on creating a corporate or family brand with strong credibility associations. Because of the often-complex nature of high-tech products and the continual introduction of new products or modifications of existing products, consumer perceptions of the expertise and trustworthiness of the firm are particularly important. In a high-tech setting, trustworthiness implies longevity and "staying power." Second, because of the rapid introduction of new products, there is a tendency by tech companies to over-brand their offerings. This is a mistake because it taxes consumers' memory and causes confusion. Consider the brand name, Intel Pentium Pro MMX; it entails many ideas and sub-ideas. Finally, high-tech marketers should use branding selectively to differentiate major new product introductions from minor extensions. For example, Intel Itanium is a new generation 64-bit microprocessor (compared to Pentium), while Intel Pentium 4 represents a performance improvement over the same generation Pentium 3 32-bit microprocessor.

In addition to having a solid understanding of traditional brand-building activities such as advertising, building brands in the high-tech arena, with its rapid change and ambiguity, requires additional considerations.

Developing a Strong Brand

So, how does a firm develop a strong brand? At a minimum, *customers expect marketers of strong brands to supply a steady stream of innovations* in exchange for their loyalty. And it is important that strong brands deliver the value they promise. The price/performance ratio must not be perceived as inequitable for the exchange, be the customer an OEM, an enterprise (business) user, or a consumer. Additional issues in branding in the high-tech environment are shown in Table 10-2 and discussed next.

TABLE 10-2 Strategies for Branding in the High-Tech Environment

Create a steady stream of innovations with strong value proposition

Emphasize traditional media advertising and PR rather than sales promotion

Influence the influencers and stimulate word of mouth

Brand the company, platform, or idea

Rely on symbols or imagery to create brand personality

Manage all points of contact

Work with partners (cobranding and ingredient branding)

Use the Internet effectively

Traditional Media Advertising and PR Tools

Advertising with a strong brand message focused on brand equity (versus price) is a vital ingredient in the branding mix. On the other hand, traditional sales promotions, with their focus on price, tend to erode brand equity. For example, when Northern Telecom (Nortel) decided in 1999 to launch a corporate-branding campaign to move beyond its roots as a maker of telephone switches, it developed ads to suggest a hip, cool image—one that was not associated with the Nortel of the past. Linking its products with the ability to bring the power of the Internet to the corporate networking market, the new tag line was "How the world shares ideas." Northern Telecom believed that mass-market advertising created value for its technology-based company, both in terms of selling more product and in identifying the company as a good investment.[40]

Due to the economic downturn from 2000 to 2002, many high-tech companies cut back heavily on their advertising budgets. Then, in 2003, many blue-chip technology companies started advertising again.[41] Hewlett-Packard launched its "Everything Is Possible" brand advertising campaign to mend the company's image after its acquisition of Compaq. HP also launched a new "You + HP" campaign focusing on digital imaging. This consumer campaign portrayed HP as the complete provider of all digital tools including cameras, printers, computers, inks, and paper. The advertising budget estimated for HP in 2003 was $400 million. Microsoft also planned to spend $400 million for its "Realizing Potential" TV ad campaign, depicting a boy in many possible professions, including an astronaut. IBM's ad budget for 2003 was estimated at $600 million. Media rates had come down and these companies felt the timing was right to stimulate demand once again.

Influence the Influencers and Word of Mouth[42]

Many high-tech companies rely on an "influence the influencers" program to generate publicity and favorable word-of-mouth endorsements. This is also a form of buzz marketing (discussed on p. 324). Public relations personnel must work hard to court the experts that influence the masses. For example, both Palm and Sony gave products away for free or at substantially reduced prices to technology experts, enthusiasts, and opinion leaders. Some journalists specialize in reviewing technology products, like Walter Mossberg for the *Wall Street Journal* and Stephen H. Wildstrom for *BusinessWeek*. Their reviews have a big impact on new product sales. As a result, these experts can become advocates for some products, with corresponding credibility greater than advertising. Third-party endorsements from top companies, leading industry or consumer magazines, or industry experts may help to achieve the necessary perceptions of product quality.[43] To gain such endorsements requires demonstrable differences in product performance, suggesting the importance of innovative product development over time.

Because brands in the high-tech arena are not built up over decades but over months, the frequent use of product giveaways helps to stimulate word of mouth, public relations and publicity, and brand familiarity. The Internet can accelerate word of mouth influence, a practice called viral marketing discussed on p. 324.

Brand the Company, Platform, or Idea[44]

Given the rapid obsolescence of high-tech products, it would not only be prohibitively expensive to brand new products with new names; it would be equally difficult for customers to develop product or brand loyalty. Therefore, firms should rely on *family names* taken from either *the company* or *the underlying technology platform*. Typically, names for new products are given modifiers from existing products—Windows 2000, for example, or Microsoft Word, Microsoft Works, Microsoft Explorer, and so forth. Alternatively, the company might choose to *brand the idea* behind the product, as in "Powered by Cisco" or Apple's "Think Different" campaign.

When a company does use a new name for a new product, it should be done to signal to customers that this new generation is a major departure and significantly different from prior versions of the product. Regardless, high-tech firms should not introduce new sub-brands too frequently.[45]

Rely on Symbols

Associations related to brand personality or other imagery may help establish a brand identity, especially in near-parity products.[46] For example, Kinetix, a multimedia company owned by AutoDesk, developed an animation software package. Part of the demo package was a funky dancing baby character, which become indelibly linked to the Kinetix brand.[47] Similarly, Napster's use of a silent, inscrutable cat, part of its attempt to build a hip, rebellious, edgy image, was successful, and its new owner (Roxio) stated its desire to continue using that icon given its high awareness level and positive associations with the Napster brand.

Manage All Points of Contact

Customer service is equally important in building strong relationships to the brand. If customers cannot get the service they need across all their touch points with the company, it negatively affects the brand.

Work with Partners

Moreover, high-tech brands are targeted not only to consumers and the financial community but also to partnerships and alliances. **Cobranding** is based on the idea of synergy: The value of two companies' brands, when used together, is stronger than that of one brand alone. Intel has relied on a variation of this strategy extensively, based on the idea of creating a brand identity for a component or ingredient used in the customer's end product, or *ingredient branding*,[48] discussed subsequently. Yahoo has also relied on this strategy, co-branding certain areas of its Web site with its content partners.

Use the Internet

High-tech companies must also consider the role of the Internet in their branding strategies. The notion of online branding seems to make sense for companies whose value proposition is intimately linked to the Web, as it is for Dell Computer. At a minimum, an effective branding strategy needs coordinated on- and off-line campaigns.

Rather than being a passive viewing experience (as in the media in which traditional branding has been most successful), the Internet shifts power and choice of viewing to the customer. Hence, branding tactics that work in the physical realm don't always translate online. The Internet is as powerful an embodiment of the company as is the product or a store. The Web must be more than brochure-ware, or "gratuitous digitization,"[49] and used to create customer value. Communicating with and helping a customer at the same time build brands online.

Indeed, on the Internet, the *customer's experience* at a particular Web site communicates brand meaning.[50] A traditional television ad might drive customers to the Web, and the Web then turns that into an experience. This dual strategy, of first driving traffic to the Web site and then ensuring that the technological creation and delivery of value at the Web site create a meaningful customer experience, creates online brand messages. However, one of the valuable lessons of the dotcom boom and bust is that branding is a necessary, but not sufficient, condition for market success. Pets.com built a successful brand with its advertising campaign that created the endearing "sock puppet" mascot for the company. But good branding cannot compensate for a poor business model or user experience.

Ingredient Branding

An **ingredient branding** strategy pulls demand from end users through the distribution channel to the OEMs, which feel pressure to use the branded ingredient in the goods they make. Business-to-business marketers will recognize this strategy as one designed to stimulate **derived demand:** Demand for the component, or ingredient, is derived from the end customers' demand for the products in which the components are used.

Supplier of ingredient or component ➜	OEM manufacturer ➜	Retail ➜	Customers
Intel	Hewlett-Packard	CompUSA	You

The way these strategies typically work is that the ingredient supplier provides cooperative advertising dollars to OEM manufacturers, who when they advertise, feature the ingredient in their own (OEM's) ads. Using this campaign, Intel's awareness level increased from 22 percent in 1992 to 80 percent in 1994.[51] Cooperative advertising dollars are typically provided as a percentage of product purchased. For example, at one time, Hewlett-Packard's cooperative advertising program was to set aside 3 percent of a dealer's purchases of HP products into a cooperative advertising (co-op) account. Then, when the dealer ran an ad that also featured HP product, the dealer would submit to HP a copy of the ad along with the invoice from the media in which the ad was run, and HP would then cut the dealer a check from his or her co-op account.

Advantages and Disadvantages of Ingredient Branding.
As shown in Table 10-3, ingredient branding does have its advantages and disadvantages. On the upside, the brand awareness created by the strategy creates a competitive advantage in the marketplace for the ingredient supplier. It establishes brand preference among end users, allowing the firm to fend off growing competition and stake out its own turf or identity. However, it is a costly proposition to develop a strong brand name for an ingredient and to provide the necessary cooperative advertising dollars to make it work. Moreover, if one of the OEMs that is participating in the branding program experiences product/performance problems, a halo effect occurs in which the supplier's reputation is tarnished also. (Note that this halo effect can work both ways: If it is the supplier's ingredient that has performance problems, the OEM's image will be tarnished as well.) And, as discussed subsequently, this strategy can create conflict with a supplier's large OEM customers.

TABLE 10-3   Pros and Cons of Ingredient Branding

	Pros	*Cons*
Supplier	Creates competitive advantage	Costly
		Possible risk if OEM has product problem
		Conflict with large OEMs
Large OEM		Erodes ability to differentiate
		Risky if supplier's product has performance problems
		If it doesn't cobrand, consumers might question product quality
		Worry about supplier forward integrating
Small OEM	Lends credibility to its product	Risky if supplier's product has performance
	Gets advertising support	problems

From a small OEM's perspective, using a cobranding strategy can lend credibility to its brand, making it more competitive with stronger brands in the market. Smaller companies can benefit from the sharing of advertising expenses through the cooperative advertising funds.

However, for large OEMs, the strategy can erode the ability of top-end brands in the market, say, IBM, for example, to differentiate its products. Customers in the marketplace often assume that industry leaders have something special in their products, but, with a cobranding strategy, customers begin to realize that the industry leader's product has the same ingredients as do lesser-known brands. This can cause conflict between the ingredient supplier and key OEMs in the marketplace, making the strategic relationship difficult to manage.

An ingredient branding strategy can also cause problems when the ingredient supplier and the OEM have different goals in the marketplace.[52] For example, the ingredient supplier, say, Intel, would like end users to demand the latest chips—those that sell at the highest margin. To drive this engine, ingredient suppliers typically invest huge expenditures in research and development. And the best scenario from the ingredient supplier's perspective is to have a large number of OEMs all competing on the basis of price.

But the OEM may have a very different strategy. For example, Compaq in the early- to mid-1990s wanted to develop a large mass market for relatively inexpensive PCs, which could best be done by using inexpensive rather than cutting-edge semiconductor chips. (The price that the OEM pays for the chip from the chip supplier, say, Intel, accounts for 20 to 25 percent of the cost of the personal computer.) An OEM with this strategy is less interested in working with a supplier that has developed a strong brand for its ingredient and charges a premium price for that brand. So, large OEMs may actively try to cultivate ties with other suppliers, to try to avoid being so dependent on the branded-ingredient supplier only.

And, if an OEM sells products without relying on the cobranding arrangement, products without the ingredient's logo (say, "Intel Inside") might arouse suspicion among consumers in the marketplace. Last, if OEMs fear that the ingredient supplier will attempt to enter their marketplace in the future (say, if Compaq fears that Intel will try to develop its own line of personal computers), they don't want to participate in the cobranding strategy, which will only serve to develop a strong brand for a company that may become a future competitor.

Reconciling these tensions may be next to impossible, but at least being aware of them helps firms to make more educated choices.

Branding for Small Business

Small businesses often face resource constraints in building a brand. A low-cost way of entering any market is with a no-brand strategy. If the small business is able to find one or more large business customers who resell the product under their own brands, this is a good way for the small business to get started. Samsung started out like this in the semiconductor business (see Box 10-1). If the small business wants to build and own direct relationships with end users, then branding is critically important. Creativity in marketing programs has to compensate for a low budget.[53] A small business needs to focus on one or two brands with one or two key associations. Creative messages can attract attention and get people to try the product but product quality and the customer experience are critical in building brand preference or loyalty over time. Google was co-founded by two graduate students in Computer Science from Stanford University, Larry Page and Sergey Brin. Their offering was a search engine with a simple, clean interface that produced the most relevant search results. The user experience reinforced this association. Superior search results got them free publicity early on in *USA Today*, *Le Monde*, and *PC Magazine*.

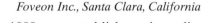

TECHNOLOGY EXPERT'S VIEW FROM THE TRENCHES

INGREDIENT BRANDING: FOVEON X3 *

ERIC ZARAKOV, VICE PRESIDENT OF MARKETING

Foveon Inc., Santa Clara, California

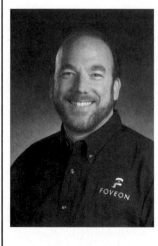

Between 1999 and 2002, the scientists and engineers at Foveon made a breakthrough in the fundamental design of an *image sensor*. Image sensors are the semiconductor chips that detect light and replace film inside of a digital camera. The new sensor had multiple benefits over the incumbent image sensor technology including: significantly sharper images, elimination of color artifacts, and better overall color quality. A cornerstone of our marketing strategy was to establish an ingredient brand for our image sensors in order to develop a premium for our products and rise above the established products. Our goal was to establish a relationship between the Foveon ingredient brand and the assurance of the highest level of image quality regardless of the host brand of the digital camera.

We launched our new image sensor in February 2002 and received worldwide visibility. The press coverage extended to almost all of the available photography media, key business press, and highly visible consumer press. Overall, our introduction resulted in more than 100 feature stories about Foveon and our X3 technology. The coverage we received included: the front page of the *New York Times* Business Section, front page of the *Wall Street Journal* Business Section, *Time Magazine*, *Newsweek Magazine*, fifteen seconds of primetime television coverage with ABC World News Tonight with Peter Jennings, and exposure to 150 million people through a radio interview with the BBC. This brought an enormous amount of visibility to Foveon. Through this news-based media introduction, we created interest in our technology by our potential customers (digital camera makers) and sparked the foundation for developing end-user pull that would reach through the digital camera makers to generate demand for new Foveon products.

As I look back to that time, I believe the success of our launch can be attributed to a number of factors. First, the timing of our story aligned with the explosive growth of digital photography. A second contributing factor was the visibility of Dr. Carver Mead, our company's founder. Dr. Mead is a world famous physicist who co-developed the very first methods for designing computer chips using Very Large Scale Integrated (VLSI). He is well known and followed by key industry analysts due to his association with many breakthrough technologies. Lastly, our visibility was high due to the obvious breakthrough nature of our technology and products. While all of the above were significant contributions, we would not have achieved our visibility if we had not planned our strategy and prepared carefully for the product launch.

PLANNING THE LAUNCH STRATEGY

As a foundation for the launch, our new technology needed a name, a logo, and simple, eye-catching visuals that we could collectively use to develop a brand identity. After considering many alternatives, our final decision was to use "X3." The "X3" name had multiple benefits. It was two syllables, which made it short and memorable. It reinforced the fact that our technology has three layers. The competing image sensors have just one layer. The X3 name also worked easily as a suffix which would minimize the potential identity tension between "X3" as an ingredient brand and the product host brand with which we would co-exist. For example, if

Nikon developed a digital camera with our technology, a potential name such as Nikon D100 X3 sounded good without imposing on the well-established Nikon brand.

Having selected a name, we needed to develop a logo for our brand. The current design was selected because of its simplicity, scalable output size, color match to our corporate identity and its ability to work as a logo on the front of a camera. In the "X3" logo that we developed, we included our corporate name, Foveon, underneath the stylized "F" to bridge the familiarity that Foveon had achieved through previous technology announcements. The logo "X3" was designed, however, to eventually work without the corporate name. Color versions of the logo and images can be found at www.foveon.com.

In addition, we developed a single dominant visual that was easy to understand,

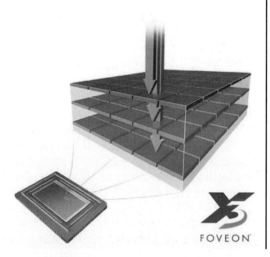

would catch attention, and communicate the essence of the technology. This single illustration was used consistently throughout our product launch and is still used in virtually all of our current marketing communications.

The third element of our visual portfolio was a fifteen second video that animated the basic concept of X3. The animation shows light separating into color layers. A Foveon X3 logo watermark was positioned in the bottom right corner of the animation to support the brand association. This effort paid off dramatically when we received a phone call from ABC World News Tonight saying they wanted to cover the Foveon X3 sensor. We offered the animation, and they ran it as a backdrop, with the visible "X3" brand, while Peter Jennings gave a fifteen second overview of X3. The animation was also used on local television stations throughout the country. The ABC news story alone gave us exposure to over 10 million prime time viewers. The video cost us less than $5,000.

While we had no certainty that our product and technology would receive the visibility that it did, we had made sure that we were prepared. Based on this experience, I can offer the following themes that resulted in a successful introduction of a brand campaign:

1. ***Get crystal clear on product positioning.*** Without this clarity, the foundation of the marketing communications plan will not hold together. Our positioning drove our strategy, marketing communication asset development and press communications.

2. ***Spend the time to work out the details.*** We spent countless hours on details of the illustrations, wording, and messages. We knew it was worth the time to get it right. The foundation communication elements would ultimately been seen by millions of people, and once it was launched, there was no going back.

3. ***Work small.*** Ultimately, all of the key marketing communications work was done with a very small internal team of two to three people. Large agencies cost a lot of money, need education on your products (at your expense), and take time to get up to speed. Make sure your team has balanced points of view so that you do not steer yourself down a blind alley of enthusiasm. Seek out specialized consultants if needed. This approach requires more hands on management—but you get what you need.

4. ***The press is your friend.*** Make their job as easy as possible. They constantly get pitched stories. Use your specific elements to make your story easy to understand and have a catch. Try to anticipate what might be needed—visuals, sound bites, testimonials, etc. These will be gold when you need them, and you have them ready.

5. ***Create mystique.*** Most people like to be on the inside of a secret. By creating mystique, you get people intrigued and create an atmosphere of excitement and inside knowledge.

6. ***Create critical mass in the press.*** I never to this day turn down an interview, no matter how small. You never know where it might lead. Once a story starts to spread, the press start coming to you and you need to keep it alive to build momentum. Each exposure generates more inquiries.

7. ***Keep a secret.*** Of all the product launches in which I have participated, the best results were achieved when NO news leaked out in advance. News that leaks out through dribbles has no impact when the formal announcement is made.

---

*Author's note: While I headed and directed the launch of the X3 technology, I worked side-by-side with a co-worker, Brian Behl, who was a key contributor to our brand identity. Brian's relentless perfectionism, keen eye for design, and multifaceted skills significantly influenced the elegant look and feel of Foveon's branding and communications.

Google is now one of the most established brands on the Internet. To read about how a small technology startup went about introducing a new brand, see the Technology Expert's View from the Trenches.

NEW-PRODUCT PREANNOUNCEMENTS

Gamespot's Vaporware Hall of Shame (www.gamespot.com/features/gs/vaporware) reads something like this:

> How would you feel if you'd seen dozens of ads for a new product, but when you went to buy it, you discovered that the product wouldn't be out for several months? Or, what if you read a glowing review of another new product, but were told by the salesperson at the store that it hadn't yet been shipped—and, in fact, would be months before it did?

Scenarios such as these happen regularly in the world of PC games, and in that industry, it is called **vaporware:** products that are announced before they are ready for market.[54]

More formally referred to as **preannouncements**[55]—or formal, deliberate communication before a firm actually undertakes a particular marketing action (say, shipping a new product)—they are a form of market signaling that conveys information to competitors, customers, shareholders, employees, channel members, firms that make complementary products, industry experts, and observers of the firm's future intentions. Because of their versatility, preannouncements are a very appealing tool for strategic marketing communications.

High-tech firms routinely preannounce new products. On April 28, 2003, Apple launched its iTunes service for music downloads. Initially, this was available only for the Apple Mac OS X operating system. At the time of this launch, Apple preannounced that it would come up with an iTunes service for Windows before the end of the year. On October 9, Roxio announced that it would launch its Napster music download service for Windows on October 29. Not wanting to be upstaged by Napster, Apple launched iTunes for Windows on October 16, getting a two-week lead and PR win over Napster.[56]

Advantages and Objectives of Preannouncements

Firms choose to preannounce new products for many reasons. In order to maximize their value, firms must have a clear intent for their preannouncements. As shown in Table 10-4, by preannouncing, firms can potentially reap a *pioneering advantage,* creating barriers to

TABLE 10-4   Pros and Cons of Preannouncements

Pros	Cons
Pioneering advantage: Preempt competitors	Cue competitors
Stimulate demand	Product delays damage reputation or jeopardize firm survival
Encourage customers to delay purchase	Cannibalize current products
Help customers plan	Confuse customers
Gain customer feedback	Create internal conflict
Stimulate development of complementary products	Generate antitrust concerns
Provide access to distribution	
Pursue a leadership position	

entry to later entrants. By announcing products before they are fully available, a firm can preempt competitive behaviors. For example, networking equipment maker Alteon Web-Systems Inc. leaked news of its new product a full year early to "freeze" the market for comparable gear from competitors. Its CEO Selina Lo said this was one more bullet she used to gain competitive advantage.[57]

Preannouncements can also *stimulate demand.* By helping develop word of mouth and opinion leader support, preannouncements can accelerate the adoption and diffusion of innovation when the product does hit the market. In addition to building interest for the product among channel members and customers, another factor related to consumer behavior is to *encourage customers to delay purchasing* until the announcing firm's new product is available. This latter reason is primarily used for big-ticket items that are purchased rather infrequently.

Businesspeople say that announcing products before they are ready for market is a "valuable tradition" in the software industry, divulging future product plans to customers. Such preannouncements can be beneficial when they *help customers to plan for their future needs,* or when they *allow customers to have input* in order to develop a more useful product. Users often need to know a software maker's plans early, because the software is so critical for their own businesses. So, alpha and beta versions (prototypes used to test and refine a program) may be sent to lead users months before the program is made commercially available. This gives customers, channel members, and OEMs time to prepare their operations for the new product and gives the manufacturer valuable market feedback.

Other advantages of and reasons for using preannouncements might include to *stimulate the development and marketing of complementary products* and to *provide access to distribution.* For example, Barco Projection Systems used preannouncements to keep its dealer channel interested in the vitality of its product line. Ultimately, one of the key drivers of a firm's propensity to preannounce is its *pursuit of a high-profile leadership position* within its industry.[58]

Disadvantages of Preannouncements

The benefits of a preannouncement strategy must be balanced against the costs of preannouncing. Preannouncements can *cue competitors* to what is coming down the pike, allowing them the opportunity to react to the new product.[59] For example, Storage Technology Corporation took a proactive strategy in preannouncing a new innovation in massive disk-storage systems used by large corporate clients. However, delays in its development turned into an advantage for its competitors, which were able to beat it to market. EMC Corporation and IBM were both able to bring out similar systems and capitalize on Storage Technology's preannouncements. Indeed (and despite the desire to use preannouncements to reap a pioneering advantage), the risk of cuing competitors is one reason why a firm's propensity to be a pioneer is *negatively* related to a firm's use of preannouncements.[60] Concerns about competitive retaliation and the risk of delays combine in such a way that firms seeking a pioneering advantage may actually avoid preannouncements.

Indeed, preannounced products may turn into vaporware and never materialize. Because the development process of high-tech products is so complex, *delays* are sometimes inevitable. For example, boo.com hyped the launch of its fashion and shopping Web site for June 1999. However, the launch didn't actually happen until mid-November.

(Moreover, when the site was finally launched, many technical glitches, including slow-loading graphics and consumers being bumped off the site, prevented users from buying anything.)[61] Such difficulties in delivering the preannounced product can be *damaging to the firm's reputation.*

Preannouncements can result in *cannibalization* of the firm's current product line, caused when customers delay purchases of current products in anticipation of the new ones. At the extreme, the combination of cannibalization and delays in meeting delivery can prove to be *catastrophic to a firm,* as Storage Technology found out in the early 1990s. Its new product, code-named "Iceberg," was formally preannounced in January 1992, with anticipated deliveries in one year's time. However, due to development problems and "buggy" software, beta testing did not begin until early 1994, and shipments didn't occur until later that year. In the ensuing time period, irate customers canceled orders, the firm lost approximately $189 million, and its stock dropped 76 percent from $78.00 to $18.50 a share.[62] Other disadvantages include the risk that preannouncements might also *confuse customers* who try to buy the product thinking that it is already available. Preannouncements might also *cause internal conflict* between departments, frustrating the efforts and objectives of another group. For example, engineering might want secrecy, but financial officers may want to send signals to the market early.

Finally, *antitrust concerns* can lead a firm to avoid preannouncing strategies, particularly for dominant firms. Vaporware can have detrimental effects in the marketplace when a firm has no intention of following through on its announcements, which are used merely as a competitive tactic in the marketplace to harm competitors. In an early investigation of Microsoft, the Justice Department decried the frequent use of vaporware, which businesspeople know is deceitful.[63] When the preannouncement is for the sole purpose of causing consumers not to purchase a competitor's product, then the predatory intent is inferred to be anticompetitive and subject to regulation under antitrust laws. Some believe that dominant players in a market—such as Microsoft—must be held to a higher legal standard because of the likely harm of their preannouncements to competitiveness in a marketplace.[64]

Tactical Considerations in the Preannouncement Decision

In making decisions about new-product announcements, firms must consider several tactical factors, shown in Figure 10-3. These include the timing of the announcement, the nature and amount of information, the communication vehicles used, and the target audience(s).[65]

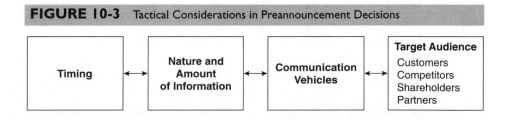

FIGURE 10-3 Tactical Considerations in Preannouncement Decisions

Timing

The timing of preannouncements must consider many factors, including the advantages and disadvantages of preannouncing relative to the

- Innovativeness and complexity of the new product.
- Nature of customer switching costs and the length of the buying process.
- Timing of final determination of the product's attributes.

Earlier preannouncements (farther away from the actual product launch) are particularly useful when complements to the new product are necessary to its success, for highly novel or complex products that will engender buyer uncertainty, for products that have a long buying process, or for those in which buyer switching costs are high. *Later preannouncements* (closer to the actual product launch) make more sense when the firm needs to keep information about the new product from potential competitors, when product features are frozen late in the process, and when the firm seeks to minimize risks of cannibalization. Regardless, preannouncements must be *timed to coincide with the purchase cycle of customers.* For example, if customers take approximately six months to decide on a new purchase, a preannouncement with a six-month lead time would be acceptable.[66]

Information

Firms must consider how much and what kinds of information to include in the preannouncement. Some new-product preannouncements contain information on the attributes of the new product, how the product works, and how it compares to existing products in the market, whereas others contain very limited information. Information about pricing and delivery date may also be important. Communication vehicles might include trade shows, advertisements, press releases, or press conferences. Target audiences for the information might include customers, competitors, shareholders, or partners.

Other Considerations

In their study of new-product preannouncing behavior, Eliashberg and Robertson found that preannouncements are useful under the following conditions:[67]

- For firms with low market dominance, which face lower cannibalization risks.
- For smaller firms, which face fewer antitrust concerns.
- When a firm believes that competitors are not likely to respond preemptively to the preannouncement. This might be likely in an industry in which R&D and technology are specialized (e.g., the pharmaceutical industry in which specialization around therapeutic category reduces the number of competitors who can react). Reliance on patents may also reduce competitive response.
- If a product requires substantial customer learning or if customers face switching costs, preannouncing can be advantageous to advance the learning process. Preannouncements can encourage advance planning for customers to switch to the new technology and can help them standardize around key specifications and operating systems.

On the other hand, firms may *not* benefit from preannouncing when

- A firm has a strong portfolio of products, and preannouncing may encourage customers to postpone purchases. (As an aside, strong brands may also be used to counter competitive preannouncements.)
- Large firms have more risk due to inferences of predatory behavior.

THE ROLE OF MARKETING COMMUNICATIONS IN CUSTOMER RELATIONSHIPS

Another important role for advertising and promotion tools is to develop and maintain relationships with customers. *Customer relationship marketing* allows a firm to create long-term mutually beneficial relationships with customers that result in greater loyalty and improved sales.

As noted in Chapters 3 and 9, although customer relationship marketing is an important tool, not all customers warrant the cost- and time-intensive efforts that relationship marketing entails. As a result, many firms rate their customers based on customer volume and profitability to the firm.[68] Identifying, recognizing, and rewarding the best customers to sustain their profitable behavior make sense. A recent book notes that the top 20 percent of a firm's customers generate almost all the profit ("angel customers"), while the bottom 20 percent actually destroy value ("demon customers").[69]

For example, using a customer database, a telecommunications company can identify which customers ("the good") buy many services at a high level of profitability (i.e., proportionately fewer services and marketing investments) compared to those who spend just as much but cost more to keep ("the bad"). Customers who spend little and likely won't spend more in the future ("the ugly") don't receive as high a level of service or marketing attention, which brings the firm's costs down.[70] By sifting through information about each customer's calling patterns, demographic profiles, the mix of products and services used, and related data, a company can identify the less profitable customers who have the potential to be more profitable in the future. The company can also determine a maximum amount to spend on marketing to a particular customer before it becomes an unprofitable venture.

Discussed subsequently, sophisticated customer relationship management software is used to this end. Firms are using customer databases to compare the mix of marketing and services that go into capturing and retaining each individual customer to the revenue that customer is likely to bring in. Also referred to as **database marketing,** or **one-to-one marketing,** these efforts allow marketers to target individual customers with pinpoint accuracy.

Categories of Customers

As shown in Figure 10-4, at a minimum, a company can identify four segments of customers, based on the company's share of customer purchases in a category and the customer's relative consumption level of the category.[71]

1. *Low share–low consumption.*   These are low-consumption customers with which the firm has a low share of their purchases. Absent some compelling reason to invest in these customers (Are they opinion leaders? Might their buying status change in the near future, for example, a college student?), a firm should avoid these customers by providing a low base of reinforcing communications and services.

   At the extreme, a firm may decide to weed them out of the customer base entirely—for example, if these customers consume more resources than they return. Although such a notion is anathema to many new high-tech startups (with salaries to pay, rent that is due, and infrastructure investments to make), a firm

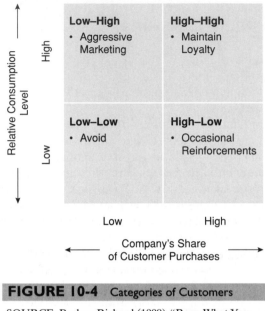

FIGURE 10-4 Categories of Customers

SOURCE: Barlow, Richard (1999), "Reap What You Reward," *The DMA Insider*, Fall, pp. 35–38.

must shepherd its resources wisely. Indeed, the hardest part of being an entrepreneur can be learning to turn down business. However, it may be the only way to survive and grow, if customers take a firm away from its mission, values, and areas of expertise. In addition, saying yes to the wrong business can damage employee morale.[72]

How does a firm say no? It can politely recommend a competitor whose value proposition and capabilities more closely match the customer's needs. Or, based on the pricing principles discussed in Chapter 9, it can charge a sufficiently high price that turns the account into a more profitable one—assuming the customer doesn't defect.

Of course, there are risks to this strategy. If done incorrectly, alienating customers can cost a company negative word of mouth. Moreover, if a company rejects the wrong customers, who prove to be major spenders in the future, it may be a nonrecoupable decision. Despite these risks, serving unprofitable customers does not make good business sense.

2. ***High share–low consumption.***   These customers are relatively loyal to the firm, but spend relatively little in the category. Although they can be reasonably profitable, if they have little growth potential, the firm should attempt to maintain them with a minimum level of marketing investment. Communicating often enough to sustain a sense of specialness and giving occasional bonuses should be sufficient to sustain their current level of activity.

3. ***Low share–high consumption.***   Although the firm has a low share of these customers' purchases, they represent a major opportunity, given their high consumption in the category. Many firms lack an understanding of how much their customers spend in a category. For example, if a company knew that 15 percent of its best customers purchased nearly twice as many products from a competitor, think

of how it would change the way it markets to them. Once these high-opportunity customers have been identified, the goal is to grow the firm's share of their business aggressively by convincing them to switch and cementing their loyalty. This may require significant value-added services to overcome their current loyalties.

4. ***High share–high consumption.*** These customers are a company's bread and butter and represent, to a competitor, the customers whom it would like to lure away. The goal is to maintain these customers' current spending levels while increasing their commitment to the firm.

How can a company make the ideas in this section actionable? Let's take a hypothetical situation. Say a wireless telecommunications company knows, for example, that it costs, on average, $350 to acquire a new customer and about $35 per month to service a customer. Given average revenues of $60 a month per customer, customers must be retained for at least 14 months to be profitable. Moving beyond averages into the customer tiering advocated here, this company may discover that a specific customer is using wireless technology for personal safety needs, is susceptible to churn (competitive switching based on price offers), and frequently calls the customer service department with questions. With additional data, this company may discover that this customer is in the low–low category and decide not to engage in the typical telemarketing or direct-mail campaigns used with its customer base. On the other hand, another customer may be using wireless technology in place of landline phones for both personal and business use and relying on the Internet for customer service needs. The greater types of usage make switching costs higher for this customer. Depending upon its relative consumption in the category, this customer would be in either the third or fourth category. Such a customer definitely warrants specialized attention and intelligent marketing to cross-sell other products.

The vital message here is that it is entirely appropriate—indeed, it is imperative—that a company segment its customers on the basis of profitability. Different strategies are required to be successful with different types of customer accounts. Moreover, as explored in the next section, in fast-paced environments with a rapid pace of innovation, strategies for customer relationship management can include more than the standard focus on capturing lifetime customer value.

Strategies for Customer Relationship Management

Strategies to manage customer relationships proactively go beyond mere data mining and database marketing. As shown in Table 10-5, Rashi Glazer identifies six strategies that are useful in turbulent market environments.[73]

Capture the customer is based on the objective of realizing as high a share as possible of the customer's total purchases over time (lifetime customer value). Using interactive communications, a firm can use information gained during previous encounters and transactions to cross-sell and up-sell customers appropriate products. In this case, once a company has acquired the customer, any acquisition costs can be amortized over many different product/service transactions. For example, Oracle Software's telemarketing sales effort is driven by (what else?) a sophisticated relational database from which an Oracle representative interacting with a prospect can call up all relevant information on the product in question, competitors' offerings, and all previous interactions with the caller (including other people at the prospect's firm who have interacted with Oracle). The system has documented success rates of over 90 percent, prospects who after a single call have either purchased a product or been converted into highly qualified sales leads.[74]

TABLE 10-5   Strategies for Customer Relationship Management

Capture the Customer	Similar to database marketing. Uses information from past interactions with a customer to tailor future offers to their particular needs. Objective is to capture as high a share as possible of the customer's total purchases over time (lifetime customer value).
Event-Oriented Prospecting	Based on information about customers that might trigger a purchase, marketing is tailored to a particular life cycle event (graduation, first job, etc.).
Extended Organization	Dissolves functional boundaries between firms, allowing one firm to manage the activities of another in its value chain. Creates high switching costs for customers.
Manage by Wire	Guides interactions with customers based on a combination of human decision making and expert computer systems (a type of artificial intelligence).
Mass Customization	Flexible manufacturing that tailors products to individual customer needs.
Yield Management	Tailors pricing to different customers' price sensitivities in order to maximize returns.

SOURCE: Reprinted from "Winning in Smart Markets" by Rashi Glazer, MIT Sloan Management Review, Summer 1999, pp. 59–69, by permission of publisher. Copyright © 1999 by Massachusetts Institue of Technology.

Event-oriented prospecting is based on a firm's ability to store information about customers that might trigger a purchase. For example, when a child graduates from high school, a computer firm might send an ad to the family with an offer to purchase a computer as a graduation gift for the child to take to college.

Extended organization strategies dissolve functional boundaries between firms and allow one firm to manage the activities of another in its value chain. For example, Federal Express has developed sophisticated software systems that allow it to manage the shipping and logistics functions for its customers. This creates high switching costs for customers.

Manage by wire strategies guide customer interactions, based on a combination of human decision making and expert computer systems (a type of artificial intelligence). For example, the healthcare industry is moving to a model in which, when customers call a health maintenance organization for questions, the person on the phone is guided by an elaborate computer script, updated continuously based on hundreds of customer records, about how to treat a specific problem.

Mass customization, or flexible manufacturing, tailors products to individual customer needs. For example, Dell Computer has based its business model around mass customization. Although this model typically requires a significant initial investment in fixed costs, customized units can be made with little marginal cost. The firm can also typically avoid heavy inventory carrying costs under this model.

Yield management (also known as smart pricing, or dynamic pricing, covered in Chapter 9) is based on tailoring pricing to different customers' price sensitivities. Pioneered in the airline industry, which uses sophisticated pricing algorithms to determine seat pricing that will maximize returns, the conditions under which this technique is appropriate are similar to those for mass customization: high fixed-cost assets, low marginal cost of providing an additional unit, and the product cannot be inventoried.

Customer Relationship Management Software

The primary tool used to implement the sophisticated strategies used in customer relationship marketing, which are premised on gathering and using detailed information about each customer's interaction with a company, is customer relationship management (CRM) software. This industry is expected to grow from $10 billion in 2002 to $14.5 billion by 2007, according to data from AMR.[75]

At a broad level, these software programs are used to capture data about customers from any contact within the enterprise. These one-to-one marketing applications provide the ability to track profitability per customer; to detect customers' dissatisfaction before they leave; or to improve product selling and retention, loyalty, and revenue. There is some evidence that a 1 percent increase in customer satisfaction can lead to a 3 percent increase in the market capitalization of a company.[76] Independent surveys have shown that business customers of Siebel software, one of the largest vendors of CRM software, report a 13 to 24 percent increase in contentment of their customers after installing Siebel CRM software.[77]

More specifically, customer relationship management software includes the following:

- Sales force automation, which allows sales reps to track accounts and prospects
- Call-center automation, used to create customer profiles, provide scripts to help service reps answer customer questions or suggest new purchases, and to coordinate phone calls and messages on Web sites
- Marketing automation, helping marketers to analyze customer purchasing histories and demographics, design targeted marketing campaigns, and measure results
- Web sales, to manage product catalogs, shopping carts, and credit card purchases
- Web configurators, to walk a customer through the process of ordering custom-assembled products
- Web analysis and marketing, uscd to track online activities of individual shoppers, and offer them merchandise they're likely to buy based on past behavior and special, on-the-spot prices for specific customers, and to target marketing via emails to individual customers

When a customer calls a company that has installed CRM software, the representative at the other end of the line can immediately access that customer's information on historical interactions culled from a variety of databases,[78] quickly solve customer problems, and make new offers to sell them additional products and services. For example, salespeople can use CRM software to figure out how much a prospect is probably authorized to spend, profile a product against competition, or find an advocate within a customer's organization. Customer service reps can use the data in these programs; the knowledge enables a meaningful, intelligent dialogue with the individual customer. These programs provide a strong value proposition not only for businesses but also for their customers. While a custom CRM implementation can be quite expensive for a company, a new trend is to use an application service provider like Salesforce.com to provide similar benefits through a simple Internet browser at a low annual per-user fee.

SUMMARY

This chapter has provided four specific tools to help high-tech managers make decisions about advertising and promotion strategies. Beginning with the idea of the advertising and promotion pyramid, this chapter initially addressed how a firm can leverage a variety of advertising and promotion tools in a cost-efficient manner. Second, the chapter provided coverage of why and how a high-tech firm can develop a strong brand name, including ingredient advertising or cobranding. Third, the chapter addressed the strategic issues in making product preannouncements, with explicit coverage of the objectives, risks, and tactical considerations. Finally, the chapter concluded with a discussion of the role of marketing communications in managing customer relationships. In addition to discussing the categories of customers and strategies to manage customer relationships, an overview of customer relationship management software was provided.

Advertising and promotion issues are particularly salient in the high-tech marketing environment. What can work particularly well is an integrated marketing campaign that makes use of mass media such as TV, radio, print, billboards, and online communication tools like banner and rich media ads, paid search, permission-based email, and viral marketing. Although high-tech marketers may be either unfamiliar or uncomfortable with these strategies, the importance of a strong brand, the use and timing of new-product preannouncements, and the development and maintenance of strong customer relationships, through the leveraging of the tools in the advertising and promotion pyramid, are vital.

DISCUSSION QUESTIONS

1. What is the logic behind using the advertising and promotion pyramid as a coordinating device?
2. What are some of the critical issues in each of the tools in the advertising and promotion mix?
3. Find examples of the various types of online ads. Examine their strengths and weaknesses.
4. What are the advantages of a strong brand to firms? To customers?
5. How does a firm develop a strong brand?
6. What is the role of the Internet in branding strategies? How does the Internet change the branding environment? What are the major implications?
7. What is the logic behind ingredient advertising (also known as cobranding and derived demand advertising)?
8. What are the advantages and disadvantages of ingredient advertising? Be sure to consider the viewpoints of both the supplier and the OEM.
9. Under what conditions might ingredient advertising be most useful or most likely?
10. What are preannouncements? What are the pros and cons?
11. What is vaporware? What are the ethics involved? (Be sure to consider the perspectives of the various constituencies involved, including customers, competitors, and the firm.)
12. What factors affect the timing (early or late) of preannouncements?
13. When is preannouncing most likely to be effective (from the company's perspective)? When is it most likely to cause problems?
14. What is database, or one-to-one, marketing?
15. On what basis should companies identify customers who are good prospects for customer relationship management? What are the resulting categories?
16. What six strategies can be used for customer relationship management?
17. What functions can customer relationship management software perform?

GLOSSARY

Advertising. A paid, nonpersonal form of corporate communication using mass-media channels such as print, radio, television, billboards etc.

Brand equity. Power of a brand to positively affect consumer response to marketing activity.

Branding. The strategy of creating a strong, recognizable, familiar brand name in a target market.

Buzz marketing. Harnessing word-of-mouth communications to generate interest and excitement in a company's product; often done in a subtle (even covert) fashion by marketers, who seek out trendsetters in particular areas, who are then compensated in some fashion to "talk up" the product to friends.

Cobranding. When the value of two companies' brands, when used together, is stronger than one brand alone, synergy is created in developing ties between the two companies' brands.

Cost per thousand (CPM). The cost of reaching a thousand different people through an advertising medium. Good for comparing the efficiency of different media.

Database (one-to-one) marketing. Using a customer database to capture data on which to rate customers, based on customer volume and profitability to the firm. Allows a firm to compare the mix of marketing and services that go into capturing and retaining each individual customer to the revenue that customer is likely to bring in. Allows marketers to target individual customers with marketing programs geared to their profitability and volume level.

Derived demand. Demand for the component, or ingredient, is derived from the end customers' demand for the products in which the components are used. For example, the demand for chips is derived from end users' demand for computers, which is pulled up the distribution channel to retailers, then to computer manufacturers, and finally, to chip suppliers.

Ingredient branding. The strategy of creating a brand identity for a component or ingredient used in the customer's product.

Integrated marketing communications. The use of different promotional tools including various forms of advertising, public relations, events and the Internet in a planned coordinated campaign to deliver a clear and consistent message to a target audience.

Permission marketing. Ask customers or prospects to opt-in to receive marketing messages on-line.

Preannouncement. Formal, deliberate communication before a firm actually undertakes a particular marketing action (say, shipping a new product); a form of market signaling that conveys information to competitors, customers, shareholders, and firms that make complementary products.

Public relations. The activities a firm undertakes to develop goodwill with its customers, the community in which it does business, stockholders, and other key stakeholders.

Search engine optimization. The practice of designing Web pages so that they rank high when users do a Web search with specific keywords.

Vaporware. Products that are announced before they are ready for market and that may never materialize.

Viral marketing. Use of the Internet to spread word-of-mouth influence among members of a target audience.

ENDNOTES

1. Ames, B. Charles and James D. Hlavacek (1984), *Managerial Marketing for Industrial Firms,* New York: Random House.

2. Raine, George (2003), "New ad agency to focus on tech," *San Francisco Chronicle,* April 24.

3. Bellizzi, Joe and Jakki Mohr (1984), "Technical Versus Nontechnical Wording of Industrial Print Advertising," in R. Belk et al. (eds.), *AMA Educators' Conference Proceedings,* pp. 171–175, Chicago: American Marketing Association.

4. Frook, John Evan (2001), "Cisco scores with its latest generation of empowering ads," *B to B,* 86 (15) (August 20), page 20.

5. Poletti, Therese (2003), "Intel finds glitch in new chips," *San Jose Mercury News,* April 15.

6. Peterson, Kim (2003), "Comdex 2003: Glitz Takes a Back Seat to Court Serious Buyers," *Seattle Times*, November 16.

7. www.comdex.com

8. Klein, Karen (1999), "Show and Sell," *Business Week Frontier*, August 16, pp. F.26–F.30.

9. www.comdex.com

10. Klein (1999), op. cit.

11. Klein, Karen (1999), "Trade Shows Are Indispensable . . . but You've Got To Prepare Ahead," *Business Week Frontier*, August 16, pp. F.21–F.24.

12. Krupnick, Matt (2003), "Seniors hold out hope for no-call list," *Contra Costa Times*, October 2, page 1.

13. Auerbach, Jon (1996), "Cutting-Edge EMC Sells the Old-Fashioned Way: Hard," *Wall Street Journal*, December 19, p. B4.

14. Mangalindan, Mylene (2003), "Web Ads On The Rebound After Multiyear Slump," *Wall Street Journal*, August 25.

15. Elliot, Nate, David Card, Andrew Peach, and Gary Stein (2003), "Online Advertising Through 2008: Paid Search Drives Modest Recovery," *Jupiter Research*, August 28.

16. M2 Presswire (2003), "Doubleclick: Doubleclick's third quarter Ad Serving Report reveals that rich media continues to make gains; Newer, very large units taking off; Older, smaller formats continue to decline; Report breaks out international trends for the first time," October 28, page 1.

17. Gumbel, Peter (2001), "Ads Click," Special Report: E-Commerce, *Wall Street Journal*, October 29.

18. Kuchinskas, Susan (2003), "Rich Media Growth Trend Continues," October 27, cyberatlas.internet.com/markets/advertising/article/0,,5941_3099401,00.html

19. Mangalindan, Mylene (2003), "Playing the Search-Engine Game," Special Report: E-Commerce, *Wall Street Journal*, June 16.

20. Gumbel (2001), op. cit.

21. Borzo, Jeanette (2003), "On Point: New services promise to deliver ads to Web sites that are a lot more relevant–and a lot more lucrative," Special Report: E-Commerce, *Wall Street Journal*, October 20, page R4.

22. advertising.wsj.com/online/rates/, downloaded November 3, 2003.

23. "Interactive Audience Measurement and Advertising Campaigning Reporting and Audit Guidelines, *Internet Advertising Bureau*, January 2002, www.iab.com/standards/measurement.asp.

24. Rosen, Emanuel (2001), *The Anatomy of Buzz: How to Create Word of Mouth Marketing*, Khermouch, Gerry. (2001, July 30). "Buzz Marketing." *Business Week*, 50–56. Dye, Renée (2000). "The Buzz on Buzz." *Harvard Business Review*, November–December, pp. 139–146.

25. Gladwell, Malcom (2000), *The Tipping Point: How Little Things Can Make a Big Difference*, Boston: Little, Brown, and Company.

26. Godin, Seth (1999), *Permission Marketing: Turning Strangers into Friends, and Friends into Customers*, New York: Simon & Schuster.

27. Parker, Pamela (2003), "House Lists Generate Best Email ROI," October 15, news.earthweb.com/IAR/article.php/3092211

28. Borzo, Jeanette (2002), "Dialing for Dollars," Special Report: E-Commerce, *Wall Street Journal*, April 5.

29. "The Top 100 Brands," Special Report, *Business Week*, August 4, 2003, 72–78.

30. Aaker, David A. and Robert Jacobson (2001), "The Value Relevance of Brand Attitude in High-Technology Markets," *Journal of Marketing Research*, 38 (4), November, 485–493.

31. Gunther, Marc (1998), "The Internet Is Mr. Case's Neighborhood," *Fortune*, March 30, pp. 69–80.

32. Vise, David A. (2003), "Time Warner Sheds 'AOL' From Its Name," *The Washington Post*, October 17, page E1.

33. Morris, Betsy (1996), "The Brand's the Thing," *Fortune*, March 4, pp. 73–86

34. Ibid.

35. Ibid.

36. Duffy, Jim (2003), "Cisco's loss is Juniper's gain," *Network World Optical Networking Newsletter*, February 26, www. nwfusion. com/newsletters/optical/2003/0224optical2.html.

37. "Cisco Systems Announces Agreement to Acquire The LinkSys Group, Inc," Cisco Press Release, March 20, 2003, newsroom. cisco.com/dlls/corp_032003.html

38. O'Connell, Vanessa (2003), "In Hopeful Sign for Ad Firms, Cisco Begins New Campaign," *Wall Street Journal,* February 18.

39. Keller, Kevin Lane (2003), *Strategic Brand Management,* Second Edition, Upper Saddle River, NJ: Prentice Hall.

40. Mehta, Stephanie N. (1999), "Northern Telecom Plays Down Phone Roots, Embraces 'I' Word,"*Wall Street Journal,* April 14.

41. Tam, Pui-Wing (2003), "Tech Companies Are Spending on Advertising Once Again," *Wall Street Journal,* May 15.

42. Winkler, Agnieszka (1999), "The Six Myths of Branding," *Brandweek,* September 20, p. 28.

43. Keller (2003) op. cit.

44. Winkler (1999), op. cit.

45. Keller (2003), op. cit.

46. Ibid.

47. Winkler (1999), op. cit.

48. Arnott, Nancy (1994), "Inside Intel's Marketing Coup," *Sales and Marketing Management,* February, pp. 78–81.

49. Used by Tim Smith, Red Sky Interactive, San Francisco.

50. "Branding," (1998), *Business 2.0,* November, pp. 69–84.

51. Morris, Betsy (1996), op. cit.

52. Kirkpatrick, David (1994), "Why Compaq Is Mad at Intel," *Fortune,* October 31, pp. 171–176.

53. Keller (2003), op. cit.

54. Yoder, Stephen Kreider (1995), "Computer Makers Defend 'Vaporware,'" *Wall Street Journal,* February 16, pp. B1, B6.

55. Except as noted, this section is drawn from Eliashberg, J. and T. Robertson (1988), "New Product Preannouncing Behavior: A Market Signaling Study," *Journal of Marketing Research,* 25 (August), pp. 282–292; Lilly, Bryan and Rockney Walters (1997), "Toward a Model of New Product Preannouncement Timing," *Journal of Product Innovation Management,* 14, pp. 4–20; and Calantone, Roger and Kim Schatzel (2000), "Strategic Foretelling: Communication-Based Antecedents of a Firm's Propensity to Preannounce," *Journal of Marketing,* 64 (January), pp. 17–30.

56. Evangelista, Benny (2003), "Singing a different iTunes—Apple introduces Windows version of popular music service," *San Francisco Chronicle*, October 16.

57. Reinhardt, Andy (2000), "'I've Left a Few Dead Bodies,'" *Business Week,* January 31, pp. 69–70.

58. Calantone, Roger and Kim Schatzel (2000), op. cit.

59. Additional detail about incumbent competitors' reactions to new product preannouncements can be found in Robertson, Thomas, Jehoshua Eliashberg, and Talio Rymon (1995), "New Product Announcement Signals and Incumbent Reactions," *Journal of Marketing,* 59 (July), pp. 1–15.

60. Calantone and Schatzel (2000), op. cit.

61. "Boo Who," (2000), *Business 2.0,* February, p. 105.

62. Schifrin, Matthew (1993), "No Product, No Sale," *Forbes,* June 7, pp. 50–52; Ambrosio, Johanna (1993), "Users Cooling to Oft-Delayed Iceberg," *ComputerWorld,* November 1, p. 4.

63. Yoder (1995), op. cit.

64. Ibid.

65. Lilly, Bryan and Rockney Walters (1997), "Toward a Model of New Product Preannouncement Timing," *Journal of Product Innovation Management,* 14, pp. 4–20; Calantone, Roger and Kim Schatzel (2000), op. cit.

66. Eliashberg, J. and T. Robertson (1988), op. cit.

67. Ibid.

68. Judge, Paul (1998), "Do You Know Who Your Most Profitable Customers Are?" *Business Week,* September 14.

69. Selden, Larry, Geoffrey Colvin, Larry Seloen (2003), *Angel Customers and Demon Customers: Discover Which is Which and Turbo-Charge Your Stock.*

70. Judge (1998), op. cit.

71. Barlow, Richard (1999), "Reap What You Reward," *The DMA Insider,* Fall, pp. 35–38.

72. Bishop, Susan (1999), "The Strategic Power of Saying No," *Harvard Business Review,* (November–December), pp. 50–64; and Davids, Meryl (1999), "How to Avoid the 10 Biggest Mistakes in CRM," *Journal of Business Strategy,* November–December, pp. 22–26.

73. Glazer, Rashi (1999), "Winning in Smart Markets," *Sloan Management Review,* Summer, pp. 59–69.

74. Ibid.

75. Insightexec (2003), "First year of negative growth for CRM market," *Customer Management Zone*, July 30, www. insight exec.com/cgi-bin/item.cgi?id= 115756.

76. Fryer, Bronwyn (2001), "Tom Siebel of Siebel Systems: High Tech the Old-Fashioned Way," *Harvard Business Review*, March, 119-125.

77. Ibid.

78. Ibid.

APPENDIX

Search Engine Placement[a]

UPDATED VERSION BY ANTHONY FERRINI

Acquire Marketing, Missoula, MT
*(*www.acquiremarketing.com*)*

How important are search engines to the traffic of a Web site? According to a study by IMT Strategies[b] that asked 400 consumer and business email users how they discovered new Web sites, 45.8 percent cited search engines as their top method. Word of mouth was mentioned by 20.3 percent, followed by random surfing at 19.9 percent. Banner ads (1 percent) trailed accidents (2.1 percent), newspapers (1.4 percent), and television (1.4 percent). Clearly, businesses must know how to effectively position their Web sites to appear in search engine rankings in order to gain maximum visibility.

However, search engines change rapidly. Their algorithms change. They form and dissolve partnerships and alliances daily. How can a business know how a search engine is going to rank a Web site? The first step is to understand how search engines work. The following explanation relies on an example of a specific company, snapApps.com.

How Search Engines Work

The term *search engine* is often used generically to describe both search engines and directories. They both contain a wealth of information gathered from billions of Web pages throughout the Internet. Directories and search engines differ mainly in how each compiles its database of information.

Directories

A directory such as the Open Directory www.dmoz.org *depends on people for compiling its information*. People from around the world submit their Web site URLs, such as www.myWeb site.com, to the Open Directory Add URL text box with a brief description of the content. Volunteer editors view the Web site and decide whether it is appropriate for the Open Directory, and then place it in a category. Web surfers who visit the Open Directory can either browse through the categories to find what they want or conduct a keyword search.

---

[a]Original version (2001) by Dallas Neil, MBA, University of Montana; President, Kinetic Sports Interactive.

[b]Source: CyberAtlas, IMT Strategies
http://cyberatlas.internet.com/markets/advertising/article/0,1323,5941_870711,00.html

Search Engines

A search engine such as Altavista.com *compiles its information automatically.* No human interaction takes place with the Web sites submitted. Search engines have three major elements.

1. *The Spider (also called the crawler).* The spider visits a Web page, reads it, and then follows links to other pages within the site. This is what it means when someone refers to a site being "spidered" or "crawled." The spider returns to the site on a regular basis, such as every month or two, to look for changes.

2. *The Index (also called a catalog).* Everything the spider finds goes into the second part of a search engine, the index. The index is like a giant digital book containing a copy of every Web page that the spider finds. If a Web page changes, then this book is updated with new information. Sometimes there is a time lag for new Web pages to enter the index; a Web page may have been "spidered" but not yet "indexed." Until a Web page is added to the index, it is not available to those searching with the search engine.

3. *The Software.* Search engine software is the third part of a search engine. This is the program that sifts through the millions of pages recorded in the index to find matches to a search and ranks them in an order the specific search engine deems most relevant.

Hybrid Search Engines

Many search engines maintain a directory *and* have search engine results for keywords that are not in the directory. This may seem confusing, but think of typing the term *nuclear missile* into a hybrid search engine. First, the hybrid search engine will look in its directory for the page under one of the many categories. If no page is found in the listing, then the hybrid search engine will default to the index (catalog) of a large search engine. Examples of hybrid search engines are Yahoo.com, lycos.com, and America Online.

One of the most popular default search engines is the Inktomi database. Inktomi is a technology-based company; one of its main products is the Inktomi search engine. The Inktomi database does not have a Web page that someone can search; instead it sells the use of its index to other search engines, such as Yahoo!. Inktomi licenses its search engine out to other companies that want their own search engines without having to build them from scratch. HotBot, launched in May 1996 and owned by Wired Ventures at the time, was Inktomi's first customer. Inktomi powers MSN Search and other search engines, and it provides supplementary results to Yahoo!, which acquired Inktomi in 2003.

What Spiders Look for When They Crawl a Web site

Imagine typing "digital camera" into a search engine. The search engine will check its index of pages and find the Web page with the highest relevance. How does the spider decide which page is the most relevant? This is a very important step in the success of search engine positioning.

When a spider searches a page, it generally looks for items referred to as HTML[**] tags (refer to Figures 10A-1 and 10A-2). The relationship between what the user actually sees (Figure 10A-1, the Web page) and the HTML code that creates it (Figure 10A-2, the HTML code) is explained here.

1. *Title Tag*—<**title**> Inexpensive Digital Camera Software</title>

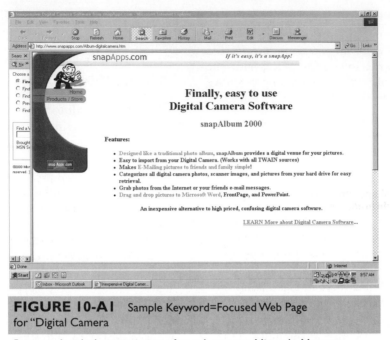

FIGURE 10-A1 Sample Keyword=Focused Web Page for "Digital Camera

Reprinted with the permission of snapApps.com, Missoula, Montana.

a. This tag is often used as the title in the search engine results. Often the text is transparent to the user viewing the page. It is highlighted in boldface in the HTML Code in Figure 10A-2 and denoted by the <title> tag.

2. *Meta Tags*—Meta tags are invisible to those who view the Web page. They are written in the HTML code right after the title and are used for providing keywords and a description of the page contents. Search engines often use meta tags to sort and rank Web pages according to the keywords entered into the HTML code.

   a. *Keyword Meta Tags*—**<META NAME = "keywords" content = >**

      These keywords are chosen to replicate what a user would enter into a search engine to find that particular site.

b. *Description Meta Tags*—**<META NAME = "description" content = >**

   Usually this description is located below the title on search engine results and offers an explanation of what the Web site contains.

3. *Link Tags*—**href=**"http://www.snapapps.com"

   a. Link tags are URL addresses to other Web sites. One of the link tags on this page is to www.snapapps.com, snapApps' homepage. Many search engines follow the links to other pages and find out how many times the keyword comes up on the corresponding Web pages. By following the hyperlinks to other pages, search engines can ensure that the Web site's content is devoted to the specific keyword that is entered into the search engine.

```
<!DOCTYPE HTML PUBLIC "-//W3C//DTD HTML 4.0 Transitional//EN">
<html>
<head>
<title>Inexpensive Digital Camera Software</title>
<META NAME = "keywords" content = "digital cameras, DIGITAL CAMERA,
cameras, digital photos, digital camera software">
<META NAME="description" content="snapAlbum 2000 works like a digital photo
Album and stores pictures from any digital camera.">
</head>
<body leftmargin=0 topmargin=0>
<map name="buttons">
<area shape="rect" coords="2,3,141,16" href="http://www.snapapps.com">
<area shape="rect" coords="2,23,141,36"
href="https://commerce.snapapps.com/index.cfm?page=store</map>
<table cellspacing="0" cellpadding="0" border="0">
<img src="images/snap_left_options2.gif" width=158 height=141 ALT="Digital
Cameras" border="0"
usemap="#buttons"><br>
<!—This is where the body text goes—>
<!—This is a comment and cannot be seen by users: note the exclamation mark—>
<h1 align="center"> This is a heading tag and carries the most weight compared to h2, h3
<br>Finally, easy to use Digital Camera Software</center></h1>
<h2 align="center"><font color="#000080">snapAlbum 2000</font></h2>
<!—More content for the page and then the final link at the bottom—>
href="https://commerce.snapapps.com/index.cfm?page=detail&amp;ProductID=16&amp
;x=45&amp;y=30" target="_top">LEARN More about snapAlbum</a> . . . </b></p>
</body>
</html>
```

FIGURE 10-A2   The HTML Code for the Web Page Shown

Reprinted with the permission of snapApps.com, Missoula, Montana.

4. *ALT Tags—ALT="Digital Cameras"*

   a. An ALT tag is the description of an image. Each image on the Web page can have one ALT tag. The ALT tag tells browser software to show specified text (ALT Tag) while the images are loading, or the text will be used as an alternative to the image if a person's browser has "Graphics turned off."

5. *Comment Tags—<!—This is a comment and cannot be seen by users—>*

   a. Denoted by an exclamation mark in the HTML code, the comment tag consists of HTML notes that the developer can see, but are not shown to the user viewing the HTML page.

6. *Heading Tags—<H1 align="center">*

   a. Web pages can have different-sized text. These are referred to as heading tags. Most search engines give a higher relevance to larger headings. H1 is the largest heading and the size decreases with H2, H3, and H4.

   Example (H1)—Refer to the Web page in Figure 10A-1 and the text "Finally, easy to use

Digital Camera Software." This can be found in the corresponding HTML code in Figure 10A-2 as **<h1 align="center"> Finally, easy to use Digital Camera Software </center></h1>**

How Search Engines Determine Ranking

Visiting the popular Google search engine today (December 2003) and conducting a search for "digital cameras" yielded 7,200,000 results. Within a matter of a few seconds, Google even sorted these millions of results and presented the Web pages that most closely matched the search. So how does a search engine determine which of the 7 million Web pages comes first?

Each search engine has a unique algorithm for ranking Web sites, but a Web site can follow a few general guidelines to generate successful results across the engines. In general, search engines look at two major factors when determining how to rank Web sites:

1. The *relevance* of the Web page content in relation to the keyword search.

2. The *link popularity* of the Web page.

Relevance

Relevance is the term used to describe how a search engine determines which Web pages match the user search. When a user conducts a search, all of the results that are returned are deemed to be relevant to the search. The results that appear in the top ten are considered to be more relevant than the others.

Each search engine uses a different scoring method for determining which pages are more relevant than others, which accounts for why different search engines rank Web pages differently. Essentially, the scoring method is determined by counting the number of times the user search term appears in the Web page, and the placement of those keywords in the page. This is called *keyword density*. If the search term appears over and over again throughout the page

content, this page will have a high keyword density for that search term.

It is important to note, however, that overwhelming a Web page with keywords is considered "spamming" the search engines. This will create a high keyword density, but the search engines will penalize ranking. Search engines identify spamming by noting if keywords have been forcefully added in inappropriate places, or if the content has lost its readability for a human user. A better approach is to craft a Web page for human readers that includes keywords where appropriate.

Link Popularity

Creating keyword-rich Web pages is a strong foundation to increase relevance in the search engines for specific searches, but another very important factor in search engine ranking is link popularity. In the scoring process, once a search engine determines which Web pages are relevant for a certain search, the ranking process is further refined by determining which pages have the highest link popularity.

In the search engine ranking process, *link popularity* is considered to be a natural voting process on the Internet. When Web site A creates a link to Web site B, search engines define that link to be a vote for site B. Beyond the sheer number of links to a Web site, search engines place greater importance on the quality of a link. If site A has high link popularity, the link from site A to site B will create additional link popularity for site B. Also, if site A is relevant to the search term, it is determined to be a more valuable link.

In our example search for "digital cameras," this is precisely how Google determined which Web pages would be presented first. Google first determined that 7,200,000 Web pages in its index were relevant for a search on "digital cameras," ranked them based on page content, and then further refined the ranking based on link popularity.

Figure 10A-3 shows the various features that each search engine crawls (or indexes) in determining the ranking of pages.

FIGURE 10-A3 Determining Ranking

Features Crawled*	Yes	No	Notes
Deep Crawl	AllTheWeb, Google, Inktomi	AltaVista, Teoma	
Frames Support	All	n/a	
robots.txt	All	n/a	
Meta Robots Tag	All	n/a	
Paid Inclusion	All but...	Google	
Full Body Text	All	n/a	Some stop words may not be indexed
Stop Words	AltaVista, Inktomi, Google	FAST	Teoma unknown
Meta Description	All provide some support, but AltaVista, AllTheWeb, and Teoma make most use of the tag		
Meta Keywords	Inktomi, Teoma	AllTheWeb, AltaVista, Google	Teoma support is ìunofficialî
ALT text	AltaVista, Google, Teoma	AllTheWeb, Inktomi	
Comments	Inktomi	Others	

Source: www.searchenginewatch.com (December 2003)

*Deep crawl: This essentially means that submitting only the domain name (www.yoursite.com) is necessary for search engines to index a Web site. A deep crawl will follow the links of a site, starting with the homepage. Engines that do not support deep crawl must have pages submitted individually.

Frames Support: Frames refer to parts of the website that remain in place when a user scrolls through the content. In the past, it was hard for search engines to index Web pages that used frames. Now, all of the major search engines now support the use of HTML frames. Using frames within a Web page will not hurt search engine placement.

Robots.txt: In some cases, certain Web pages should not be indexed by a search engine (i.e., in the case of a login page to a secure Web site). The use of robots.txt files specify a complete Web site that should not be indexed by a search engine. All of the major search engines support the use of a robots.txt file.

Meta Robots Tag: This tag specifies an individual Web page that should not be indexed.

Paid Inclusion: This specifies whether payment to the search engine will guarantee inclusion in the index. This shouldnít be confused with a paid placement, as with pay-per-click advertising (discussed later in this Appendix).

Full Body Text: All of the search engines scan the body text of a Web page and include these words in the search database. These are the words that determine how relevant page content is to a keyword search.

Stop Words: Some commonly used words (and, a, an, is, the, etc), referred to as ìstop wordsî by search engines, will be ignored.

Meta Description: The meta description is used in the head of an HTML document, and generally contains a short description of a Web page. Some search engines use the meta description in the results of a user search, others donít display the description but index the keywords. All of the engines utilize the meta description tag in some form.

Meta Keywords: The meta keywords tag is used in the head of an HTML document, and contains a list of keywords that the Web page creator deems relevant to the content.

ALT Text: The ALT text tag is utilized by some search engines. If a Web page is created in proper form, it should already include ALT text (the descriptive text that is assigned to images used on a Web page). The main purpose of ALT text is to supply an alternative method for users to visualize an image. If a user is visually impaired, a text-reader will read the ALT text and help the user understand the visual elements of a Web page. Some search engines have deemed these as relevant content, and will index these keywords in a relevance search.

Comments: Comments are included in the HTML code of a Web page, but are not included in the visual display of a Web page. Comments are used in the HTML code to make programming notes by a Web developer. Only a few search engines deem this content relevant to a keyword's search.

Search Engine Placement

Now that we have a general understanding of how search engines determine ranking, how might a Web site go about gaining a top-ten position for a keyword search?

Research Target Keywords

The first step is to determine which keywords to target for search placement. What would customers type in a search engine to find a Web site? These search terms are the target keywords. Guessing the words customers might type into a search engine is a good start. One can also research the keywords competitors use in their html code (View Source). Experts advocate researching actual searches on the Internet.

There are several methods available on the Internet to help assist in this process. One method is to type a search term into an engine that provides "related" or "refined" suggestions. Several engines provide this service to help users refine their search, and this is the same data that provides valuable marketing keywords. Teoma and Altavista currently (as of December 2003) provide this service.

Other tools offer valuable data by taking a single search term and suggesting a list of related search terms. These tools also list the number of actual user searches for that search term. One such tool is provided by a company called WordTracker (www.wordtracker.com); it offers a subscription service that provides related keywords and search counts for the major search engines.

Overture, a pay-per-click search engine that Yahoo acquired in 2003, offers a similar tool for free called the "Search Term Suggestion Tool." Table 10A-1 shows the first eight results for a search performed on "digital camera." It shows related keywords and the number of actual user searches for each term.

This type of keyword research, using actual user search data on keywords and related terms, is an important first step in a

TABLE 10-A1	Search Term Suggestion Tool*
Count	Search Term
1852438	digital camera
90048	digital camera review
47816	sony digital camera
44741	digital camera memory
37099	canon digital camera
35303	digital camera accessory
25513	olympus digital camera
22190	digital camera comparison

SOURCE: www.overture.com
*Searches done in October 2003

search engine placement campaign. Target the wrong keywords, and Web site traffic might be disappointing.

Position Keywords

Once a list of targeted keywords has been defined, the next step in a search engine placement campaign is to position keywords across the Web site. Targeted keywords must be matched appropriately with Web pages that contain related content. Including keywords with matching content will create pages that are of interest to the user, while providing increased placement on the search engines (due to increased relevancy). No more than a few targeted keywords should be positioned on each page to prevent dilution of the keyword density. Creating focused Web pages with specific keywords will generate the best results.

There are three main areas within a Web page in which positioning keywords will provide the best results: head tags, body text, and link text. Use every effort to include targeted keywords in all of these areas.

1. *Head tags:* The head tags are at the top of the HTML of a Web page and include the meta keywords, meta description, and title tag. Keep in mind that both the title and description tags will be seen by users, and these tags contain the data that search engines display

directly in their search results. So it's important to write a title and description rich in keywords, but also written in a compelling way to draw users to the Web site from search results.

2. ***Body text:*** The body text is the main content of a Web page. It is all the plain text that is found in an HTML document between the body tags. Include keywords throughout the body, with special attention to the top of the page. Placing a keyword closer to the top of a Web page, with text defined as a large heading "<H1>," will provide greater scoring in the search engines.

3. ***Link text:*** Link text is text that appears in the body of a Web page but is defined as a link—usually blue and underlined. For example, when a Web page has a link with the words "digital camera" in the link text, that page is determined to be more relevant for "digital camera" than a page that has those words only as plain text in the body.

Once the keywords have been carefully selected and incorporated into the Web site, the next step is to submit the site to the search engines.

Search Engine Submission

Submission to the search engines and directories is an essential step in a placement campaign. There are thousands of search engines on the Web, so determining which engines to submit to is the first step in the submission process. It is unnecessary to submit a site to every search engine on the Web, since the majority of searches occur on only a small list of popular engines.

Generally, submission to the engines with the greatest audience reach will yield

TABLE 10-A2   Which search engines provide the greatest audience reach?

Search Engine	%
Google	32 %
Yahoo	26 %
AOL	19 %
MSN	17 %
Ask	2 %
Overture	1 %
Altavista	1 %
Lycos	0.4 %
CNET	0.2 %
AllTheWeb	0.2 %
Others	2 %

August 2003
SOURCE: SearchEngineWatch.com, comScore/Media Metrix http://searchenginewatch.com/reports/article.php /2156431

the greatest results. Table 10A-2 shows the reach of the various search engines (August 2003).

After determining which search engines to submit to, the approach to actually submit the Web pages can vary. Each search engine or directory generally publishes a link on the homepage to "add a url" or "submit a site." Since each search engine has a unique submission process, hand-submission according to each format and set of rules is the safest way to ensure that a site will be indexed.

It is not necessary to submit a Web site over and over again to the same search engine. The proper approach includes submitting the Web site once; if the site doesn't appear in the search engine index within two to four weeks, then try submitting again.

Pay-per-Click Search Engine Submission

Another type of search engine submission is pay-per-click advertising. This service uses an auction-style model for determining placement for specific search terms. The top positions go to the highest bidders. With pay-per-click search engines, bidding on a search term (keyword) is the only method to guarantee placement in the top results of search engines.

The main advantage to using pay-per-click is that Web sites have control over what, where, and when a listing appears. The Web site is charged only when users click, or actually visit, the listed Web page. The disadvantage to using pay-per-click is that it can be expensive as a long-term solution.

Pay-per-click advertising is available on all major search engines, although the procedure to get listed varies slightly. The major facilitators for pay-per-click advertising are Overture and Google. These companies have emerged as the leaders of pay-per-click advertising by developing a network of partners that display sponsored listings in their search results. As of December 2003, these are the pay-per-click network partners for Overture and Google:

- Google partners: America Online, CompuServe, Netscape, AskJeeves, AT&T Worldnet, Earthlink, Excite.
- Overture partners: MSN, CNN, AltaVista, InfoSpace.

Submission to a pay-per-click search engine involves an account setup procedure, generally requiring a minimum deposit. Once an account is created and funds have been deposited, a Web site can bid on keywords. It can be difficult to ascertain what a top bid for a particular keyword might be. Contacting the search engine is a requisite first step. In any case, the top bidders for keywords will see their listings appear across the pay-per-click network (as in Figure 10A-4). Each time a visitor clicks the listing, the bid amount is deducted from the Web site's account.

If a Web site currently has little visibility in search engine results, needs to achieve visibility quickly, or is waiting for regular placement, pay-per-click placement is a great way to achieve immediate listings at the top of the results in the major search engines. Figures 10A-5 and 10A-6 show the placement of ads on the basis of pay-per-click search engine positioning for Google and Overture, respectively.

Monitoring Search Engine Placement and Statistics

After submitting a Web site to search engines, it's important to begin tracking which search engines have indexed the Web site. One way to track whether a Web site has been indexed is by visiting the search engine and selecting the "advanced search" feature. Most search engines have an advanced search tool that will give users the ability to select search options—including the ability to search for only a specific Web address. These results will identify if the site has been included in the index.

Another option for tracking search engine submissions is to examine more

---

FIGURE 10-A4   Viewing a bid for the search term "digital camera."

HP Digital Cameras - HPshopping.com

The official Hewlett-Packard store featuring the complete line of HP digital cameras, including point and shoot, as well as photo enthusiast digital models. Free shipping on items over $250. *www.hpshopping.com*

(Advertiser's max Bid: $1.02)

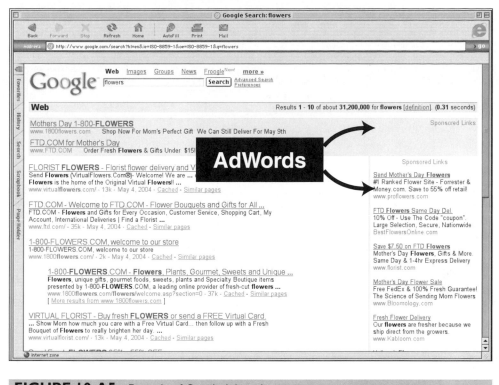

FIGURE 10-A5   Example of Google Adwords:

SOURCE: http://www.google.com/ads/

advanced statistics about the users who come from search engines. Almost all servers that host Web pages provide some sort of Web statistics to the Web sites hosted by that server. These statistics are generated by logging the activity of all visitors to a Web site, and include vital information that allows for statistical analysis. Many different services are available on the Web to provide a simple user interface that displays graphical charts of Web site statistics (e.g., Hitbox). This data can be extremely useful to determine which search engines and keywords have been most effective in driving traffic to a Web site.

Building Link Popularity

As discussed earlier, links are an important factor in how search engines determine ranking. Remember: It is the quality of the links, not the quantity, that determines ranking. Soliciting links from a long list of random sites doesn't provide a good solution. Instead, take the time to research Web sites with similar content. Search engines will reward a Web site that has links from many other sites with similar content.

A good method for researching Web sites and quality links is to simply use the search engines. Visit a major search engine and perform a search on the target keywords. The resulting Web sites are all prime candidates for quality links. This is evidenced by the fact that the search engines themselves rank these pages so well.

With a list of prime link candidates, then next step is to visit each Web site individually and request a link. Some of these Web sites might be competitors and, obviously, will not provide a link. However, there will be many Web sites with relevant

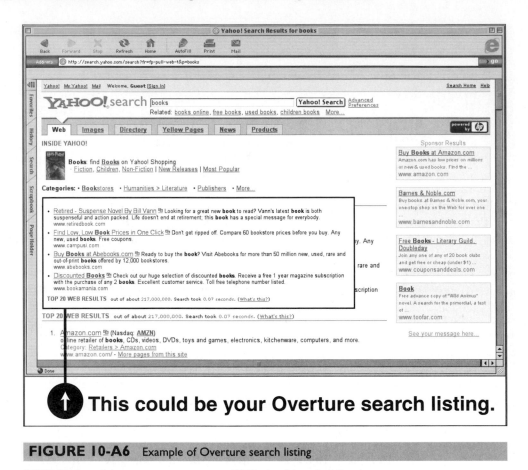

FIGURE 10-A6 Example of Overture search listing

SOURCE: http://www.content.overture.com/d/USm/ays/bjump/yah.jhtml

content that are not competitive. The most successful approach for soliciting links from these sites is by requesting a "reciprocal link." Each site will create a link to the other, a win-win solution that improves the visibility for both Web sites.

There are many other creative ways to create quality links with other Web sites. Ensure that links are included with any content or services shared with other sites on the Web. For instance, if a company writes press releases or articles for syndication, including a link back to the company's Web site can provide quality link opportunities. Similarly, affiliate programs (see Chapter 10) have become very popular on commercial Web sites. These programs provide a network of affiliate Web sites that present a prime opportunity to build links.

Limitations of Search Engine Positioning

Even though search engine positioning offers many advantages, there are some limitations. The first and foremost limitation is understanding that attracting consumers to a Web site does not necessarily mean they are going to purchase anything. The traditional tools of advertising and promotion and a convincing message are still very important. An excellent source for understanding online selling can be found at www.sitesell.com. This digital book outlines the key essentials on how to not only increase the number of visitors but also increase sales.

The second limitation of search engine positioning relates to the nature of consumer search. How does a company target a person who doesn't know the correct word to type in to find the appropriate Web site? For example, snapApps.com sells a product that organizes user names and passwords in a simple computer software application. Although it offers a useful service, this product, called snapSafe, has been very difficult to sell through search engine positioning. People don't generally search for a product to organize their passwords. Finding interested customers means using keywords that are not directly associated with the product, such as the keyword *Web surfer*. Web surfers generally access many different Web sites and therefore would need to store many user names and passwords. If on one hand, snapApp.com focuses its Web page on generating traffic from the keyword *Web surfers*, and if on the other hand, its Web page is about organizing passwords, that could be characterized as spamdexing. Search engine positioning is generally limited to consumers that are informed rather than uninformed about the product offering. Uninformed consumers need to be reached in other environments and advertising media.

Third, the whole process of (re)designing the Web site and (re)engineering the pages so that they are keyword focused is a time-consuming process. Furthermore, Web page submissions take a significant amount of time. After the pages have been submitted to an engine, the pages must be frequently checked to see if they have been indexed. One option for search engine positioning is to use the services of a third-party vendor. Outsourcing search engine positioning can save a significant amount of time and money. For example, third parties, such as Morevisibility.com or Positionsolutions.com, can usually deliver rankings in the top 10 for each keyword and are accustomed to understanding how the online environment for search engines is changing.

The potential value of high rankings on search engines makes engaging in search engine positioning difficult to ignore. However, the process to get to the top of search engine rankings is not a simple one. The various techniques used to perform effective search engine positioning were explored in detail in this Appendix to help marketers and business decision makers understand the concepts and processes associated with search engine positioning that will aid in making an educated and informed decision.

CHAPTER 11

E-Business, E-Commerce, and the Internet

In 1888, David Goodman Croly, a newspaper columnist known as "Sir Oracle," predicted that:

A "national information computer-utility system" would take form, "with tens of thousands of terminals in homes and offices 'hooked' into giant central computers providing library and information services, retailing, ordering and billing services, and the like."

(Croly's predictions appeared in an 1888 volume called Glimpses of the Future; *he asked that the predictions be evaluated in the year 2000.)*

From *The Book of Predictions*

All Business Is E-Business at GE

Thomas Alva Edison invented the light bulb around 1880 and formed the General Electric Company in 1892. In 1896, GE invented the first equipment for producing X-rays, based on the discovery by Marie and Pierre Curie. GE got into engines for the fledgling U.S. aviation industry around 1917, and its first plastics department was created in 1930. In 1962, the company built a superconducting magnet, which led to the development of magnetic resonance imaging (MRI) for medicine. From power generation to plastics, and jet engines to medical imaging, GE has come to be recognized as an innovative high-tech company.

Jack Welch became GE's CEO in 1981. He restructured GE's diverse portfolio by closing more than 200 businesses that were not the number one or two competitors in their industries. The restructuring was effective in reducing costs and bureaucracy. In the late 1980s, Welch shifted his focus to changing GE's corporate culture. He launched major corporate initiatives in the years that followed to change GE's cultural norms to develop a learning organization that excelled at setting and reaching challenging performance targets, and he incorporated cutting-edge practices to boost organizational productivity. These cultural changes transformed GE and propelled it to great heights in the closing years of the twentieth century.

Consistent with this organizational culture, in January 1999 Welch announced his first corporate e-business initiative, "Destroy Your Business.com" (dyb.com). Each business unit, under an e-business leader and its own dyb.com task force, was asked to imagine how a hypothetical "dotcom" might take away business and to devise countermeasures. By mid-1999, all business units had developed Web sites that facilitated online transactions and provided value-added services. After a year, it became clear that although e-business was not going to result in a major sales boost, it was possible that it could be extremely useful in cutting costs. So, many business units began exploring online purchasing or e-procurement as a way to cuts costs and enhance profitability. In December 1999, GE Transportation Systems demonstrated an e-auction software tool it had developed for $20,000. Within 30 days, all business units had adopted e-auctions, resulting in significant savings in purchasing costs. By March 2000, GE had a Web-based report for each business unit on the performance of "e-sell" and "e-buy." By May 2000, GE undertook the "e-make" initiative, a productivity push. The idea was to digitize and bring business processes online that had earlier required human intervention, thereby eliminating the latter and making processes efficient and effective. For example, all travel arrangements throughout GE were now done only online, resulting in tens of millions of dollars in savings.

In 2000, 5 percent of GE's sales and 25 percent of its 30,000 suppliers were online, and the e-make initiative had resulted in $1 billion of savings. With e-business now well established as a part of business operations, GE was well positioned to reap even greater rewards in the future.

SOURCES: Bartlett, Christopher A. and Meg Glinska (2003), "GE's Digital Revolution: Redefining the E in GE," Harvard Business School case #9-3-2-001, August 18;

Koprowski, Gene J. (2003), "General Electric's Tech: Past, Present and Future," *TechNewsWorld,* August 26.

By lowering transaction costs, the Internet fundamentally changes the economics of transactions, in ways that benefit both consumers and companies. Consumers gain convenience; timely access to products, service, and information; and a variety of choices not available in more traditional contexts. The Internet matches buyers and sellers online while minimizing the need for face-to-face interaction. For companies, **e-business,** or the application of Internet technologies to streamline business processes such as procurement and supply chain management, human resource management, or new product development, can squeeze out costs, minimize inventory, and shorten reaction times, resulting in greater efficiency and effectiveness.

In addition to the myriad ways in which businesses are using the Internet to streamline business processes, individual consumers also use the Internet for a variety of purposes. In 2003, 66 percent of U.S. households surfed the Web compared to 40 percent of European households. New users in the United States today are more likely to be women, older, lower income, and less experienced in technology, compared to earlier users. Thus, Internet users have begun to reflect the mainstream population in the United States. More than one-third of U.S. households shopped online in 2002.[1] **E-commerce** is the subset of e-business activities that enables and supports customers to do online transactions on sellers' Web sites.

The volume of e-commerce transactions with consumers (or *B2C*) reached $95 billion, and the volume of e-commerce transactions between businesses (or *B2B*) reached $2.4 trillion, in 2003. These were in line with or exceeded the 1999 estimates made by Forrester Research of $108 billion for the consumer market and $1.3 trillion for the business market.[2]

As shown in Table 11-1, many technologies have evolved over time to allow such widespread use of the Web. The earliest e-commerce companies, started around 1995,

TABLE 11-1 History of Development of Internet Technologies

1969	Arpanet	• A decentralized network of computers commissioned by the U.S. Department of Defense to maintain communications in the event of catastrophe. • Limitations on the amount of data traffic it could carry. • Scientists, engineers, and universities used the network to communicate about research projects. • Evolved into the global Internet.
1974	TCP	• Vince Cerf developed Transmission Control Protocol (TCP), which could transmit large amounts of traffic over long distances.
1980s	Ethernet Protocol	• Developed by XEROX PARC. • Allowed stand-alone PCs to be used for local area networking.
1982	TCP/IP	• TCP evolved into the current Internet protocol.
1990	HTML	• Tim Berners-Lee, a computer programmer at CERN, Switzerland, started work on distributing and accessing documents across servers on a network using Hyper Text Markup Language (**HTML**) and a primitive browser.
1992	Open access to the Internet	• Congress approved commercial use of the Internet, which for the first 20 years was only for scientists, engineers and administrators.
1993	Mosaic	• Mark Andreeson and his colleagues at the University of Illinois released Mosaic, a user-friendly browser that built on the work of Tim Berners-Lee.
1994	Netscape Navigator	• Netscape released the first commercial browser.

were *pure-play dotcoms,* like Amazon and eBay, which operated online with few offline assets. *Portals* like Yahoo also emerged in 1995, providing a variety of content such as news, weather, stock quotes, as well as a directory of other sites on the Web.

During this early period of the Internet, many companies published informational Web sites that provided company and product information without offering transactional capability. Oftentimes referred to as "brochure-ware," some technology enthusiasts derided this as "gratuitous digitization" (meaning that such a strategy did not fully capitalize on the potential that the Internet could offer in terms of interactive capabilities). Email was the "killer app" (short for "application") that ultimately resulted in droves of people utilizing the capabilities of the Internet. Hotmail, the first free Web-based email service, signed up 11 million subscribers in two years; Microsoft recognized the potential of this market and purchased Hotmail for $400 million in 1999.

Seeing the initial success of the dotcoms, many traditional businesses got into e-business as a defensive move (see vignette on GE on page 362), in order to counter the threat posed by these new online competitors. Different approaches to realizing revenue from online business models proliferated. The years 1995 through 2000 were a boom period for Internet-based business models, followed by a bust period until 2003.

The purpose of this chapter is to review the lessons learned from the recent Internet boom-and-bust cycle, to glean insights about what works and what doesn't work in this environment. An organizing framework for business models is presented and discussed. This is followed by a discussion of the characteristics that must be considered for effective Web site design. Next, the ways in which the Internet is used in personal (i.e., consumer) and business (organizational) contexts are discussed. Finally, some trends are reviewed that could help or hinder the ability to fully capitalize on the potential the Internet offers.

LESSONS FROM THE DOTCOM BOOM AND BUST

The monthly NASDAQ composite index, comprised primarily of technology startup stocks, had a wild ride up from about 750 in January 1995 to 5000 in March 2000. This bull market was driven by thousands of **dotcom** companies that rode the wave of initial public offerings (IPOs). But from its peak in March 2000, a bear market brought the NASDAQ down to a low of 1250 around October 2002. During these thirty months, the market capitalization of all NASDAQ companies declined about 75 percent. Similar declines were observed in stock exchanges through most of Europe and Asia. In the United States, thousands of dotcoms closed due to lack of funding or were bought out by stronger companies.

Among the many reasons underlying the dotcom crash are three key concerns: (1) the strategies and operations of the dotcoms, (2) the venture capitalists that funded them, and (3) competition. We discuss each of these in turn.

1. *Strategies and Operations of the Dotcoms.*   theglobe.com was an idea for a global online community that came to Stephan Paterno and Todd Krizelman, both 20, when they were undergraduate students at Cornell University in 1994. Their goal was to have a huge global bulletin board where members could meet online and discuss topics of interest such as cooking, gardening, or golf. Their idea was for their online business to capitalize on network externalities. By offering free

service to individual members, the company would follow a "get big fast" strategy. After signing up millions of members, the value of the Web site would support a revenue model based on reselling "eyeballs" to advertisers for a fee.

The President of Cornell introduced the two young men to Michael Egan, a businessman who had recently sold Alamo Rent-a-Car. After a five-hour lunch with Stephan and Todd, Egan decided to invest $20 million in theglobe.com in 1996. After two years of hard work developing the Web site, theglobe.com released an IPO on November 3, 1998. The stock, which was issued at $9 per share, rose to $97 per share by the end of the day, giving the co-founders, both now 24, stakes of $75 million each in the company. theglobe.com set the record for the largest first day gain in stock price in history—and this, even before the budding young company had seen its first profit.

In the months and years following the IPO, theglobe.com was not successful in attracting a critical mass of members or advertisers. Its stock price declined rapidly until May 2001 when it was delisted by NASDAQ. On August 15, 2001, the company finally had to close its doors.[3]

theglobe.com's experience is fairly typical of many dotcom failures during the boom and bust of the late 1990s. Many of the founders lacked business experience, or their Web sites lacked a compelling value proposition for users. Instead of a clear vision and long-term strategy, these startups kept experimenting with the underlying value propositions of their online businesses. Many over-spent on their customer acquisition strategies. When coupled with overly optimistic forecasts of customer retention rates and lifetime value estimates, the losses these online pioneers sustained were staggering. Moreover, because barriers to entry were low, and underlying sources of competitive advantage were easily imitated, the competitive environment was intense. Despite their hugely successful IPOs, they eventually ran short of cash and had to close down.

2. **Venture Capitalists.**   Venture capitalists (VCs) that funded the dotcoms were also responsible, in part, for the boom and bust. Many did not do adequate due diligence on these startups before financing them. Instead, they imitated their own competitors' funding decisions for fear of being shut out of lucrative market opportunities. While many VCs in Silicon Valley had experience with launching hardware and software technology products, they were unfamiliar with the consumer marketing required for many of these new Web sites. Moreover, the VCs were understaffed, taking so many dotcoms into their portfolio that they were spread too thin. Not recognizing their limitations, they also micromanaged the startups, even those that hired experienced managers. They put pressure on the startups to do IPOs even before they turned a profit. This frenzy of easy venture money available in the late 1990s was followed by a massive write-off of venture investment during the bust. Many of the well-known venture capitalists during the boom and bust continue to face legal actions, even in 2003, from investors whose money they were entrusted to manage.

3. **Competition.**   Initially, the competitors to the dotcoms were dotcoms themselves, all behaving as reactive, venture-funded startups. For example, in one category alone, the pet food category, there were nearly a dozen Web sites including pets.com, petopia.com, petsmart.com, and petstore.com. As noted previously, low

barriers to entry led to overcapacity on the Web and little meaningful differentiation among Web sites. Then came formidable competition from the traditional bricks-and-clicks businesses.

In the e-commerce arena, Barnes and Noble, Wal-Mart, and Sears started e-commerce Web sites to compete with the likes of Amazon and other dotcoms. In the e-business arena, Cisco, General Electric, Grainger, and other traditional companies adopted e-business strategies to compete with online electronic marketplaces like PlasticsNet, Verticalnet, and FreeMarkets. With their well-established brand names, strong cash positions, and bricks-and-clicks value propositions, they were able to retain existing customers and prevent many dotcoms from reaching critical mass.

The dotcom boom and bust was a speculative bubble similar to the rise and fall of railroad company stocks in the United States in the 1840s and "tulipmania" in Holland in the 1630s (in which a temporary shortage of tulip bulbs resulted in a huge run-up in prices and company stocks, only to be followed by the inevitable crash). A key lesson for technology entrepreneurs is to rely on a solid foundation of business strategy. Recall from Chapter 2 that the objective of starting a new business should be to deliver long-term value to customers and shareholders rather than get rich quick through an early sellout or IPO. There must be a demonstrable path to profitability in the business plan, and it must be executed with integrity and hard work.

There is evidence that the worst of the shakeout and consolidation during the bust is over. Of the publicly held dotcom companies that survived the shakeout, about 40 percent had become profitable by the fourth quarter of 2002. This percentage was expected to increase to 50% by the end of 2003.[4] These profitable companies have survived because:

- They capitalized on the inherent properties of the Internet, by matching their online business to the unique characteristics of the online environment.
- They had strong underlying business models.

Each of these issues is discussed in turn.

Capitalizing on the Unique Characteristics of the Online Environment

The Web has transformed many industries where consumers have migrated to the Web to perform functions previously performed in an offline environment. For example, travel industry Web sites like Expedia, Hotwire, Travelocity, Orbitz, and Priceline have helped close down some 13 percent of traditional offline travel agencies.[5] Music fans who got used to downloading and sharing files with peer-to-peer networks have caused a 31 percent revenue decline in recorded music sales from mid-2000 to mid 2003.[6] Now legitimate music download sites like iTunes and Napster are catering to consumer needs for convenience and flexibility. Direct Web sales of computers have resulted in increased operating margin for Dell, with HP and Gateway trying to catch up. While online shopping remains less than 5 percent of total U.S. retail sales, eBay is positioned to become the fifteenth largest retailer in the United States and Amazon the fortieth.[7]

On the other hand, online shopping for high-end fashion apparel does not appear to have much traction with consumers to date. By 2003, Burberry, Gucci, and Prada still did not have e-commerce Web sites. And Prada says it has no plans to provide this service to consumers. Prada tries to combine fashion with architecture and design in its offline retail stores. The retailer believes online shopping may work for basic apparel but not high fashion.[8]

As these examples suggest, there are some common features of businesses and industries that are being transformed by the Web, industries that can capitalize on the benefits the Internet can truly deliver. Primary among these benefits is a *lower cost of communication.*[9] Any industry that depends heavily on the flow of information stands to be radically altered by the Internet. The financial services industry has shown that enhanced access to information via online channels has done more than merely cannibalize offline trades; it has actually resulted in a net increase in the number of trades performed. The entertainment (movie and music) businesses that rely on digitized content of information as their products are being similarly transformed, particularly as broadband technologies are being more widely adopted. The large information component in the healthcare industry makes it a natural to shift claims processing and other aspects of medical care to the Web. Similarly, government and education are primarily in the business of information sharing and should see major shifts in how business is done.

Across a different set of industries, however, communication and information (while important) tend not to be the critical deliverables. In retailing, for example, physical goods are ultimately what people are purchasing. As such, the logistics of inventory, order fulfillment, shipping, and returns of the physical goods mean that there is a natural limit to the extent to which the Internet will transform the industry. In addition, when customers want to examine the goods prior to purchase, or when the shopping experience is part of the value proposition, online models may not be a compelling advantage.

Other industries and contexts whose value propositions go well beyond communication and information are similarly finding the Internet to be less revolutionary than evolutionary. For example, manufacturing still requires that goods be made on a production floor. And, even in the transportation industry, despite the movement to buying tickets online or tracking orders online, one still must get on a physical plane to be "transported" or have the goods delivered. In these areas, the Internet is being used behind the scenes, as an enabler of business processes. These organizational/business uses of the Web are discussed later in this chapter.

Business Models

A second characteristic determining which online businesses have staying power is based on their underlying business model. The term "business model" appeared as a buzzword during the boom years of the Internet and has persisted through the bust. What it really means is having a sound business plan for an online business venture.[10]

Framework for Business Plan

An organizing framework for all the difficult decisions that have to be made to develop an online business plan or model is presented in Figure 11-1. As presented in Chapter 2, a good business strategy must conceptualize both strategy and execution.

Strategy.   The strategic aspects must include segmentation and selection of target markets, decisions regarding how those target markets will be served in terms of a company's online efforts, and positioning of the customer value proposition.

FIGURE 11-1   Framework for Developing Online Business Strategies

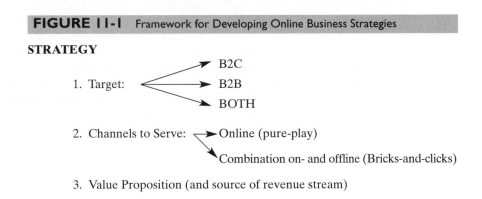

STRATEGY

1. Target:
  - B2C
  - B2B
  - BOTH

2. Channels to Serve:
  - Online (pure-play)
  - Combination on- and offline (Bricks-and-clicks)

3. Value Proposition (and source of revenue stream)

EXECUTION

1. Infrastructure:
  - Technology
  - People
  - Physical assets

2. Financial projections (path to profitability)

3. Startup financing

In its selection of *target markets,* at an aggregate level, an online venture can target consumers (B2C) or businesses (B2B) or both. As noted previously, the volume of e-commerce transactions with consumers (or B2C) reached $95 billion, and the volume of e-commerce transactions between businesses (or B2B) reached $2.4 trillion, in 2003. For instance, eBay started out targeting only consumers online, but is now also targeting businesses online through a new category, Business and Industrial. Any company's efforts must follow sound segmentation procedures, as described in Chapter 6.

After the selection of the target market to serve, a critical question is *to what extent the selected customers' needs can be met via an online offering.* Customers could be targeted exclusively online (pure play) or reached through both online and offline methods **(bricks-and-clicks)**. For example, Cisco has adopted a bricks-and-clicks model in which it serves corporate customers through its Web site, while its value-added resellers are serviced by Cisco's salesforce. As noted in Chapter 8, issues surrounding channel conflict and channel boundaries must be addressed if a company chooses to use a bricks-and-clicks model.

In making these decisions, a company must ask:

- What are the needs of this target market?
- Are there unmet needs not being served by the competition?
- Where are the greatest opportunities for serving potential customers online?

The answers to these questions assist in developing the online venture's value proposition.

After identifying target customers and the combination of on-and offline channels to serve them, the online venture must *address the value proposition* it offers to those customers and, correspondingly, the source of revenue from customers (or other partners) to support the costs of running the Web site. Many Web sites provide a value-adding service to users at no cost to the user. For instance, Yahoo started out providing content for free. In contrast, AOL has always charged monthly subscription fees for Internet access and content. To assist with this step in developing the online business plan, basic questions to ask include:

- How will this Web site address the needs of the target market?
- What are the specifics of our value offering?

If one purpose of the online venture is to generate revenue, many possibilities exist, including:

- Charging membership subscriptions or usage fees for content or services.
- Bringing in advertising revenue.
- Earning a margin or a percent of the transaction for e-commerce.
- Getting referral fees for generating traffic to other sites.

For example, Google is a Web search service based (in 2003) on an advertising revenue model. It is a fast, relevant search engine for users. But it accepts advertising only for paid keyword searches.

Once the basic strategic elements have been decided, the online venture must then move to its plan to execute, or implement, those decisions.

Execution.   In terms of execution, the business plan or model needs to discuss the supporting infrastructure, projected financial results (including estimates of fixed and variable costs of operation), and financing.

Infrastructure includes technology (e.g., hardware and software, site design), people required to run operations, and physical assets including office space, warehouses, and vehicles.

It is interesting to note that many companies believe that if they decide to add an Internet channel to their existing distribution channels, they will gain a larger portion of the margin they might otherwise give up to their channel members. In reality, however, the costs of supporting infrastructure, including the necessary hardware and software to enable e-commerce, the need for order fulfillment of individual orders, a mechanism to handle returns and customer inquiries, and other activities, can quickly eat up the margin. It is important to be realistic in estimating the costs of all the requisite infrastructure and activities.

Second, in terms of execution, the *projected financial results* must demonstrate a realistic path to profitability. As noted previously, this requires estimates regarding the costs of customer acquisition, estimates of customer loyalty and repurchase rates, returns, and other costs. Amazon's early experience demonstrates that it is easy to be overly optimistic in these estimates.

Finally, the plan must address *how the online venture will finance* itself initially (prior to reaching breakeven), either by bootstrapping cash flow from internal operations, requesting monies from a corporate parent, or seeking external funding from angel investors or venture capitalists.

By having a complete definition of all these various elements of the online business model, the odds of success are more likely. Importantly, each of these elements must "hang together" in a cohesive fashion. Within this broad framework of the necessary elements for an online business plan or strategy, there are some distinct differences between different types of online business models, as explored in the next section.

Types of Online Business Models

The previous components of an online business model must be addressed by all types of online businesses. Despite these commonalities, there are several distinct business strategies to achieve online profitability, and these distinct models have been given identifiable labels in practice.

1. **Portals** are information Web sites that are users' browser gateways into the World Wide Web. Content is usually given away free to attract "eyeballs," which are then sold to advertisers. They also direct users to other useful Web sites. Yahoo and AOL are examples of consumer portals.

2. **"Market makers"** are intermediaries that bring buyers and sellers together to facilitate transactions. They also go by other names such as vertical markets, electronic marketplaces, Net markets, or online exchanges (recall Chapter 8's discussion of online vertical markets). These online businesses have higher domain specific knowledge than portals about products, services, or transaction format. Since they don't take title or physically handle the goods, they are also called infomediaries.

   EBay is an example of an **infomediary** that facilitates auctions for consumers as well as businesses. For example, the Business and Industrial category on eBay—featuring equipment and supplies for restaurants, printing, construction, and healthcare, among other things—is one of the newest and fastest growing categories. Liquidators buy excess inventory offline and sell it on eBay. Buyers are mainly small and medium businesses with less than 100 employees who have less bureaucratic procurement processes.[11] In the auto industry, online auctions of used cars are big business among dealers through Web sites like AutoTradeCenter. Even General Motors has entered this business through its SmartAuction Web site for sale of its off-lease cars. GM sells about 300,000 cars, or 40 percent of its off-lease fleet through SmartAuction.[12] Auctions are a popular B2B e-commerce format because they permit sellers to convert assets into liquidity while providing good value to buyers. In addition, to the extent that these auction sites amass a large number of buyers, they are able to maximize the likelihood that the seller will realize maximum value for their goods.

3. Product/service providers can use the Internet to engage in direct transactions/ exchanges with their customers.[13] These include:
   a. **seller storefronts,** from companies like Dell or Cisco,which enable online transactions and let customers make online purchases directly on the Web site
   b. **e-tailers** like Amazon that resell other companies' goods online
   c. **e-procurement**, where a large buyer such as General Electric invites suppliers to bid for its business on its own private Web site (often called a virtual private

network or exchange) (see vignette on page 362). GE's Global Exchange Services started out as its e-procurement Web site. (However, it was sold to an independent company, and now operates like a business market maker.)

In an influential article, Michael Porter decried the use of the term "business models" instead of the terms "strategy" and "competitive advantage."[14] Clearly, a business model must include more than just the mechanism by which revenues will be generated; an online business venture must be able to create sustainable economic value, defined as true price minus true cost.

Importantly, Porter argues that deployment of Internet technology by the players in an industry inherently reduces average profitability because the Internet provides better information to buyers and sellers, reduces transaction costs, and levels the playing field for all who use the technology. Therefore, the only way to create sustainable economic value in an industry with reduced average profitability is either to lower costs through operational efficiencies or to charge premium prices by competing in a distinctive way. Dell, GE, and Wal-Mart provide examples of companies leveraging e-business technologies to gain operational efficiencies. EBay differentiates itself through product assortment, user experience, and community. Thus, the strategic principles used to create competitive advantage for online business ventures are the same as those that apply to regular business.

Regardless of what type of business model is used, all Web sites must be crafted with any eye to enhancing the user experience. The next section of the chapter explores best practices in the development of Web sites.

EFFECTIVE WEB SITE DESIGN AND MANAGEMENT

Web Site Design

What makes an effective Web site? As shown in Figure 11-2, pure-play and bricks-and-clicks Web sites need to provide an excellent user experience through a well-designed online customer interface, consisting of context, content, community, connection, customization, communication, and commerce.[15]

Context refers to a Web site's design and layout, sometimes termed "look and feel." Some sites are more aesthetic, intended to create a mood or image with lots of graphics, sound, or video. Others are functional, largely text-based, like Google. The context of a Web site must be designed with an eye to the target customers and the value proposition.

Content includes all the digital material on a Web site including information, products, and services. Content pertains to what is presented while context is about how it is presented. It is imperative that a Web site include the relevant content a user will need, in a design and layout that is user-friendly and easily navigable.

Community refers to the ability of users on a Web site to interact with each other, build bonds, and share common interests or values. The community itself is made up of the people who participate in the activities of the site on a regular basis and are willing to share information with those who share their interests. Community is a useful tool for a Web site to build relationships with and among its users. EBay has a vibrant community of sellers who help each other with tips and

FIGURE 11-2   Elements of Effective Web Design*

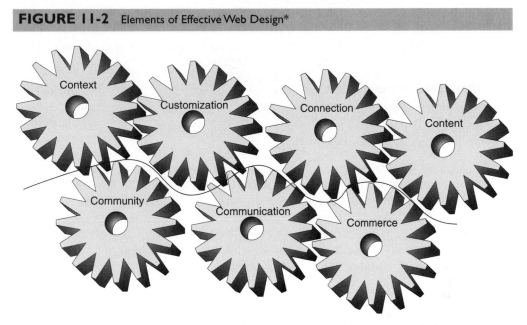

*All elements have reinforcement (consistency) and fit (meet needs of the target market)

information and provide useful feedback on company policies. Amazon has an active community of book, music, and other product reviewers.

Customization is the ability of a Web site to present flexible information tailored to different users or to be modified by users to suit their unique needs. For example, MyYahoo allows users to choose what types of information they would like to receive when they log into this portal. Customization is an important feature that delivers personalized value to individual visitors.

Communication refers to the means by which the Web site permits interaction between itself and its users—e.g., live chat, email, or telephone. This communication could be one-way (the company or Web site to the user) or two-way (company or Web site to the user, and the user back to the company or Web site); studies show that two-way communication contributes to a more satisfying user experience.

Connection refers to how well the Web site is linked with other useful Web sites. If a user cannot find what is needed on a Web site, it adds value to the user experience to be referred to a Web site where the needs can be met. The desire to build connection explains why Amazon has integrated a Google search box into its Web site. Building connections also presents opportunities to earn revenue from referrals to affiliated Web sites.

Commerce refers to a Web site's capability to execute commercial transactions between buyers and sellers. It requires features such as a shopping basket, invoicing, secure credit card processing, and order acknowledgment.

Rayport and Jaworksi highlight the importance of reinforcement and fit in designing effective Web sites.[16] *Reinforcement* refers to the degree of consistency between the seven characteristics discussed above. If a Web site wants to build a vibrant community but does not provide easy two-way communication tools, it would be low on reinforcement. *Fit* refers to the match between the Web site design and the needs of the target market. For example, Yahoo's Web site has an attractive look and feel, as well as useful content to draw users and keep them; therefore, it is high on fit.

Web Site Management

After the Web site itself is designed, it must be effectively managed to leverage its potential. There are two key aspects to doing so: building site traffic and managing online, customer relationships.

Build Site Traffic

In terms of building site traffic, it is important to promote the Web site through all of the various tools discussed in Chapter 10. Importantly, many Web sites have followed a "get big fast" strategy, designed to focus on customer acquisition as the key to success. It is noteworthy, however, that most companies don't consider when such a strategy makes sense. There are three conditions supporting a "get big fast" strategy (i.e., grow the number of visitors as quickly as possible):[17]

- *Network effects:* Recall that network effects exist when adoption by a customer increases the value of that product for other customers already using it, because existing customers can connect with additional people. Network effects are minimal for e-tailers (most shoppers do not communicate with each other). Although chats and customer recommendations do benefit from network externalities, these features are only of secondary importance. Portals, on the other hand, as well as market makers, can see benefits in the area of aggregating buyers and sellers, online chats, communities, etc.
- *Scale economies:* When a company can spread its fixed costs over a growing revenue base, it makes sense to get big fast. Some Internet companies have only fixed costs (i.e., portals), and so, operating margins can increases dramatically with growth. However, online retailers have a large share of variable costs (i.e., cost of goods sold), and so are less likely to benefits from scale economies.
- *Retention rates:* The real payoff from get-big-fast customer acquisition strategies are found in customer lock-in. If a first-mover can cement its advantage, then a get-big-fast strategy can yield benefits. Retention rates tend to be high in the presence of network effects, solid competitive position, and high customer switching costs. Note, however, that retention rates are quite fickle for e-tailers. Online e-tailers say they can build loyalty based on customer service or knowledge of customer preferences. However, this is highly variable by segment.

As this third point (customer retention) suggests, management of ongoing customer relationships can be key to online success.

Manage Online Customer Relationships

Online ventures must manage the customer experience so that customers have an incentive to return. Conceptualizing, implementing, and successfully managing any online venture requires an intimate understanding of customer behavior, and how the Web will deliver compelling value to those customers. The next two sections of this chapter examine, in turn, the online behavior of individual consumers, and the online behavior of organizational or business customers.

CONSUMER BEHAVIOR AND THE INTERNET

When the Internet was introduced for commercial purposes, early adopters were curious about the content available on the World Wide Web. They would spend time exploring content on their favorite Web sites and also surf or browse new Web sites. As noted previously, email was the first killer application and people started using it in large numbers with the availability of free email services from Hotmail and Yahoo.

Individuals use the Internet for a variety of purposes: online shopping (e-commerce), communication and information, online communities to support their hobbies, interests, and lifestyles. In fact, search of all kinds, product or content, has become the most performed task on the Internet, after email and instant messaging.[18] Not all of a consumer's uses of the Internet are inherently positive. Examples of negative use include identity theft, pornography, and gambling.

Thousands of B2C e-commerce Web sites proliferated during the boom, competing for consumers' attention. Well-funded startups engaged in national TV advertising to build their brands. The marketing costs to acquire new users or consumers escalated.[19] Unfortunately, consumers did not embrace online shopping as anticipated. The grocery industry is an interesting case study of how consumer behavior affects the success of online business ventures.

Peapod vs. WebVan

Peapod was the first online grocery delivery service on the World Wide Web. Peapod was merely the online Web site on which customers could place orders for groceries; however, they did not carry any inventory themselves. Rather, they partnered with regional supermarkets like Andronico's in the San Francisco Bay Area to fill customer orders. More specifically, orders placed on the Peapod Web site were sent electronically to the physical store of the retail partner nearest to the customer's address. Store personnel filled the order and delivered it in their trucks to the customer. The electronic information system was so primitive that there was no feedback from the store to Peapod about out-of-stock items. Therefore, orders were often only partially filled and customer satisfaction suffered.[20]

In contrast, Louis Borders, the well-known retailer who started Borders Books, wanted to provide consumers a better online grocery shopping experience than Peapod. So he founded a well-funded startup, WebVan, in 1999. WebVan had its own automated warehouses well stocked with inventory, one for each large, metropolitan market. The warehouses were so automated with conveyor belts that no employee had to move in any direction more than 19 feet (6 meters) to handle the merchandise. WebVan also invested in its own fleet of trucks and hired its own drivers instead of relying on any partner. When consumers placed orders on WebVan, they had to choose a 30-minute window for scheduled delivery in advance. WebVan's drivers were brand ambassadors for the company, well trained in customer service. They would offer to load groceries into the consumer's refrigerator at the time of delivery. Delivery service was free to customers who ordered more than $50 in groceries. WebVan also had blue-chip top management, headed by CEO George Shaheen, formerly the worldwide head of Andersen Consulting.[21]

WebVan managed to generate a small but loyal customer base after its 1999 high-profile launch supported by TV advertising. Despite this, it ran short of cash and had to shut down in July 2001. Many analysts attribute WebVan's failure to the high fixed cost of its infrastructure. It is estimated that each warehouse cost about $100 million to build.[22] Further, it did not achieve the critical mass of consumers needed to break even on their huge fixed costs. The consumer households that WebVan targeted were affluent, busy people who did not have time to shop, or did not enjoy shopping, in stores. Yet the system demanded the discipline of advance planning from these people who were required to go to the Web site, place their order, and then schedule a delivery time in advance when they would be home.

Inhibitors of Consumer Use of the Web

The failure of WebVan highlights how difficult it is for Web sites to change the behavior of consumers online when they are used to shopping in a different way offline. These and other factors either can inhibit or facilitate consumers' use of the Internet as shown in Table 11-2.

One of the obstacles to migrating consumer behaviors to the online world is *access to the technology*. This issue is sometimes framed as the **digital divide**, or the disparity in access between technology "haves" and "have-nots." It has been estimated, for example, that the average Internet user in Africa has access to 20 times less bandwidth (or the capability to receive data online) than the average European user and 8.4 times less than the average North American user.[23] Even within the United States, only 30 percent of Internet users have access to broadband access through DSL or cable modem compared to 70 percent in South Korea, primarily due to differences in pricing and government policy.[24] Although some predict that as the price of technology declines over time, and as new technological solutions such as wireless and satellite are rolled out, the digital divide will narrow, it is an issue that has important social and economic implications (see Chapter 12).

TABLE 11-2   Inhibitors and Facilitators of Consumer Use of the Web

Inhibitors

Not compatible with consumer behavior
Access to technology
Spam
Viruses, hackers, and fraud
Concerns about privacy

Facilitators

Empowerment (via access to information and transaction cost efficiencies)
Bricks-and-Clicks channels
Capability to join online communities
Peer-to-peer computing
Broadband technologies
Wireless technologies

Another obstacle to the consumers' embrace of the Internet is **spam,** or unsolicited email. Spam accounted for about 50 percent of email sent in 2003 and was estimated to go up to 80 percent in 2004. Because spam is often offensive or fraudulent, consumers are becoming hostile to unsolicited email and becoming less trusting of the Web in general.[25]

Other obstacles to consumer usage of the Internet include the vulnerabilities to **viruses, hackers,** and fraud. In a survey of 39,000 Internet users, nearly one-third were pessimistic about Internet security. It is estimated that by 2006, 25 percent of all identity theft will be from information obtained online, up from 5 percent in 1998.[26] Both spam and fraud are being fought with technology, regulation, and the concerted efforts of the major industry players. These efforts will intensify in the future.

A final obstacle to wider consumer usage of the Internet is consumers' concern over *privacy*. When an individual surfs the Web, most Web sites leave a small bit of information on the hard drive of that individual's computer (a "*cookie*"). Then, the next Web site the individual visits accesses the cookies stored on the hard drive in order to compile a profile of that individual's surfing habits and interests. When coupled with any information volunteered by the individual at various Web sites (such as address, gender, activities, interests, or opinions), marketers have a wealth of data to better target that individual's interests. Indeed, online technologies allow marketers to personalize the ads that are delivered (based on access to cookies). The ostensible benefit is that offers and ads are better targeted to the individual surfer's interests. A dark side, however, is that companies often don't limit how they use customer information, transferring it to partners without telling their customers. Such practices ultimately erode consumer trust. Companies must follow the best practices of opt-in permission marketing (see Chapter 10) to maintain consumer trust.

While required changes in behavior and privacy and other concerns can act as constraints on consumer adoption of online business offerings, other factors can facilitate consumers' online experiences.

Facilitators of Consumer Use of the Web

A positive force facilitating adoption is the *empowerment* of consumers. As noted previously, due to enhanced efficiencies in transaction costs, the Internet changes the dynamic between buyer and seller. Importantly, through enhanced access to information, the ability to have a greater selection of vendors beyond one's immediate geographic location, and anytime/anywhere ("24/7") shopping, the Internet can empower customers, relative to the companies from whom they purchase. This empowerment, when coupled with transaction cost efficiencies from bringing together a large number of buyers and sellers, is also one of the driving factors in the growth of consumer auction sites such as eBay.

Also, technological tools available to consumers on the Internet can provide more and better information about price and availability. For example, **shopping bots** are Web sites that surf the Web, looking for pricing information for a specific product (www.botspot.com). With the availability of Web sites such as MySimon.com, Yahoo Shopping (http://shopping.yahoo.com), Froogle (http://froogle.google.com) and Shopping.com, consumers can find information on product availability and prices from a plethora of e-tailers and bricks-and-clicks stores by visiting just one Web site.

Consumer acceptance of the Web is also getting a boost with the appearance of multiple offline and online channels from the same retailer, also known as *bricks-and-clicks* ("harmonized" or integrated) channels. Many consumers feel comfortable doing research on the Web but are uncomfortable providing their credit card information online. Some consumers like the convenience of ordering and paying on the Web, but they want to avoid paying shipping and handling charges. When a deadline looms, it may be faster to order online and pick up offline rather than wait for home delivery. Consumers may also want to physically examine the goods before purchase. For all these reasons, some online shoppers prefer to pick up their purchases offline. The offline channel also provides a convenient way for consumers to exchange or return merchandise they are not satisfied with. At stores like REI, Sears, and Circuit City, the percentage of online shoppers opting for in-store pick-ups is reported to be 33, 40, and 50 percent respectively.[27] Further, Sears has found from customer surveys that one in ten major appliance purchases done in its offline stores, or sales of about $500 million per year, is influenced by research done on Sears.com.[28] These integrated, or harmonized, channels provide a seamless shopping experience for consumers.

Another factor contributing to enhanced consumer use of the Internet is the capability to join communities of interest with individuals who share common interests. Whether used for hobbies, health conditions, online dating, family genealogy, or the plethora of other human interests, the capabilities to connect like-minded individuals is one of the key enabling factors of the Internet.

Finally, the rise in peer-to-peer computing that facilitates file sharing between individuals, faster transmission speeds (via access to broadband technologies, such as digital subscriber lines—or DSL—and cable modem), and wireless surfing (via Wi-Fi networks) will also continue to transform the ways in which consumers use the Web. This chapter's Technology Tidbit examines the use of the Web in predicting trends; it features a new Internet phenomenon, blogging.

E-BUSINESS AND ORGANIZATIONAL (BUSINESS) BEHAVIOR

As shown in Figure 11-3, this section of the chapter explores how business organizations and industries have modified their strategies and processes in capitalizing on the capabilities of the Internet.

Changing Distribution Channels

The earliest benefit of the Internet was to let businesses maintain direct contact with their customers without the need to rely on channel intermediaries, a phenomenon called **disintermediation**. Disintermediation potentially allows companies to reach buyers faster and cheaper. This has happened most extensively to date in the travel industry, in which airlines continue to move toward bypassing travel agents entirely and allowing customers to buy their tickets directly from the airlines' Web sites. So the Web challenges intermediaries to consider carefully how they add value to the customer. Some intermediaries are trying to reinvent themselves. Traditional intermediaries are consolidating or offering improved services of their own. For example, Grainger, a traditional distributor of maintenance, repair, and operating (MRO) supplies with 370 branch offices, offers its

A TREND AWAY FROM TRENDS?

A major social phenomenon of our times is the preoccupation with trying to predict all kinds of trends. From stock prices and technology to political campaigns and the economy, those who dare to make predictions (right or wrong) are accorded the status of experts, interviewed by the media and paid big bucks. Now comes the Internet, the great social leveler. Web sites like Google Zeitgest (www.google.com/press/zeitgeist.html), JeevesIQ (sp.ask.com/docs/about/jeevesiq.html), Yahoo! Buzz Index (buzz.yahoo.com/weekly/), and Lycos 50 (50.lycos.com/) make news, analysis, and trends available to the average Joe or Jane. These trend-predicting Web sites are based on the assumption that Internet web searches correlate to the things that people are interested in and, therefore, reflect trends in society. For example, the *Atkins Diet* was among the top 50 searched items on the search engine Lycos for 50 straight weeks, making it a hot trend in 2003.

Special search engines, like Daypop (www.daypop.com), index weblogs on a daily basis. Weblogs or *blogs* are personal journalism Web sites where individuals publish their opinions or daily happenings in their lives. As such, blogs provide a peek at the Web's collective consciousness. For example, the Daypop Top 40 identifies the most popular Web sites that Web loggers are linking to, while Word Burst identifies the most used words in weblogs in the past couple of days. Technorati (www. technorati.com) is another index that keeps track of who's linking to what URLs in the weblog community. Thus, weblog indexes like Daypop and Technorati can also be used to find out what the leading thinkers in particular fields are talking about.

So what happens when Internet technology makes available all this information on trends? Everyone becomes an expert. To quote Lee Gomes of *The Wall Street Journal:*

> Gone will be the hateful distinctions that have divided people for generations—like "trend setter" and "trend follower." We'll all be together online like a giant school of fish, everyone moving in the same direction, everyone changing course simultaneously.

entire catalog of 220,000 items on its Web site; it also offers FindMRO.com, a service that helps customers find items that Grainger does not carry.[29]

In the area of logistics and distribution, a new type of intermediary—called the **online category manager**—has emerged. These are offline wholesalers that have expanded to offer services to online retailers in specific product categories. Take the case of CircuitCity. For DVDs, CDs, and videos, typically 500 to 3,000 titles appear in their offline stores. However, on CircuitCity.com, Web shoppers expect 55,000 titles including older, slow-moving items. So CircuitCity negotiated a deal with Alliance Entertainment Corporation. Alliance, an entertainment products wholesaler based in Coral Springs, Florida, maintains 300,000 square feet of inventory. It has a Web site customized for CircuitCity for these products. CircuitCity controls the retail prices on this site. When a

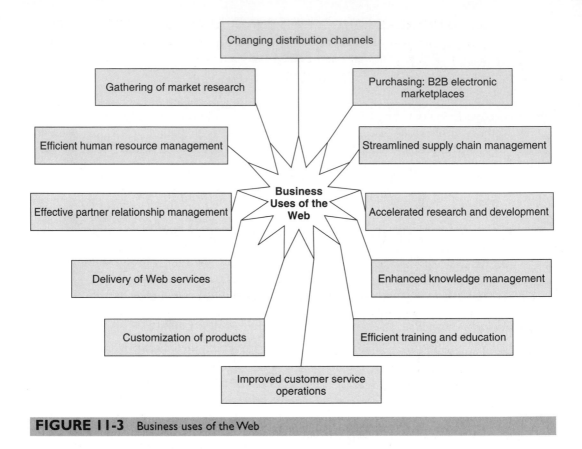

FIGURE 11-3   Business uses of the Web

customer places an order, Alliance processes the order, ships the goods to the customer's address in packaging that has only CircuitCity's brand on it, and bills Circuit City only upon shipment.[30] For online retailers like CircuitCity, the benefits of using an online category manager include minimum inventory investment, ownership of the customer relationship, and efficient outsourcing of the order processing and fulfillment functions to the online category manager. For the customer, the arrangement delivers a seamless Web experience. The online category manager gets a one-time fee for integration of the e-commerce Web site plus additional revenue for every product shipped. Amazon has a similar arrangement with Ingram Book Group for books and Kmart with Global Sports Inc. for sporting goods. These examples demonstrate the win–win nature of how traditional wholesalers can benefit by embracing e-business as online category managers.

As a third way in which distribution channels are affected by the Internet, the Web has created a whole new class of intermediaries, in a phenomenon called **reintermediation**. Because the information explosion on the Web has resulted in an unwieldy and inefficient marketplace, these new online intermediaries or infomediaries are needed as facilitators, functioning as brokers, helping people gather and decipher the vast quantity of information on the Web, bringing together buyers and sellers, and providing value by offering trusted advice, personal service, or other benefits. This use of the Internet to broker information in online marketplaces has transformed the purchasing function in many businesses.

New Purchasing Approach: Utilization of B2B Electronic Marketplaces

Business-to-business online intermediaries, also known Net markets, B2B hubs, e-marketplaces, or online exchanges, have an interesting history and have evolved in terms of some of the following dimensions:

- Affiliation
- Industry focus
- Private versus public
- Business model

Affiliation refers to whether the online exchange is independent or allied with either buyers or sellers on the exchange. **Independent exchanges** were the first versions of online B2B marketplaces, affiliated with neither the buyer nor the seller.

Industry focus refers to whether the marketplace serves the needs of a specific industry or offers goods and services across multiple industries.

- *Vertical,* or industry-specific, hubs, such as PlasticsNet.com, offer industry-specific solutions.
- *Horizontal,* or functional, hubs supply goods and services across multiple industries such as MRO.com, iMark.com (for used capital equipment), and Employease.com.

Both types bring buyers and sellers together and facilitate exchange without taking title to the goods. Early revenue models tended to be based on advertising or listing fees and a percent of the transaction. However, most of these independent exchanges did not reach critical mass and failed for a number of reasons. Sellers were reluctant to surrender margin to these infomediaries. They were also concerned that their product specifications and prices would be transparent to their competitors. Buyers were unaware of, or unwilling to trust, new, untested intermediaries. These independent exchanges tended to lack domain-specific knowledge and did not have the offline relationships to motivate enough buyers and sellers to join. Eventually, these exchanges either closed down, got acquired, or morphed into suppliers of software for exchanges.

The second iteration of B2B exchanges gave rise to those sponsored by the leading players in the industry, called **consortia,** or coalitions. Buyer-initiated consortia, like Covisint for auto parts procurement, were initially sponsored by Ford, General Motors, and Daimler-Chrysler. Another example in consumer-packaged goods is Transora, initially sponsored by Coca-Cola, Procter and Gamble, Colgate Palmolive, and others. These consortia have the power to get suppliers to participate in their Web sites to lower the buying companies' procurement costs. They also have domain-specific knowledge and industry relationships. But they have not been very successful in attracting a large number of buyers beyond the original sponsors. This is because prospective buyers are suspicious that the sponsors will use their clout to get preferential treatment from suppliers on these sites. Since buyers are all competitors in the industry, the lack of trust among them acts as a barrier to membership in these consortia.

One of the most common formats for these B2B marketplaces is a *reverse auction*. Unlike a standard auction, in which *buyers* bid sequentially higher prices in order to win the auction, reverse auctions have *sellers* bid sequentially lower prices in order to "win" the right to supply the requested good to the buyer (lowest bid wins). Freemarkets.com is an example of such a B2B market where buyers put their purchase specifications up for bid and sellers compete for the business. This model favors buyers, who get lower prices on the supplies they must buy. Studies show that reverse auctions are effective in procuring commodity items. However, a critical concern is the *long-term impact on buyer-supplier relationships*. To what extent will the use of reverse auctions undermine such relationships? Will this squeeze vendors to the breaking point? The long-term impact on suppliers remains to be seen.[31]

Because of concerns such as these, a third—and in late 2003, the most popular—iteration of the B2B hubs is the *private exchange* (or virtual private network). In this version, a large buyer invites a trusted group of suppliers and partners to collaborate on a secure **extranet**. Extranets allow suppliers, distributors, and customers the ability to view company data on a restricted, secure basis. Unlike public B2B exchanges, there is no sharing of sensitive information with unknown parties, and buyers retain control of their supply chain. For example, Ace Hardware has 5,100 retail stores in the United States supplied by 14 Ace distribution centers and 9 suppliers' distribution centers. Using software from a company called E3 in Atlanta (now part of JDA Software), Ace integrated its own distribution centers with those of its suppliers so that Ace buyers and their suppliers could look at the same data on sales, inventory, and forecasts. This led to more streamlined order planning and is estimated to have reduced suppliers' distribution costs by 28 percent and their transportation costs by 18 percent.[32]

As another example, in August 2002, Wal-Mart told its suppliers they would have to start sending and receiving electronic data over the Internet. By September 2003, more than 98 percent of Wal-Mart's exchanges with suppliers were being done over the Internet using AS2, a software package from iSoft in Dallas. AS2 is an Internet EDI (electronic data interchange) system, much cheaper and easier to implement than the 1960s-era, proprietary EDI. Coty Inc., a cosmetics and fragrance manufacturer that depends on Wal-Mart for 30 to 40 percent of its sales, installed an AS2 system for about $22,000. In six months it almost paid for itself because Coty's long distance bill was reduced by approximately $1,000 to $5,000 per month.

Predictions are that private exchanges will continue to see increased adoption rates in B2B e-commerce because of their potential to realize supply chain efficiency.[33]

Streamlined Supply Chain Management

Many companies are using the Internet as a way to manage their supply chain, or the flow of products from suppliers to the factory and, ultimately, to channel intermediaries and customers. This allows a company to track sales on a timely basis, get instant feedback from customers, keep inventories to a minimum, and simplify complex, costly transactions (see Chapter 8 for additional discussion). The Internet is making huge changes in the supply chain, facilitated by powerful software tools that automate the flow of information. These software programs sharpen a company's view of parts and products moving through the supply chain so it can better grasp what inventory is available and how demand may be shifting. Companies like GE (see vignette on page 362), Ace Hardware, and Coty (see previous discussion on extranets) are cutting costs and generating savings by implementing supply chain e-business initiatives.

This chapter's technology expert speaks to some of these initiatives.

TECHNOLOGY EXPERT'S VIEW FROM THE TRENCHES

DEVELOPING AN E-BUSINESS STRATEGY

CHRIS COUTURE

Former Vice President & Chief Information Officer
Dell Financial Services, Round Rock, TX
Current President, Round Rock Advisory
Services, Round Rock, TX

In developing an e-business strategy, companies must first define their strategic objective for creating an Internet presence. In today's highly competitive business world, where direct interactions with the customer are rewarded and middlemen find it harder and harder to define their value proposition, "the Internet could be considered the ultimate extension of a direct business approach." [a] It is the ability to leverage the Internet in order to conduct business directly with customers and suppliers that will separate the average company from the truly exceptional company.

In order to implement an e-business strategy, a company may choose one of three primary courses of action:

- Incremental changes in its current business model
- Replacement of its current business model
- Innovation/incubation of a completely new business model

Those companies who choose an *incremental strategy* will take the best parts of their current business and simply "Internet enable" them in order to take advantage of new opportunities. This is the lowest risk approach to developing an e-business and allows a company to leverage its existing strengths in order to take advantage of new Internet opportunities. The downside is that it may take companies a long time to migrate their traditional business to the Web because they are forced to migrate old business rules as well. Additionally, companies may have a business model that does not necessarily lend itself to transition to an e-business. These two issues offset the low risk of an incremental business strategy with potential low rewards.

With a *replacement strategy,* a company decides to throw away its current business model in favor of a completely new one. This strategy is extremely risky and can be presumed to fail in all but a very few circumstances. Companies that abandon their existing business structures may also abandon their existing customers. Customers have different speeds at which they are willing and able to adjust to a new way of doing business. A complete replacement strategy, without a migration path for existing customers, will open up a huge window of opportunity for the competition to gain market share.

The last e-business strategy is referred to as *innovation or incubation*. This occurs when a completely new Internet-enabled business opportunity is discovered, and a company allows a team of innovators to develop and pursue the opportunity without being encumbered by existing business rules and thought processes. This methodology allows businesses to develop breakthrough products and services that can take advantage of all that the Internet has to offer, while still maintaining the existing business structure in order to serve the current set of customers. A company that follows this methodology can attract new customers to its new e-business model, while giving current customers a migration path from the traditional model to the new.

One of the great mistakes companies have made in developing an Internet strategy was to assume that a new sales channel was the same thing as a new business model. The best examples of this are the multitudes of Internet retailers that were high fliers during the late 1990s, only to go out of business after the crash in 2000. Many startup companies were launched, and billions of dollars were lost, in trying to provide customers a way to shop for goods and services online without paying attention to the "old-fashioned" parts of running a business like inventory management and logistics. Traditional businesses that have implemented an incremental strategy to incorporate e-commerce into their model are the ones that have been successful. The lesson learned is that just having an Internet sales channel does not lead to a profitable e-commerce business. There are a handful of e-commerce industries truly born of innovation (online auction, travel, and digital entertainment to name a few) that will be seen as the survivors of the initial e-commerce boom because they developed a completely new way of conducting business, not just a new way to shop.

Regardless of which e-business strategy (incremental, replacement, or incubation) that a company chooses, there are really only three key success factors:

- Enhanced information sharing
- Efficiency
- Customer experience

Enhanced information sharing reduces both the time and distance between a business and its customers and suppliers. In the case of supply chain management, allowing a company's suppliers instant access to inventory levels, without having to physically visit a warehouse, will dramatically aid in the supplier's ability to meet demand in an accurate and timely manner. Old business thinking meant that companies would protect their internal business data. E-business thinking requires companies to open up their vaults of corporate data and to share the information with customers and suppliers in order to make everyone in the value chain more effective.

In an e-business world, *efficiency* is more than internal cost savings to a company. It means finding ways to make it easier and more cost effective for your customers to do business, which in turn translates into internal cost savings for the company. For example, in the case of a typical procurement function, purchase orders are filled out and sent to a supplier. The supplier then enters these purchase orders into its fulfillment system. Companies that leverage the Internet are able to offer direct links that identify a customer's standards and specific pricing, allow a customer to build its own quote, automate the purchase approval process, and submit a paperless purchase order. By focusing on the customer to eliminate the expense of a purchase order process on their end, the supplier is able to eliminate the expense of an order-processing step on its end.

The final success factor of an e-business, *customer experience,* is the most important. The Internet allows consumers to comparison shop between many different companies in a matter of seconds. This has had the effect of turning many products and services into simple commodities. Ultimately then, the success of any e-business is based on its ability to retain its customer base and ensure repeat business. The creation of an experience that is unique and customized for each customer is the differentiator between those e-business models that will survive and those that will fail. "Quality of customer service (62%) is three times more important to repeat online sales than product or price (19%).[b] Those companies who are focused on mass customer segments, and compete on products and prices offered, will be most vulnerable to the competition. Those e-businesses that offer customers personalized access to their own account information can make the Web site better for the customer with each visit and can personalize services to every customer. These are the companies that will be around for the long term and will be the true leaders of the Internet economy.

[a] Dell Computer Corporation CEO Michael Dell (1997), *Austin-American Statesman,* February 26.
[b] Hanrahan, Timothy (1999), "Price Isn't Everything," *Wall Street Journal,* July 12, p. R20.

A new e-business initiative in distribution is the Auto-ID Center project, a partnership between the Massachusetts Institute of Technology, Gillette, Wal-Mart, and 85 other retailers and manufacturers.[34] Its goal is to put radio-frequency identification (RFID) microchips on every product package, from razors to soda, and track these through a smart network. Each item will be tracked from the time it leaves a factory until it is picked off the shelf and scanned by a shopper. As supplies dwindle on the shelf, an alert will be relayed over radio waves to a central computer, which can signal a store clerk to restock shelves. The system can also monitor inventory at the store level and send an alert to a manufacturer when shipments are needed. Wal-Mart is working with its suppliers to tag every box and pallet with RFID chips by 2006. This system will help reduce out-of-stock items as well as "shrinkage" (product loss due to theft or supply chain problems). Analysts estimate Wal-Mart could save $8.35 billion annually by using RFID.[35] However, suppliers will have to incur additional costs in adopting RFID. As of December 2003, each chip costs about $2. The prediction is that when this price comes down to a penny per chip, then RFID will become ubiquitous in retail distribution.

Accelerated Research and Development

Companies have begun to use the Web in creative ways in the R&D process. In 2001, pharmaceutical giant Eli Lily and Company had about 7,500 employees in R&D. By 2003, thanks to the Web, they almost tripled that number without adding headcount. How was that possible? Lily created an online scientific forum, InnoCentive Inc. (www.innocentive.com), where it posts difficult problems in chemistry for anyone to solve. It offers rewards of $2000 to $100,000 for correct answers to these problems. In this way, Lily is able to get solutions to problems that have stumped its own employees without major investments in R&D. Now companies like Procter and Gamble and Dow Chemical have also started using InnoCentive to cut down R&D costs.[36]

Some companies are using the Web as a corporate **intranet** to speedily distribute information among internal departments. Intranets have secure firewalls so that only authorized users have access to the information. For example, the Ford Motor Company uses its corporate intranet to share strategic information about product designs. Its intranet connects 120,000 computers around the world, letting engineers, designers, and suppliers carefully document and share in real time the thousands of steps in product development, including testing and design. Sharing such information has helped Ford reduce the time it takes to turn new car models into full products from about 36 months to 24 months.[37]

Enhanced Knowledge Management

Use of the Internet to facilitate information dissemination within a company is a good way to proactively manage a firm's knowledge bases and to effectively use that knowledge to produce good decisions. Note that effective knowledge management requires the tearing down of walls and barriers between departments, functions, and individuals, both inside and outside the company, in order to better share and use information. Many companies are harnessing the power of the Web to generate, store, and disseminate knowledge and information to improve employee productivity.

Efficient Training and Education

Companies are finding that Internet capabilities allow their employees to receive more training and education opportunities (i.e., "webinars") at a fraction of the cost they used to pay to send employees to seminars and other training. At IBM, "web jams" using Web conferencing can bring together employees in small or large groups to brainstorm on challenges. This has cut down IBM's travel expenses by about $20 million per year. In 2002, Kinko's replaced 51 U.S. training centers with a $2.5 million e-learning network that lets workers take online courses covering everything from products to policies. This use of the Web allowed Kinko's to cut its training expenses to $5.5 million, about one-third of what they spent in 2001. Moreover, this training allowed increased revenues as well: For a new service on making banners and signs, Kinko's stores that provided online course offerings saw revenues rise 27 percent compared to an 11 percent revenue increase in stores that did not.[38]

Improved Customer Service Operations

Service-oriented organizations, such as those in healthcare and transportation, can find that the Internet allows them to better manage their operations. For example, a formidable challenge in managing service operations is how to ensure consistent quality of the delivered service—particularly when demand can vary over time. Sutter Health, a Sacramento, California, hospital chain, invested $20 million on a Web system, "eICU," which combines patient monitoring with digital record keeping. In-room cameras and monitoring devices feed a patient's vital statistics such as pulse and blood pressure to doctors, who track patient progress and make needed interventions. Patient blood-clot rates have dropped from 25 percent to nearly zero, stress-ulcer rates have dropped from 14 percent to near zero, and average ICU stays have also declined.[39]

Another challenge for service operations is to effectively manage demand. BostonCoach, a nationwide limousine service, uses a wireless network to match supply and demand for its fleet. Each driver hits buttons on a wireless phone to let a dispatcher know the vehicle's location. At the control center, dispatchers track two Web dashboards developed by IBM. One lets them know the location of drivers based on their mobile phone signals. The other shows who needs to be picked up, where, and when. Software for mapping and matching tells the dispatcher which driver is best to pick up which passenger. Since the system was installed, BostonCoach has squeezed 20 percent more rides from its cars, adding $10 million in annual revenue.[40]

Apart from pure service organizations, the Internet can also be used by product companies to automate existing customer service routines that are currently handled in an offline manner. For example, many companies are using the Internet to answer customer questions in a timely, cost-efficient manner. E-service savings can be huge: Web service costs companies just four cents per customer on average for a simple Web page query, compared to $1.44 per live phone call. Shifting service to the Net lets companies handle up to one-third more service inquiries at 43 percent of the cost.[41] Companies using the Internet for customer service can see cost savings of 30 percent or more.[42]

Customization of Products

The Web is being used to analyze preferences, customize offerings, and enhance value to customers. German automaker BMW is using the Web to allow buyers to custom-order cars without compromising production efficiency. Instead of choosing from a pool of dealer-purchased cars, buyers can design their own from 350 model variations, 500 options, 90 exterior colors, and 170 interior trims. Eighty percent of cars bought in Europe and 30 percent of cars bought in the United States are built to order. After entering the buyer's order into BMW's Web ordering service, the dealer receives the precise date of delivery within about 5 seconds. The cars usually arrive within 11 to 12 days, one-third of the time it took before the Web ordering service was in place.[43]

Delivery of Web Services

Another use of the Web by businesses is the delivery of software solutions. **Web services** are business and consumer software applications, delivered over the Internet, that let users access and share data across devices, computers, databases, applications and organizations. By using a set of shared protocols and standards, these applications allow disparate systems to share data and services without requiring human intervention.[44] These standards allow a company's software to link up and communicate with another software application that sits on a different company's computer.[45] One of the standards used widely to enable Web services is Extensible Markup Language **(XML)**, which "tags" digital content in standardized formats. For example, a document can be tagged to identify it as an invoice, and a numeric data field can be tagged as a price, to enable data exchange. A simple application of a Web service is a credit card payment done on an e-commerce Web site like eBay. A buyer may enter credit card information on the Web site and submit it for payment. The data is exchanged from the eBay server to the server of a credit card transaction company like Verisign and an acknowledgment is sent back to the buyer when the process is completed. It is seamless and convenient for the consumer.

The earliest forms of Web services were browser-based software programs from **application service providers (ASPs)**. Companies like Corio and Usinternetworking "served up" software on their Web sites from companies like Peoplesoft and Siebel for a per-user fee. Renting software became more popular when companies like Intuit and Salesforce offered it directly to their customers on their own Web sites. This method of software distribution, which relies on an ASP model, is especially well-suited to small and medium businesses without large IT departments who don't want to incur the investment or the hassle of large-scale in-house implementations. In the future, it is likely that software from vendors will be stitched together and made available as Web services to customers who want integrated solutions.[46]

More sophisticated Web services are implemented in supply chain extranets of manufacturers or customer relationship management extranets involving dealers or distributors. Companies like HP, IBM, Microsoft, Oracle, Sun, and others are developing Web services. The potential benefits to businesses from Web services include better collaboration with business partners and cheaper, faster, and more flexible implementation.[47] However, much more work remains to be done in terms of ensuring security and getting industry agreement on standards before the full potential of Web services is realized.

Effective Partner Relationship Management

Customer relationship management, sales force management, and partner/vendor relationship management initiatives are all facilitated by the use of integrating data from disparate functions across the organization. Because these programs have been discussed in Chapter 10, their use is merely noted in this section.

Efficient Human Resource Management

Many companies are using a corporate intranet to assist with traditional human resource functions, such as recruiting and management of employee benefits. Again, the cost efficiencies to be gained from using a standard interface to manage a repetitive function can be significant.

Gathering of Marketing Research

The Internet is also enabling companies to facilitate the marketing research process in an efficient manner. Web-based surveys, tabulated instantly; email communications analyzed for common themes and patterns; and click-stream data all enable marketers to be more effective in tailoring their programs to meet customer needs.

Clearly, the Internet allows firms to streamline business processes to gain enhanced effectiveness and efficiency. Importantly, companies that choose not to leverage the Internet for these capabilities do so at their own peril.

REALIZING THE INTERNET'S FULL POTENTIAL

The Internet and the World Wide Web, to date, have seen unprecedented innovation in new technologies and their application. There has been considerable experimentation with various online business models, and the dotcom bust helped separate the wheat from the chaff. Email, instant messaging, and search have been established as killer applications widely used on a daily basis.

Many believe that the many uses of the Internet are still in their infancy. Regarding the Internet's future potential, Jerry Yang, co-founder of Yahoo, likes to use a baseball metaphor, "We are only at the bottom of the second inning."[48] Where is innovation going in the future to realize the full potential of the Internet? The following trends (Figure 11-4) are likely to assist in such realization.

New Devices for Access

Diffusion of Broadband

Semantic Web

Overcoming Other Barriers:
• Quality/reliability
• Privacy
• Sociocultural/legal questions
• Internet taxes

FIGURE 11-4   Trends and Considerations in Realizing the Full Potential of the Internet

New Devices for Access

One discernible trend is a move away from the PC platform to access the Internet. Handheld devices of all kinds, from cellphone handsets to personal digital assistants (PDAs) to hybrids (combination devices that function as both cell phone and PDA), are becoming enabled for Web access. While the "form" factor (i.e., tiny design) limits their user-friendliness, the lower cost (compared to a PC) encourages adoption and helps bridge the digital divide. For example, the worldwide installed base of cell phones is about 1.3 billion, compared to about 750 million PCs. The annual sales of new cellphones worldwide is about 400 million, compared to about 150 million PCs, so the gap in installed base is going to increase.[49] In China, there are about ten times the number of cellphones in use compared to PCs. Indeed, teenagers and young adults in Japan leapfrogged PCs to adopt NTT DoCoMo's iMode Internet access service, which provided them access to email, useful information, and entertainment content on their mobile phone handsets.[50] Other devices that are being designed to provide Internet access in the future include kitchen appliances, cars, airplanes—essentially any place that people spend time that could be used for communications and connectivity.

Diffusion of Broadband

Another trend is the diffusion of broadband access at affordable prices to greater sections of the population. As noted earlier in this chapter, the United States lags certain countries in Asia and Europe on this count and needs to achieve higher adoption rates of broadband technologies in order to realize the full potential of the Internet in the future. With widespread availability of broadband, the Internet will be a better medium to deliver applications such as entertainment, education, and communication. Music downloads are already popular; movie downloads will be the next popular application through Web sites like movielink.com. The major broadcast networks are already contemplating video-on-demand for primetime shows and these could be delivered on PC or non-PC platforms.[51] Playing online computer games is another growth opportunity in the consumer market.[52] Portals like AOL, Yahoo, MSN, and Real Networks are offering access to online games for a per-use download fee or monthly subscription. Some games allow the user to compete online against people from all over the world. Going beyond the young male target audience, these games are attracting adults, especially women.[53]

In addition to broadband access, continued development of Wi-Fi networks, affording wireless connectivity to the Internet via "hotspots," can also allow greater leverage of the Internet for both personal and business productivity.

Semantic Web

There is radical development work being done on the Semantic Web by Tim Berners-Lee and his colleagues at the Worldwide Web Consortium (W3C). The Semantic Web will be a smart network that can understand human language and make computers as easy (or difficult?) to work with as humans. It would understand not only the meaning of words and concepts but the logical relationships among them. The heart of the Semantic Web will be XML, the language that has been accepted as a standard for all kinds of Web services. Intelligent agents could be sent out to explore Web sites and bring back relevant information to users. The Semantic Web faces formidable technical and political challenges, including acceptance by the 400 corporate members of W3C. While it cannot be predicted when it will materialize, this would be of great value to businesses for a whole range of applications.[54]

Overcoming Other Barriers

In order for e-business to achieve its potential, a series of hurdles remain and must be addressed:

- *Quality and Reliability.* While Internet infrastructure has improved over time, occasionally a major vendor reveals vulnerability to hackers in its products.[55] There is need for more thorough and continuous testing of products to keep hackers at bay. Fighting viruses and worms requires continuous vigilance of software firms and security agencies on a global scale. For example, Microsoft offered cash rewards of $250,000 to each person with information leading to the conviction of people responsible for the SoBig virus and Blaster worm in 2003.[56]
- *Privacy.* The success of electronic commerce depends on the ability to keep personal data safe from abuse. Privacy policies that are strictly adhered to, in which users can see how different sites will use their personal information, are vital.
- *Social/cultural/legal questions.* The community of Internet users is beginning to represent a broad cross-section of society. Along with legitimate uses, the Internet has been used for criminal activities such as terrorism, illegal drug dealing, and pedophilia. If the latter activities are not kept in check, this can become a threat to the legitimate uses of the Internet. As of the writing of this edition, legislation is being considered to make spam illegal. The volume of spam emails has threatened to choke the Internet; clearly, solutions to this must be offered and addressed.
- *Internet taxes.* Businesses and consumers in the United States continue to worry about when the government will begin to levy new taxes on the use of the Internet, especially in an environment of huge state budget deficits. If this happens, it may have a negative backlash on e-businesses.

On balance, despite the above issues, the trend is clear that progressive societies and organizations will continue to expand their use of the Internet and e-business. In closing, in a lighter vein, see Box 11-1, "Proverbs for the Millennium."

BOX 11-1

PROVERBS FOR THE MILLENNIUM

1. Home is where you hang your @.
2. The email of the species is more deadly than the mail.
3. A journey of a thousand sites begins with a single click.
4. You can't teach a new mouse old clicks.
5. Great groups from little icons grow.
6. Speak softly and carry a cellphone.
7. C:\ is the root of all directories.
8. Oh, what a tangled Web site we weave when first we practice.
9. Pentium wise, pen and paper foolish.
10. The modem is the message.
11. Too many clicks spoil the browse.
12. The geek shall inherit the earth.
13. There's no place like home(page).
14. Don't byte off more than you can view.
15. Fax is stranger than fiction.
16. What boots up must come down.
17. Windows will never cease.
18. Virtual reality is its own reward.
19. Modulation in all things.
20. Give a man a fish and you feed him for a day; teach him to use the Net, and he won't bother you for weeks.

—Author unknown.

SUMMARY

This chapter started with a historical review of the technologies that have facilitated the explosive growth of the Internet and are used to enable e-business applications. The reasons behind the dotcom boom-and-bust cycle were examined, and two critical success factors were identified:

- Understanding of the characteristics of the online environment, and how those can be used to transform entire industries.
- Knowledge of the essential elements of e-business strategies and business models.

Regardless of the specific underlying business model, an online venture must be supported by business fundamentals and deliver long-term value to customers and shareholders. Moreover, all Web sites must have effective design and layout. Effective Web site design principles discussed were context, content, community, customization, communication, connection, and commerce.

Issues relating to consumer behavior on the Web—and in particular, online shopping behaviors—were explored. Another important lesson to be learned from the dotcom bust is to realize how difficult it is to change the behavior of consumers online compared to their shopping behavior offline. Online consumer shopping behavior is facilitated by the use of search engines to find relevant information and by the availability of multiple channels, or bricks-and-clicks. However, online consumer usage of the Web is inhibited by access concerns (the digital divide), spam, fraud, and privacy concerns.

Business uses of the Internet continue to drive the primary ways in which the Internet is affecting our economy. Important issues discussed included disintermediation and reintermediation, the evolution of electronic marketplaces, and the myriad other ways in which businesses are using the Internet to streamline and manage their business processes.

Finally, trends that could help or hinder the realization of e-business' full potential were presented.

DISCUSSION QUESTIONS

1. What are the major historical technological innovations that made e-business possible?
2. Analyze the reasons for the failure of theglobe.com. What lessons for e-business success can be learned from this and other examples of the dotcom bust?
3. What is your understanding of the five basic business models: seller storefront, e-tailer, infomediary, portal, and e-procurement? Find a real Web site example of each and explain it in terms of the framework in Figure 11–1.
4. Is there a viable business model for online grocery shopping? Discuss.
5. From a consumer perspective, why are bricks-and-clicks channels better than pure-play channels? If a pure-play startup without deep pockets wants to move in the direction of bricks-and-clicks, what should it do?
6. Pick one of your favorite e-business Web sites. Comment on each of the seven aspects of design presented in Figure 11–2. Further, comment on the concepts of reinforcement and fit for this Web site.
7. Explain the evolution and innovation that has ocurred in the area of online B2B intermediaries. Which of these are here to stay?
8. What other applications of e-business are being used to support business beyond e-commerce? Look at the current organization where you work or study. What are the promising areas of e-business in this organization?
9. What are Web services? Can you describe a futuristic consumer and business application of Web services?
10. Do you envision any technological trends not discussed in this chapter that may help or hinder the realization of the full potential of e-business? Discuss.

GLOSSARY

Application service provider (ASP). A business that provides application software on its Web sites that users can run in an Internet browser, usually for a per-use or subscription fee.

Bricks-and-clicks. A business that has both online and offline operations visible to customers. Also known as "clicks and mortar."

Consortia. Web sites that bring buyers and sellers together to facilitate exchange. They are not independent but sponsored by some large buyers or sellers in the industry. Also known as "coalitions."

Digital divide. This term refers to the disparity in access to digital technology by different sections of the population.

Disintermediation. Bypassing of channel intermediaries in favor of going directly to customers using the Internet.

Dotcom. The term used for a business that had only an online operation visible to customers during the Internet boom. Also known as *pure play*.

E-business. The application of Internet technologies to make business processes more efficient or effective.

E-commerce. The subset of e-business that supports and enables customers to conduct online transactions on seller Web sites.

E-procurement. A large buyer's Web site to make online purchases from suppliers who are invited to participate on that Web site.

E-tailer.   A pure-play retailer with a Web site that lets customers make online purchases; it has no offline presence or stores.

Extranet.   A secure network based on Internet technology that allows a company's business partners access to sensitive or restricted information via secure login.

Hacker.   A person who gains unauthorized electronic access to a company's computers or data.

HTML.   Hypertext Markup Language, the authoring language used to create documents on the World Wide Web.

Independent exchange.   An independent Web site (typically B2B) that provides information and services to business buyers and business sellers to facilitate transactions.

Infomediary.   See market maker.

Intranet.   A company's secure internal network based on Internet technology.

Market maker.   An independent Web site (B2B or B2C) that provides information to buyers and sellers to facilitate exchange without physically handling the product. Also known as an *infomediary*.

Online category manager.   A wholesaler that provides order processing and fulfillment services to an online retailer's cus-tomers as an extension of the retailer's Web site in specific product categories.

Portal.   An informational Web site; content is usually given away free to attract "eyeballs," which are then sold to advertisers.

Reintermediation.   The opportunity for new types of intermediaries to help users navigate the Web; an opposite trend to disintermediation.

Seller storefront.   A Web site of a manufacturer (not a retailer) that lets customers make online purchases directly on the Web site.

Shopping bot.   A Web site that lets shoppers search for and compare prices of a product across many e-commerce Web sites.

Spam.   Unsolicited and unwelcome email in a user's inbox.

Virus.   A malicious software program created by a hacker that can invade a user's computer and destroy programs or data.

Web services.   Programs and services delivered over the Internet that let users access and share data across devices, computers, databases, applications, and organizations.

XML.   Extensible Markup Language, the cornerstone language of Web services, that uses standardized tags to identify documents and fields for business purposes.

ENDNOTES

1. Totty, Michael (2003), "The Masses Have Arrived," Special Report: E-Commerce, *Wall Street Journal,* January 27.

2. Mullaney, Timothy J., Heather Green, Michael Arndt, Robert D. Hof, and Linda Himelstein (2003), "The E-Biz Surprise," *Business Week,* May 12.

3. Dugan, Ianthe Jean and Aaron Lucchetti (2001), "After Becoming Stars of the Dot Com Boom, TheGlobe.com Founders Find Fame Fleeting," *Wall Street Journal,* May 2.

4. Mullaney, et al (2003), op. cit.

5. Ibid.

6. Evangelista, Benny (2003), "RIAA Decries Drop in CD Sales," *San Francisco Chronicle,* September 3, http://sfgate.com/ cgi-bin/ article.cgi?f=/chronicle/a/2003/09/03/BU24 9534.DTL.

7. Mullaney, et al (2003), op. cit.

8. Beatty, Sally (2003), "Fashion Tip: Get Online," *Wall Street Journal,* October 31.

9. Mandel, Michael and Robert Hof (2001), "Rethinking the Internet," *Business Week,* March 26, pp. 117–122.

10. Applegate, Lynda (2001), "Emerging E-Business Models: Lessons from the Field," Harvard Business School, 9-801-172.

11. Kopytoff, Verne (2003), "Businesses click on Ebay: Web Site Sees Growth Potential in Companies' Shopping Trips," *San Francisco Chronicle,* July 28, http://www.sfgate.com/ cgi-bin/article.cgi?file=/chronicle/archive/ 2003/ 07/28/BU255473.DTL.

12. Angwin, Julia (2003), "Used-Car Auctioneers, Dealers Meet Online," *Wall Street Journal,* November 20.

13. Mahadevan, B. (2000), "Business Models for Internet-Based E-Commerce: An Anatomy," *California Management Review,* 42 (4) (Summer), 55–69.

14. Porter, Michael E. (2001), "Strategy and the Internet," *Harvard Business Review* (March), 63–78.

15. This section is based on the 7 Cs Framework in Rayport, Jeffrey F. and Bernard J. Jaworski (2004). *Introduction to e-commerce,* (2nd ed.), New York: McGraw-Hill/Irwin.

16. Ibid.

17. Eisenmann, Thomas. (2000, December 15). "Online Portals." *Harvard Business School.;* Eisenmann, Thomas. (2000, December 11). "Online Retailers." *Harvard Business School*

18. Mangalindan, Mylene, Nick Wingfield and Robert A. Guth (2003), "Rising Clout of Google Prompts Rush By Internet Rivals to Adapt," *Wall Street Journal,* July 16.

19. Hoffman, Donna L. and Thomas P. Novak (2000), "How to Acquire Customers on the Web," *Harvard Business Review,* May–June, 179–185.

20. Tapscott, Don, David Ticoll and Alex Lowy (2000), *Digital Capital,* Boston: Harvard Business School Publishing.

21. McCaffee, Andrew (2001), "WebVan," Case # 9-602-037, Boston: Harvard Business School Publishing.

22. Ibid.

23. *E-Commerce and Development Report* (2003), United Nations Conference on Trade and Development, New York and Geneva.

24. DiCarlo, Lisa (2003), "What's The Problem With Broadband?" *Forbes.com,* November 13, http://www.forbes.com/home/2003/11/13/cx_ld_1113roadblocks.html.

25. Blakeley, Kiri (2003), "Spam: It's Even Worse Than You Think," *Forbes.com,* November 11, http://www.forbes.com/home/2003/11/11/cz_kb_1111spam.html.

26. Miller, Matthew (2003), "Flaws In The System," *Forbes.com,* November 12, http://www.forbes.com/home/2003/11/12/cz_mm_1112fraud.html.

27. Xiong, Chao (2003), "Online Stores Try New Pitch: Fetch It Yourself," *Wall Street Journal,* November 19.

28. Gumbel, Peter (2001), "Ads Click," Special Report: E-Commerce, *Wall Street Journal,* October 29.

29. Feder, Barnaby J. (2000), "Nuts, Bolts and Bumps in the Road," *New York Times,* September 20.

30. Sechler, Bob (2002), "Behind the Curtain," Special Report: E-Commerce, *Wall Street Journal,* July 15.

31. Jap, Sandy, (2003), "An Exploratory Study of the Introduction of Online Reverse Auctions," *Journal of Marketing,* 67 (July), pp. 96–110.

32. Harris, Nicole (2001), "'Private Exchanges' May Now Allow B-to-B Commerce to Thrive After All," *Wall Street Journal,* March 16.

33. Zimmerman, Ann (2003), "To Sell Goods To Wal-Mart, Get on the Net," *Wall Street Journal,* November 21, B1.

34. Keenan, Faith (2003), "If Supermarket Shelves Could Talk," *Business Week,* March 31, http://www.businessweek.com/print/magazine/content/03_13/b3826049.htm?mz

35. Boyle, Mathhew (2003), "Wal-Mart Keeps the Change," *Fortune,* November 10, 46.

36. Mullaney, et. al (2003), op. cit.

37. Cronin, Mary (1998), "Ford's Intranet Success," *Fortune,* March 30, p. 158.

38. Green, Heather (2002), "The Web Smart 50," *Business Week,* November 24, 82–106.

39. Ibid.

40. Ibid.

41. Stepanke, Marcia (1999), "You'll Wanna Hold Their Hands," *Business Week* e.biz, March 22, pp. 30–31.

42. Burrows, Peter (1998), "Instant Info Is Not Enough," *Business Week,* June 22, p. 144.

43. Green (2002), op. cit.

44. Hagel, John, III (2002), "Edging into Web Services," *The McKinsey Quarterly,* 4, 4–13.

45. Ismail Ayman, Samir Patil, and Suneel Saigal (2002), "When computers learn to talk: A Web services primer," *The McKinsey Quarterly,* 4, 14–21.

46. Clark, Don (2003), "Renting Software: Is It the Next Big Idea?" *Wall Street Journal,* June 3.

47. Hagel (2002), op. cit.

48. Yang, Jerry (2000), Remarks made at the National Press Club, April 11, www.pcworld.com/news/article/0,aid,16241,00.asp.

49. Gurley, J. William (2003), "The comeback of the mobile Internet," Cnet News.com, July 17, http://news.com.com/2010-1071-1026742.html.

50. Moon, Youngme (2002), "NTT DoCoMo: Marketing i-Mode," Case # 9-502-031, July 17, Harvard Business School Publishing.

51. Nelson, Emily and Martin Peers (2003), "As Technology Scatters Viewers, Networks Go Looking for Them," *Wall Street Journal,* November 21, A1.

52. Loftus, Peter (2003), "Pay and Play," Special Report: E-Commerce, *Wall Street Journal,* June 16.

53. Xiong, Chao (2003), "Where the Girls Are," *Wall Street Journal,* October 28.

54. Port, Otis (2002), "The Next Web," *Business Week,* March 4, 97–102.

55. Roberts, Paul (2003), "Microsoft Warns of Widespread Windows Vulnerability," *Computerworld,* July 16.

56. Pope, Charles (2003), "Microsoft after virus writers: International effort offers cash rewards," *San Francisco Chronicle,* November 6.

CHAPTER 12

Realizing the Promise of Technology: Societal, Ethical, and Regulatory Considerations

If the automobile industry had made as much progress as the computer industry in the past 50 years, a car today would cost a hundredth of a cent and go faster than the speed of light.
—RAY KURZWEIL, *The Age of Spiritual Machines*, 1999

"The wealthiest 1 billion people in the world are pretty well served by information technology companies," says Lyle Hurst, director of Hewlett-Packard's e-inclusion initiative in 2001. "We're targeting the next 4 billion."[1] HP's e-inclusion initiative, begun in 2000, is designed to help bring the benefits of technology to the world's most impoverished citizens. Among the various efforts in the initiative are:

- India: Studying how to provide technology to assist the rural poor with access to government records, schools, health information, crop prices, and the like.
- Costa Rica and the Dominican Republic: Establishing telecenters in remote villages equipped with PCs, satellite hookups, and solar generators to allow villagers to surf the Internet, swap emails with overseas relatives, search for information, and watch health and agricultural training videos.
- Senegal: Joining with aid groups to give seed capital and staff for a program to train 600 Senegalese to use the Internet, with the idea of developing future businesses that leverage the Net.

HP is also developing variations of its handheld computers to process soil test samples and small business loans.

HP's motivations are driven, in part, by social concerns, bringing the many benefits of technology to the world's poor, such as creating jobs, improving education, and providing better access to government services. HP is also driven by economic concerns. First, it hopes to profit by selling equipment for the ventures, since governments and private foundations are assisting with the funding. Second, HP's managers hope that spreading information technology throughout developing countries will result in new businesses unimagined today and trigger demand for simple and economical computer products. For example, the centers in Costa Rica already are starting a newspaper and coffee-trading business. In Senegal, the aid group will help entrepreneurs start e-businesses that allow overseas Senegalese remit funds and communicate with relatives at home. In India, village photographers, using solar-powered digital cameras and printers, can take pictures for government ID cards, saving residents a trip to the city. Indirectly, building the HP brand and developing relationships are also likely to be beneficial to HP in the long term.

HP's initial successes in its forays into these developing countries are due, in large part, to its willingness to work with grassroots social groups, aid agencies, and local governments to learn what people in these low-income nations need. Their research, for example, revealed demand for low-price, simple IT devices and ways to connect them to the Internet. After understanding these local needs, HP then uses that knowledge to build a business model around them. By figuring out what applications work in one area, the company hopes to replicate it in thousands of others.

HP's belief is that by "doing good" (in a social sense), it will "do well" (in a traditional business sense). It also believes that, given the scale of the need of these impoverished peoples, only programs that sustain themselves can be successful over the long term. At a minimum, projects such as this can build goodwill for the HP brand and establish relationships that may prove to be beneficial in the future. At best, these projects provide a win–win solution for both society and the company. They can help the company discover new, profitable lines of businesses, while helping to alleviate poverty.

SOURCES: Engardio, Pete, (2001), "Smart Globalization," *Business Week,* August 27, pp. 132–137; Gunther, Marc (2003), "Tree Huggers, Soy Lovers, and Profits: Some of America's biggest corporations believe that the best way to make money is by saving the world. And guess what? They just might be right." *Fortune,* June 23, p. 98; Murphy, Cait (2002), "The Hunt for Globalization that Works," *Fortune,* October 28, pp. 163–176.

Technological developments offer promises of a better life, from making us more efficient, to entertaining us, to keeping us in touch with each other, to making us healthier. At the same time, however, technological advances have fallen far short of their promises, creating problems and unforeseen hazards. For example, air bags were touted as a space-age fail-safe measure to protect people who refused to wear seat belts. But the early air bags, which deployed at speeds approaching 200 miles per hour, ended up killing 141 people, mostly children or small adults.[2] Similarly, breakthroughs in antibiotics early last century inspired predictions about the eradication of disease, but now we are faced with drug-resistant microbes and are running out of antibiotics.[3]

The purpose of this chapter is to address some of the issues that can pose obstacles to realizing the promise of technology; these obstacles are overviewed in Figure 12-1. A major obstacle arises from the unintended consequences that accompany technology development. Whether explicitly acknowledged or not, these unintended consequences of technology can inhibit adoption and utilization of new technologies. This chapter first explores the paradoxical effects that technology presents to potential adopters, juxtaposing the enabling benefits of technology against the drawbacks and unintended consequences.

Then, the chapter continues with a discussion of how high-tech firms can resolve the ethical dilemmas posed when introducing their technologies to the marketplace. Next, the chapter provides an overview of the ongoing debate regarding corporate social

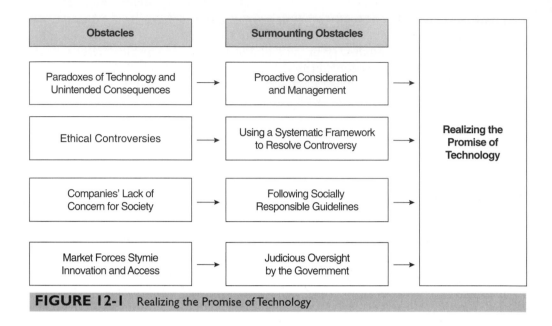

FIGURE 12-1   Realizing the Promise of Technology

responsibility and how that debate can play out in technology-focused businesses. Finally, the chapter concludes with a discussion of market forces that may stymie innovation and access to technology and the debate over the role of the government in addressing these complicated social issues. Within each section, a decision-making framework is provided to assist high-tech marketers in grappling with these complicated issues.

The Technology Expert's View from the Trenches provides insights on challenges to realizing the promise of biotechnology.

THE PARADOXES OF TECHNOLOGY[4] AND UNINTENDED CONSEQUENCES

In the 1930s, the fierce, biting South American fire ant entered the United States as a ship stowaway. Although it was targeted for eradication with DDT and super-pesticides, after three decades, the pesticides had done more damage to the ants' predators than to the invaders. The chemicals actually ended up helping to increase the ant population rather than to eradicate it.[5]

As this example shows, the best-laid technological plans often go awry. Many people are susceptible to overinflated enthusiasm about the possible benefits of high-tech solutions. Yet technology is inherently paradoxical, with every positive quality potentially countered by an opposing negative quality. Although technology enthusiasts would have us believe in their promises that technology can transform our lives in positive ways, the pace, complexity, and unintended consequences of technological development are all too apparent. Yes, technology has provided people freedom, control, and efficiency; but it also has degraded the environment, undermined the social fabric of our lives, and, at the extreme, put us on the brink of obliteration.

TECHNOLOGY EXPERT'S VIEW FROM THE TRENCHES

BIOTECHNOLOGY: PROMISES AND CHALLENGES
RALPH E. CHRISTOFFERSEN, PH.D.

Partner, Morgenthaler Ventures
Boulder, Colorado

The last thirty years have been described by many as the "age of biology." Cloning, from proteins to sheep, has provided unprecedented opportunities for understanding and treating diseases. Sequencing of the human genome has expanded the possibilities dramatically and allows for the long-term possibility of diagnosing and treating diseases with specificity previously unimagined. Indeed, many diseases thought to be untreatable now have treatments thanks to biotechnology advances, and many others are being studied with the potential for treatment in the future. With such significant advances and future potential, it has been thought by many that the decades ahead will bring major advances in therapeutic design and development, and the practice of medicine will be changed dramatically. Not surprisingly, this potential has spawned a new industry, the "biotechnology industry," driven by entrepreneurs who believe that not only scientific advances but new business paradigms are possible that can accelerate the process.

At the same time, these significant promises have been mitigated by a number of serious challenges. Historically, development of a new therapeutic agent has been highly risky. Depending somewhat on the nature of the disease, development of a new therapeutic agent has taken approximately ten years, required approximately $200 million or more in cash, and had an overall probability of success of approximately 10 percent. Given these odds, many potential investors passed on the biotechnology opportunity, claiming that "the odds of success were probably better in Las Vegas." However, entrepreneurs believed that new business models, along with new scientific paradigms were possible, which would improve the probability of success substantially.

The result has been, in large part, an enormously successful experiment. Several very successful companies have been built with products meeting critical medical needs, including Amgen, Genentech, Biogen, and Chiron. Furthermore, the next generation of companies is developing rapidly and promises to continue both the business and scientific revolutions that have begun. The result has been creation of thousands of new, high-paying jobs along with a competitive advantage for the United States in which the environment has been nurtured. At the same time, traditional pharmaceutical companies have discovered the power of the new technologies and nontraditional business models and have embraced them through partnerships, internal efforts, and acquisitions to improve their traditional record of success.

With such successes, it would be expected that the biotechnology industry would be burgeoning with new entries, each following the paradigm for success set down by the leaders. In some ways this has happened, since approximately 1,450 biotechnology companies now exist in the United States, of which approximately 350 have successfully become public companies. On the other hand, a number of significant challenges to development of the industry remain that make starting or developing a biotechnology company a very risky business.

Among the challenges and barriers are the product development time and life cycle. While patent protection can, in principle, provide about twenty years of proprietary commercial sales for a successfully developed product, regulatory requirements for product approval can use up ten or more years of the patent protection time. When compared with other product development times and commercial opportunities (e.g., information technology products and services), investors frequently choose other options. In addition, the amount of money needed to develop each product presents a daunting challenge. No single person or group of investors is likely to invest $200 million in a ten-year development program at the outset. The result typically has been initial investment by venture capital investors in the $10-to-$15 million range, with the hope that subsequent private investments, an IPO, and successive public investments will fund the remainder of product development costs. This means that a biotechnology company is almost always on the verge of running out of money, which has been exacerbated further by both venture capital and public investor expectations of

results sooner than science or funding could reasonably be expected to produce. This makes recruitment and retention, as well as maintenance of morale, of employees a significant challenge. Furthermore, with the specter of price controls on the horizon in the United States, the potential for increased investor reluctance due to decreased potential profit is very real.

More subtly, a number of other internal challenges exist within individual companies. One of the most important is having a technology that has broad therapeutic application. All too often, companies have been formed based on development of a single product, which raises the potential for failure significantly. Another challenge relates to finding appropriate personnel. The need for persons with drug discovery, development, and commercial experience from large pharmaceutical companies is evident, but finding such persons that also possess entrepreneurial skills and risk tolerance is frequently a very difficult task. Adding to these are uncertainties associated with patents (will they issue, will they have broad claims, will they be challenged, etc.?), political and ethical issues (price controls, fears of "eugenics" or the attempt to control one's hereditary qualities, the ways in which human genetic information is gathered and used, etc.), and the status of competitors.

As a result, biotechnology is not for the faint of heart. The benefits, both to society and individuals in the field, can be enormous, but the risks and frustrations can make success quite challenging to achieve.

Research conducted by David Mick and Susan Fournier identified eight paradoxes regarding consumers' views of technology.

Technology Paradoxes

As Figure 12-2 shows, eight paradoxes can be identified that characterize the relationship between consumers and technology.

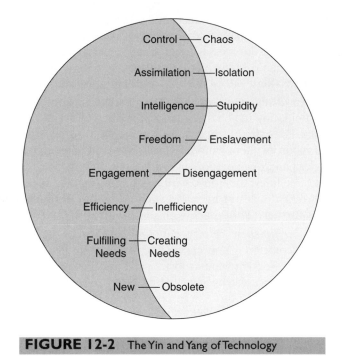

FIGURE 12-2   The Yin and Yang of Technology

1. ***Control–chaos.***   Technology is supposed to help bring order to our lives and businesses. However, once purchased, technology can disrupt our everyday routines. For example, as a society, we have become so dependent on computers that if they fail, we can no longer do our work. Indeed, in this wired world, we are more susceptible than ever to crashes from viruses and worms that bring work to a standstill. In July 2003, the most prevalent worm was the Klez.E worm, accounting for 19.2 percent of computer virus infections in that month.[6]

2. ***Assimilation–isolation.***   Many technologies help us connect with other people, either by facilitating communication or by providing shared experiences. But these technologies have also become a substitute for face-to-face communication, personal contact, and other social activities. Ultimately, when we log on to virtual communities, we sit alone at our computers. For example, research shows the more hours people spend on the Internet, the more depressed, stressed, and alienated they feel, even if most of their time is spent sending messages or "interacting" in chat rooms. "We've become so removed from reality that we don't even know how to feed our increasing hunger for intimacy."[7] Moreover, 6 percent of Internet users suffer from some form of addiction to it, such as compulsive Internet use, online gambling, accessing porn/sex sites, or stock trading.[8] False feelings of intimacy, timelessness, and lack of inhibition contribute to the addictive force of the Internet.

3. ***Intelligence–stupidity.***   Sophisticated products are supposed to help us be smarter by allowing us to perform complex tasks. But if the products are difficult to master or cause us to lose old skills, we actually feel inadequate or inferior in mastering new technologies.

4. ***Freedom–enslavement.***   Products that offer freedom can end up creating new restrictions. For example, voice mail gets us away from our offices, but makes us feel obligated to check our messages constantly. Cellphones give us freedom, but now we are always available. Many fear that wireless access to the Internet (via Wi-Fi technology) means there will be no safe haven from the ubiquitous reach of technology.

5. ***Engagement–disengagement.***   Although some technologies are designed to make participating in activities easier, they may detract from the quality of the experience. For example, instead of smelling and squeezing lemons for ripeness, we can now order them from a Web site. We watch food heat under plastic wrap behind glass microwave doors instead of carefully stirring pots and inhaling aromas that have filled the house with comforting smells.[9] The very things that are supposed to simplify our lives and allow us to stay in touch with each other are putting us further out of reach with sense-dulling "conveniences."

6. ***Efficiency–inefficiency.***   New technology can help us perform tasks faster, but also creates time-consuming new chores. For example, in the late 1990s, the concern was that U.S. office workers were drowning in email. Indeed, the average number of email messages received each week by U.S. office workers in 1998 was 179, according to eMarketer.com. For those in Internet firms, the number was higher, at 233 messages per week.[10] In the 2003/2004 timeframe, the concern focused more on the inefficiencies arising from the large proportion of spam email, email that was both unsolicited and unwanted. In July 2003, spam volume accounted for 50 percent of all email messages. CyberAtlas estimated that spam emails cost corporate America $9 billion in 2002, based on loss of worker productivity, consumption of bandwidth, and the use of technical support time, or "about $10 per user per month."[11]

   Additionally, many companies invested in information technology with the expectation that such investments would make their businesses more productive. Although most statistics do suggest that the increases in economic productivity are, in large part, due to investments in information technology, at the personal level, inefficiencies arising from such things as continued efforts to keep current with software upgrades, the use of computers for personal activities (such as email and Internet surfing), and so forth, cause frustration.[12]

7. ***Fulfilling needs–creating needs.***   Many technology products make us aware of new needs even as they fulfill others. For example, although we might be able to use software to perform more sophisticated tasks, we also now need to attend training programs to learn how to use the software to do those tasks. Similarly, instant messaging has become a popular tool not only among consumers, but among corporate/business users as well. Yet, network security managers said that instant messaging poses the most dangerous security risk to their networks, and as a result, network security managers are having to develop and implement new technologies to handle the security threat.[13]

8. ***New–obsolete.***   Although people are excited by owning cutting-edge products, their excitement is chronically undermined by a fear of falling behind. This paradox creates both the simultaneous desire to purchase new technologies and to delay purchase in order to wait for the latest/best technology.

   These paradoxes highlight the fact that technological developments do have unintended consequences. Of course, the unintended consequences are not always negative. For example, the Internet has resulted in a much more efficient utilization of energy and resources. Online retailers with warehouses have eight times the number of sales per square foot as a physical store. If, as predicted by the Organization for Economic Cooperation and Development, the Internet makes 12.5 percent of retail space superfluous, it would save $5 billion worth of energy every year.[14] Similarly, a book purchased online costs about one-sixteenth the energy of one bought in the store. And, one minute spent driving uses the same amount of energy as 20 minutes of shopping from home. By putting reading materials online, the Internet is predicted to reduce worldwide demand for paper by about 2.7 million tons by 2003.[15]

Marketing Implications of Technology Paradoxes

How do these paradoxes affect the way people view technology and, ultimately, their purchase strategies and behaviors? Potential purchasers of technology create coping strategies to lessen the presence of the paradoxes. Understanding such coping strategies can give marketers insight into segments of technology consumers and the differing types of marketing messages that may be useful, given consumers' concerns. For example, buyers of computers may place limits on how often they use them so that they don't take over their lives. Indeed, some Silicon Valley executives decidedly reject the use of technology during their personal time, refusing to own computers and other electronic gadgets.[16] For others, yoga, massage, or Tai Chi can reduce the stress caused by the technology paradox.

At its extreme, the backlash against technology can be seen in the Luddite movement. **Luddites** are people opposed to technology and changes brought about by technology. (The etymology of the word possibly comes from Ned Ludd, an eighteenth-century Leicestershire worker who destroyed machinery [circa 1811],[17] or from a boy named Ludlam, who, to spite his father, broke a knitting frame.) Regardless, to protest unemployment caused by the industrial revolution in the early nineteenth century, English workers known as Luddites resorted to a campaign of breaking machinery, especially knitting machines. The Luddites signed their destruction "General Ludd," "King Ludd," or "Ned Ludd." The government dealt harshly with the Luddites—fourteen were hanged in January 1813 in York. Although sporadic outbreaks of violence continued until 1816, the movement soon died out. The contemporary equivalent of the Luddite movement is found in people who consider themselves "neo-Luddites." There are alternative music bands by that name, there is a folk opera dedicated to Ned Ludd, and (paradoxically enough) plenty of Web sites dedicated to Luddism. Even some prominent contemporary writers such as social critic Neil Postman can be counted as proponents of this informal movement. For example, one of the targets of neo-Luddism is a category of food products that the protesters have dubbed "frankenfoods," with obvious reference to Mary Shelley's 1818 novel (written at the end of the Luddite movement) depicting the catastrophes that ensue when science goes too far in its quest for knowledge. Frankenfoods are, of course, genetically engineered foods, a category that includes a large and increasing variety of both plant and animal products. Kirpatrick Sale's book (1996), *Rebels Against the Future: The Luddites and Their War on the Industrial Revolution: Lessons for the Computer Age* also highlights another perspective of the current application of Luddites.

Unfortunately, technology marketers, with their enthusiasm for the products they market, are sometimes blind to the dual impacts of technology from a customer's perspective. At a minimum, marketers must be alert to the presence of the paradoxes, try to actively consider what the unintended consequences of a new technology might be, and develop contingency plans for their marketing efforts. To its credit, the recently passed Nanotechnology Research and Development Act requires, as part of its funding, the study of societal and ethical consequences of nanotechnology, including environmental and health effects.[18] The inaugural issue of *Nanotech Briefs* (October 2003) announced that the National Science Foundation has given two grants in excess of $1 million each to study the societal implications of nanotechnology and its potential for unintended consequences. NSF Director Rita Colwell states, "we can't allow the societal implications to be an afterthought. The program has to build in a concern for those implications from the start."[19] Technologies that don't do this have a way of coming to grief later on. More information on nanotechnology appears in this chapter's Technology Tidbit.

TECHNOLOGY TIDBIT

NANOTECHNOLOGY

A futuristic robot device ("nanobot") retrieves plaque and bacteria from the lining of a blood vessel.

Nano- is a prefix used in science meaning "a billionth of," and in technology, it means building things based on atomic and molecular structures. Its components are single devices that are a fraction of the size of DNA molecules. They work at the individual atomic or molecular level, with devices ranging in size from one nanometer (one-billionth of a meter) to 100 nanometers. Predicted to have as significant an impact in the twenty-first century as antibiotics, the many potential applications include:

Molecular manufacturing is the anticipated ability to inexpensively fabricate complex devices, both large and small, with precise control over the arrangement of the individual atoms that constitute the device. For example, nanotechnology may allow bricks to be built with molecules instructed to repair themselves when cracks appear.

Molecular computing is predicted to be the next generation of computing, bringing a giant leap in computing beyond the current generation of transistors (currently based on electric impulses), based on molecular reactions.

Another use will be based on biomolecular reactions. Medical nanomachines will operate at the cellular level, say, to identify and attach to individual cancer cells. Nanoparticles will also help prevent immune reactions against implants with prosthetic limbs and artificial organs. Medical diagnostics can also use DNA detection based on small amounts of blood, which can be screened for numerous diseases.

MEMS, or microelectromechanical systems (MEMS), are micromachines that integrate mechanical elements, sensors, and electronics to allow the manufacture of extremely small devices.

Nanotechnology, as it is expected to develop during the next thirty years, offers the potential to cure all physical diseases, reverse aging and prolong or restore vital youth, provide abundant and almost limitless wealth, open the frontiers of space, and in general empower individuals to seek fulfillment of their wildest dreams, within, of course, the limits imposed by the laws of physics and chemistry.

SOURCES: www.nano.gov; www.nano.org.uk; www.nanotech-now.com; www.nanotechbriefs.com

Ultimately, firms must be aware that blind pursuit of the technological imperative can potentially be viewed as threatening and interfering with sociological needs for safety and human dignity. Because customers will ultimately make the decision whether to buy, firms must pay attention to the tension these paradoxes pose. The controversy surrounding genetically modified food provides an example.

Example of Backlash Due to Fears Arising from Technology: Protests over Genetically Modified Foods

What are genetically modified foods? To answer this question, one must understand the function of DNA. The code of all life-forms is written in deoxyribonucleic acid (DNA), which takes the form of a double helix (long spiral staircase). The rungs linking the two sides of the staircase are composed of pairs of nucleotides: adenine and thymine, or cytosine and guanine. Importantly, these base pairs contain the complete set of instructions for an organism's biological processes required to live and reproduce, its "**genome**." As an example, there are three billion base pairs for the human genome.[20]

So, genetically modified (GM) foods are grown from seeds that reweave the strands of DNA to alter the instruction set. For example, modifications have rewoven genes to make crops withstand frost, herbicides, and even produce their own pesticides. Monsanto includes DNA from a common bacterium in its seeds that makes plants toxic to insects, but not to humans. These modifications have proven popular with farmers: Increases in crop yields mean an increase in profits.

Sales of genetically modified seeds rose from $75 million in 1995 to $1.5 billion in 1998. Predictions are that genetically modified crops will surpass natural crops in acreage planted by 2020, with 100 percent of crops coming from genetically modified organisms by 2020.[21]

In the United States, genetically modified components are viewed as additives and do not require Food and Drug Administration (FDA) approval. The FDA says foods with GM ingredients are perfectly safe. However, others argue that there are profound differences between GM and traditional crops and that problems have arisen in the past. For example, pollen from some strains of corn with built-in pesticides kills the larva of the Monarch butterfly. There is passionate concern over modifying the genetic code of other species. For example, genetic modifications to salmon have produced fish that grow twice as fast, resist disease, and out-mate competitors.[22] Critics worry that this is a biological time bomb that could destroy the remaining natural salmon populations.

The rise of companies marketing products guaranteed to be free from genetic alterations, and selling such foods at premium price, is evidence of the backlash against this technological advance. Firms that have invested hundreds of thousands of research dollars into GM foods inevitably question the value of their investments when the resulting products sell at a lower rather than higher price.

Public fears about GM food are very acute in Europe. Whereas 90 percent of Americans believe the U.S. Department of Agriculture's statements on biotechnology, only 12 percent of Europeans trust their national food regulators. The European Union had a five-year moratorium on GM modified corn from 1998 through 2003 that was being reconsidered in December 2003. David Byrne, the European Union commissioner in charge of consumer protection said that the EU had an effective set of rules to guarantee the traceability and labeling of GM products, and so could "move on."[23]

Marketing and Business Implications

What are some of the marketing and business lessons that can be learned from this controversy? People's instincts often are to concentrate not on potential benefits, but on potential accidents and abuses. These instincts are particularly pronounced in the case of GM foods because, rather than making food better tasting, cheaper, or safer, these technological discoveries have helped primarily with the production and distribution of food.[24] Because the benefits are not obvious to consumers, it is natural that fears about the technology supercede desires for it. As a result, probably the biggest lesson—and challenge—is to *manage the public's misgivings* about the altered products. The best way to manage public misgivings is to educate with information. Marketers must believe in and practice full disclosure. But marketers will always be constrained by the limits of scientific evidence and the confusion of conflicting empirical studies on new phenomena.

Another implication comes in the area of *labeling*. Research by Konstantinos Giannakas and Murray Fulton shows that if consumers are averse to genetic modification and don't see the benefits of genetic modification (say, in lower prices), to the extent that they trust the regulatory system, labeling makes sense. However, if consumers don't trust the regulatory system, an outright ban may occur. Because consumer awareness about GM products and their preferences or aversion for them affect purchase decisions, public policy makers must consider the effects of labeling on consumer demand.[25]

In terms of other business implications, these scientific discoveries are *erasing boundaries between businesses*. Agricultural seed companies compete with chemical companies—the new seeds are direct threats to pesticides and herbicides—and with pharmaceuticals (when the foods are designed with medicinal value). Future discoveries with important implications for human health may now come from agricultural and chemical labs. This convergence of industries arising from developments in the life sciences is just beginning to take effect. Ripple effects are predicted to be felt in healthcare, consumer products, and cosmetics companies. For example, the ratio of doctor bills to pharmaceutical costs, which is now 9:1, is predicted to shift to 1:1 in the next 25 years.[26] Other implications will be felt in the energy business, which will use renewable plant sources for energy, textiles, mining, and environmental remediation. Concerns over unintended consequences also highlight the fact that many technological breakthroughs can create ethical dilemmas.

ETHICAL CONTROVERSIES SURROUNDING TECHNOLOGICAL ADVANCES

Many technological breakthroughs are outpacing society's capacity to deal with the ethical dilemmas that arise. For example, scientists have identified a gene that enhances memory in mice. The gene, NMDA receptor 2B, or NR2B, directs production of a nerve protein that helps the brain recognize when two things are linked. The gene normally becomes less active in middle age, as memory declines. But when an extra copy of the gene programmed to remain active in old age is inserted into mouse embryos, the enhancement is permanent and is passed on to offspring.[27] Such discoveries raise many ethical concerns. Who will have access to the genes that improve cognitive performance?

Only those who can afford to pay for the procedures? Can families now order "designer babies," specifying the gene enhancements that go beyond simply correcting abnormalities to increasing human capacities above those currently deemed normal?

Similarly, in 1995, scientists discovered that women of Ashkenazi Jewish descent (of Central or Eastern European ancestry) whose close relatives had breast and ovarian cancers and who had inherited a particular mutated gene called BRCA1 had as much as a 90 percent risk of developing breast cancer and a 40 to 50 percent risk of developing ovarian cancer. Further study revealed that as many as 1 percent of Jewish women carried this mutation.[28] Scientists and labs decided *not* to offer the test for the mutation to the general public, because questions remained about the risk posed by the gene (it was not clear if a woman with the mutated gene would get cancer), about what could be done to lessen the risk of getting cancer, and about whether testing would do more harm than good (concern over insurance coverage for these women). However, at least one lab found these reasons patronizing, saying that women have right to know, and offered the test.

Controversies in other technology arenas also arise. Technologies for sharing music files, such as those developed by MP3.com and the original Napster, allow for the violation of intellectual property rights and copyright laws (these issues are addressed subsequently in this chapter). The parties whose rights are violated have taken their fight to the courts, but at its heart, such developments can also boil down to issues of ethics.

Controversy inevitably surrounds many technological developments, and high-tech firms cannot simply ignore these issues when they arise. The risk of possible negative publicity and anti-technology backlash can have a very detrimental effect on the commercialization and success of new technologies. Because of the many ethical dilemmas presented by the development and marketing of high-technology products and innovations, this chapter offers a framework for addressing such dilemmas, as shown in Table 12-1. The use of this framework will be demonstrated in the context of the pharmaceutical industry with the situation faced by Merck Pharmaceutical in the 1980s.

Framework to Address Ethical Controversies: Merck, Ivermectin, and River Blindness[29]

In 1978, a disease called river blindness plagued at least 85 million people throughout Africa and parts of the Middle East and Latin America. The cause of the disease is a parasitic worm carried by a tiny black fly that lives and thrives along fast-moving rivers. When the flies bite people, the larva of the parasitic worm enters the human body, eventually growing to more than two feet in length and causing grotesque but relatively innocuous nodules in the skin. The health problems caused by the worms begin when the adult worm reproduces, releasing millions of microscopic offspring, which swarm through the body tissue causing terrible itching, so terrible that some past victims committed suicide. After several years, the microfilariae cause lesions and depigmentation of the skin. Eventually, they invade the eyes, often causing blindness.

Indeed, the disease was so prevalent in some areas that the children assumed blindness was simply part of growing up. The World Health Organization labeled river blindness as a public health and socioeconomic problem, because of the burdens of the illness coupled with attempts of people to avoid flies, when they abandoned fertile ground near rivers, moving to poorer land with decreased food production.

TABLE 12-1   Framework for Addressing Ethical Dilemmas

1. Identify all stakeholders who are affected by the decision.

2. For each stakeholder group, identify its needs and concerns, both if the decision *is* implemented and if the decision is *not* implemented.

3. Prioritize the stakeholder groups and perspectives.

4. Make a decision.

In Merck's labs, a scientist stumbled upon the possibility that one of its drugs used to eliminate insect-borne parasites in cows—ivermectin—might actually have properties that would enable it to eliminate the parasite that caused river blindness. However, from Merck's perspective, this was not necessarily good news. For starters, the process to identify which discoveries to pursue was difficult. For every pharmaceutical compound that became a "product" candidate, thousands of others fell by the wayside. Moreover, if Merck did make the decision to invest in further research for this drug, including conducting field trials in remote areas of the world, it was plagued by many critical issues. First, the population that would benefit from this discovery was relatively small. Second, the population lacked the means to pay for the medication. Third, there was no infrastructure in place to deliver the medication and oversee the administration of the series of treatments that was required. Fourth, if a human derivative of ivermectin proved to have any adverse health effects when used on humans, it might taint its reputation as a veterinary drug. Fifth, there was concern that a human version of the drug distributed to the Third World might be diverted to the black market, undercutting sales of its veterinary product. Clearly, Merck faced an ethical dilemma in how to proceed with this scientist's discovery. How would the use of the framework help in resolving this dilemma?

Step 1: Identify All Stakeholders Who Are Affected by the Decision
In Merck's case, the stakeholders might include the following:

- Shareholders
- The public at large
- The Third World population affected
- The government
- Employees

Step 2: For Each Stakeholder Group, Identify Its Needs and Concerns
This step requires identifying the results both if the decision is implemented and if the decision is *not* implemented. What were each stakeholder group's needs and concerns if the decision to explore the drug further were to proceed, and if the decision to explore the drug further were curtailed?

From the *shareholders' perspective*, shareholders were concerned about the likely negative impact if this discovery were pursued. Studies showed that, on average, it took 10 years and $200 million to bring a new drug to market. So, if the company pursued this development, it would be costly, regardless of the outcome. And, even in the most positive scenario, the costly development and drug trials, coupled with the lack of a target market's ability to pay for the drugs, indicated that the company might lose money in proceeding further with this discovery.

On the other hand, the company might face negative publicity if it withheld useful medications from people. Possible repercussions from such negative publicity might include competition gaining a stronghold.

From the *public at large's perspective,* if the drug were further pursued, the company might be viewed favorably for its efforts and gain more loyal customers. If the drug were shelved, the public might begin to question Merck's motives in developing drugs to address human suffering. In choosing not to develop drugs that have the potential to enhance significantly the quality of life for people, the resulting skepticism could cause problems in managing customer relationships. Moreover, given the presence of knowledge spillovers in high-tech markets, in which innovations and developments in one area have the potential to lead to new innovations in other areas, there was a legitimate question related to what other human diseases this discovery might ultimately lead to, discoveries that might never be made if the drug were not pursued.

From the *perspective of people afflicted with the disease,* if the drug were developed, and if it proved to be effective in combating river blindness, the quality of life of this population would be greatly enhanced. This could lead to other improvements, in that any support currently provided by relief agencies or local governments to assist this population with ongoing sustenance and survival could be redirected to other needy populations. There was also the risk that if the drug were developed, some unknown side effects might occur (which hopefully would be uncovered during additional development and prior to administration of the drug to this population). One would have to weigh (as with any drug) the benefits the drug could deliver relative to possible side effects.

If the drug were not developed, not only would this population continue with the grim situation, but the placement of profits over people could lead to a general backlash against corporations in developing countries, with, at a minimum, ensuing negative publicity or possibly even local protests with resulting violence.

From the *government's perspective,* if the drug were further pursued, the U.S. government might choose to expedite the lengthy approval process in order to hasten Merck's ability to relieve human pain and suffering. If the drug were not pursued, although the U.S. government typically does not legislate drug development, a sufficient outcry might cause intervention, requiring Merck to make its formularies for the bovine medicine available on a licensing basis to other competitors so they could pursue its development for river blindness. Issues related to infrastructure and drug administration would also ultimately be government concerns.

From the *employees' perspective,* if the drug were developed and created financial losses, Merck employees might experience possible belt tightening and layoffs. Still, the scientists and people involved would feel confident that their work had purpose and meaning. If the drug weren't developed, employees might not face negative revenue implications, although the decision not to pursue potentially lifesaving discoveries might be demoralizing.

Step 3: Prioritize the Stakeholder Groups and Perspectives

How did Merck prioritize the various stakeholders? It is vital that this step not be considered simply a debate over people versus profits (which is frequently what happens with ethical business dilemmas). Framed in such a manner, the discussion typically becomes one based on whose view is "right" or which views come from more powerful people in the firm, rather than leading to insights about priorities. Instead, priorities should be based on a company's mission and a long-term perspective. A real benefit of this step comes in the clarification of values and the explication of implicit assumptions that underlie decision making in a company.

In this case, Merck had, over the years, deliberately fashioned a corporate culture to nurture the most creative, fruitful research. Its scientists were among the best paid in the industry and were given great latitude to pursue intriguing leads. Moreover, they were inspired to view their work as a quest to alleviate human disease and suffering worldwide. Employees found inspiration in the words of George W. Merck, son of the company's founder and its former chairman, that formed the basis of Merck's overall corporate philosophy:

> We try never to forget that medicine is for the people. It is not for the profits. The profits follow, and if we have remembered that, they have never failed to appear. The better we have remembered it, the larger they have been.[30]

At this step, a company can also look for creative solutions that reframe the ethical debate, looking for win–win solutions that do not require pitting different stakeholder groups' needs against one another. It is vital that a company remain in touch with the underlying ethical dilemma, however, and not attempt to resolve the debate in a moral vacuum.

Step 4: Make and Implement a Decision

In wrestling with the dilemma, Merck explicitly recognized that its success in the pharmaceutical market was, in large part, due to the efforts of its scientists in making discoveries just like this one. If its scientists did not believe that their discoveries would be used to their fullest capabilities, then not only would they possibly become demoralized, Merck might also have difficulties recruiting, attracting, and retaining the best scientists. On this basis, it decided to proceed with further study on the drug, which was eventually released for human purposes in 1987.

Box 12-2 on the Merck case provides additional information.

Benefits of the Framework

The benefits of using this framework to guide the resolution of ethical dilemmas are threefold. First, the framework makes explicit the various issues that the company will need to wrestle with, regardless of the resulting outcome. For example, in the Merck case, the framework makes explicit the need to manage possible shareholder concerns about a loss of profitability against the need to manage scientists' incentives for discovery. Second, the framework brings into stark relief the various perspectives of the affected stakeholders. By highlighting the various perspectives and their respective "stakes," a firm gets a better sense of the magnitude of the controversy it will encounter, regardless of the decision it makes. Third, the framework leads to a heightened sense of commitment to the resulting outcome. In an ethical dilemma, a company is exposed to criticism, regardless of which side of the dilemma the resulting decision supports. By using the framework, company spokespeople can

ADDITIONAL INFORMATION ON THE MERCK CASE

After a seven-year clinical research program, Merck & Company, Inc., in collaboration with the World Health Organization (WHO), demonstrated that a single oral dose of Mectizan, taken once per year, could prevent blindness and alleviate skin disease. In 1987, Merck decided to donate Mectizan free of charge to all people affected by river blindness, for as long as necessary. Merck approached William Foege, M.D., then executive director of the Carter Center, for assistance with the global distribution of Mectizan. Together, they created the Mectizan Donation Program (MDP) and housed it at the Task Force for Child Survival and Development, an independent partner of the Carter Center. MDP acted as the liaison between Merck, nongovernmental development organizations (NGDOs), affected countries' ministries of health, and United Nations organizations such as WHO.

The Mectizan Donation Program is currently advised by the Mectizan Expert Committee, an independent body established by Merck, which is comprised of public health and parasitic disease experts. These experts must approve all applications for the drug that are submitted by river blindness treatment programs worldwide. Applicants must demonstrate effectiveness in addressing such issues as storage and transport of Mectizan, dosage and administration of the drug, medical support personnel, and patient selection.

The statue of a boy leading a blind man has become recognized as a symbol of the fight to eliminate river blindness. The statue commemorates the success of three WHO–Pan-American Health Organization–led pro

Photograph reprinted with the permission of Merck & Co., Inc.

grams: the Onchocerciasis Control Programme in West Africa (OCP), operating in 11 countries; the African Programme for Onchocerciasis Control (APOC), covering 19 countries outside West Africa; and the Onchocerciasis Elimination Program for the Americas (OEPA), present in 6 countries. OCP, after 25 years, has practically eradicated the disease in West Africa, APOC became operational in 1996 and will continue its support until 2007; and OEPA aims at eliminating the disease in the Americas by 2007.

THE GIFT OF SIGHT

Due to these efforts, the disease is on the verge of elimination in West Africa. An estimated 12 million children have grown up without the risk of being blinded by onchocerciasis. On the 25 million hectares of land now reclaimed for cultivation near river areas, enough food is produced to feed 17 million people per year.

SOURCE: www.who.int/ocp/ (Onchocerciasis Control Programme of West Africa); www.merck.com/about/cr/mectizan/; www.cartercenter.org.

feel greater confidence in their decisions that come under attack. The systematic approach to considering all stakeholders involved, their relative stakes in the decisions, and using some solid basis to prioritize their relative needs leads to a sense that the decision is not a knee-jerk tendency about "people over profits" (or vice versa as the case may be), but rather, is one that is well grounded in explicit consideration of all parties' perspectives. In communicating the issues to concerned third parties (e.g., the media or boards of directors), the thoroughness of the process used to make the decision becomes clearer.

Some may say that Merck's decision, based on its desire to keep talented, driven scientists, was fundamentally based on profits and self-serving; therefore, Merck could be criticized for a lack of altruism. However, it is important to recognize that the most obvious decision to support profitability in this case would have been for Merck to forego the drug, rather than to proceed. Indeed, given the lack of a profitable market opportunity, a decision based solely on profits would have led to a very different outcome.

The unfortunate reality faced by companies wrestling with such ethical dilemmas is that they often find themselves in a catch-22: They are "damned if they do and damned if they don't." Even if a company engages in an action that ultimately reflects some level of social responsibility, it can be criticized by those who believe that "people above profits" should have been the underlying basis for the decision.

In order to explore the issue of socially responsible corporate behavior further, additional information on social responsibility and business is provided.

SOCIAL RESPONSIBILITY AND BUSINESS DECISIONS

Seventy-eight percent of American adults say they are more likely to buy a product from a company associated with important social causes, and 84 percent say that such "cause-related marketing" creates a positive image of a company.[31] The ethical scandals in business (e.g., Andersen Consulting, Enron, WorldCom) have heightened the public's sensitivity to corporate behaviors. As a result, companies are paying increased attention to corporate social responsibility.

Corporate social responsibility refers to a company's simultaneous focus on economic profitability ("the bottom line," or "doing well") and societal needs and concerns ("doing good"). There are a variety of ways that companies can engage in socially responsible business practices. They can align with a nonprofit organization to support a worthy cause or make corporate donations in support of a cause. For example, Apple Computer, Microsoft, Hewlett-Packard, and other high-tech companies have long been recognized for their product donations to education. Companies can also develop sustainable (environmentally friendly) business practices as part of their model of corporate social responsibility. For example, DuPont has recently adopted a corporate mission based on sustainable growth, focused on generating wealth while helping to save the planet. As a result, it has spun off oil-and-gas units and is focused on technology developments that use renewable materials, such as using corn for a new stretchable fabric called Sorona (rather than oil-based textiles such as nylon and polyester). DuPont's goal is to increase the percentage of its revenues generated from renewable resources from 14 percent in 2003 to 25 percent by 2010.[32]

Although the belief that business has obligations to society that go beyond making money is not new, the debate surrounding this issue has taken on increased intensity in recent years. On one side of the debate are those who say that business should mind its own business—which is turning a profit for shareholders—and leave social problems for others to wrestle with. Typically espoused by business conservatives, the thought is that business's concern for social causes leads to spending and expenses that do not contribute to profitability. People on this side of the debate believe that when companies donate company funds to charities, for example, they are giving away shareholders' money that, if not going directly into the business, should be the shareholders' decision regarding how to spend it.

On the other side of the debate are those who believe that business economics and social responsibility are not mutually exclusive, but rather, businesses that are socially responsible are actually more successful. As exemplified by Hewlett-Packard in the opening vignette, the thought is that socially responsible companies will reduce their risks, energize their employees, and build a stronger emotional connection with their customers and investors by taking on social causes and issues that matter. Indeed, many believe that it is directly in a business's interests to attack social problems. As stated by Paul Tebo, DuPont's corporate vice president for safety, health, and environment, "The closer we can align with social values, the faster we'll grow."

There are ways to assess business decisions with an eye to social responsibility. Michael Porter and Mark Kramer[33] offer a set of guidelines for companies who believe that pursuit of social causes can strengthen their competitive position in the market. They believe that business and social concerns are simultaneously addressed when a company's philanthropic investments occur in four key domains:

1. *Supply/input conditions:*   Investments in social causes that either directly or indirectly develop a company's human resources, capital resources, physical infrastructure, natural resources, scientific or technological infrastructure, and so forth will deliver both economic and social benefits.

2. *Demand/customer conditions:*   Philanthropic investments that can develop local markets, improve the capabilities of local customers, provide insights into needs of emerging customers, or develop product standards will be in a firm's strategic interests. The opening HP vignette also provides an example of investments designed to stimulate local customer markets.

3. *Competitive context:*   Any investments a firm makes to facilitate policies to reduce corruption, encourage fair competition, protect intellectual property, and in general, support an attractive business environment can be beneficial to both society and the individual business. Microsoft has long been involved with efforts in support of recognizing intellectual property rights. This is clearly an example of a cause that has both social and economic benefits.

4. *Supporting infrastructure:*   Investments that bolster supporting industries (e.g., services, suppliers) can foster the development of vibrant industry clusters, which can become an engine for economic development, in line with the spirit "a rising tide floats all boats."

Cisco's Networking Academy provides an example of the powerful links that exist between a company's philanthropic strategy, the four domains that can be affected by or targeted with such investments, and the resulting economic and social benefits.[34] Cisco (maker of networking equipment and routers used to connect computers to the Internet) found that many of its customers faced a chronic shortage of qualified network administrators. Although Cisco was already engaged in a type of cause-related marketing (that included donations of networking equipment to a high school near its headquarters), the company decided to formalize not only the donation of equipment, but also a training program to train teachers at the schools (and students) on how to build, design, and maintain the networks. This program grew into a Web-based distance-learning curriculum. At the suggestion of the U.S. Department of Education, the company began to target schools in "empowerment zones," areas designated by the federal government as the most economically challenged communities in the country. The program was expanded to include developing countries as well. Cisco has added a worldwide database of employment opportunities for academy graduates. Other companies have joined the effort, donating Internet access and other needed computer hardware and software. And, rather than reinventing their own training infrastructure, other companies such as Sun Microsystems and Adobe Systems have expanded the Academy's curriculum by sponsoring courses in other areas. After five years (as of December 2002), the Academy was operating 9,900 programs in all 50 states and 147 countries and it continues to grow rapidly. Cisco's $150 million investment has brought technology careers, and technology itself, to people in some of the most economically depressed areas of the world. More than 115,000 students have graduated from the two-year program, and one-half of the 263,000 students were outside the United States. Cisco has found a pool of talented employees, improved the sophistication of new customers, attracted international recognition, generated pride and enthusiasm among its employees and partners, and is known as a leader in corporate philanthropy.

To guide socially responsible decision making, companies should examine their goals, mission, and potential liabilities. In certain cases, a company may decide to engage in socially responsible behavior either because its corporate values suggest such behavior, because key personnel believe that it is the right thing to do, or for a host of other reasons. When based on a corporate mandate, a company may worry less about the profit impact of the decision and more about convincing other company personnel or stakeholders about the merits of the decision. In other cases, the potential motives for and benefits of ties to nonprofit causes, charitable giving, or socially responsible practices are less clear. A company can use three criteria to help it trade off the benefits versus the costs of social responsibility or, more broadly, sponsorship of events in general (see Table 12-2).

TABLE 12-2 Considerations in Social Responsibility

Does the mission of the company match the mission of the cause?

Does the target market of the company have some vested interest in the cause?

Will the socially responsible behavior yield goodwill among key stakeholder groups, and will the company benefit from positive exposure the behavior generates?

- **Does the Mission of the Company Match the Mission of the Cause?**   To the extent that the mission of the company has an explicit tie to the social cause, the investment of corporate funds makes sense. For example, a telecommunications company's investments in broadband access to rural communities makes sense, because it is part of the company's raison d'être. However, based on this factor, such a company's sponsorship of environmental reclamation in these communities may make less sense, because environmental reclamation is unrelated to its mission.

- **Does the Target Market of the Company Have Some Vested Interest in the Cause?**   This consideration requires that the company explicitly address whom it is targeting with its products and services, and whether that target market has a vested interest in the cause. In this case, if a telecommunications company's primary target market, new users of telecommunications services, was women-owned small businesses, then sponsorship of a breast-cancer awareness campaign or support of educational programs for young girls in math and sciences would match this criterion. Note that in this case, attempting to connect to the target market and to build competitive advantage via causes in which the target is interested may take the company into socially responsible activities somewhat removed from its primary business.

- **Will the Socially Responsible Behavior Yield Goodwill among Key Stakeholder Groups, and will the Company Benefit from Positive Exposure the Behavior Generates?**   This consideration is based on the fact that socially responsible behavior is ultimately one way to maintain a positive image among key stakeholder groups. For example, a telecommunications company might donate money to a local food bank in some of its primary communities. In this case, the cause is not directly tied to its mission, nor specifically targeted to its primary customers. Even though the direct link to possible sales is not explicit, such an activity can create goodwill in its local communities.

    Further, socially responsible behaviors increase the likelihood that the company will benefit from publicity. To what extent will the company's behaviors be known to the broader public? If the event is marketed or televised, or if the company plans on advertising its affiliation to the cause, then some exposure will be generated. Such exposure can be used to generate brand-name awareness. If, however, the event is small and not marketed or televised, then the exposure will be more limited.

    Companies that operate at only this third level of social responsibility run great risk if they are merely engaging in socially responsible behaviors in one area to compensate for socially irresponsible behaviors in others. As in the ethics section, socially responsible behaviors can be viewed skeptically when people question the motives behind business behavior; this is more likely to be the case if companies' actions are not perceived as genuine. Companies that exhibit social responsibility in areas more closely related to their businesses, and that are genuine in their efforts, are less likely to be accused of hypocrisy.

Whether these corporate social responsibility initiatives are more than mere window-dressing, a public relations ploy designed to "divert attention from corporate rapacity and corruption,"[35] remains to be seen. Social responsibility advocates strongly argue that, if reputation and brand matter, companies with a mission that goes beyond making money will do better when it comes to recruiting, retaining, and engaging their workers and attracting loyal customers. The alternative—a focus on short-term profits that comes from polluting the environment (or avoiding investments in sustainability) or mistreating workers—can be profitable in the near-term, but may prove costly over the long term. Ultimately, the debate over corporate social responsibility will be resolved by comparing the performance of companies that adhere to "business as usual" to companies that embrace socially responsible initiatives. If it leads to profit and growth, corporate social responsibility will prove to be a lasting business strategy.

Social Responsibility and Innovation

There are some who believe that social responsibility can actually support a company's pursuit of breakthrough technologies.[36] New technologies are needed to address the social and environmental challenges associated with economic growth and the saturation of mature markets. Although corporate strategic philanthropy can be initiated in any geographic location, the opening vignette on Hewlett-Packard highlights the attention that developing countries are receiving in this area. Not only can companies expand their markets by doing business at "the base of the pyramid" where the world's poor are desperate to join the market economy, they can also invent new products to serve the needs of the world's most impoverished people whose basic needs often go unmet. For example, four billion people earn less than $1,500 annually in purchasing power. Three billion lack reliable telecommunications service, in large part due to the cost of extending wire-line infrastructure from urban to rural areas.

The need for power generation and distribution in developing countries provides an example of the impetus for breakthrough innovations in developing countries. More than two billion people in the world lack access to dependable electric power; as a result, rural poor spend what little income they have on candles, kerosene, and diesel to have periodic lighting at night. Breakthrough technologies to solve this problem will be developed where they can be profitably deployed—in markets where they do not compete against established systems. "Distributed generation" of power takes advantage of renewable fuels to generate small quantities of electricity near the actual point of use (avoiding the need for expensive distribution infrastructure). For example, Iowa Thin Films Technologies makes a type of solar photovoltaic cell that can be used for rural, off-grid power needs. Roll-to-roll technology can produce solar cells in inexpensive factories located directly in the countries they serve.

At its heart, the business of innovation, and the decisions of innovative businesses, are human endeavors very much shaped by human forces. Ethics and values do underlie the push to innovate, as Greg Simon speaks to so eloquently in the Technology Expert's View from the Trenches.

REALIZING THE PROMISE OF TECHNOLOGY
GREG SIMON

President,
Center for Accelerating Medical Solutions
Washington, D.C.

Over five hundred years ago in Germany, a goldsmith who had bungled a surefire money-making venture by getting a crucial date wrong was looking for a way to mollify his business partners. He decided to use his goldsmithing skills to mold what became known as movable type, and he used the type to print the one book he knew would sell, the Bible, in this case, the Gutenberg Bible. About a hundred years later in Antwerp, another entrepreneur began to print Bibles with maps of the Holy Land in the back, leading to the development of printed cartography that could be updated with each new voyage. The first fanatics who used these new and improved maps for their ventures, the Dutch, soon began returning from voyages with profits of 500 percent or more.

This got everybody's attention, to say the least, and led to a demand for cash to invest that was so great that new ways were needed to allow for borrowing against land, which was the source of wealth at the time. The result was land registers and secured mortgages, the bulk of which went into paying for ships. These new ventures needed some protection, referred to as insurance, and limited joint stock companies sprang up to pool resources for both investing and insurance, leading to the creation of the stock market. Naturally, something was needed to coordinate all this—national banks were the answer, and they brought along credit (and, unfortunately, credit agencies). With further refinement of the printing press, institutions arose to manage specialized information in standardized formats. Because of this standard, more reliable data, individuals and companies could take more risks, leading to massive economic expansion and development.

The only management system capable of holding all this together turned out to be the business contract, which eventually made its way into social theory and then, with the help of those like John Locke, into political theory. The best example of the social contract is the contract between state and citizen contained in the Constitution of the United States. All this, thanks to a goldsmith's mistake.

The history of the United States shows how the forces that shape communication shape our economic, social, and political well-being. The history of its democracy is closely entwined with the history of communication. The United State's revolutionary battle cry reflected the failure of communication with its mother country. "No taxation without representation" was the colonial protest against one-way, centralized communication from England that did not provide for any response or interactivity. And it is no accident that the United State's First Amendment to the Constitution is the right of free speech, the right to communicate.

Thomas Jefferson put very well why information is a peculiar resource, one that grows as people share it. He said, "He who receives an idea from me receives instruction himself without lessening mine. He who lights his taper with mine receives light without darkening me."

Throughout history, people's desire to improve the means of communication has helped us grow and prosper. Again, in the United States, the earliest laws regarding radio frequencies and communications spectrum were provoked by the tragedy of the *Titanic,* in which the failure of radio operators to man their stations and the inability of the *Titanic* to communicate its SOS signal on a clear frequency led to the death of thousands.

Morse was also a famous portrait artist in the United States. His portrait of President James Monroe hangs today in the White House. While Morse was working on a portrait of General Lafayette in Washington, his wife, who lived about 500 kilometers away, grew ill and died. But it took seven days for the news to reach him. In his grief and remorse, he began to wonder if it were possible to erase barriers of time and space, so that no one would be unable to reach a loved one in time of need. Pursuing this thought, he came to discover how to use electricity to convey messages, and so he invented the telegraph.

The transcontinental railroad, the telephone, Marconi's wireless radio, the television, the interstate highway, and now the Internet both fulfilled people's dreams and set us to dreaming anew.

But today's dream is not about technology. It is about breaking the barriers of how we know our world, our neighbors, and ourselves. It is about millions of individual journeys to explore the frontiers of knowledge, whether it is a child's email conversation with a scientist at the South Pole or a tour of the Louvre from one's living room.

Some things have to be seen in order to be believed. But some things have to be *believed* in order to be seen. The revolution that sprang from the development of the Gutenberg press was not simply the result of the innovation by a goldsmith with a need to raise money. It succeeded because the world was ready to receive and nurture the idea Gutenberg set forth. After all, movable metal type had been invented in Korea 200 years before Gutenberg invented it again. But conditions conspired to keep that first movable typeface from spreading. Confucianism prohibited the commercialization of books, and the Korean royal presses would print only classical Chinese literature, not the more popular Korean literature.

By Gutenberg's time, there were better conditions: better paper, better metals, and eyeglasses. And Europeans were ready for a cheaper way to copy books than using scribes who charged for one copy what a printer would charge for a thousand.

We have a similar challenge today to create the commercial, technical, legal, and social conditions that will produce the foundation for a global information infrastructure. We have, each of us and each nation, a job to do to bring the world the next generation of technology. The work we do to cross our common oceans to build a global information infrastructure is not in the service of wires or satellites but is in the service of a global vision that can be realized in every neighborhood of the world.

The technology that so fascinates and occasionally dominates us is a mute guest in our lives. It cannot, left alone, speak of values or a vision. It cannot yet tell the difference between the bricks of a church and the bricks of a prison. But we can and must speak of values and we can and must recognize the moral choices in our lives. We cannot choose to delay or deny the future, we must make ready for it. But there is no better way to predict the future than to create it.

THE ROLE OF THE GOVERNMENT

What is the role of government in realizing the promise of technology? Depending upon one's perspective, the government can play several pivotal roles. For example:

- It can support science and education through research and outreach programs to maintain innovativeness. In the 1980s and 1990s, the United States was ranked at the top on an overall innovation index based on R&D funding and other factors. By 2005, because of their spending on basic research and education and percentage of technical workers, Japan, Finland, Denmark, and Sweden are expected to be at the top.[37]

    As an example of this role, in November 2003, the U.S. Senate passed the Nanotechnology Research and Development Act, setting up a National Nanotechnology Initiative and authorizing $3.7 billion in funding over four years. The bill provides government grants to industry, the development of research centers, education, and training. Supporters believe the bill is a vital catalyst for the development of what will become a $1 trillion piece of the global economy.[38]

- It can initiate legislation to enhance competition, as in the Telecommunications Act of 1996, designed to open the telecommunications industry to industry newcomers.

- It can introduce measures to ensure and protect consumer welfare, as in the proposed "can-spam" legislation, and rules regarding privacy.

- It can help with dialogues over standards (or legislate them if the industry players cannot agree among themselves, as in the case of digital TV standards).[39]

Some worry that, rather than helping to realize the promise of technology, the government actually interferes with innovativeness and the support of technology in the economy. On the one hand, the role of the government is to offer policies and legislation that provide competitive markets and ensure consumer welfare. On the other hand, the government must also offer policies that provide incentives to undertake the risks of technology innovation, including protecting companies' intellectual property rights. Clearly, the government faces a delicate balancing act in addressing the needs of established companies and industries, new technology startups, and consumers.

In particular, three arenas of government activity are particularly salient in today's high-tech environment, as shown in Figure 12-3. The government must (1) continue efforts to update its antitrust models to reflect the characteristics of a business economy that is driven in large part by technology-focused goods and services; (2) continue to modify and update rules regarding intellectual property to reflect changes in the economy, particularly as they relate to digital rights management; and (3) consciously monitor issues related to society's access to technology.

Updating Antitrust Models

Many of the guiding principles used for antitrust thinking were formed in the late eighteenth century, in an era when the economy was driven by the Industrial Revolution and the development and production of physical (tangible) goods. This not only meant that firms were concerned with the factors of production and limitations on supply; it also implied that a good could be consumed by only one individual.[40] However, in an

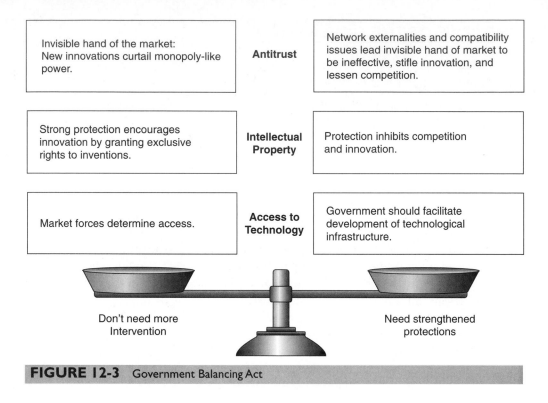

FIGURE 12-3 Government Balancing Act

economy driven by digitization and technology, information can be consumed by many people at once. Indeed, due to network externalities, the more people who adopt a common good, the more valuable it becomes. For example, consumers want to buy what everyone else is using, so their equipment is compatible with other users. At its extreme, the most popular products are widely used, creating monopoly-like economic power, as in the case of Microsoft's Windows operating system.

The question is whether the characteristics of a high-tech environment, such as unit-one cost structures, network externalities, and knowledge spillovers, render traditional antitrust models moot. Some technology enthusiasts say "yes." In favor of relaxed antitrust restraints, they argue that while an economy driven by information and technology-based products may lead to more monopoly-like economic power, such concentrations will be temporary and quickly overthrown by new innovations that are superior. As shown on the left-side of Figure 12-3, reliance on the invisible hand of the market will allow new innovations to curtail monopoly-like economic power.

Others say "no," that because of the unique characteristics of a high-tech environment, society needs traditional antitrust protections more than ever. Proponents of this school believe that monopoly-like economic power in technology markets can be particularly problematic. For example, Microsoft's control of the operating system that is crucial to running over 90 percent of the world's computers can create issues of access to the market that, if limited, can stifle competition and innovation.[41] Moreover, once a technology standard is set, consumers feel a need to buy products based on the standard, even if they're not the best. Even temporary monopolies can stifle innovation, inhibit competition, and slow the pace of innovations, highlighting the need for strengthening traditional antitrust protections (the right-hand side of Figure 12-3).

Consistent with the view that a vibrant technology-based economy does require effective antitrust enforcement,[42] after a nineteen-month trial, Microsoft was found guilty in July 2000 of violating state and federal antitrust laws by using its monopoly power in the market for operating systems software to crush rivals and extend its dominance to Web browsers. Microsoft's alleged restrictive and discriminatory business agreements were a key element in the legal case. Appeals to higher courts failed to overturn the decision. Although disputed by some as being "too soft," the settlement reached in the penalty phase of the case called for Microsoft to give customers the option to purchase the operating system without the integrated Web browser. It also required that Microsoft allow competitors access to its communications protocols so that they could write programs that worked seamlessly with its Windows operating system.[43]

Regardless of one's perspective on how the rules for antitrust ought to be applied, a technology-based economy does have unique characteristics that highlight the need for a dialogue regarding the impact of antitrust regulations.[44]

Examining Intellectual Property Models

The government's efforts in enforcing intellectual property rights through patents and copyrights also has a profound impact on technological developments—and has been impacted by technology developments.

Patent Protection

Some worry that the patents being granted for technological breakthroughs such as genes, prime numbers, and even lab animals (a Harvard mouse) will lead to excessive control of information that will stymie possible knowledge spillovers that could build on these developments in profound, but unknown, ways.

The Internet and the human genome are two such breakthroughs. The Chapter 7 discussion on intellectual property protection provided a summary of the debate with respect to patenting Internet business models. With respect to the life sciences, a single company, Human Genome Sciences, had received patents on 106 complete human genes, including some that are crucial to treating osteoporosis and arthritis, and in 2000, had patents pending on more than 7,500 genes.[45] The rush to patent genes raises profound ethical and social questions. Will such patenting allow breakthroughs to be shared with the broad scientific community? Will advances that could improve consumer welfare be restricted? For example, in genetically modified foods, five U.S. farmers and one French farmer filed an antitrust lawsuit against Monsanto, accusing the company of conspiring to control markets for corn and soybean seeds; Novartis, DuPont, and seven other companies were named as co-conspirators.[46]

The balancing act that the government must strike in the case of patent protection is providing inventors with an incentive to undertake the risks of innovation (via patent protection) against stifling innovation and competition with such protection. Although costly research and development does merit recompense, some middle ground between the right to exclude others from using the innovations and sharing that knowledge may be desirable.[47]

Copyright Protection

At least two forces have converged that are making it difficult for "content" owners to retain control over their copyrighted works:

1. Digitization of content
2. The development of information technology to facilitate sharing of digital content:
   a. Hardware, such as CD or DVD burners
   b. Software, such as *peer-to-peer computing services;* these file-sharing services allow users to search one another's PC hard drives and copy digital files such as movies, videogames, software, images, and text
   c. Broadband connections that allow faster downloads of data-intensive media, such as movies

These two forces have created a schism between the rights of entertainment companies and technology providers. Copyright laws grant copyright holders (often, the entertainment companies) a monopoly over the right to reproduce and distribute their works during the term of the copyright. Technology providers make devices that enable consumers to reproduce and distribute copyrighted works (photocopying machines, TiVo, DVD burners, to name a few). The public finger-pointing between the CEOs of Disney (Michael Eisner) and Apple (Steve Jobs) epitomized the acrimony between the entertainment and technology companies. Eisner charged that Apple's slogan ("Rip, Mix, Burn"), touting its products' multimedia capabilities, endorsed digital piracy. Entertainment companies criticized technology companies for dragging their feet in establishing technical standards to prevent digital piracy. In a very strong lobbying effort by content providers, lawmakers initially proposed that computer and consumer electronics manufacturers embed copyright protection technology in all of their products. At its heart, the question is: Are technology providers facilitating copyright infringement by their customers?

The government has attempted to address the issue of what has become known as **digital rights management** (designed to curb digital piracy) in several different ways. For example, the 1992 Audio Home Recording Act granted consumers the right to unlimited private use of legally purchased music, videos, books, and other media content. These rights were somewhat compromised with the passage of the 1998 Digital Millennium Copyright Act, which made it a crime to circumvent copy protection.

Lawsuits have also guided the development of legal policy in this arena. The landmark case in this arena came in 1984, the Betamax case (*Sony Corp. of America v. Universal City Studios*), hailed as the "Magna Carta of the technology age," defining the rights of entertainment companies and technology providers. In this case, Sony was *not* held to be liable for facilitating copyright infringement, even though it knew that some consumers would use its VCRs illegally to make copies of copyrighted materials. The critical issue was that because the VCR's primary uses were "noninfringing," the courts should not stifle potentially beneficial technologies before their usefulness might be fully understood.

Importantly, at the time of the Sony ruling, the studios had not been able to show any actual revenue loss attributable to VCRs, even though they had been in circulation for almost a decade. In contrast, providing fuel on the fire in the battle between the music industry and technology companies, between 1999 and 2003, unit sales of recorded music, and record industry revenues fell precipitously.[48]

The Sony case has been used to analyze the rights of the music industry with respect to downloading music from the Internet. In December 1999, the Recording Industry Association of America (RIAA) successfully sued Napster for contributing to copyright infringement. After a lengthy trial and appeal process, Napster was shut down in July 2001. Because Napster was a service—which meant that the company had an ongoing relationship with its users—the courts held that the Sony ruling didn't apply. The VCR was a product that, once sold, Sony had nothing to do with any longer; as a result, Sony could not track what its users did. However, Napster had an ongoing relationship with its users and could track what users were trading.

Other peer-to-peer networks developed in the wake of the Napster closure. Smart technology developers modified their approach to file-sharing services and developed stand-alone software products rather than an ongoing service. Disaggregated file trading did not have a centralized index, but rather, stored information on files across the network. Unlike Napster (the first-generation file-sharing service), these second-generation file-sharing services (such as Morpheus, Grokster, Kazaa, Gnutella, LimeWire, and BearShare) were mere providers of software tools that enabled their users to search one another's PC hard drives and copy digital files, with no ability to know what people searched for and downloaded. Despite being sued in October 2001, Morpheus and Grokster were protected from liability by the Sony ruling. The legal cases in this arena deal with complicated issues and continue to receive significant attention.

Perversely, the outcome of the more recent cases implied that although commercial operators could not be prosecuted, the actual users of the file-sharing services could be sued (users engage in copyright infringement when they upload or download copyrighted files). The ruling meant that, if the RIAA was going to enforce its rights as copyright holders, it was faced with the prospect of suing individual users, which it did in September 2003. Using sophisticated, parallel computing technology that searched popular songs, and in turn, the hundreds and often thousands of people offering to share them, the RIAA was able to compile evidence against individuals. Along the way, Internet Service Providers were subpoenaed for the names of people using file-sharing services.[49]

Between September and December 2003, RIAA sued nearly 400 individuals for copyright violations. Initial evidence suggested that the lawsuits had the desired effect. Prior to the lawsuit, 29 percent of Internet users downloaded songs to their computers; in the months following RIAA's legal strategy, 14 percent of Internet users reported downloading of songs to their computers.[50]

Some believe that the entertainment industry's attempts to restrict information sharing, rather than to develop new business models, exhibit a desire to control intellectual property in ways that may be incompatible with an information-based economy.[51] It is instructive to note that studios, fearful of new technology, initially fought the advent of the VCR, nearly asphyxiating the home-video market that today provides up to 50 percent of their revenues from sales of videotapes and DVDs. Some believe the moral to the story is that, if the entertainment industry were to embrace the new file-sharing technology, it could end up making more money than ever before.[52] In other words, maybe it's time to accept technology and deal with the Digital Age.

Rather than searching for ways to defend their current business models, entertainment companies might reframe the issue as how to maximize revenues in the Digital

Age and still protect the value of their intellectual property. As many say, "content yearns to be free." Paradoxically, this implies that one way to maximize revenue may be to abandon efforts to protect digital rights at all costs. Rather, entertainment companies may begin to think nontraditionally about how to distribute content in order to increase their options for creating value. For example, instead of simply selling a book, CD, or movie via the Net, entertainment companies could focus on selling a variety of complementary products and services, such as merchandise or tickets to special events with artists and authors. With such a strategy, music companies could deliver many forms of additional services that their customers would be willing to pay for—and have a direct connection to the consumer.[53] Consistent with the notion of creative destruction, established incumbents at the top of the industry likely have to reinvent themselves—with some degree of pain—or face being "amazoned."[54]

Missing from the intense rhetoric is the voice of the consumer. Is it possible to reframe this debate with a win/win/win lens, for technology companies, entertainment companies, and consumers? Although strong protections for intellectual property are essential for promoting continued innovativeness, such protections also shouldn't stifle competition and access. A Consumer Technology Bill of Rights has been developed by a consumer advocacy group, DigitalConsumer.org,[55] formed by Silicon Valley businesspeople who opposed the erosion of consumer rights and of technological innovation. The Bill of Rights proposes that once consumers have legally purchased digital content, they have the right to "time-shift" (record for later playback); "space-shift" (copy to blank CDs or portable players); make backup copies; use content on any platform (PC, MP3 player, etc.); and to translate content into different formats. These ideas are also supported by legal scholars, such as Lawrence Lessig (creativecommons.org).[56]

Assisting with Access to Technology

A final area where the government has been walking a tightrope in balancing various constituencies' needs is with respect to access to technology. Some believe that access to technology is an "essential facility," from which, like education, all should have the opportunity to benefit. Therefore, as in the development of the interstate highways, phone lines, and electricity, in order to fully capitalize on the promise of technology, the government has a responsibility to its citizens to develop a widely available network of computer technology.

The debate in this area has become known as the **digital divide**, referring to the disparity in access to technology between technology "haves" and "have-nots." Disparity in access is considered to be an important issue, because "if the technology revolution leaves some behind, all will suffer."[57]

This disparity in access can manifest itself in myriad ways:

- Between different socioeconomic groups (affluent versus poor)
- Between different geographic areas, say, between urban and rural users, or between inner-city and suburban users
- Between different ethnic groups, say between Caucasians and African Americans or Hispanics
- Between different countries (developed versus developing versus least developed).

Disparity in Access for Rural Areas

Small towns and rural areas, particularly in sparsely populated areas of the western United States, simply have not had access to broadband technology that will allow high-speed Internet connections. Both the distance and the sparse population contribute to the problem. Consider that in New Jersey, the average distance between a customer and phone company's nearest switching facility is about 2.6 miles; in Wyoming, the distance is twice as far, making the cost to the phone company of reaching a customer twice as high. Moreover, parts of the rural West have as few as half a dozen households per square mile, compared with thousands in urban and suburban areas.[58] Some phone companies say they should not have to serve customers in those areas if that's where they choose to live. As stated by Garry Betty, president and CEO of Earthlink.net: "To tell me I've got to serve someone at a certain speed regardless of the cost because he chooses to live in the far reaches of Montana is not fair. Let them pay for it themselves."[59]

If the economics of the situation do not support businesses developing digital access for rural communities, the government may choose to play a role in a manner similar to the role it played with the subsidies provided to develop the telephone and the interstate freeway infrastructures. Because the long-term health and survival of rural America is, to a large extent, dependent on its ability to attract and retain economic development and activity—which, in turn, requires access to current technologies—this type of support is considered by many to be vital.

Disparity in Access between Different Ethnic Groups

Table 12-3 shows the rate of Internet penetration by U.S. household ethnicity. Despite growing 22 percent since 2001, the number of African Americans online in 2003 represented only 8 percent of the Internet population (compared to accounting for 12.7 percent of the U.S. population).[60] Despite the slightly lower penetration rate, African Americans' rate of broadband adoption is higher than the overall Internet population (43 percent for African Americans compared to 36 percent for the overall Internet population). Moreover, African Americans purchase more clothing online than the general Internet population (48 percent compared to 41 percent) and more music and videos (44 percent vs. 39 percent). Marketers should note that African American Internet users read online ads, and 46 percent find them informative (vs. 26 percent of the general population).[61]

TABLE 12-3   Internet Penetration by U.S. Household Ethnicity

	2001	2007**
Caucasian and other	62%	81%
African American	45%	69%
Hispanic	45%	68%
Asian American	63%	82%

**Predicted
SOURCE: Greenspan, Robyn (2003), "African-Americans Create Online Identity," *CyberAtlas,* September 26. From Jupiter Research.

These statistics suggest that the continued debate over the digital divide is more likely to be found in terms of socioeconomic indicators rather than ethnicity per se. Indeed, online marketers are catering to these various ethnic groups. Nissan has designed an African American–themed online component for its "Shift" campaign. With "Shift-Respect," Nissan highlighted African American achievements and influential icons during Black History Month. Moreover, Office Depot, Toyota, Ford, and Nissan have launched Spanish language sites.

Disparity in Access between Developed and Developing Countries

Based on the belief that technology provides access to more diffuse communications, information, and economic opportunity, at least some believe that providing technology access for developing countries could stimulate a push to democracy and economic development.

> The developing world's first goal in a technology "Marshall Plan" should be to ensure that poor countries have elementary access to communications. . . . Properly used, technology in its many guises can raise living standards and quality of life.[62]

Indeed, the Digital Opportunity Task (DOT) Force was created in 2001 at the meeting of the G8 developed countries to develop international strategies for bridging the so-called digital divide.

Solutions to Bridging the Digital Divide

Government efforts to facilitate access to technology are one possible way to bridge the digital divide. So, too, are the efforts of many high-tech companies mentioned in the section of this chapter on social responsibility. Another solution to the digital divide can also be found in new technological developments, such as satellite and other types of high-speed wireless access, that do not require wiring "the last mile." Although not specifically intended for rural markets, Motorola and Cisco jointly contributed $1 billion to create wireless, high-speed Internet networks, and AT&T is experimenting with cellular-like services that compress data and bring high-speed Web access into homes. Moreover, the efforts of the Association for Competitive Technology to lobby for the government's approval of the Echostar/DirecTV merger was, in part, based on their statements that satellite delivery of broadband could help to overcome the digital divide. The digital divide will get narrower over time as prices of new technology decline and become more affordable.

As these solutions suggest, many people seem to believe that providing access to technology for disadvantaged populations is the solution to bridging the digital divide and that such access will bring economic prosperity and wealth to disadvantaged populations. Although *access* to information technology is clearly a necessary condition to bridging the digital divide, it is certainly not the only factor that will help integrate technology into people's lives. An overlooked factor is the *willingness* of the people (to whom such technology is brought) to fully embrace and utilize the technology to their advantage.[63] Hence, efforts to facilitate the effective utilization of technology must also be undertaken.

CONCLUDING REMARKS: REALIZING THE PROMISE OF TECHNOLOGY

Many of the vital technologies in the world today were literally inconceivable two centuries ago. The steamship was not an evolutionary development of the sailing ship; the automobile is not an evolutionary development of the horse and buggy; the transistor did not evolve from improvements in vacuum tubes; and the personal computer did not evolve from the mechanical calculator.[64] Technological development requires creativity, resources, perseverance, and serendipity. Indeed, the time lag between an innovation and product development may be up to forty years. The market may not be ready for it, government regulators may not know how to facilitate it, the costs may be too high, or some other development may come along that supersedes it. Innovators must be tireless crusaders regarding their inventions. Although they may not immediately find a viable commercial application, they must persevere, hoping that it one day may be useful.[65]

Ultimately, technology is a tool meant to serve human purposes. However, the promises that new technologies offer can be realized only if inventors take careful steps in the commercialization and marketing of their ideas. A good idea alone is insufficient. Smart marketing based on systematic consideration of critical issues is necessary to allow innovations to benefit society in the ways in which they are intended. Without effective marketing of high-technology products and innovations, the benefits such innovations can yield will remain elusive.

SUMMARY

This chapter explored factors related to realizing the promise of technology, including paradoxes and unintended consequences, ethical dilemmas and social responsibility, and the role of government. Although technology is not the panacea that can solve all human problems—including food shortages, health problems, transportation snafus, and business inefficiencies—neither is it essentially incompatible with human values, the source of all woes in our society, including the spread of nuclear weapons and environmental degradation. Proactive consideration of the paradoxes technology poses for users, as well as its unintended consequences, will enable firms to better anticipate obstacles so that the promises of their innovations can be realized. Similarly, proactive consideration of ethical dilemmas can go a long way toward minimizing possible negative effects. Use of the framework on wrestling with ethical dilemmas as well as explicit attention to the considerations in socially responsible business decisions will help a firm navigate through potential controversies. The government's role is to balance carefully the need to foster a climate that promotes innovation while ensuring a competitive marketplace.

DISCUSSION QUESTIONS

1. What are the eight paradoxes technology can pose from a user's perspective? Give an example of each one. What are the marketing implications?
2. Identify and describe one unintended consequence arising from technology (not mentioned in this chapter). What are the marketing implications of this (i.e., how should a marketing manager use knowledge of this unintended consequence)?

3. Identify and describe an ethical controversy arising from technological developments (not mentioned in this chapter). Use the framework presented in the chapter to make a decision on how to deal with this ethical controversy.

4. Summarize the debate over corporate social responsibility. Which side of the debate do you fall on? Why?

5. Find an example of a high-tech company's efforts towards social responsibility.

   a. Evaluate this company's efforts in terms of the four domains of corporate social responsibility (identified by Porter and Kramer).

   b. Evaluate the company's efforts based on the three considerations for social responsibility presented in Table 12-2.

   c. Are there other causes or areas where this company could usefully put its philanthropic efforts to be more effective? Which ones? Why?

6. How can social responsibility facilitate the development of radical innovations?

7. For each of the following three issues, summarize the pros and cons for government involvement:

   a. Active intervention in the market with respect to antitrust concerns

   b. Active intervention in the market with respect to intellectual property concerns

   c. Active intervention in the market with respect to access to technology concerns.

GLOSSARY

Corporate social responsibility. Refers to a company's simultaneous focus on economic profitability ("the bottom line," or "doing well") and societal needs and concerns ("doing good").

Digital divide. The disparity in access to technology between different groups in society, including different socioeconomic groups, geographic locales, different ethnic groups, and different countries.

Digital rights management. Efforts by entertainment/media companies, technology companies, and the government to recognize and enforce the rights of copyright holders and to curb digital piracy.

Genome. The complete set of instructions for an organism's biological processes required to live and reproduce, as written in two pairs of nucleotides (comprised of adenine and thymine, or cytosine and guanine) in its DNA (deoxyribonucleic acid).

Luddites. People who are opposed to technology and changes brought about by technology.

ENDNOTES

1. Engardio, Pete (2001), "Smart Globalization," *Business Week,* August 27, pp. 132–137.

2. Ball, Jeffrey (1999), "High-Tech Air Bags Are Lacking in Grasp of Human Dimensions," *Wall Street Journal,* August 5, pp. A1, A6.

3. Dibbell, Julian (1996), "Everything That Could Go Wrong . . . ," *Time,* May 20, p. 56.

4. Except as noted, this section is drawn from Mick, David Glen and Susan Fournier (1998), "Paradoxes of Technology: Consumer Cognizance, Emotions, and Coping Strategies," *Journal of Consumer Research,* 25 (September), pp. 123–143.

5. Dibbell, Julian (1996), op. cit.

6. Greenspan, Robyn (2003), "The Deadly Duo: Spam and Viruses, July 2003," *CyberAtlas,* August 11.

7. Donato, Marla (1999), "Sense-Dulling Conveniences Creating Alienated World," *Missoulian,* January 12, p. A8, from *Chicago Tribune.*

8. Down, Jeff (1999), "Plugged in to Excess," *Missoulian,* August 23, p. A1, Associated Press.

9. Donato, Marla (1999), op. cit.

10. Bryan, Rebecca (2000), "Two-Timing the Clock: Trying to Slow the Raging River of Progress? Take a Step Back and Disconnect from the World," *Business 2.0,* February, pp. 227–230.

11. Morrissey, Brian (2003), Spam Cost Corporate America $9B in 2002," January 7, *CyberAtlas.*

12. Siegel, Matt (1998), "Do Computers Slow Us Down?" *Fortune,* March 30, pp. 34, 38; Landauer, Thomas (1995), *The Trouble with Computers: Usefulness, Usability, and Productivity,* Cambridge, MA: MIT Press; Brynjolfsson, Erik (1993), "The Productivity Paradox of Information Technology," *Communications of the ACM,* 36 (December), pp. 67–77.

13. Woods, Bob (2003), "IM Use a Big Security Threat," June 5, 2002, *CyberAtlas.*

14. Cox, Beth (2000), "E-commerce Said to Be Eco-Friendly," *InternetNews.com,* January 11.

15. Taylor, Chris (2000), "Why Mother Nature Should Love Cyberspace," *Time,* April–May, p. 82.

16. Tam, Pui-Wing (2000), "Taking High Tech Home Is a Bit Much for an Internet Exec," *Wall Street Journal,* June 16, p. A1.

17. www.m-w.com/cgi-bin/dictionary, downloaded January 12, 2004.

18. *Chemical Week* (2003), "Senate Passes Nanotechnology Bill," November 26, p. 44.

19. *Nanotech Briefs* (2003), "NSF Grant to Study Nano's Societal Impact," October, p. 4.

20. Enriquez, Juan and Ray Goldberg (2000), "Transforming Life, Transforming Business: The Life-Science Revolution," *Harvard Business Review,* March–April, pp. 95–104.

21. The Futurist (2003), "Genetically Modified Crops May Surpass Natural Crops in Acreage Planted by 2020; (November/ December), 37 (6), p. S2; Kluger, Jeffrey (1999), "Food Fight," *Time,* September 13, pp. 42–44.

22. Golden, Frederic (2000), "Make Way for Frankenfish," *Time,* March 6, p. 62.

23. *The Financial Times* (2003), "EU moves toward lifting ban on genetically modified food," December 5, p. 10.

24. Enriquez, Juan and Ray Goldberg (2000), op. cit.

25. Giannakas, Konstantinos and Murray Fulton (2000), "Consumption Effects of Genetic Modification: What If Consumers Are Right?" working paper, University of Nebraska, Department of Agricultural Economics, Lincoln, NE.

26. Enriquez, Juan and Ray Goldberg (2000), op. cit.

27. Weiss, Rick (1999), "Smarter Mice Run Ethical Maze," *Missoulian,* September 2, pp. A1, A9, from *Washington Post.*

28. Kolata, Gina (1996), "Breaking Ranks, Lab Offers Test Assessing Breast Cancer Risk," *Boulder Daily Camera,* April 1, p. A3.

29. "Merck & Co., Inc." (1991) in David Held, *Property, Profit, and Justice,* The Business Enterprise Trust.

30. Ibid.

31. Kadlec, Daniel (1997), "The New World of Giving," *Time,* May 5, pp. 62–64.

32. Gunther, Marc (2003), "Tree Huggers, Soy Lovers, and Profits: Some of America's biggest corporations believe that the best way to make money is by saving the world. And guess what? They just might be right." *Fortune,* June 23, p. 98.

33. Porter, Michael and Mark Kramer, (2002), "The Competitive Advantage of Corporate Philanthropy," *Harvard Business Review,* (December), pp. 5–16.

34. Porter, M. and Kramer, M. (2002), op. cit.

35. Gunther, Marc (2003), op. cit.

36. Hart, Stuart and Clayton Christensen (2002), "The Great Leap: Driving Innovation from the Base of the Pyramid," *Sloan Management Review,* 44 (Fall), pp. 51–57; Kirkpatrick, David (2003), "Two Ways to Help the Third World," *Fortune* October 27, pp. 187–196.

37. "U.S. Innovation Ain't What It Used to Be," (1999), *Business Week,* March 22, p. 6.

38. *Chemical Week* (2003), op. cit.

39. The reason that digital TV is considered an issue of consumer welfare is that the switch to digital signals allows broadcasters to squeeze more video and data into existing channel space, which can be used to provide more options for consumers.

40. Murray, Alan (1999), "Pushing Adam Smith Past the Millennium," *Wall Street Journal,* June 21, p. A1.

41. Murray, Alan (2000), "In the New Economy, You've Got Scale," *Wall Street Journal,* January 17, p. A1.

42. Murray, Alan (1997), "Antitrust Isn't Obsolete in an Era of High-Tech," *Wall Street Journal,* November 10, p.A1.

43. *Los Angeles Times* (2002), "Microsoft Investigation Could Kill Antitrust Agreement," October 24.

44. Murray, Alan (1997), op. cit.

45. Enriquez, Juan and Ray Goldberg (2000), op. cit.

46. Ibid.

47. Shulman, Seth (1999), "We Need New Ways to Own and Share Knowledge," *The Chronicle of Higher Education,* 45 (24), p. A64.

48. Parloff, Roger (2003), "The Real War Over Piracy," *Fortune,* October 27, pp. 148–156.

49. France, Mike (2003), "Striking Back," *Business Week,* September 29, pp. 94–96.

50. Associated Press (2004), "Survey: Music downloading declining after crackdown," January 5.

51. Griffin, Jim (2000), "The Digital Delivery of Intellectual Property Is Our Generation's Nuclear Power," *Business 2.0,* February, p. 212.

52. Parloff, Roger (2003), op. cit.

53. Yang, Catherine, (2003), "A Rising Chorus of Music Downloaders?" *Business Week,* September 29, p. 96.

54. Black, Jane (2002), "The Freebie Road to Digital Riches," *Business Week,* May 13.

55. Mossberg, Walter (2002), "Consumers Must Protect Their Freedom to Use Digital Entertainment," *Wall Street Journal,* March 14, B1.

56. Black, Jane (2002), "Lawrence Lessig: The 'Dinosaurs' Are Taking Over," *Business Week,* May 13.

57. Crocket, Roger (1999), "High Tech's Next Big Market? Try the Inner City," *Business Week,* December 20, p. 48.

58. O'Malley, Chris (1999), "The Digital Divide," *Time,* March 22, pp. 86–87.

59. Ibid.

60. Morrissey, Brian (2003), "Black Online Population Narrows Adoption Gap," *CyberAtlas,* February 28.

61. Greenspan, Robyn (2003), "African-Americans Create Online Identity," *CyberAtlas,* September 26.

62. Kirkpatrick, David (2001), "Tech into Plowshares," *Fortune,* October 15, pp. 211–213.

63. Albert, Terri and Charles L. Colby (2003). "The Technology Readiness of Vulnerable or Impacted Groups and Public Policy Considerations: A Cross-Cultural Research Program." Paper presented at the 2003 AMA Public Policy Conference, Washington, D.C., May; Albert, Terri and Jakki Mohr (2004), "Technology Readiness of Vulnerable Consumers: Implications for the Digital Divide," Working Paper, University of Montana, Missoula.

64. Ayres, Robert (1994), "Technological Trends," *National Forum,* 74 (Spring), pp. 37–43.

65. Bronson, Gail (1987), "Technology: Songs the Sirens Sing." *Forbes,* July 13, pp. 234–237.

Author Index

Subject Index